Game-Theoretical Models in Biology

Covering the major topics of evolutionary game theory, *Game-Theoretical Models in Biology, Second Edition* presents both abstract and practical mathematical models of real biological situations. It discusses the static aspects of game theory in a mathematically rigorous way that is appealing to mathematicians. In addition, the authors explore many applications of game theory to biology, making the text useful to biologists as well.

The book describes a wide range of topics in evolutionary games, including matrix games, replicator dynamics, the hawk-dove game, and the Prisoner's Dilemma. It covers the evolutionarily stable strategy, a key concept in biological games, and offers in-depth details of the mathematical models. Most chapters illustrate how to use Python to solve various games.

Important biological phenomena, such as the sex ratio of so many species being close to a half, the evolution of cooperative behaviour, and the existence of adornments (for example, the peacock's tail), have been explained using ideas underpinned by game theoretical modelling. Suitable for readers studying and working at the interface of mathematics and the life sciences, this book shows how evolutionary game theory is used in the modelling of these diverse biological phenomena.

In this thoroughly revised new edition, the authors have added three new chapters on the evolution of structured populations, biological signalling games, and a new chapter on evolutionary models of cancer. There are also new sections on games with time constraints that convert simple games to potentially complex nonlinear ones; new models on extortion strategies for the Iterated Prisoner's Dilemma and on social dilemmas; and on evolutionary models of vaccination, a timely section given the current Covid pandemic.

- Presents a wide range of biological applications of game theory
- Suitable for researchers and professionals in mathematical biology and the life sciences, and as a text for postgraduate courses in mathematical biology
- Provides numerous examples, exercises and Python code.

Chapman & Hall/CRC Mathematical Biology Series

Series Editors:
Ruth Baker, Mark Broom, Adam Kleczkowski, Doron Levy, Sergei Petrovskiy

About the Series
This series aims to capture new developments in mathematical biology, as well as high-quality work summarizing or contributing to more established topics. Publishing a broad range of textbooks, reference works, and handbooks, the series is designed to appeal to students, researchers, and professionals in mathematical biology. We will consider proposals on all topics and applications within the field, including but not limited to stochastic modelling, differential equation modelling, dynamical systems, game theory, machine learning, data science, evolutionary biology, cell biology, oncology, epidemiology, ecology and more.

An Introduction to Physical Oncology
How Mechanistic Mathematical Modeling Can Improve Cancer Therapy Outcomes
Vittorio Cristini, Eugene Koay, Zhihui Wang

Introduction to Mathematical Oncology
Yang Kuang, John D. Nagy, Steffen E. Eikenberry

Stochastic Dynamics for Systems Biology
Christian Mazza, Michel Benaim

Systems Biology
Mathematical Modeling and Model Analysis
Andreas Kremling

Cellular Potts Models
Multiscale Extensions and Biological Applications
Marco Scianna, Luigi Preziosi

Quantitative Biology
From Molecular to Cellular Systems
Edited By Michael E. Wall

Game-Theoretical Models in Biology, Second Edition
Mark Broom, Jan Rychtář

For more information about the series, visit: https://www.routledge.com/ Chapman--HallCRC-Mathematical-Biology-Series/book-series/CRCMBS

Game-Theoretical Models in Biology

Second Edition

Mark Broom

City, University London, UK

Jan Rychtář

Virginia Commonwealth University, USA

CRC Press

Taylor & Francis Group

Boca Raton London New York

CRC Press is an imprint of the
Taylor & Francis Group, an **informa** business

A CHAPMAN & HALL BOOK

Second edition published 2022
by CRC Press
6000 Broken Sound Parkway NW, Suite 300, Boca Raton, FL 33487-2742

and by CRC Press
4 Park Square, Milton Park, Abingdon, Oxon, OX14 4RN

© 2022 Taylor & Francis, LLC

First edition published by CRC Press 2013

CRC Press is an imprint of Taylor & Francis Group, LLC

ISBN: 978-0-367-45668-9 (hbk)
ISBN: 978-1-032-30870-8 (pbk)
ISBN: 978-1-003-02468-2 (ebk)

DOI: 10.1201/9781003024682

Typeset in CMR10
by KnowledgeWorks Global Ltd.

Publisher's note: This book has been prepared from a camera-ready copy provided by the authors.

To Monica and Dewey and Walter

Contents

Preface

When we wrote the first edition of our book, we were aiming to produce an extensive book focused on the static, as opposed to dynamic, aspects of game theory and its applications to biology, which was both rigorous enough for the mathematical reader whilst containing enough biological applications to interest the biological reader. This original edition contained twenty chapters covering a wide range of both general theoretical models and particular application areas of biology. We focused on models that were of interest to us, explaining their motivation, giving (at least some) mathematical detail of the models and also discussing some of their important conclusions.

We believe that the book was at least reasonably successful in its aim, and we have both had interesting conversations with readers of the book over the years. Nine years have passed, however, and as this is a fast-moving area of research, a lot of important work has happened since. We were asked by Callum Fraser at Taylor & Francis to produce a second edition and felt that the time was right to do this.

As well as refreshing the book in general, we have added new sections on areas of work that have seen significant development since the original book. We have been selective in the areas we have chosen for development, and much of the core content of the book is not significantly changed from before, even though we know that in many of these, there has been a lot of new work. In particular, we have developed three new chapters and some new sections to existing chapters. We describe the main changes below.

Chapter 13 is a new chapter on the evolution of structured populations. This has taken a little content from the old Chapter 12, which now focuses on games in finite populations and on graphs, but is mostly new content which goes beyond the original evolutionary graph models representing structured populations (though Chapter 12 also has some interesting new work in this still very relevant area). Chapter 13 looks at models of general structures, includes games that are often multi-player in character and also looks at some models involving games on evolving structures.

Chapter 18 is a new chapter on biological signalling games, starting with classical models of the handicap principle, which featured in the first edition as part of Chapter 16 (Chapter 17 in the new edition). We have added some signalling models, which include additions to the general theory that offer alternative or complementary explanations for the widespread use of signalling within the biological world and also a specific application to plant pollination.

Chapter 22 is a totally new chapter which considers evolutionary models of cancer. Here cancer is treated as an ecological population evolving within an environment, which is a person or animal. This area has seen rapid development since our original book, and we discuss a number of model types, including adaptive therapies, where the physician can use evolutionary knowledge to deliver treatments more effectively, especially for patients with very serious conditions.

We have added new sections to: Chapter 7 on games with time constraints that convert simple games to potentially complex nonlinear ones; Chapter 15 (the old Chapter 14) adding new models on extortion strategies for the Iterated Prisoner's Dilemma and on social dilemmas and multi-player games of cooperation versus defection; and Chapter 21 (the old Chapter 19) on evolutionary models of vaccination, a timely section given the current Covid pandemic.

We believe that in this new version of our book, we have provided some updated, interesting models for our readers to consider whilst leaving the fundamental character of the book unaltered. We hope that you enjoy reading this new edition of our book.

As before, the book is principally aimed at final-year students, graduate students, and those engaged in research at the interface of mathematics and the life sciences. The level of mathematical content is set so that it can be read by a good graduate-level or final-year undergraduate mathematics student without any biological background, and graduate biologists with significant mathematics short of the level of a mathematics degree and a willingness to put in sufficient work. Different sections will inevitably appeal to different parts of this audience, but it is our aim that every part of the book will be accessible to the target audience described.

The production of this book has inevitably involved many people. Some of the people listed were involved only in helping with the original edition, some with only the second, and some with both. We would, in particular, like to thank: Ben Allen, Steve Alpern, Chris Argasinski, David Basanta, Nick Britton, Joel Brown, Chris Cannings, Ross Cressman, Meghan Fitzgerald, Christoforos Hadjichrysanthou, Michal Johanis, Vlastimil Křivan, Mike Mesterton-Gibbons, Hans Metz, Rob Noble, Sebastian Pauli, Luigi Pistis, Jon Pitchford, David Ramsey, Graeme Ruxton, Paulo Shakarian, Peter Sozou, Kateřina Staňková, Shanmugathasan Suthaharan and Arne Traulsen for discussions and providing critical comments on earlier drafts of the manuscript.

This work was supported by funding from the European Union's Horizon 2020 Research and Innovation Programme under the Marie Skłodowska-Curie Grant Agreements No. 690817 and No. 955708.

We also wish to thank our editors at Taylor & Francis, Callum Fraser, Mansi Kabra and their production team.

Inevitably there will be some mistakes in the text. Please let the authors know of any errors by emailing us at mark.broom@city.ac.uk or rychtarj@vcu.edu; a list of corrections will be maintained on the Taylor & Francis, website at http://www.routledge.com/9780367456689

Authors

Mark Broom is a professor of mathematics at City, University London. For over 30 years, he has carried out mathematical research in game theory applied to biology. His major research themes include multi-player games, patterns of evolutionarily stable strategies, models of parasitic behaviour (especially kleptoparasitism), the evolution of defence and signalling, and evolutionary processes in structured populations. He earned his PhD in mathematics from the University of Sheffield.

Jan Rychtář is a professor of mathematics at Virginia Commonwealth University. He works on game theoretical models and mathematical models of kleptoparasitism. His recent research interests include models of disease prevention and elimination. He earned his PhD in mathematics from the University of Alberta.

Chapter 1

Introduction

Unlike physics, which has been inextricably linked to mathematics since its foundations, the relationship between biology and mathematics has not been as strong, and until relatively recently, the application of mathematics in the life sciences, in general, has been very limited. Mathematical biology, however, has been rapidly advancing on a number of fronts with both the rise of powerful computing methods and the understanding of complex low-level biological structures, in particular the sequencing of the human genome, allowing the modelling of complex molecular systems and the development of early-stage multiscale models. Important early mathematical developments in the modelling of organisms include the predator-prey modelling of Lotka (1925) and Volterra (1926), and the development of population genetics (e.g. Hardy, 1908; Weinberg, 1908; Wright, 1930 and Fisher, 1930), but arguably the most influential modelling in this area concerns the development of the theory of evolutionary games, which is the subject of this book. Both the type of dynamics used in predator-prey models and the core concepts of population genetics have had important influences on evolutionary games, but the origins of evolutionary games go back a long way with ideas developed for a very different type of scenario, and we begin by discussing games and game theory more generally.

1.1 The history of evolutionary games

People have played games for thousands of years, and there has always been an interest in how to play them in the best way, to maximise the chances of victory. One example, though it has evolved during this time and only reached its current form in the fifteenth century, is the game of chess. Chess is an example of a game of perfect information (see Chapter 10), where both players are aware of the precise state of the game. The players take turns to play, and there is (effectively) a finite number of sequences of play, and with sufficient computing power, it would be possible to decide the result with optimal play. A simpler game with the same properties, but where this is more obvious, is noughts and crosses (also known as tic-tac-toe). Here there are 9! sequences (in fact, far less due to symmetries, and the fact that sometimes

DOI: 10.1201/9781003024682-1

1

games are completed without all squares being filled) and with best play the game ends in a draw. In fact, the strategy is sufficiently obvious, that a game between two players with some knowledge of the game will effectively always be drawn; see Figure 1.1. The game of bridge, another popular and complex game, is a game of imperfect information, as the cards of the players remain hidden from each other (common property of card games) and are inferred by both the bidding and the play.

How are such games to be solved? In fact, solving problems where there is uncertainty and/or there are decisions to be made has a long history. Probability theory, which can be thought of as the mathematical theory of random events, has much of its origins in the solution of gambling problems. For instance, in the 17th century Chevalier de Mere, a member of the French court and a gambler, played a lot of dice games and brought a problem to the attention of mathematicians Pascal and Fermat; see Haigh (2003). The aim of Chevalier de Mere was to understand the odds of different games better, so that he could know when the odds favoured him and avoid games when they did not. In general, when faced with a situation where there are a number of choices, and potential outcomes are uncertain, we are faced with a problem of optimisation (which might be simply to choose whether to play the game or not). The scientific methodology of operational research, for example linear programming, came to prominence during the Second World War, when complex problems of military logistics involving troop and equipment deployments were analysed. Similar problems also feature in modern industry, and such techniques are now very widespread, see e.g. Walsh (1985); Luenberger and Ye (2008). In general, there are many ways of performing a set of tasks, and the aim is to find the best (or as close to the best as you reasonably can). The situations described above are examples of decision theory and have in common an individual faced with a range of options, and possibly also by various possible random acts of nature (and so also relying on probability theory). However, there is no other agent whose choices need to be taken into consideration.

Game theory is a mathematical theory which deals with the interactions of individuals in some (conflict) situation. Each individual has choices available to him or her, and the eventual outcome of the game depends upon the choices of all of the players. This, in turn, determines the rewards to all of the players. Every player thus has some influence on the eventual result of the game and the reward that he or she receives. The key feature of most interesting games is that the best choice depends upon the choices of the other players, and this is what differentiates game theory from optimisation theory. We define what we mean by a game in the evolutionary context more precisely in Chapter 2.

1.1.1   Early game playing and strategic decisions

In a game such as chess where players alternate moves but where there is (effectively) only a finite number of game sequences, there will be a determined

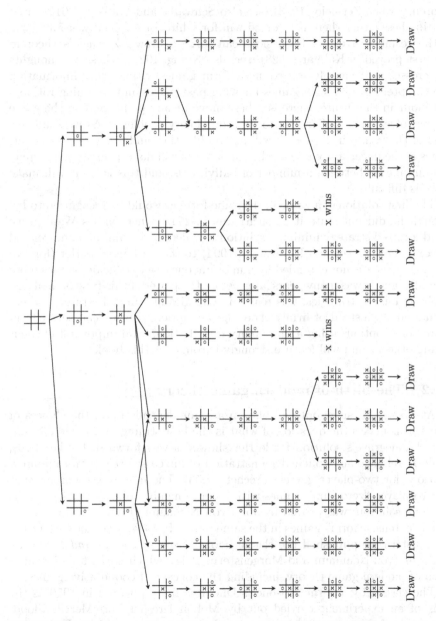

FIGURE 1.1: Game of noughts and crosses. Noughts starts; crosses can always at least draw. We follow noughts and assume that crosses play optimally in response. Isomorphic or forced moves not shown.

outcome with best play (usually assumed to be a draw in chess). Thus if a player is sufficiently intelligent, he or she could work out the best play. This is formalised in perhaps the original theorem of game theory (Zermelo's theorem), when Zermelo (1913) (see also Schwalbe and Walker, 2001) stated that in chess either there is a forced win for white, or a forced win for black, or the result is a draw (both sides can force a draw). Zermelo's theorem was first proved in Kalmár (1928) (see also König, 1927). Chess and noughts and crosses are examples of extensive-form games with perfect information (see Chapter 10). For such games knowing past moves and assuming rational behaviour in the future, there is a best move at any point, and so the game can be solved in a similar way to decision theory problems. As we shall see, most of the games in this book will not be like this, and a key feature of our games is the uncertainty about what opponents will do, especially where play is simultaneous between a number of individuals, and this is what will make analysis difficult.

The first solution of a game in the same form as would be recognised today (though he did not state it as such) was in 1713, when James Waldegrave found a mixed strategy minimax solution (for details on minimax and related concepts see Osborne and Rubinstein, 1994) to the card game le Her (for two players). This was not extended by him or anyone else to consider games more generally, and it was many years later before the power of this type of analysis, involving mixed strategies, was realised. The idea of a mixed strategy, where an individual, instead of firmly choosing one option, can instead select from a number of options using a probability distribution, is an important element of game theory and will feature strongly throughout this book.

1.1.2   The birth of modern game theory

At first sight, since optimal play in a game depends upon the choices of others, this makes the question of what is the best strategy impossible to answer. Waldegrave's solution for le Her showed a way forward. In the 1920s, Emile Borel gave an explicit demonstration of mixed strategies and minimax solutions for two-player games (Fréchet, 1953). The minimax theorem, that for two-player "zero sum" games with a finite number of strategies, there is a unique solution, was proved by John von Neumann (1928) who also introduced "extensive form" games in the same paper. In 1944, together with Oskar Morgenstern, he published the classic book *Theory of Games and Economic Behavior* (von Neumann and Morgenstern, 1944), which is the foundation of much of modern game theory, including the concept of cooperative games.

The classical game the Prisoner's Dilemma first appeared in 1950 as the basis of an experiment carried out by Melvin Dresher and Merrill Flood. The name of the game and its associated story was invented by Tucker, see Straffin Jr (1980); Tucker and Straffin Jr (1983). We see in Chapters 4 and 15 that the Prisoner's Dilemma is important in a number of study disciplines and that these experiments were the start of a huge experimental and theoretical industry. The Prisoner's Dilemma, including the repeated (or iterated) version, is still central to experiments investigating cooperation to this day.

One of the key developments in the early theory of games, and one which makes frequent appearances in this book, is the concept of the Nash equilibrium. This elegant solution to the problem of finding the "best" strategy was developed by Nash (1950, 1951), together with other important results such as the Nash bargaining solution. The Nash equilibrium is still the basis of all of the more elaborate solutions that have followed it in the various branches of game theory, and we see this clearly for evolutionary games in Chapter 3. For an insight into the life and times of John Nash, the reader is referred to the book *A Beautiful Mind* (Nasar, 1998) and, to a lesser extent, the film of the same name.

In the 1960s, two other great contributors to game theory came to prominence. Reinhard Selten (1965) introduced the idea of (subgame) perfect equilibria. Shortly afterwards, John Harsanyi (1966) developed the modern distinction between cooperative and non-cooperative games, where (essentially) cooperative games have binding agreements and non-cooperative ones do not. He followed this work with a series of papers, e.g. Harsanyi (1967, 1968a,b) developing the theory of games of incomplete information, as mentioned above (see Chapters 10 and 19 for biological examples), which has become important in economic theory. Selten (1975) introduced the concept of trembling hand perfect equilibria, which has been central to much subsequent theory (see Chapters 4 and 8 for examples of this idea).

Game theory is used widely in economics, in a variety of areas such as auctions and bargaining; both non-cooperative and cooperative games are common, and games can be pairwise or include multiple players. The vital importance of game theory to the study of economics was recognised in the awarding of the Nobel Prize for Economics in 1994 to Nash, Selten and Harsanyi "for their pioneering analysis of equilibria in the theory of non-cooperative games", and the prize has been awarded to game theorists several times since then.

1.1.3 The beginnings of evolutionary games

It was only more recently that it was realised that game theory could be applied to biology in a systematic fashion, where the reward is represented by a gain in Darwinian fitness and natural selection makes an excellent substitute for rationality. In fact, it can be argued that animal behaviour is more reasonably modelled by game theory than human behaviour because of this. Evolutionary games model large populations, where games are (usually) played between members of the population chosen at random, and this leads to different types of solutions to conventional games, based upon stability considerations, as we see in Chapter 2.

The first (implicit) game-theoretical argument in evolutionary biology is a lot older and comes from Charles Darwin (1871), who explains why natural selection should act to equalise the sex ratio (see Chapter 4). More mathematically rigorous explanations for the same problem were given by Dusing in 1884 (Edwards, 2000) and by Fisher (1930) in terms that would be recognisable to a game theorist, although it was the explicit discussion of this problem by Hamilton (1967) that presented one of the earliest works on game theory. As well as being an important problem in its own right, it is the classic example of a playing the field game, an important class of evolutionary game. We look into this type of problem, including the sex ratio problem itself, in Chapter 4 (see also Chapter 7).

The first explicit application to evolutionary biology was by R. C. Lewontin (1961) in "Evolution and the Theory of Games". Note that this involved a game played by a species rather than an individual against nature, which is generally accepted as the wrong starting point by modern evolutionary biologists. Even if a particular strategy is beneficial to a species as a whole (such as Cooperate in the Prisoner's Dilemma, see Chapter 4), if it can be invaded by an alternative strategy (such as Defect) then that strategy cannot persist in the population. It was argued (e.g. Wynne-Edwards, 1962) that spatial dispersal and the formation of groups could allow the groups with the strategies best for the group as a whole to prosper, "group selection". Whilst Maynard Smith (1964) considered this theoretically plausible in some circumstances, and he did not believe that this would occur commonly in nature. More recently, interest in the idea of group selection has been revived with the concept of multi-level selection (see Section 15.7), where selection is influenced at a number of different levels such as the gene, the cell and the individual, and in particular the group also has a role to play, see Wilson (1975); Eshel (1972); Eshel and Cohen (1976); Boyd and Richerson (2002) and Chapter 5.

Another key strand of evolutionary games was started in the 1960s by Hamilton (1964) and by Trivers (1971) in their work on relatedness and altruism. The idea of inclusive fitness, where an individual's fitness is not just dependent upon the number of its own offspring but also on those of its relatives, provided a convincing explanation for much cooperative behaviour in nature. We discuss such cooperation in general, together with considering the Prisoner's Dilemma above, in Chapter 15.

The next contribution, and the one that did most to establish an independent theory of evolutionary games, was by Maynard Smith and Price (1973). They introduced the central concept of evolutionary game theory, the Evolutionarily Stable Strategy (ESS), a development on a par with the Nash equilibrium in terms of importance. This idea is not only key to the mathematical theory of evolutionary games but is a vital component of myriad applied models across evolutionary biology and is the central idea throughout this book. Maynard Smith and Price (1973) and Maynard Smith (1974) established some of the classical games that are commonly used today, in particular, the Hawk-Dove game and the war of attrition (see Chapter 4). John Maynard Smith in particular, as well as jointly creating the basic tool for analysing biological problems, had a great interest in a variety of real problems to apply the games to, and it is to him more than any other figure that the popularity of evolutionary games can be attributed (see in particular Maynard Smith, 1982).

From the early pioneering work of Maynard Smith, Hamilton and others, there has been a vast explosion of game theory in biology, both in the mathematical developments and particularly in the range of applications. The most striking results often relate to behaviour or biological features, which are at first sight paradoxical from the conventional evolutionary standpoint. Examples are the classical results of how altruism can evolve, through kin selection and through reciprocal altruism (Chapter 15), and the explanation of biological ornaments, such as the peacock's tail, as signals of mate quality (Chapter 17). Whilst the nature of the games used in biology and those used in economics are rather different, the evolutionary approach has also fed back into models from economics, see e.g. Kandori (1997); Sandholm (2010).

We should note that there has been some criticism about the use of evolutionary games to model behaviour. Typically strategies are chosen that represent visible behaviours, phenotypes, without taking the underlying genetics into account, and we discuss this to some extent in Chapter 5. In particular, the concept of a restricted repertoire, where genetics only allows a certain subset of strategies to occur, or different strategies are controlled from the same genetic locus, was considered by Maynard Smith (1982). We certainly acknowledge the importance of genetics and that we effectively neglect its influence throughout most of the book, in common with most evolutionary game models (but see Section 5.3.3 for a nice argument for why this neglect is often hopefully not too problematic).

For people who want to read some more about the history of game theory, a valuable resource, from which much of the information in this chapter was obtained, is Walker (2008).

1.2 The key mathematical developments

Following the introduction of evolutionary games, a number of important pieces of work were published. These established the core of the mathematical theory around the static concept of the ESS, introduced the equally important area of evolutionary dynamics and discussed a number of real biological behaviours using the theory of games to explain apparently paradoxical behaviours and features of the natural world. Below we discuss some of the key areas from their origins to some modern developments, considering similar developments together, irrespective of year of origin.

1.2.1   Static games

Following the early work of Maynard Smith and Price (1973), the general theory of matrix games (an example of which is the Hawk-Dove game) was developed. Here populations play independent two-player games against randomly chosen opponents so that the game can be summarised by a single matrix (see Chapter 6). Haigh (1975) showed how to systematically find all of the ESSs of a matrix game (but see Abakuks, 1980). An important result on the potential co-existence of ESSs was proved by Bishop and Cannings (1976). Further developing this theme, in a series of papers starting with Vickers and Cannings (1988b), Cannings and Vickers (1988), Cannings, Vickers and co-workers created the theory of patterns of ESSs. For a review of the work of Chris Cannings, one of the leading early contributors to evolutionary game theory, especially of the static variety, see Bishop et al. (2020).

A key book bringing together a lot of the key conclusions of this early work is Maynard Smith (1982). Written from a biological rather than mathematical perspective but still with significant mathematical content, this remains a very important reference point today.

The theory of matrix games requires both participants to occupy identical positions. If each participant can be identified in a particular role, such as owner and intruder, the game changes and must be analysed differently, with each individual (potentially) having a different strategy in each role. Models of this type were introduced by Maynard Smith and Parker (1976) and further developed by, e.g. Hammerstein and Parker (1982), with important theoretical developments due to Selten (1980).

Whilst much of the mathematical theory of games is based upon linear games where rewards are governed by a set of independent contests, most models of actual biological situations are nonlinear in character. From the early work of Hamilton (1967) on the sex ratio and the patch foraging models of Parker (1978), nonlinearity has featured. The theoretical developments on general nonlinear evolutionary games are still limited, although some important work was carried out by Bomze and Pötscher (1989). Work on multi-player evolutionary games, which are related to nonlinear games, was first developed by Palm (1984), and this was followed by Haigh and Cannings (1989) for the multi-player war of attrition and by Broom et al. (1997b) for multi-player matrix games. A type of game more commonly used in conventional game theory and with applications in economics is the extensive form game. Extensive form games allow us to model a sequence of interactions between two or more players. The original work was from von Neumann (1928) and developed to its modern form by Kuhn (1953); other important results are given in Selten (1975). This is less used in biological modelling but is invaluable for modelling complex interactions, and we see a few examples throughout this book.

When modelling biological populations, the usual starting point is to consider an effectively well-mixed infinite population of identical individuals. Real populations may contain individuals of different sizes as considered by

Maynard Smith and Parker (1976). More generally, Houston and McNamara (1999) consider a methodology for modelling a population where the state of individual (e.g. size, hunger level) could significantly affect their strategy; see also McNamara et al. (1994) for an example. Evolution in finite populations was considered in Moran (1958, 1962), and this was developed to consider games in finite populations in Taylor et al. (2004). In real populations, various factors, including geographical location, mean that interactions between certain individuals are more likely than others. The evolution of a population with explicit structure was popularised with the concept of cellular automata in the game of life, see Gardner (1970) (though cellular automata had been in existence since the 1940s), and this idea developed to consider a more general structure using evolution on graphs in Lieberman et al. (2005) and Nowak (2006a).

1.2.2 Dynamic games

Conventional game theory modelling is concerned with what is the best strategy, and rational individuals can change their strategy in light of experience. In the static theory of evolutionary games, we also assume that the population has found a given strategy by some means and analyse it for stability against invasion. But populations must evolve gradually, so how do they change and how can stable strategies be reached? This is the subject of evolutionary dynamics. We note that our book mainly covers static games, but dynamics are certainly as important and are the focus of a number of significant existing books, as we discuss in Chapter 3.

The most commonly used dynamics in evolutionary games is the continuous replicator equation introduced by Taylor and Jonker (1978), though the term replicator dynamics first appeared in Schuster and Sigmund (1983). These track how the composition of a population changes through time. Individuals cannot change strategy, but their offspring copy their strategy, so the more successful strategies spread. Important work relating the developing dynamical theory to the static theory was given by Zeeman (1980, 1981); Hines (1980). In particular, it was shown by Zeeman (1980) that any ESS is an attractor of the replicator dynamics, so some set of initial population values will converge to it. An example game with two ESSs, together with the trajectories of the evolving population, is shown in Figure 1.2 (see Section 3.1.1.2 for an explanation of this figure). Results relating to stability properties of the replicator dynamics were shown by Bomze (1986) and Cressman (1990). The key results are still perhaps best summarised in the classic books by Hofbauer and Sigmund (1998) and Cressman (1992). The discrete replicator dynamics was introduced by Bishop and Cannings (1978). Evolutionary dynamics, especially the replicator dynamics, have been applied to most of the game types described in Section 1.2.1, including multi-player games (Palm, 1984), extensive form games (Cressman, 2003) and games with an infinite number of strategies (Bomze, 1991).

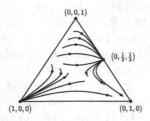

FIGURE 1.2: Replicator dynamics for a matrix game with payoff matrix $A = \begin{pmatrix} 1 & 0 & 0 \\ 0 & -1 & 2 \\ 0 & 0 & 1 \end{pmatrix}$ and two ESSs $(1,0,0)$ and $(0,1/2,1/2)$.

A related dynamic is the replicator-mutator dynamics of Page and Nowak (2002). A different dynamic is the imitation dynamics of Helbing (1992), where individuals can change strategy based upon others that they meet. Similarly, the best response dynamics (Matsui, 1992) allows individuals to be rational and update by picking the best play in a given population.

Under the replicator dynamics, the set of strategies does not change (except that unsuccessful strategies can be lost). However, for evolution to occur, new strategies must be introduced. Adaptive dynamics studies how a population might change when new strategies very close to existing ones can appear (infrequently) within a population. Important early work was due to Eshel (1983), and the term adaptive dynamics was first used in Hofbauer and Sigmund (1990). Key theoretical developments appeared in Geritz et al. (1998), see also Metz et al. (1996). Adaptive dynamics has much in common with the static analysis, which is the focus of our book, and we consider it in Chapter 14.

1.3   The range of applications

Game theory has been used to model a wide range of different application areas. In this book, we focus on six main areas, each of which has a chapter devoted to it. The evolution of cooperation, especially in humans, has been modelled using a number of developments from the original models. Models either follow the altruism models of Hamilton (1964), for example West et al. (2007b), or more commonly are based upon the Prisoner's Dilemma, important examples of which are Axelrod's tournaments (Axelrod, 1984) and the reputation model of Ohtsuki and Iwasa (2004). The modelling of interactions in animal groups is also another important area for game theory. Stable social groups often form dominance hierarchies, and particular important models are the reproductive skew models introduced by Vehren-camp (1983), and the winner-loser models introduced by Landau (1951a,b); see also Dugatkin (1997a). Other important group behaviour involves how the existence of groups influences how individuals respond to the threat of predators, see e.g. McNamara and Houston (1992).

The two areas which have received the most attention in terms of modelling animal behaviour are models of mating and models of foraging. Competition for mates has long been the most important single area of modelling, and the Hawk-Dove game is based upon this scenario. More subtle competition than just straight contests of the Hawk-Dove type include the signalling processes introduced by Zahavi (1975) to explain ornaments like the peacock's tail, and modelled mathematically by Grafen (1990a,b), and sperm competition modelling, for example in Parker (1982). A second commonly modelled area is that of foraging. When animals only compete with each other indirectly, the central theory is that of the ideal free distribution introduced by Fretwell and Lucas (1969); see also; Parker (1978). Whilst the ESS solution was intuitive, a formal mathematical proof only appeared in Cressman et al. (2004). Two different models of direct foraging competition are the kleptoparasitism models starting with Ruxton and Moody (1997), Broom and Ruxton (1998a), and the producer scrounger models of Barnard and Sibly (1981), Caraco and Giraldeau (1991).

A further important area of behaviour is that of parasitism, where one in-dividual exploits another. Animals do this in a number of ways; for instance, predators exploit prey. Whilst most predator-prey models do not involve game theory, its use is becoming more common; see, for example, Brown and Vincent (1992) and Křivan and Cressman (2009). A well-known relationship between exploiter and exploited involves cuckoos and cowbirds, which parasitise other birds by tricking them into raising their chicks (see Planque et al. (2002) for a game-theoretical model). Similarly, humans and other animals are host to myriad diseases, and this can be thought of as another form of parasitism. Modelling the evolution of epidemics is also an important use of mathematics. Such models, starting with Kermack and McKendrick (1933), predate evolu-tionary game theory, but evolutionary game theory has made an important contribution, in particular through the modelling of the evolution of viru-lence, as in Nowak et al. (1990), Nowak and May (1994) and to strategies for vaccination (Bauch and Earn, 2004).

In this book, we will see a range of applications across the areas described above, but these are only a few of the models that have been developed. It is important to realise that there is a huge variety of other models of related behaviour; for example, game theory has been applied to many distinct types of mating and foraging behaviour. There are also models that do not neatly fall into the categories described above, and more generally models which are game-theoretical but do not use evolutionary games, for example in terms of modelling fisheries policy as in Sumaila (1999). Thus in terms of applications in particular our book provides a subjectively chosen subset of available models rather than anything more definitive.

1.4   Reading this book

There are a large number of chapters within this book, and to understand
the ideas within a chapter it is not necessary to read every preceding chapter.
However, for those without significant prior knowledge, some earlier chapters
are particularly important. The content of Chapters 2–4 is useful for all readers
(although it is perhaps not necessary to read Chapter 2 in great detail on a
first reading, as this chapter is aimed more at introducing the mathematical
methods rather than providing an intuitive understanding of game theory,
which can best be gained from the other two chapters). Similarly, the main
content at the start of Chapter 6 is useful for most of the later chapters, but the
second half of this chapter is for more specific mathematical interest. Figure 1.3
shows various distinct paths through the book for those with specific interests
(often not all of a given chapter's content is necessary for later chapters on
a path). For example, readers interested in evolutionary games in structured
populations in Chapter 13 should read up to Chapter 4 but can then jump
straight to Chapter 12. Readers particularly interested in the mathematical
models should read from the start of the book, from Chapters 2 through to
Chapter 8 in particular (possibly missing out Chapter 5, which discusses some
biological issues). Readers primarily interested in biological applications may
go directly to Chapters 15 and higher and visit Chapters 2–8 only if needed
to refresh the most relevant underlying game-theoretical concepts.

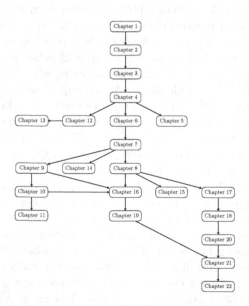

FIGURE 1.3: Recommended order of chapters for reading.

Chapter 2

What is a game?

In this chapter we describe the terms that are used for the vast majority of evolutionary game models in this book and elsewhere. As stated at the end of Chapter 1 this chapter is primarily aimed at introducing some important mathematical methods, and it is left to Chapters 3 and 4 to see how we use these tools to model populations using evolutionary games.

"Game" is used in a wide variety of contexts, most of which have nothing to do with the content of this book. For game theorists and especially those working in applications of game theory in biology, economics or social sciences, the word "game" means a mathematical model of a situation where several entities interact (directly or indirectly) with one another and where each entity acts in its own interest (which can be potentially in conflict with the interests of other entities, although there can also be common interests).

Most games, such as the Prisoner's Dilemma discussed in Chapters 4 and 15, have indeed been developed to model and analyse such conflict situations. There are also games that have existed for a long time, played by children and adults of all ages, and only after the development of game theory have these games been adapted to model situations from real life. Before we introduce the mathematical concept of a game, let us look at one such game, the Rock-Scissors-Paper game.

Example 2.1 (The Rock-Scissors-Paper (RSP) game). Two players put their hands behind their backs and at a given signal produce one of three forms simultaneously, with no knowledge of the other's play: a clenched fist (a "Rock"), two parted fingers (a pair of "Scissors") or a flat palm (a sheet of "Paper"). The convention is that Rock beats (blunts) Scissors, Scissors beats (cuts) Paper and Paper beats (wraps) Rock. If you win you gain a point, and if you lose you lose a point, with no points if both produce the same symbol.

The RSP game has been used to model some biological phenomena, for example Coliform bacteria (see Exercise 2.2), or mating strategies in the common side-blotched lizard (*Uta stansburiana*) as described below.

Example 2.2 (RSP game in lizards, Sinervo and Lively, 1996). The common side-blotched lizard *Uta stansburiana* exhibits a throat-colour polymorphism. Males with orange throats are very aggressive and defend large territories. Males with dark blue throats are less aggressive and defend smaller territories. Males with yellow stripes do not defend any territories but they

DOI: 10.1201/9781003024682-2

look like females and use a sneaking strategy when mating. It was observed in Sinervo and Lively (1996) that (a) if blue is prevalent, orange can invade, (b) if yellow is prevalent, blue can invade and (c) if orange is prevalent, yellow can invade.

As is the case in the RSP game, and which is typical for most interesting games, the result of the game depends not only on the action of a particular player, but on the actions of other player(s) as well. Much of game theory is devoted to answering a question any player of any game has to answer—what is the best action one can take? The difficulty lies in the fact that the best play generally depends upon the choice of the other player(s). This is the fundamental concept of game theory. We will see in subsequent sections how game theory deals with the definition of "the best action" and how we search for such an action.

2.1 Key game elements

Considering the example of the RSP game, we see some of the key concepts we need in game theory. We must specify the rules of the game. Some of the most important factors are who plays the game (two players), what actions these players can make (choose Rock, Scissors, or Paper) and what is the outcome that follows a specific set of actions. There are also some other factors, necessary for a complete mathematical description of a game, which are often implicitly assumed, without explicitly stating them. Examples of such factors include the requirement that the players make their choices simultaneously (i.e. without knowledge of the choice the other player will make) or in a given order; that the game is played once or several times, etc.

Below, we introduce the concept of a game in normal form. An alternative game concept will be introduced in Chapter 10.

2.1.1 Players

A game scenario may have any number of players, who make strategic decisions and obtain rewards accordingly. A particular game can have any finite number, or even an infinite number, of players. The most common number of players is two, as in the RSP game, and this scenario is covered by much of the literature on the subject.

As described in Chapter 1, it is very common in biology that the game is defined to have two players only, but it can still be "played" within a population of more than two individuals. In fact, very often the population is considered effectively infinite. We may think about the population as a pool of potential players that engage in pairwise contests playing a particular (relatively simple) game between two players rather than every individual being a player of a (relatively complicated) game. We can argue about the realism of these assumptions (see Chapters 4 and 9) but this is one of the simplifications modellers use to make their model mathematically tractable. As we will see in this book, there are enough mathematical difficulties even with this simplification.

The popularity of pairwise games is not simply due to their relative simplicity, but also to the wide applicability of this idea. However, real populations contain many individuals, and their interactions cannot always be reduced to a sequence of independent pairwise contests. Sometimes, larger groups of individuals may meet and simultaneously fight for a resource; this is considered in Chapter 9. Similarly, games may be pairwise, but the result of one may influence which individual will be the next opponent, or the likelihood of success in a subsequent game, thus generating a structure of non-independent pairwise games. An example of this is in the modelling of dominance hierarchy formation discussed in Chapter 16 and also in Chapter 9. A different situation again is where an individual must employ a strategy in its general behaviour which is not used against any single individual or group of individuals, but whose reward is affected by the play of everyone in the population. Such a game is termed "playing the field", and a classical example of this is the sex ratio game which we look at in detail in Chapter 4.

2.1.2 Strategies

Once we have decided upon the players, we have to specify what choices the players can make. There may be a number of points where a player has a choice to make, and the possible choices at each point are often referred to as *actions*. Mathematically, a *strategy* can be defined as a complete specification of the chosen action in every possible position in a game that could be encountered. From the biological side, a strategy can be thought of as a phenotype, such as skin colour in lizards or toxin production in bacteria, or a behaviour such as the brood parasitism of cuckoos, which is determined genetically. There is generally a distinction between a *pure strategy* and a *mixed strategy*. This distinction is, mathematically, very often only a matter of definition. However, biologically, the distinction is often important.

2.1.2.1 Pure strategies

A *pure strategy* is a single choice of what strategy to play. There can be a finite or infinite number of pure strategies for a particular game. In the RSP game, the strategy is a choice in an interaction, such as "play Rock", "play Scissors", "play Paper". If the game is modified so that the players play the RSP game until there have been three decisive rounds (i.e. rounds with a winner, with the overall winner being the one who wins at least two out of three rounds), the pure strategies can get increasingly complex. A pure strategy in such a case specifies what to play in every round, conditional on every possible sequence played previously. This means that even in very simple scenarios where every choice is one of two, the number of pure strategies can turn out to be very large. Biology plays an important role in trimming the set of pure strategies in such cases. For example, if players have no memory at all, then even in the multiple-round RSP game, a pure strategy can still only be one of "play Rock", "play Scissors", "play Paper" (all the time). If the players have a short-term memory, a strategy can be a rule like "start with Rock and then play whatever would beat the opponent in the previous round".

If biology does not help with trimming the strategy set, we often deal with such games in a different way. We introduce the idea of Extensive Form Games in Chapter 10 and cover repeated games in more detail in Chapter 15 (see also Section 4.2.5).

2.1.2.2   Mixed strategies

If there are finitely many pure strategies, given by the set $\{S_1, S_2, \ldots, S_n\}$, then a mixed strategy is defined as a probability vector $\mathbf{p} = (p_1, p_2, \ldots, p_n)$ where p_i is the probability that in the current game the player will choose pure strategy S_i. For example, in the RSP game, a player may choose to play each of Rock and Scissors half of the time, but never play Paper, which would be represented by the vector $(1/2, 1/2, 0)$.

Definition 2.3. *The* support *of* $\mathbf{p}$, $S(\mathbf{p})$, *is defined by* $S(\mathbf{p}) = \{i : p_i > 0\}$, *so that it is the set of indices of pure strategies which have non-zero chance of being played by a* $\mathbf{p}$*-player.*

For example, the support of the above strategy $(1/2, 1/2, 0)$ is $\{1, 2\}$.

A pure strategy can be seen as a special case of a mixed strategy; a strategy S_i can be identified with a "mixed strategy" $(0, \ldots, 0, 1, 0, \ldots, 0)$ with 1 at the ith place meaning that the player using (mixed) strategy S_i uses (pure) strategy S_i with probability 1. The set of all mixed strategies can be represented as a simplex in $\mathbb{R}^n$ with vertices at $\{S_1, S_2, \ldots, S_n\}$. One advantage of this approach is that one can see a mixed strategy as a convex combination of pure strategies,

$$\mathbf{p} = (p_1, p_2, \ldots, p_n) = \sum_{i=1}^{n} p_i S_i. \tag{2.1}$$

If this linearity is carried over to the rewards from the game, the payoffs (see Section 2.1.3) of the game, it can be exploited further (as shown in Chapter 6 on matrix games).

In the example of the RSP game, pure strategies can be seen as three points in 3D space, with $(1, 0, 0), (0, 1, 0), (0, 0, 1)$ representing strategies "play Rock", "play Scissors", "play Paper". The set of mixed strategies can then be seen as points in an equilateral triangle with vertices at the pure strategies; see Figure 2.1. The notion of a mixed strategy is naturally extended even to cases where the set of pure strategies is infinite. For example, if the set of pure strategies can be seen as a set $[0, \infty)$ (such as in the war of attrition investigated in

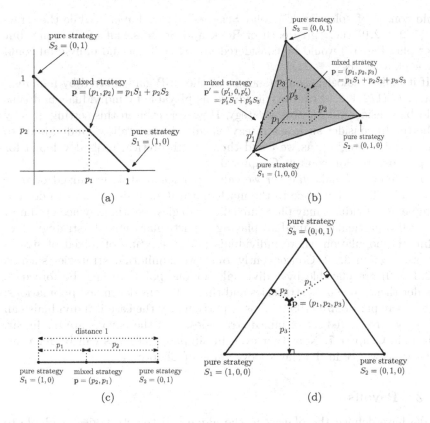

FIGURE 2.1: Two ways to visualise pure and mixed strategies. Mixed strategies of a game with two pure strategies can be seen in (a) $\mathbb{R}^2$, or (c) $\mathbb{R}^1$. A game with three pure strategies can be seen in (b) $\mathbb{R}^3$, or (d) $\mathbb{R}^2$.

Section 4.3) then a mixed strategy is given by a probability measure $\mathbf{p}$ on $[0, \infty)$ such that, for any (measurable) $A \subseteq [0, \infty)$, $\mathbf{p}(A)$ is the probability that in the current game a player will choose a strategy x belonging to A.

2.1.2.3 Pure or mixed strategies?

When modelling a real biological scenario, what is defined as a pure and what as a mixed strategy depends solely on the choice of the modeller and there are mathematical and also biological factors that must be taken into consideration. From a mathematical perspective, a rule of thumb would be that the smaller the set of pure strategies, the easier the analysis. From the biological perspective, a pure strategy is a phenotype, i.e. it should be something an individual can really play. For example, if there is strong evidence that, for the RSP game, individuals can only play one symbol all the time (as in the case of the lizards from Example 2.2), then the set of pure strategies would consist of "play Rock", "play Scissors", "play Paper", while the strategy $(1/2, 1/2, 0)$ (i.e. "play each of Rock and Scissors half of the time, but never play Paper") would be considered mixed as in fact no individual could play such a strategy.

If it makes biological sense (such as in the RSP game played by humans), a strategy $(1/2, 1/2, 0)$ can be considered as playable by individuals and thus could be considered as a pure strategy. However, even in this setting, it may be better to consider such a strategy as mixed (and "allow" individuals to play mixed strategies), as we could then use theoretical tools developed for this purpose; see for example Chapter 6.

On the other hand, whether we call a particular strategy mixed or pure often makes little difference to the mathematical analysis we have to do. For example, if individuals may play "mixed" strategies, we still may need to know how well a particular individual playing a particular "mixed" strategy does within the population where individuals play a mixture of mixed of strategies; see Section 3.2.2. Consequently, once we admit that strategies such as $(1/2, 1/2, 0)$ are playable by individuals, at one point we may be forced to consider them to be pure strategies and then still consider mixed populations; see for example Bomze (1990). This is particularly the case if individuals can only play a restricted set of mixed strategies, or if the payoffs are nonlinear, such as in Chapter 7. Note that even in simple games this can be relevant, however, as we see in the Hawk-Dove game of Chapter 4.

2.1.3　Payoffs

We have defined the players in the game and the strategies available to them, but so far we have omitted a vital aspect of the game for any player which provides the criteria for which strategies are successful—payoffs, i.e. rewards and/or punishments. In biology, it is often assumed that fitness is a measure of how many offspring that reach adulthood an individual produces (see Chapter 5), and thus natural selection leads to the spread of strategies that are associated with the highest fitness. We assume that the payoff of a game is some kind of contribution to the individual's fitness; very often, the fitness of the individual is considered to be a payoff of the game plus some additional background fitness (the same for all individuals). The background fitness is biologically relevant as the interactions between individuals are very often not the only source of resources. We will see (for example in Chapters 3 and 6) that in many cases, the background fitness is essentially irrelevant to the analysis.

If a game has m players, and $\mathbf{S_i}$ is the set of pure strategies available to player i, then the payoff to a player i is a (usually real) function f_i from $\mathbf{S_1} \times \mathbf{S_2} \times \cdots \times \mathbf{S_m}$.

For the RSP game, there are two players, each one has three strategies and there are thus nine pairs of payoffs, defined by the $3 \times 3 = 9$ combinations of the choice of player 1 and player 2. The payoffs can thus be seen as a bimatrix,

$$
\begin{pmatrix}
(\ 0,\ \ 0) & (\ \ 1,-1) & (-1,\ \ 1) \\
(-1,\ \ 1) & (\ \ 0,\ \ 0) & (\ \ 1,-1) \\
(\ \ 1,-1) & (-1,\ \ 1) & (\ \ 0,\ \ 0)
\end{pmatrix},
\tag{2.2}
$$

where the first (second) number of the ij entry $(f_1, f_2)_{ij}$ represents the payoff to the first (second) player if the first plays strategy S_i while the second plays strategy S_j.

The RSP game is an example of a *zero sum game*, where the interests of the players are diametrically opposed, so that the total reward will be zero. This is a very common concept in game theory, and there is a significant body of theory on zero sum games. This is often logical for financial transactions where there is a fixed amount of money to be divided between the players. In fact, this concept is much less frequently used when game theory is applied to biology, and there will not be much use made of it throughout the book, since animal interactions are rarely zero sum.

2.1.3.1 Representation of payoffs by matrices

In general, payoffs for a game played by two players with each having only finitely many pure strategies can be represented by two matrices. For example, if player 1 has available the strategy set $\mathbf{S} = \{S_1, \ldots, S_n\}$ and player 2 has the strategy set $\mathbf{T} = \{T_1, \ldots, T_m\}$, then the payoffs in this game are completely determined by the pair of matrices

$$
A = (a_{ij})_{i=1,\ldots,n;j=1,\ldots,m}, B = (b_{ij})_{i=1,\ldots,m;j=1,\ldots,n},
\tag{2.3}
$$

where a_{ij} and b_{ji} represent rewards to players 1 and 2, respectively, after player 1 chooses pure strategy S_i and player 2 chooses pure strategy T_j. We thus have all of the possible rewards in a game given by a pair of $n \times m$ matrices A and B^T, which is known as a bimatrix representation. Entries are often given as a pair of values in a single matrix form as seen in (2.2) for an RSP game.

Sometimes (as in the case of the RSP game), the choice of which player is player 1 and which player 2 is arbitrary, and thus the strategies that they have available to them are identical, i.e. $n = m$ and (after possible renumbering) $S_i = T_i$ for all i. Further, since the ordering of players is arbitrary, we can switch the two without changing their rewards, so that their payoff matrices (2.3) satisfy $b_{ij} = a_{ij}$, i.e. $A = B$. We thus now have all of the possible rewards in a game given by a single $n \times n$ matrix

$$
A = (a_{ij})_{i,j=1,\ldots,n},
\tag{2.4}
$$

where in this case, a_{ij} is a reward to the player that played strategy S_i while its opponent played strategy S_j. The payoff matrix for the RSP game is thus given by

$$\begin{pmatrix} 0 & 1 & -1 \\ -1 & 0 & 1 \\ 1 & -1 & 0 \end{pmatrix}. \tag{2.5}$$

In fact, we often generalise the RSP game so that it is no longer a zero-sum game. This is the case in the lizards game from Example 2.2; see Sinervo and Lively (1996) for the creation of the payoff matrix in that game. In general, for an RSP game involving players who choose different pure strategies, where the unchosen strategy is S_i, we set the reward for a win to be $a_i > 0$ and that for a loss to be $-b_i$ where $b_i > 0$. This thus gives the payoff matrix as

$$\begin{pmatrix} 0 & a_3 & -b_2 \\ -b_3 & 0 & a_1 \\ a_2 & -b_1 & 0 \end{pmatrix}. \tag{2.6}$$

Whilst analysis of a single payoff matrix is more convenient, one reason for using and analysing bimatrix games, rather than reducing everything to a single matrix, is that we can model individuals with two distinct roles, such as owner and intruder, where the strategy set and payoffs may be different for the two players. We shall see some games of this kind in Chapter 8.

2.1.3.2   Contests between mixed strategists

It is often necessary to specify the reward to a player playing a mixed strategy $\mathbf{p}$ against another player playing a mixed strategy $\mathbf{q}$.

The problem with evaluating the reward from the contest is that player 1 plays strategy S_i with probability p_i, i.e. in most cases does not stick to a single strategy. The same is true for player 2. Consequently, nobody can predict in advance what strategy players 1 and 2 will play; thus (in most cases) nobody can predict in advance the rewards to the individual players. We can, however, evaluate the *expected payoff* to individual players.

Consider a game whose payoffs are given by a matrix A. If player 1 plays the mixed strategy $\mathbf{p}$ and player 2 plays the mixed strategy $\mathbf{q}$, then the proportion of games that involve player 1 playing S_i and player 2 playing S_j is simply $p_i q_j$. The reward to player 1 in this case is a_{ij}. The expected reward to player 1, which we shall write as $E[\mathbf{p}, \mathbf{q}]$, is thus obtained by averaging over all possibilities, i.e.

$$E[\mathbf{p}, \mathbf{q}] = \sum_{i,j} a_{ij} p_i q_j = \mathbf{p} A \mathbf{q}^{\mathbf{T}}. \tag{2.7}$$

Note that for games where the roles of players 1 and 2 are distinct, if player 1 chooses $\mathbf{p}$ and player 2 chooses $\mathbf{q}$, we would write the reward to players 1 and 2 as $E_1[\mathbf{p}, \mathbf{q}]$ and $E_2[\mathbf{q}, \mathbf{p}]$, respectively; and similarly as above, these become

$$\begin{aligned} E_1[\mathbf{p}, \mathbf{q}] &= \mathbf{p} A \mathbf{q}^{\mathbf{T}}, \\ E_2[\mathbf{q}, \mathbf{p}] &= \mathbf{q} B \mathbf{p}^{\mathbf{T}}. \end{aligned} \tag{2.8}$$

2.1.3.3 Generic payoffs

We can think of payoffs as parameters of the game-theoretical model. For example, if we consider two-player games where each player has exactly two strategies, the payoffs can be represented by a 2×2 matrix

$$\begin{pmatrix} a & b \\ c & d \end{pmatrix}, \tag{2.9}$$

and our model thus has four parameters a, b, c and d. This gives us a *parameter space*. In this case, the parameter space is $\mathbb{R}^4$. We can think of the payoffs being drawn from the parameter space based on certain probabilistic distributions dictated by the underlying biology. We might expect that each parameter may be able to take a value from a continuum of possibilities. However, for the sake of simplicity, let us suppose that there are biological reasons for all payoffs to be within the range $[0, 2]$, we have $a = d = 1$, b being (almost) normally distributed around 0.5 and c being (almost) normally distributed around 1.5 (we note that in the analysis of matrix games, fixing the value of one element in each column does not affect the generality of our analysis, as we see in Chapter 6). This would mean that in reality, the parameter space is restricted to the square $[0, 2]^2 = \{(\beta, \gamma); \beta, \gamma \in [0, 2]\}$; the point (β, γ) in the parameter space determines the payoffs $a = 1, b = \beta, c = \gamma$ and $d = 1$; moreover the point (β, γ) is drawn from the parameter space randomly so that the distributions of β and γ satisfy the biological requirements.

The *generic payoffs assumption* is that we can ignore sets of measure zero in the parameter space. More specifically, given the parameter space P, the (biologically motivated) distribution over it (i.e. a measure μ on P) and a set of measure zero on the parameter space (i.e. $A \subset P, \mu(A) = 0$), we can draw the parameters only from $P \setminus A$. In other words, the assumption says that since the probability of the parameters (and consequently the payoffs) coming from a given set A is 0, so that it is an extremely unlikely event, we can ignore this event completely.

The assumption can be used by selecting the set A as a set of roots of a function G over the parameter space (or even countably many functions $\{G_n\}$). If we do that, the generic payoff assumption can be formulated as assuming that $G \neq 0$. For example, in the game above, we may consider a function $G(\beta, \gamma) = \beta - \gamma$. This function determines the diagonal of the square, see Figure 2.2, and by the generic payoff assumption we can thus safely assume that the parameters are drawn so that $G(\beta, \gamma) \neq 0$, i.e. $\beta \neq \gamma$, which means that payoffs are such that $b \neq c$. Equivalently, it is very unlikely that a particular game scenario the modeller chooses to investigate would have payoffs such that $b = c$. In fact, as we see in Section 6.2, complications can occur when another pure strategy has the same payoff as the pure strategy against itself (i.e. $a = c$ or $b = d$ here). Thus the functions that we particularly do not want to equal 0 in this case are $\beta - 1$ and $\gamma - 1$.

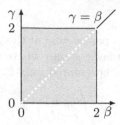

FIGURE 2.2: Example of a parameter space, for a game with payoffs given in (2.9) and restrictions $a = d = 1$, $b, c \in [0, 2]$ and $b \neq c$.

However, it is important that the set A (or the functions G_n) are chosen first and the parameters or payoffs second. Once we choose the parameters and thus the payoff matrix for the game, we cannot come up with a specific function, and claim that by the generic payoff assumption, the function is not 0. Thus the payoffs can satisfy the generic requirement and at the same time satisfy some nontrivial identities that arise from our analysis as seen in the example below. A rough analogy is in statistics, where the hypotheses must be set before collecting and analysing the data, and not afterwards.

Example 2.4. Consider a game whose payoff matrix is given by

$$\begin{pmatrix} 0 & a & b \\ c & 0 & d \\ e & f & 0 \end{pmatrix},$$ (2.10)

and a strategy $\mathbf{p} = \left(\frac{a}{a+c}, \frac{c}{a+c}, 0 \right)$. We can see that

$$E[S_1, \mathbf{p}] - E[S_2, \mathbf{p}] = \frac{ac}{a+c} - \frac{ac}{a+c} = 0,$$ (2.11)

which seems to go against the "generic" assumptions. However, we will see later (for example in the discussion which follows Lemma 6.7) that (2.11) is a special case of a result that holds for all generic payoffs when $\mathbf{p}$ is an Evolutionarily Stable Strategy (see Definition 3.4). The solution to this lies in the fact that if we define

$$G(\alpha, \beta, \gamma, \delta, \varepsilon, \zeta) = \text{``}E[S_1, \mathbf{p}] - E[S_2, \mathbf{p}]\text{''} = \frac{c}{a+c}\alpha - \frac{a}{a+c}\gamma,$$ (2.12)

it is clear that we have defined our function only after we picked our parameters (which is not allowed). If we define the equivalent function prior to parameter selection, we obtain

$$G(\alpha, \beta, \gamma, \delta, \varepsilon, \zeta) = \text{``}E[S_1, \mathbf{p}] - E[S_2, \mathbf{p}]\text{''} = \frac{\alpha\gamma}{\alpha+\gamma} - \frac{\alpha\gamma}{\alpha+\gamma},$$ (2.13)

so trivially $G = 0$ for all values for which $\alpha + \gamma \neq 0$, i.e. on a set of non-zero measure and thus this function also cannot be used. On the other hand, the function

$$G(\alpha, \beta, \gamma, \delta, \varepsilon, \zeta) = \text{“}E[S_1, \mathbf{p}] - E[S_3, \mathbf{p}]\text{”} = \frac{\alpha\gamma}{\alpha + \gamma} - \frac{\alpha\varepsilon + \gamma\zeta}{\alpha + \gamma} \qquad (2.14)$$

shows that for generic payoffs, we can assume that once $\mathbf{p}$ is a mixed strategy with $p_3 = 0$ such that $E[S_1, \mathbf{p}] = E[S_2, \mathbf{p}]$, we get $E[S_3, \mathbf{p}] \neq E[S_1, \mathbf{p}]$.

In most of the following text, and throughout the book, we will assume that payoffs are generic; we will only explicitly discuss this if a non-generic case is considered, and this choice has an important effect on analysis. Often modellers will choose payoffs which are formally non-generic, which nevertheless do not affect the analysis, for the sake of neatness; an example is the Hawk-Dove game of Chapter 4, which has a payoff matrix as above where $b = 2c$. Whether or not this assumption can be justified depends on the particular problem and also why we are solving such a problem. For a discussion of the issues around generic games, see Samuelson (1997). Non-generic payoffs create situations that can be the most mathematically complicated, with the least genuine biological insight (see Abakuks, 1980, for such a case). Thus it is generally safe to ignore non-generic payoffs when our goal is to model biological scenarios.

2.1.4 Games in normal form

In the above, we have discussed a game in so-called *normal form*. For two players, such a game is defined by

$$\{\mathbf{S}, \mathbf{T}; f_1(S, T), f_2(T, S)\}, \qquad (2.15)$$

where $\mathbf{S}$ is the set of pure strategies available to player 1, $\mathbf{T}$ is the set available to player 2 and $f_1(S, T)$ and $f_2(T, S)$ are the payoffs to players 1 and 2, respectively, when player 1 plays $S \in \mathbf{S}$ and player 2 plays $T \in \mathbf{T}$. More generally, for n players, a normal form game is defined by

$$\{\mathbf{S}_1, \dots, \mathbf{S}_n; f_1, \dots, f_n\}, \qquad (2.16)$$

where $\mathbf{S}_i$ is the set of pure strategies available to player i and f_i is the payoff function that specifies the payoff to player i for all possible combinations of the strategies of all of the players.

2.2 Games in biological settings

So far we have considered a single game contest between individuals. However, individuals are not usually involved in a single contest only. More often, they are repeatedly involved in the same game (either with the same or different opponents). Each round of the game contributes a relatively small portion of the fitness to the total reward. Of ultimate interest is the function $\mathcal{E}[\sigma; \Pi]$ describing the fitness of a given individual using a strategy σ in a given population represented by Π.

We shall represent by $\delta_\mathbf{p}$ a population where the probability of a randomly selected player being a $\mathbf{p}$-player is 1.

We note that $\mathcal{E}[\sigma; \Pi]$ can be observed by biologists only if there are players using σ in the population Π, i.e. there is some fraction of the population ε playing σ with the remainder of the population represented by Π', which we shall write as $\Pi = \varepsilon\delta_\sigma + (1 - \varepsilon)\Pi'$. However, from the mathematical point of view, even expressions like $\mathcal{E}[\mathbf{q}; \delta_\mathbf{p}]$ for $\mathbf{q} \neq \mathbf{p}$ are considered. This seems to represent the payoff to an individual playing $\mathbf{q}$ in a population entirely composed of $\mathbf{p}$-players, which may seem contradictory. In finite populations, this is true, as the population consists entirely of $\mathbf{p}$-players. However, in an infinite population there may be a sub-population (e.g. a single individual) which makes up a proportion 0 of the population, and so such payoffs are logical.

The evaluation of $\mathcal{E}[\sigma; \Pi]$ depends crucially on the exact mechanism the individuals use to play the game. Many of the scenarios covered in the early literature implicitly assume the following model.

Example 2.5 (Matrix games). A population of individuals is engaged in pairwise contests. Every single contest can be represented as a game (the same one for every contest) whose payoffs are given by a (bi)matrix as discussed above (i.e. the set of pure strategies is finite). Every game is completely independent of each other. Any particular individual can play one or several such games against randomly chosen opponents. The total payoff to the individual is taken as an average payoff of all the games it plays. In the more simple games, we assume an infinite population of absolutely identical individuals in every way except, possibly, their chosen strategy.

We will see below how to calculate $\mathcal{E}[\sigma; \Pi]$ for the above scenario. In most other cases, the evaluation of $\mathcal{E}[\sigma; \Pi]$ in full generality (i.e. for all σ and all Π) is a quite difficult, if not impossible, task. However, as we will see in Section 3.2.2, we will rarely need knowledge of $\mathcal{E}[\sigma; \Pi]$ in full generality.

Below we will formalise the notion of the population and will also show how to evaluate $\mathcal{E}$.

2.2.1 Representing the population

In most cases, the population is considered infinite and the structure of the population is described by a density of individuals playing a particular strategy. This is a special case of representing a population by a measure on the strategy space.

For any (pure or mixed) strategy $\mathbf{p}$, as described above, we let $\delta_\mathbf{p}$ denote the population where a randomly selected player plays strategy $\mathbf{p}$ with probability

1. In other words, the probability of selecting a player with a strategy $\mathbf{q} \neq \mathbf{p}$ is 0.

Let δ_i denote the population consisting of individuals playing strategy S_i (with probability 1). Similarly to the case of mixed strategies, the linear structure of the strategy simplex (in fact, the linear structure of the space of all measures on the simplex) allows us to add (or to make convex combinations) of the population structures and we can see δ_i as a column vector $S_i^{\mathbf{T}}$. For example, $\sum_i p_i \delta_i$ means the population where the proportion of individuals playing strategy S_i is p_i. In particular, if individuals can play only finitely many pure strategies, then we can visualise a particular population as a point in the strategy simplex exactly in the same way that we did for mixed strategies; see Figure 2.1.

If we are concerned about the actual density (or count) of individuals playing a certain strategy, we can just multiply δ_i by the appropriate density (or count).

Also, in the spirit of (2.1), we can write

$$\sum_i p_i \delta_i = \mathbf{p}^{\mathbf{T}}. \tag{2.17}$$

Thus a row vector $\mathbf{p}$ denotes a mixed strategy (a given individual plays strategy S_i with probability p_i) while a column vector $\mathbf{p}^{\mathbf{T}}$ denotes a mixed population where a random individual plays a strategy S_i with probability p_i.

In the earlier models, there was little mathematical difference between $\delta_{\mathbf{p}}$ (a uniform population of all individuals playing a mixed strategy $\mathbf{p}$) and $\mathbf{p}^{\mathbf{T}} = \sum_i p_i \delta_i$ (a mixed population with different individuals playing potentially different, but always pure, strategies). Thus $\mathbf{p}$ may have sometimes meant the mixed strategy, sometimes the distribution of strategies in the population. This lack of distinction is natural, because one can see $\delta_{\mathbf{p}}$ as a barycenter of the population $\sum_i p_i \delta_i$. As is the case in physics, where the motion of a rigid object is often described by the motion of the barycenter, in biological games there is sometimes no real need to distinguish between the population $(\sum_i p_i \delta_i)$ and its barycenter $\delta_{\mathbf{p}}$. This is commonly exploited in classical matrix games (see Example 2.5 and Chapter 6). In this case, a focal individual plays against a randomly chosen opponent from the entire population, and it makes no difference to the focal individual (in an infinite population; this is not true for a finite population, see Chapter 12 or Taylor et al. (2004)) if the population is described by $\sum_i p_i \delta_i$ or by $\delta_{\mathbf{p}}$. In either case, the random opponent will play strategy S_i with probability p_i.

On the other hand, sometimes the nature of the game is such that the exact structure of the population matters because it influences the "effective structure" of the population, as in the following example, and we will thus try to distinguish between the two notions. We will introduce the notion of *polymorphic-monomorphic equivalence* later in Chapter 7 by the expression (7.4).

Example 2.6 (Variation on the war of attrition). Let individuals have two options when it comes to pairwise fights—either fight for a very short time and then leave, or fight for a very large time. Individuals can enter many fights during a single day, but we assume that individuals cannot engage in any other activity while in a fight. Thus fighting individuals are effectively temporarily absent from the population. In particular, a random opponent to a focal individual in the population $\mathbf{p}^{\mathbf{T}} = p_1\delta_1 + p_2\delta_2$ will be playing S_1 (short fights) with probability higher than p_1 because many individuals playing S_2 will be engaged in long fights with each other. This is in contrast to a population described by $\delta_{\mathbf{p}}$, where any opponent will always play S_1 with probability p_1.

2.2.2 Payoffs in matrix games

Recall that we are interested in the function $\mathcal{E}[\sigma; \Pi]$, the *expected* payoff to the individual that uses a strategy σ in the population described by Π. Note again the word expected. In most cases, due to randomness in the model, one cannot evaluate the reward exactly but has to rely on the expected value.

As an example, we will evaluate $\mathcal{E}[\sigma; \Pi]$ in matrix games (see Example 2.5). Let the population be described by $\Pi = \sum_i p_i\delta_i = \mathbf{p}^{\mathbf{T}}$. Let a focal individual, using strategy σ, play k games. One cannot predict against what strategy the opponent will play. However, the expected number of plays against an individual using a strategy S_i is p_ik. Thus the expected total reward is $\sum_i p_ikE[\sigma, S_i] = \sum_i p_ik\sum_j \sigma_j a_{ji}$ and the expected average payoff from one game is

$$\mathcal{E}[\sigma; \mathbf{p}^{\mathbf{T}}] = \frac{1}{k}\sum_i p_ik\sum_j \sigma_j a_{ji} = \sigma A\mathbf{p}^{\mathbf{T}} = E[\sigma, \mathbf{p}]. \tag{2.18}$$

Note that

$$\mathcal{E}[\sigma; \delta_{\mathbf{p}}] = \mathcal{E}[\sigma; \mathbf{p}^{\mathbf{T}}]. \tag{2.19}$$

Thus the payoff function is linear in the strategy of the players as well as in the composition of the population. This gives a quadratic form for the payoff function. This provides significant simplification, and allows important results to be proved for both the dynamic (see Chapter 3 or Cressman, 2003; Hofbauer and Sigmund, 1998) and static approaches (see Chapter 6). In contrast nonlinear games are more complex (see Chapter 7).

2.3 Further reading

Very accessible introductions to game theory can be found in Stahl (1998) and Straffin (1993). A comprehensive discussion of game-theoretical concepts can be found in van Damme (1991), see also Osborne and Rubinstein (1994).

A well-written, not so mathematical, book on game theory is Fisher (2008). See also Chapter 10 for further concepts, in general, game theory.

For more books on Evolutionary Game Theory, see Gintis (2000), Weibull (1995), Hofbauer and Sigmund (1998), Samuelson (1997), Mesterton-Gibbons (2000), Friedman and Sinervo (2016) and Tanimoto (2015). The classical book from a more biological perspective is Maynard Smith (1982). A very good book on mathematical modelling for biologists is Kokko (2007) and a review on the use of game theory in biology-related disciplines is Schuster et al. (2008). An interesting and recent book is McNamara and Leimar (2020).

2.4 Exercises

Exercise 2.1. Identify pure strategies for the game where two players play the RSP game until there have been three decisive rounds (i.e. rounds with a winner, with the overall winner being the one who wins at least two out of three rounds).

Exercise 2.2 (Bacterial RSP game, Kerr et al., 2002). Model the following production of an antibiotic colicin in bacteria *E. coli* as an RSP game. There are three strains of bacteria, one strain that produces a colicin (and is also immune to it), one strain that is only immune to colicin (but does not produce it) and one strain that is neither immune nor producing.

Exercise 2.3. Two friends Paul and John are trying to meet up, but have no means of communication. They have two favourite bars, which are at opposite ends of the city, and so there is only time to go to one. Bar A is better than Bar B, but the most important thing is for them to meet. Model this scenario as a (bi)matrix game.

Exercise 2.4. Suppose that in Exercise 2.3 John wants to meet Paul, but that Paul wants to avoid John. He could do this easily by going somewhere else (which is an option he will consider), but would prefer to go to one of his favourite bars, if John was not there (avoiding John is more important than having a drink). Model this scenario as a bimatrix game.

Exercise 2.5. Identify the positions of the following strategies on the diagram from Figure 2.1 (d).
(i) (0,1/2,1/2), (ii) (0,1/3,2/3), (iii) (1/3, 1/3, 1/3), (iv) (2/5, 2/5, 1/5), (v) (1/10, 3/5, 3/10).

Exercise 2.6. Show that in a matrix game, for any mixture $\sum \alpha_j \delta_{\mathbf{p}_j}$ of mixed strategies, we get

$$\mathcal{E}[\sigma; \sum_j \alpha_j \delta_{\mathbf{p}_j}] = \mathcal{E}[\sigma; \delta_{\bar{\mathbf{p}}}], \qquad (2.20)$$

where $\bar{\mathbf{p}} = \sum_j \alpha_j \mathbf{p}_j$.

Exercise 2.7. In an RSP game with payoff matrix (2.5) find an expression for the payoff to a player playing $\mathbf{p} = (p_1, p_2, p_3)$ against a player playing (q_1, q_2, q_3). Suppose that I know the strategy $\mathbf{q}$ that my opponent will choose. Find the strategy (strategies) which maximise my payoff.

Exercise 2.8. For the payoff matrix from (2.9), which of the following restrictions on the parameters can be considered non-generic?
(i) $a > 2$, (ii) $a = 0$, (iii) $a = c$, (iv) $a > 2, b < 1, c < 1, d > 3$,
(v) $a > 2, b = 1, c < 1, d > 3$, (vi) $a > 2, b = 1, c = 1, d > 3$,
(vii) $a > 2, b = 1, c = 1, d = 3$, (viii) $ab = cd$.

Exercise 2.9. Consider a population of animals which play a two-strategy matrix game from (2.9) with payoffs $a = 1, b = 0, c = 2, d = -2$. Find the expected payoffs of all of the individuals within a population which comprises a fraction 0.4 of individuals playing (0.5,0.5), a fraction 0.3 of individuals playing (1,0) and a fraction 0.3 of individuals playing (0.2,0.8).

Exercise 2.10. For the population from Exercise 2.9, find the mean strategy of members of the population (you may have already calculated it for Exercise 2.9). Find the mean strategy of the next opponent, if this opponent is not selected at random as in matrix games, but rather with probability proportional to its probability of playing pure strategy S_1.

Chapter 3

Two approaches to game analysis

In this chapter we describe the two basic approaches to the analysis of evolutionary games, the dynamical approach and the static approach. For the dynamical approach we look at the most commonly used dynamics such as the continuous replicator dynamics and adaptive dynamics. We also briefly discuss the important issue of timescales in evolutionary modelling. For the static approach we introduce the main ideas of static theory, the Nash equilibrium and the Evolutionarily Stable Strategy. We then discuss some important issues about the composition of populations. Finally we compare the two approaches.

3.1 The dynamical approach

In this section, we are interested in describing how the strategies played by the individuals in the population change over time. There are two classical approaches in the literature, *replicator dynamics* and *adaptive dynamics*, each modelling a different aspect of evolution.

3.1.1 Replicator dynamics

Consider a population described by $\mathbf{p}^{\mathbf{T}} = \sum_i p_i \delta_i$, i.e. the frequency of individuals playing strategy S_i is p_i. To simplify notation, let $f_i(\mathbf{p})$ denote the fitness of individuals playing S_i in the population in this section. Further, for the purpose of deriving the equation of the dynamics, assume that the population has N individuals and $N_i = p_i N$ of those are using strategy S_i (this is convenient for the immediate derivations below, but often we shall assume infinite populations and only the frequencies matter).

3.1.1.1 Discrete replicator dynamics

Suppose that the population has discrete generations that are non-overlapping (i.e. all adults die before the new generation, as happens in many insect species) and asexual reproduction. We assume that each individual playing strategy S_i generates $f_i(\mathbf{p})$ copies of itself in the next generation so that $N_i(t+1) = N_i(t)f_i(\mathbf{p}(t))$ and thus get the so-called *discrete replicator dynamics* (see Bishop and Cannings, 1978)

$$p_i(t+1) = \frac{N_i(t+1)}{N(t+1)} = \frac{N_i(t)f_i(\mathbf{p}(t))}{\sum_j N_j(t)f_j(\mathbf{p}(t))}$$

$$= p_i(t)\frac{f_i(\mathbf{p}(t))}{\bar{f}(\mathbf{p}(t))}, \tag{3.1}$$

where

$$\bar{f}(\mathbf{p}) = \sum_i p_i f_i(\mathbf{p}) \tag{3.2}$$

is the average fitness in the population. In the case of matrix games where payoffs are given by matrix A, we get $f_i(\mathbf{p}) = (A\mathbf{p^T})_i + \beta$ and $\bar{f}(\mathbf{p}) = \mathbf{p}A\mathbf{p^T} + \beta$ where β is a background fitness, representing the contribution to an individual's fitness that does not come from direct conflicts with others. Thus the dynamics (3.1) becomes

$$p_i(t+1) = p_i(t)\frac{\left(A(\mathbf{p}(t))^{\mathbf{T}}\right)_i + \beta}{\mathbf{p}(t)A(\mathbf{p}(t))^{\mathbf{T}} + \beta}. \tag{3.3}$$

Note that, in contrast to the continuous dynamics below, the background fitness β can have a significant effect on the dynamics. In general, for small β evolution occurs faster, but the process is less stable than for larger β; see Exercise 3.1. Also, when $\beta \to \infty$ and the generation times tend to zero, the continuous dynamics discussed below is a limiting case of the discrete dynamics (Hofbauer and Sigmund, 1988).

3.1.1.2   Continuous replicator dynamics

If the population is very large, has overlapping generations and asexual reproduction, we may consider N_i and $p_i = N_i/N$ to be continuous variables. Population growth is given by the differential equation

$$\frac{\mathrm{d}}{\mathrm{d}t}N_i = N_i f_i(\mathbf{p}(t)), \tag{3.4}$$

and we get the so-called *continuous replicator dynamics* (Taylor and Jonker, 1978, Hofbauer and Sigmund, 1998, Chapter 7),

$$\frac{\mathrm{d}}{\mathrm{d}t}p_i = p_i\Big(f_i(\mathbf{p}(t)) - \bar{f}(\mathbf{p}(t))\Big). \tag{3.5}$$

As before, for matrix games, the dynamics (3.5) becomes

$$\frac{\mathrm{d}}{\mathrm{d}t}p_i = p_i\Big(\big(A(\mathbf{p}(t))^{\mathbf{T}}\big)_i - \mathbf{p}(t)A(\mathbf{p}(t))^{\mathbf{T}}\Big). \tag{3.6}$$

The background fitness β is irrelevant for the dynamics in this case and only the payoffs given by A matter.

It should be noted that the replicator dynamics does not allow any mutations and thus if a strategy is originally not represented in the population, it will never appear. Consequently, any pure strategy (i.e. a case where all members of the population adopt the same strategy) is a rest point of the dynamics and, similarly, once the dynamics is on a face of the mixed strategy simplex, it will always stay there. For this reason, we are mostly interested in the behaviour in the interior of the mixed strategy simplex.

As an example, consider the dynamics (3.6) for the Rock-Scissors-Paper game (2.6). In this case, there is a unique internal equilibrium given by

$$\mathbf{p} = \frac{1}{K}(a_1a_3 + b_1b_2 + a_1b_1, a_1a_2 + b_2b_3 + a_2b_2, a_2a_3 + b_1b_3 + a_3b_3), \quad (3.7)$$

where K is chosen so that the three terms add to 1. There are three qualitatively different outcomes of the dynamics.

1) If $a_1a_2a_3 > b_1b_2b_3$ there is convergence to $\mathbf{p}^\mathbf{T}$ from any other interior point.

2) If $a_1a_2a_3 < b_1b_2b_3$ then $\mathbf{p}^\mathbf{T}$ is unstable and if the dynamics starts at any other point described by $\mathbf{p}_0 \neq \mathbf{p}$, it diverges in cycles closer and closer to the boundary.

3) If $a_1a_2a_3 = b_1b_2b_3$ there is an internal equilibrium $\mathbf{p}$ given by (3.7) with closed orbits around it, so that in the long term the population neither moves towards nor away from this equilibrium (note that this is a non-generic case).

The outcomes of the dynamics are shown in Figure 3.1.

Recall that in Chapter 1 we saw another example of the replicator dynamics, Figure 1.2. In this example there are two alternative strategies that the population may converge to, depending upon the initial state.

3.1.2 Adaptive dynamics

The theory of *adaptive dynamics* was developed to allow the population to evolve with small mutations. In contrast to replicator dynamics where at any given time, different individuals could play different strategies, and only the frequencies of existing strategies were allowed to change, it is assumed that every individual plays the same strategy, but the strategy played by the population can change over time. We consider adaptive dynamics in more detail in Chapter 14. For now, let us just illustrate this for the example of two-player, two pure strategy games with payoffs given by a matrix A. Assume that almost all members of the population adopt a strategy $\mathbf{p}$ (playing S_1 with probability p). In that population a small number of mutants play strategy $\mathbf{x}$

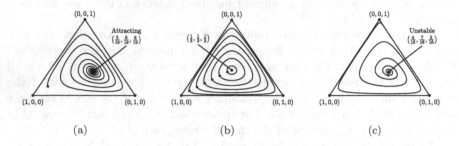

FIGURE 3.1: Continuous replicator dynamics for the RSP game given by matrix (2.6); (a) $a_1 = a_2 = b_3 = 2, a_3 = b_1 = b_2 = 1$, an asymptotically stable and globally attracting equilibrium, (b) $a_1 = a_2 = a_3 = b_1 = b_2 = b_3 = 1$, a stable (but not asymptotically stable) equilibrium with closed orbits, (c) $a_1 = a_2 = a_3 = b_1 = 1, b_2 = b_3 = 2$, an unstable and globally repelling equilibrium.

with $\mathbf{x} \approx \mathbf{p}$. Many such mutant strategies can enter the population, but only one at a time. The strategy $\mathbf{p}$ will change to the strategy $\mathbf{p} + d\mathbf{p}$, if this is fitter than $\mathbf{p}$ (against the resident population playing $\mathbf{p}$). In general, mutants with $p + dp$ on one side of p will be fitter, and those on the other side will not, so that the value of p will either increase or decrease.

The key concept of adaptive dynamics is the *Evolutionarily Singular Strategy*, which is a strategy where there are no small mutations which have higher fitness than the existing strategy. There are a number of distinct types of Evolutionarily Singular Strategy, with a rich variety of evolutionary behaviour, as we see in Chapter 14.

3.1.3 Other dynamics

The replicator dynamics is designed to model change as a result of natural selection (although it ignores some important complexities, most notably sexual reproduction; see Chapter 5). In the form of the equation (3.5) it assumes that individuals are reproduced faithfully and without any error or mutation. If we introduce a mutation matrix $Q = (q_{ij})$ where q_{ij} is the probability that the offspring of an individual using strategy S_i will be an individual using strategy S_j, the replicator equation (3.5) can be adjusted to the *replicator-mutator equation* (see Page and Nowak, 2002),

$$\frac{\mathrm{d}}{\mathrm{d}t}p_i = \sum_j p_j f_i(\mathbf{p})q_{ji} - p_i \bar{f}(\mathbf{p}). \tag{3.8}$$

If mutation rates are very small, this is closely linked to the idea of the "trembling hand" as we discuss in Section 3.2.4. It should be noted that there are various other dynamics. One example is the *best response dynamics*, used in particular for discrete processes. At every time step a new individual enters a population and chooses the strategy which maximises its payoff against the current strategy mix. An alternative version of this game, where all players update their strategies at each time step, can be used for spatial systems represented by graphs (see Chapter 13). Another example is imitation dynamics (copying the play of the fittest member of the population that you play against).

As we said in Chapter 1, static analysis is the main focus of this book, and there are a number of good books which consider evolutionary dynamics in detail. We will only occasionally consider the dynamics, generally where they are directly relevant to the discussion of the static results.

3.1.4   Timescales in evolution

When considering the evolution of populations, there are a number of different processes which must be considered. Animal behaviours, such as foraging or resting, can continuously change during the course of a day, and so behavioural dynamics, for example in Section 19.5, operate on a very fast timescale. The composition of populations, for example in terms of the numbers of different types, such as predators and prey, e.g. in Section 20.1, do not change significantly on a daily basis, but the relative numbers of prey and predators can change from year to year. Evolutionary dynamics, involving the competition between a population and a mutant strategy until potentially the mutant replaces the resident, will typically take many years. Thus behavioural dynamics is faster than population dynamics which is, in turn, faster than evolutionary dynamics (but see Argasinski and Broom (2013a) for a model where these processes run on the same timescale). In particular, when considering evolution with at least two such timescales, it is generally assumed that the faster process operates with the parameters of the slower process fixed, and that the faster process is in equilibrium (assuming that it converges to equilibrium) when we consider the slower process; see Křivan and Cressman (2009) to illustrate these ideas. A general methodology for incorporating population dynamics and evolutionary dynamics is the G-function; see Brown and Vincent (1987); Vincent et al. (1996); Cohen et al. (1999); Vincent and Brown (2005).

Similarly, when we consider the introduction of mutations to a population, we usually assume that potentially beneficial mutations (i.e. those that will not be immediately eliminated) appear sufficiently rarely, that the evolutionary competition between a previous mutant and the population has concluded before the next mutant invades. Thus mutation is considered a slower process than evolutionary dynamics. This is perhaps more questionable than the other timescale comparisons, as it is perfectly conceivable that evolution could occur sufficiently slowly for a new mutation to appear in a population involving two competing types, particularly if they are evenly matched (and the process of drift, where two equally fit types compete, is a slower process than evolutionary competition between non-equal types). However, it is plausible in many circumstances and makes the analysis more manageable, and often even if the assumption does not fully hold, there will not be a great difference in the results.

We note that models without timescale separation are becoming more common, and there are particular scenarios where timescale separation is less appropriate, see for example the cancer models in Chapter 22. We further discuss timescale separation a little in our final chapter, Chapter 23.

3.2   The static approach—Evolutionarily Stable Strategies (ESS)

A game is no more than a model and many complexities are ignored just by the construction of a game. Consequently, one can often ignore a particular underlying dynamics (which is still just another model) and focus on the game only. We will see in Section 3.3 that ignoring the dynamics and focusing on statics only does not, in many cases, significantly alter the outcomes of the analysis (from the point of view of applications in biology, for example).

Anybody playing any game will try to find a winning strategy. Defining the concept of a winning or "the best" strategy is the key to the analysis of any game. Superficially it looks like the task may be hopeless as what is the best strategy for player 1 to play will depend upon the choice of player 2, and vice versa. This is partly what makes game theory such an interesting subject. There are several definitions of "the best". The most used one for games between humans or rational entities is the so-called *Nash equilibrium*. In biological applications, the most used concept is the *Evolutionarily Stable Strategy (ESS)*. In simple terms, an ESS is a strategy such that if all members of a population adopt it, then no mutant strategy can invade the population under the influence of natural selection; see Maynard Smith and Price (1973); Maynard Smith (1982).

Most biological games model a never-ending battle to survive. Players that do not use a sufficiently good strategy are replaced by players that use a better strategy. There are many ways and variations on how to model this "replacement", and we saw some in Section 3.1. In this book, we will be looking for good strategies which satisfy some important properties, which we look at below. If we find such a strategy which is unique, the hope is that irrespective of the replacement process, it will be the strategy that will eventually be adopted by all of the individuals in the population.

3.2.1 Nash equilibria

We consider a rational (but not arrogant) individual with the following humble thought in her mind: "I am intelligent, but the other player is at least as intelligent as I am". This may be a very unfortunate thought because a repeated application of it may result in a thought loop illustrated by considering the RSP game (but which also occurs in many other instances). Player 1 is about to play Rock, having come to this choice by a sequence of logical arguments. But she knows that player 2 is intelligent as well, so player 2 replicated the same thoughts and thus knows that player 1 wants to play Rock. Naturally, player 2 chooses Paper. So, player 1, because she is intelligent and knows that player 2 is intelligent, now comes to the conclusion that she should play Scissors. But since she knows that player 2 is intelligent as well, ...

As discussed in Chapter 1, whilst economic game theorists usually assume that their players are rational, when considering biological situations it is not assumed that animals are behaving rationally. As we have described, for games a single measure of success, a reward, is needed and Darwinian fitness fits this requirement well. Animals do not behave rationally, but if strategies are faithfully reproduced in offspring, successful strategies will propagate, and evolutionary stability is thus a good substitute for rationality.

Yet, regardless of whether we assume rationality or some sort of dynamics, we get the same vicious cycle (for evolutionary dynamics this represents the real behaviour of the population). Note that this looks similar to the dynamics that we saw in Figure 3.1(b). How can one get out of this cyclic reasoning?

Static analysis does not consider how the population reached a particular point in the strategy space. Instead, we assume that the population is at that point, and ask, is there any incentive for any members of the population to change their strategies? Thus static analysis is not a priori concerned with if (and how) the population can reach a point from which there is no incentive to change.

Such a point can be found by the following reasoning in which we look at the humble thought from a different angle. Because I am intelligent, I have the capability to choose the best strategy. Because the other player is intelligent, he has the capability to choose the best strategy. Because we are both intelligent, the best strategy for me must thus be the best reply to the best strategy of the other player and the best strategy of the other player must be the best reply to my best strategy.

Definition 3.1. *A strategy S is a* best reply *(alternatively a* best response*) to strategy T if*

$$f(S', T) \leq f(S, T); \text{ for all strategies } S', \tag{3.9}$$

where $f(S, T)$ denotes the payoff to a player using S against a player using T.

This was the idea of John Forbes Nash, Jr. which was the foundation of the work that led to the award of the Nobel Prize for Economics in 1994.

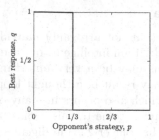

FIGURE 3.2: Best reply strategy from Example 3.2.

Example 3.2 (Best responses). Suppose that player 1 plays a two-strategy game with player 2 where payoffs to player 1 are given by the matrix

$$\begin{pmatrix} 2 & 1 \\ 4 & 0 \end{pmatrix}. \tag{3.10}$$

If player 2 plays S_1 with probability p, find the payoff of player 1 as a function of her chosen strategy q (the probability that she plays S_1) and hence find her best response q as a function of p. Plot this best response function against p.

The payoff of $\mathbf{q} = (q, 1-q)$ against $\mathbf{p} = (p, 1-p)$ is given by

$$\begin{aligned} E[\mathbf{q}, \mathbf{p}] = \mathbf{q}A\mathbf{p}^{\mathbf{T}} &= 2qp + q(1-p) + 4(1-q)p \\ &= q + 4p - 3qp = 4p + q(1 - 3p). \end{aligned} \tag{3.11}$$

The above function is increasing with q if $p < 1/3$ and decreasing with q if $p > 1/3$. Thus the best response is to play $q = 1$ if $p < 1/3$ and $q = 0$ if $p > 1/3$. If $p = 1/3$ then all values of q are best responses. A plot of this function is given in Figure 3.2.

Definition 3.3. *In a two-player game where players are allowed to play mixed strategies, a pair of strategies* $\mathbf{p}^*$ *and* $\mathbf{q}^*$ *form a* Nash equilibrium *pair, if*

$$\begin{aligned} E_1[\mathbf{p}^*, \mathbf{q}^*] &\geq E_1[\mathbf{p}, \mathbf{q}^*] \;\; \forall \mathbf{p} \neq \mathbf{p}^*, \tag{3.12} \\ E_2[\mathbf{q}^*, \mathbf{p}^*] &\geq E_2[\mathbf{q}, \mathbf{p}^*] \;\; \forall \mathbf{q} \neq \mathbf{q}^*, \tag{3.13} \end{aligned}$$

where $E_i[\mathbf{p}, \mathbf{q}]$ *denotes the payoff to the* i*th player if it uses strategy* $\mathbf{p}$ *while the other player uses* $\mathbf{q}$.

A pair $\mathbf{p}^*$ and $\mathbf{q}^*$ is a Nash equilibrium if there is no strategy which has a greater reward against $\mathbf{p}^*$ than $\mathbf{q}^*$ and vice versa, so that each strategy is a *best reply* against the other. Thus if two players played $\mathbf{p}^*$ and $\mathbf{q}^*$ against each other and now meet again, there is no incentive for either of them to change their strategy, if they assume that the other will not change. Alternatively, if all individuals in the population play $\mathbf{p}^*$ and $\mathbf{q}^*$ in the roles of players 1 and

2, respectively (or these are the average strategies of the population) then no player can do better by playing a strategy different to these. This is not true for any strategy pair that is not a Nash equilibrium, so that it would be in the interests of at least one of the players of such strategy pairs to change strategy.

The situation is somewhat easier when the roles of players are interchangeable, i.e. if $E_1[\mathbf{p}, \mathbf{q}] = E_2[\mathbf{p}, \mathbf{q}]$. In this case, the humble thought is "If I am intelligent enough to come up with strategy $\mathbf{p}$, the other player is inteligent enough to come up with the same strategy", and thus in this case being the best reply against the other player's best strategy simply becomes being the best reply to itself. Thus a strategy $\mathbf{p}^*$ is a Nash equilibrium if

$$E[\mathbf{p}, \mathbf{p}^*] \leq E[\mathbf{p}^*, \mathbf{p}^*] \ \ \forall \mathbf{p} \neq \mathbf{p}^*. \tag{3.14}$$

We can see that for Example 3.2, assuming that the payoff matrix is the same for both players, the strategy $p = 1/3$ is the only best response to itself, and thus the unique Nash equilibrium.

3.2.2 Evolutionarily Stable Strategies

In biological settings, we do not assume intelligence in our players (see Chapter 1). Yet evolution provides an equally good tool, because if one player plays a strategy S as a result of evolution, any other player's strategy will be the result of evolution as well. Consequently, the best strategy must again be a best reply to itself, as in the case of Nash equilibria. Indeed, if everybody in the population adopted a strategy S, but S was not a best reply to itself, then there would be a strategy M such that a player using M would have a higher payoff than a player using S. Consequently, strategy M could invade the population.

However, a strategy being a best reply to itself does not prevent invasion. Consider a population consisting of individuals, the vast majority of whom adopt a strategy S, while a very small number of "mutants" adopt a strategy M. The strategies S and M thus compete in the population $(1 - \varepsilon)\delta_S + \varepsilon\delta_M$ for some small $\varepsilon > 0$ (rather than in the population δ_S), and it is against such a population that S must outcompete M. This brings us the definition of an ESS.

Definition 3.4. *We say that a strategy S is* evolutionarily stable *against strategy M if there is $\varepsilon_M > 0$ so that for all $\varepsilon < \varepsilon_M$ we have*

$$\mathcal{E}[S; (1 - \varepsilon)\delta_S + \varepsilon\delta_M] > \mathcal{E}[M; (1 - \varepsilon)\delta_S + \varepsilon\delta_M]. \tag{3.15}$$

S is an Evolutionarily Stable Strategy (ESS) *if it is evolutionarily stable against M for every other strategy $M \neq S$.*

The above condition means that any mutant strategy M must do worse than the resident strategy (if mutants are present only in small numbers).

In Definition 3.4, by "strategy" we generally mean a strategy that can be played by the individuals, although as we see in Section 3.2.3 we allow mutant strategies M to be mixed even if individuals can play only pure strategies (see Chapters 6 and 7).

To illustrate the close relationship between ESSs and Nash equilibria, we will consider matrix games in more detail.

3.2.2.1   ESSs for matrix games

We have seen that in the case of matrix games we have, by (2.18), $\mathcal{E}[\mathbf{p}; (1 - \varepsilon)\delta_{\mathbf{p}} + \varepsilon\delta_{\mathbf{q}}] = E[\mathbf{p}, (1 - \varepsilon)\mathbf{p} + \varepsilon\mathbf{q}]$. By Definition 3.4, a strategy $\mathbf{p}$ is an ESS if for every other strategy $\mathbf{q} \neq \mathbf{p}$ there is $\varepsilon_{\mathbf{q}} > 0$ such that for all $\varepsilon < \varepsilon_{\mathbf{q}}$ we have

$$E[\mathbf{p}, (1 - \varepsilon)\mathbf{p} + \varepsilon\mathbf{q}] > E[\mathbf{q}, (1 - \varepsilon)\mathbf{p} + \varepsilon\mathbf{q}]. \tag{3.16}$$

For matrix games, Definition 3.5 below is an equivalent definition of an ESS (the equivalence of Definitions 3.4 and 3.5 is Theorem 6.2, which we shall see when we consider matrix games in more detail in Chapter 6).

Definition 3.5 (ESS in matrix games). *A (pure or mixed) strategy* $\mathbf{p}$ *is called an* Evolutionarily Stable Strategy (ESS) *for a matrix game, if and only if for any mixed strategy* $\mathbf{q} \neq \mathbf{p}$

$$E[\mathbf{p}, \mathbf{p}] \geq E[\mathbf{q}, \mathbf{p}], \tag{3.17}$$

$$if\ E[\mathbf{p}, \mathbf{p}] = E[\mathbf{q}, \mathbf{p}],\ then\ E[\mathbf{p}, \mathbf{q}] > E[\mathbf{q}, \mathbf{q}]. \tag{3.18}$$

If (3.18) does not hold, then $\mathbf{p}$ may be invaded by a mutant that does equally well against the majority of individuals in the population (that adopts $\mathbf{p}$) but is either getting a (tiny) advantage against them by doing better in the (rare) contests with like mutants, or is at least no worse by doing equally well against the mutants. In this latter case invasion can occur by "drift"; both types do equally well, so in the absence of selective advantage random chance decides whether the frequency of mutants increases or decreases. Thus both conditions (3.17) and (3.18) are needed for $\mathbf{p}$ to resist invasion by a mutant $\mathbf{q}$. If the conditions hold for any $\mathbf{q} \neq \mathbf{p}$, then $\mathbf{p}$ can resist invasion by any rare mutant and so $\mathbf{p}$ is an ESS.

The above reasoning relies on ε being arbitrarily small. In fact, for an ESS $\mathbf{p}$ and a potential invading strategy $\mathbf{q}$, there will be a critical value of $\varepsilon_{\mathbf{q}}$, the so-called invasion barrier (see Bomze and Pötscher, 1989), where invasion is resisted if and only if $\varepsilon < \varepsilon_{\mathbf{q}}$ and influxes of $\mathbf{q}$ in excess of this threshold will invade (note that $\varepsilon_{\mathbf{q}}$ may be equal to 1, so that a potential invader may never succeed). The situation is relatively simple for matrix games, as we shall see in Section 6.1.2, but in general, different invading mutant strategies may have different invasion barriers. The invading frequency in any population will indeed usually be small, but can only be arbitrarily small in an infinite population, so that in any real scenario an invading fraction will have some minimum frequency, and so this is a useful idea when relating assumed infinite populations back to finite reality (see Chapter 12).

Note that unlike in Definition 3.4, there is no $\varepsilon_{\mathbf{q}}$ associated with the potentially invading strategy $\mathbf{q}$.

Returning to Example 3.2 from Section 3.2.1, there is a unique candidate to be an ESS, $p = 1/3$. We saw from before that

$$E[\mathbf{q}, \mathbf{p}] = 4p + q(1 - 3p) \Rightarrow E[\mathbf{q}, 1/3] = 4/3 \tag{3.19}$$

and so (3.17) clearly holds.

$$\begin{aligned} E[\mathbf{p}, \mathbf{q}] - E[\mathbf{q}, \mathbf{q}] &= p + 4q - 3pq - (5q - 3q^2) \\ &= (p - q)(1 - 3q) = (1 - 3q)^2/3, \end{aligned} \tag{3.20}$$

which is positive unless $q = 1/3$. Thus (3.18) also holds and $p = 1/3$ is an ESS.

There is a general method of how to find the ESSs of a matrix game, and this is given in detail in Chapter 6. For our RSP example, the ESSs depend upon the values of a_i and b_i. If the values for winning and losing are independent of the strategy played so that $a_i = a$ and $b_i = b$ for all i, then if $a > b$ there is a single unique ESS $(1/3, 1/3, 1/3)$. This is what is called an *internal* or *interior* ESS, i.e., if p is in the interior of the strategy simplex. We say that $\mathbf{p}$ is an *internal strategy* if $p_i > 0$ for all i.

This internal strategy is probably what most people would think to be the natural solution to the game, due to the symmetries between the strategies. If $a \leq b$, however, it turns out that the game has no ESSs at all. To consider the long-term behaviour in this game, it is vital to look at the dynamics, which we considered in Section 3.1. The relationship between ESSs and dynamics is not straightforward however, as we see in Section 3.3. In most models, especially the models applied to more specific scenarios that we consider in the later chapters of the book, there is at least one ESS. Biological interpretations are clearest when a unique ESS is discovered, as this usually means there is a definitive solution to a given game, with correspondingly clear predictions. It is perhaps surprising how often this is actually the case in models of real behaviour. However, as mentioned above, it is possible for there to be no ESSs at all, and then to make predictions we must consider the dynamics.

3.2.3 Polymorphic versus monomorphic populations

In a population of pure strategists, there are differences if single or multiple mutations are allowed in the population. For example, consider a game shown in Maynard Smith (1982, Appendix D), given by the matrix

$$\begin{pmatrix} 1 & 1 & 1 \\ 1 & 0 & 10 \\ 1 & 10 & 0 \end{pmatrix}. \tag{3.21}$$

Strategy S_1 cannot be invaded by either of strategies S_2 and S_3 if those mutants enter the population separately. However, population δ_1 can be invaded by arbitrarily small numbers of mutants if half of them plays S_2 and the other half plays S_3. Moreover, strategy S_1 is not an ESS by Definition 3.5 as it can be invaded by a strategy $\mathbf{p} = (0, 1/2, 1/2)$, see also Maynard Smith (1982, Figure 36, p. 184). This is easy to see, since whilst $E[\mathbf{p}, S_1] = E[S_1, S_1] = 1$, $E[\mathbf{p}, \mathbf{p}] = 5 > E[S_1, \mathbf{p}] = 1$. It was therefore argued in Maynard Smith (1982) that one should test against invading mixed strategies even if individuals can play only pure strategies.

Now consider a generalised RSP game, given by the matrix

$$\begin{pmatrix} 0 & 2 & -1 \\ -1 & 0 & 2 \\ 2 & -1 & 0 \end{pmatrix}. \tag{3.22}$$

The unique ESS of this game is $\mathbf{p} = (1/3, 1/3, 1/3)$, with a payoff of $1/3$ to all members of the population; see Theorem 6.9 and Section 6.4. Here we can also see the difference between a monomorphic population $\delta_\mathbf{p}$ and a polymorphic population $\mathbf{p^T} = \sum_i p_i \delta_i$. If a small number of mutants playing $(1/2, 1/2, 0)$ enter the population $\delta_\mathbf{p}$, the mutants go extinct and the population returns to $\delta_\mathbf{p}$ exponentially fast. However, if the same mutants enter the population $\mathbf{p^T} = \sum_i p_i \delta_i$, they destabilise the population proportions with S_1 going sharply up at first, S_3 going up mildly, both rising at the expense of S_2, and also partly at the expense of $(1/2, 1/2, 0)$. The resulting equilibrium that is reached after some oscillations contains a largest group of S_3 strategists, a somewhat smaller groups of S_1 and S_2 strategists (the same proportion of each), and a small group of mutants as well, in fact in almost the same proportion as they entered the population. The resulting equilibrium in the population is still equivalent to $(1/3, 1/3, 1/3)$ in the sense that a random opponent will play strategy S_i with probability $1/3$, but the actual proportions of individuals are quite different. In any case, the mutants have invaded the population.

We have seen above that there is a difference between monomorphic and polymorphic populations. To capture the evolutionary stability in polymorphic populations, we say that a population is in an *Evolutionarily Stable State (ESState)* if, after a small disturbance, its population state (in the sense of the strategy played by a random opponent, as above) stays close to the original population state and is eventually restored by selection. Mathematically, a population state is an ESState if it is *locally superior* as described in the condition (6.16) (Maynard Smith, 1982; Hofbauer and Sigmund, 1998). As seen above the "restored state" may have a different mix of individual types than the original state. Thus any population composition which makes up an ESState is in some sense equivalent.

As we have seen, it is possible for a number of strategies to perform equally well against a given population. An *Evolutionarily Stable Set (ESSet)* is a set of strategies, which all score equally well against each other, each of which would be an ESS were it not for the other members of the set (Thomas, 1985). We will see an example of an ESSet and compare it to an ESS in Section 4.1.6 (we see a further example for foraging populations in Section 19.3.2).

3.2.4 Stability of Nash equilibria and of ESSs

In general, in game theory, strategies which are Nash equilibria (and so a rest point of the dynamics), must pass an additional test, that of the *trembling hand*, devised by Reinhard Selten; see Selten (1975) and also Myerson (1978).

It should be noted that there can be different mechanisms that cause the trembling hand, and two different mechanisms can lead to different conclusions. Trembles can be errors, if there is a (small) chance that an individual does not play the strategy it wanted because either (a) the individual made an mistake, or (b) the action was somehow modified by the environment. Alternatively, we could assume a constant supply of mutants at a low level migrating into a population, in which case the mix of mutants would be independent of the current population mixture. Mutations could occur when one type erroneously gives birth to individuals of another type, as in the replicator-mutator dynamics of (3.8), in which case the mix of mutants would depend upon the current population mixture. Another possibility is small stochastic variation in the composition of the population (when we think of our infinite population as an approximation for a large finite one).

In evolutionary games, the trembling hand is "achieved" by assuming a supply of other strategies at a low level. For example, even when all individuals in the population should play strategy S_1 (i.e. the population should be δ_1), due to mutation or other biological processes, the population will effectively be $(1 - \varepsilon)\delta_1 + \varepsilon\delta_2$.

One should prefer Nash equilibria that are not destabilised by such errors. A generic ESS resists all sensible trembles, but trembles can affect non-generic situations; see Section 4.1.6 for a case with an ESSet and the effect of potential trembles. In general, there are possible game solutions that are in some way not sensible, and we must distinguish those we would expect to see in real populations from those that we would not. We shall consider a simple example to illustrate the idea.

Example 3.6. Consider the matrix game with payoff matrix

$$\begin{pmatrix} 1 & 2 \\ 0 & 2 \end{pmatrix}. \tag{3.23}$$

There are two Nash equilibria in this game $(1, 0)$ and $(0, 1)$. The trembling hand means that if an individual tries to play a strategy $\mathbf{p} = (p_1, p_2)$ there is probability ε of an error causing the wrong pure strategy to be played, and thus it will, in fact, play a strategy $\mathbf{p}_\varepsilon = \big((1 - \varepsilon)p_1 + \varepsilon p_2, (1 - \varepsilon)p_2 + \varepsilon p_1\big)$. The best reply against $S_1 = (1, 0)$ will thus be a strategy $\mathbf{p} = (p_1, p_2)$ that maximises

$$
\begin{aligned}
E[\mathbf{p}_\varepsilon, (S_1)_\varepsilon] &= \big((1-\varepsilon)p_1 + \varepsilon p_2, (1-\varepsilon)p_2 + \varepsilon p_1\big) A (1-\varepsilon, \varepsilon)^{\mathbf{T}} \\
&= \big((1-\varepsilon)p_1 + \varepsilon p_2\big)(1+\varepsilon) + \big((1-\varepsilon)p_2 + \varepsilon p_1\big)2\varepsilon \qquad (3.24) \\
&= p_1 + \varepsilon(\dots).
\end{aligned}
$$

Thus, for small ε, the best reply against $(1,0)$ is $(1,0)$. Hence, $(1,0)$ is a Nash equilibrium even when hands are trembling. On the other hand, the best reply against $S_2 = (0,1)$ will be a strategy $\mathbf{p} = (p_1, p_2)$ that maximises

$$
\begin{aligned}
E[\mathbf{p}_\varepsilon, (S_2)_\varepsilon] &= \big((1-\varepsilon)p_1 + \varepsilon p_2, (1-\varepsilon)p_2 + \varepsilon p_1\big) A (\varepsilon, 1-\varepsilon) \\
&= \big((1-\varepsilon)p_1 + \varepsilon p_2\big)(2-\varepsilon) + \big((1-\varepsilon)p_2 + \varepsilon p_1\big)(2 - 2\varepsilon) \qquad (3.25) \\
&= 2 - \varepsilon(1 + p_2) + \varepsilon^2(\dots).
\end{aligned}
$$

Hence, the best reply in this case is a strategy with as small p_2 as possible, i.e. $(1,0)$, and thus $(0,1)$ is not a Nash equilibrium when hands are trembling.

We should note that in this particular case, a population playing $(0,1)$ can be invaded by a small group playing $(1,0)$, and so $(0,1)$ fails the usual test for being an ESS.

3.3   Dynamics versus statics

The focus of this book is mainly on the static approach. Since it may be argued (with some justification) that the dynamics approach may model the biology in a more realistic way, we point out the similarities and differences between the two approaches in this section.

Dynamic and static analyses are mainly complementary, however the relationship between the two is not straightforward, and there is some apparent inconsistency between the theories.

As the concept of an ESS as an uninvadable strategy is partially based on the same idea as that of replicator dynamics, we look at replicator dynamics.

Comparing ESS analysis and replicator dynamics, we see that the information required for each type of analysis is different. To determine whether a strategy $\mathbf{p}$ is an ESS, we just have to make sure that the minimum of a function

$$
\mathbf{q} \to \mathcal{E}[\mathbf{p}; (1-\varepsilon)\delta_\mathbf{p} + \varepsilon\delta_\mathbf{q}] - \mathcal{E}[\mathbf{q}; (1-\varepsilon)\delta_\mathbf{p} + \varepsilon\delta_\mathbf{q}] \qquad (3.26)
$$

is attained for $\mathbf{q} = \mathbf{p}$ for all sufficiently small $\varepsilon > 0$. To understand the replicator dynamics, we need to know $\mathcal{E}[S_i; \mathbf{p}^{\mathbf{T}}]$ for all i and all $\mathbf{p}$.

The above pinpoints that the main difference between static analysis and replicator dynamics is that the static analysis is concerned about monomorphic populations δ_p while replicator dynamics studies mixed populations

$\mathbf{p^T} = \sum_i p_i \delta_i$. It is therefore not surprising that both analyses can produce the same (or at least similar) results only if there is some kind of identification between $\delta_\mathbf{p}$ and $\mathbf{p^T}$ such as in the case of matrix games. It should be noted that most of the comparative analysis between ESSs and replicator dynamics assumes matrix games, so that it is not necessarily valid for other classes of games.

3.3.1 ESS and replicator dynamics in matrix games

The following two theorems are the most striking results regarding the connections between statics and dynamics.

Theorem 3.7 (Folk theorem of evolutionary game theory, Hofbauer and Sigmund, 2003). *In the following, we consider a matrix game with payoffs given by matrix A and the corresponding replicator dynamics (3.6).*

1) *If* $\mathbf{p}$ *is a Nash equilibrium of a game, in particular, if* $\mathbf{p}$ *is an ESS of a matrix game, then* $\mathbf{p^T}$ *is a rest point of the dynamics, i.e. the population does not evolve further from the state* $\mathbf{p^T} = \sum_i p_i \delta_i$.

2) *If* $\mathbf{p}$ *is a strict Nash equilibrium, then* $\mathbf{p}$ *is locally asymptotically stable, i.e. the population converges to the state* $\mathbf{p^T} = \sum_i p_i \delta_i$ *if it starts sufficiently close to it.*

3) *If the rest point* $\mathbf{p}*$ *of the dynamics is also the limit of an interior orbit (the limit of* $\mathbf{p}(t)$ *as* $t \to \infty$ *with* $\mathbf{p}(0) \in int(S)$*), then it is a Nash equilibrium.*

4) *If the rest point* $\mathbf{p}$ *is Lyapunov stable (i.e. if all solutions that start out near* $\mathbf{p}$ *stay near* $\mathbf{p}$ *forever), then* $\mathbf{p}$ *is a Nash equilibrium.*

We note that the replicator dynamics itself and the results of the above theorem rely on a certain linearity of the payoffs of the game; we will further investigate this linearity property in Section 7.1.

Theorem 3.8 (Zeeman, 1980). *Any ESS is an attractor of the replicator dynamics (i.e. has some set of initial points in the space that lead to the population reaching that ESS). Moreover, the population converges to the ESS for every strategy sufficiently close to it; and if* $\mathbf{p}$ *is an internal ESS, then global convergence to* $\mathbf{p}$ *is assured.*

Also, if the replicator dynamics has a unique internal rest point $\mathbf{p}^*$, and if the set of accumulation points of the orbit of $\mathbf{p}$ is in the interior of the strategy simplex (such as in Figure 3.1(b) but not in Figure 3.1(c)), then

$$\lim_{t \to \infty} \frac{1}{T} \int_0^T p_i(t)dt = p_i^*. \tag{3.27}$$

Thus the long-term average strategy is given by this rest point, even if at any time there is considerable variation.

The results of Zeeman (1980) and Hofbauer and Sigmund (2003) above are very helpful because it means that, for matrix games, identifying ESSs and Nash equilibria of a game gives a lot (sometimes practically all) of the important information about the dynamics. For example, if **p** is an internal ESS, and there is no other Nash equilibrium, then global convergence to **p** is assured (by Zeeman, 1980). For more complicated games the situation is less straightforward (see for example Chapter 7 for nonlinear games and Chapter 9 for multi-player games).

Yet, there are cases when an ESS analysis does not provide a complete picture. In particular, there are attractors that are not ESSs; see Taylor and Jonker (1978) and Zeeman (1981). To see this, re-consider the generalised RSP game (2.6). If $a_1 = a_2 = b_3 = 2, a_3 = b_1 = b_2 = 1$, then the replicator dynamics has a unique internal attractor, but it is not an ESS (see Section 6.4). This occurs because we can find an invading strategy for **p** where the payoffs to the different components change in such a way under the dynamics that it is inevitably forced into a combination that no longer invades. Thus if the invader is comprised of a combination of pure strategists it is beaten, but if it is comprised of mixed strategists it is not.

For the discrete dynamics, the behaviour is even more complex since an ESS need not be an attractor at all (Cannings, 1990). In general, the behaviour of the discrete dynamics will change with the value of the constant background fitness β.

3.3.2 Replicator dynamics and finite populations

Just as the infinite population assumption can have unrealistic implications for the static concept of the ESS, see discussion of the invasion barrier in Section 3.2.2.1, it can also lead to problematic conclusions in the dynamic setting which are not what you would expect in any real system. The replicator dynamics assumes a smooth change of strategy frequencies within the population until, in some cases, a rest point is reached. In real populations which are finite, there will be continued stochastic variation in the frequencies of the strategies, and this can be modelled using Markov chains (see Chapter 12), although the modelling is significantly more complex. The overall qualitative behaviour of finite models and the replicator dynamics are often the same, for instance if there is a single central global attractor, the results over the medium term will usually be consistent (though it should be noted that in any finite population stochastic variation will mean that eventually only a single pure strategy will remain, see Chapter 12, although this may take a vast time to happen, and so not be observed in a real population). Thus in our RSP example from Figure 3.1, case (a) with the stable global equilibrium will represent the finite population case well. However, in the other two cases this will not be true. For instance in case (c), the dynamics takes the population closer and closer to the boundary, with frequencies of particular strategies going down to smaller and smaller values as time goes on. Stochastic fluctuation will inevitably eliminate one strategy by chance at some point, after which the dominant remaining strategy will move to a proportion 1, and there will be a single pure strategy remaining in not too long a time frame. In case (b) too, where the dynamics leads to closed cycles, random movements will continuously move the population either closer or further from the centre, and stochastic drift will eventually move the population further from the centre on average.

3.4 Python code

In this section we show how to use Python to get numerical solutions to the replicator dynamics of an RSP game.

```
 1   """
 2   Numerical solution of the replicator dynamics
 3   for the Rock-Scissors-Paper game
 4   """
 5
 6   ## Import basic packages
 7   import numpy as np
 8   from scipy.integrate import odeint
 9   import matplotlib.pyplot as plt
10
11   ## Define payoff  matrix
12   A=[ [0, 1,-1],    # Payoffs to Rock
13       [-2, 0, 2],   # Payoffs to Scissors
14       [2,-1, 0]]    # Payoffs to Paper
15
16   # Give initial proportion of R, S, P
17   y0 = [0.9, 0.09, 0.01]
18
19   ## Define replicator dynamics
20   def dynamics(y,t):
21       " Defines the dynamics "
22       dydt = y*(np.dot(A,y) - np.dot(y,A).dot(y));
23       # dy/dt[i] = y[i] * ((A y^T)[i] - yAy^T)
24       return dydt
25
26   # Specify the time interval over which we will solve
27   t = np.linspace(0, 30, 101)
28
29   # Numerically solve the dynamics
30   Sols = odeint(dynamics, y0, t)
31
32   # Unpack the solutions (R = Sols[:,0]) etc
33   R, S, P = Sols.T
```

```
34
35  ## Plot results
36  plt.plot(t,R, label='Rock')
37  plt.plot(t,S, label='Scissors')
38  plt.plot(t,P, label='Paper')
39  plt.legend(loc='best')
40  plt.xlabel('Time')
41  plt.ylabel('Prevalence')
42  plt.show()
```

3.5   Further reading

There are a number of very good books investigating evolutionary dynamics. We recommend, in particular, the classic books by Hofbauer and Sigmund (1998) and Cressman (1992) as well as Nowak (2006a), Weibull (1995), Cressman (2003), Vincent and Brown (2005), Samuelson (1997) and Sandholm (2010). For evolutionary dynamics in continuous strategy spaces, see Cressman and Křivan (2006); Oechssler and Riedel (2002). For a short and concise overview of different dynamics see Page and Nowak (2002), Nowak and Sigmund (2004) and also Křivan (2009) and references therein.

Nash equilibria were introduced in Nash (1951). The concept of an ESS was introduced in Maynard Smith and Price (1973). The classical book on ESSs is Maynard Smith (1982). A more mathematical book on static concepts in game theory is Bomze and Pötscher (1989). A relatively short overview of game theory for evolutionary biology can be found in Hammerstein and Selten (1994). ESSs in polymorphic populations are discussed in Hammerstein and Selten (1994, p. 950) and references therein. The differences and similarities between dynamics and statics are discussed in Hofbauer (2000), Hofbauer and Sigmund (1990), Zeeman (1980), Cannings (1990) and Hammerstein and Selten (1994, Section 5) and references therein. There are a large number of stability concepts that have been introduced in the study of evolutionary games; a good summary of many of these can be found in Apaloo et al. (2009).

Models which do not involve separation of timescales are considered in Argasinski and Broom (2013a) who use classical-style animal interactions, in particular, aggressive contests for mates and territories (see also Argasinski and Broom, 2013b, 2018a). Models on smaller scales, such as for the evolution of cancer in Chapter 22, for example in Vincent and Gatenby (2008), and Section 23.3.3 also often consider unified timescales for evolutionary and ecological processes.

3.6 Exercises

Exercise 3.1. Consider the 2×2 matrix below and study how the population dynamics governed by equation (3.3) depends on the background fitness β,

$$\begin{pmatrix} 0 & 1 \\ 1 & 0 \end{pmatrix}. \tag{3.28}$$

Hint. Start the population near one of the pure strategies for a range of values of β including 0 and a large value such as 100.

Exercise 3.2. Repeat Exercise 3.1 for the matrix

$$\begin{pmatrix} 1 & 2 \\ 2 & 1 \end{pmatrix}. \tag{3.29}$$

Exercise 3.3. Consider a 2×2 matrix A and a strategy $\mathbf{p} = (p, 1-p)$. Write down the continuous replicator dynamics and show that this reduces to

$$\frac{\mathrm{d}}{\mathrm{d}t} p = p(1-p)h(p), \tag{3.30}$$

where $h(p) = \mathcal{E}[1; \delta_{(p,1-p)}] - \mathcal{E}[2; \delta_{(p,1-p)}]$ is the so-called *incentive function*. Show that $\mathbf{p} = (p, 1-p)$ for $p \in (0,1)$ is an asymptotically stable point of the replicator dynamics if and only if $h(p) = 0$ and $h'(p) < 0$.

Exercise 3.4 (ESS and Lyapunov function, Hofbauer and Sigmund, 2003; Hofbauer et al., 1979). Let $\mathbf{x}^*$ be an internal ESS of a matrix game and let $L(\mathbf{x}) = \prod_{i=1}^{n} x_i^{x_i^*}$. Show that L is an increasing function along trajectories of the replicator dynamics (3.5), and that $L(\mathbf{x}^*) \geq L(\mathbf{x})$ for every mixed strategy $\mathbf{x}$.

Hint. For the second part, take the logarithm and use Jensen's inequality.

Exercise 3.5 (Křivan, 2009, Proposition 4). Let $A = (a_{ij})_{i,j=1}^{n}$ be a payoff matrix. Show that if there exists i such that $a_{ii} \geq a_{ji}$ for all $j = 1, 2, \ldots, n$, then pure strategy S_i is a Nash equilibrium. Show that if $a_{ii} > a_{ji}$ for all $j \neq i$, then pure strategy S_i is a strict Nash equilibrium.

Exercise 3.6. Plot the best response functions for each of the following matrix games:

$$(a) \begin{pmatrix} 1 & 1 \\ 3 & 2 \end{pmatrix}, (b) \begin{pmatrix} -1 & 2 \\ 1 & 1 \end{pmatrix}, (c) \begin{pmatrix} 1 & 1 \\ 0 & 2 \end{pmatrix}. \tag{3.31}$$

Exercise 3.7. Find the Nash equilibria and ESSs of the games from Exercise 3.6.

Exercise 3.8. Consider the repeated RSP game from Exercise 2.1. Is the strategy "Play Rock in round 1 and then play whatever would beat the opponent in the previous round" an ESS?

Hint. Identify the best reply as a strategy in a similar form.

Exercise 3.9. Consider a game given by a 2×2 matrix A. Show that any trembling hand Nash equilibrium is a Nash equilibrium; see also Example 3.6.

Exercise 3.10. Show that there is a 1-1 correspondence between ESSs and the stable rest points of the replicator dynamics in a 2×2 matrix game.

Exercise 3.11. Show that dynamic and static analysis can yield different results for 3×3 matrix games.

Hint. See Figure 3.1 and Section 6.4.

Chapter 4

Some classical games

4.1 The Hawk-Dove game

The Hawk-Dove game is perhaps the best known game modelling biological behaviour. Originating in Maynard Smith and Price (1973), it attempts to explain the relative rarity of the use of dangerous weaponry in some contests over valuable resources involving heavily armed animals, where at first sight it would seem logical that individuals should fight to the maximum of their ability. An example is the highly ritualistic displays involved in contests between red deer stags during their rut.

We will only describe the basic Hawk-Dove game in this chapter, but will revisit it on a number of occasions throughout the book, as it is often the starting point for more complex models.

4.1.1 The underlying conflict situation

Two male animals often compete for a territory. The classical example is of stags during the breeding season, where possession of a good territory will mean significant mating opportunities with a number of females.

In populations of red deer *Cervus elaphus*, when two stags meet they often go through a series of rituals to decide which will control the territory (Clutton-Brock and Albon, 1979). The males can be thought of as exhibiting one of two available types of behaviour, to *display* or to *escalate*. These are represented by two distinct (pure) strategies in the model below. The strategy display, or play *Dove*, means a willingness to enter into the ritualistic contest, but not to enter into a violent confrontation. These ritualistic contests can include merely assessing each other's size, roaring, walking in parallel lines, and pushing contests with their antlers. At any point during a sequence of such behaviours, an individual can back down without injury and eventually one individual will do so. If one is significantly larger than the other, then this may happen early in the sequence, whereas evenly matched individuals may persist for a longer period of time. Stags' antlers contain sharp points and are potentially dangerous weapons, however. An individual may start to threaten to use, and then actually use, its antlers to inflict injury, with the severity of such contests increasing in intensity if neither gives way. The strategy

DOI: 10.1201/9781003024682-4

escalate, or play *Hawk*, means commencing a violent altercation. In the game we assume that a Dove individual will immediately concede the territory against a Hawk, and so will escape injury, but two Hawks will fight on until one of them is injured and cannot carry on.

4.1.2   The mathematical model

We now formulate the Hawk-Dove game in the formal mathematical terminology of matrix games. In this basic game it is assumed that individuals are of identical size, strength and other attributes. Each individual plays one or more contests against other randomly chosen individuals. A contest contains two equally able players competing for a reward (e.g. a territory) of value $V > 0$. Each of the contestants may play one of two pure strategies, Hawk (H) and Dove (D). If two Doves meet, they each display, and each will gain the reward with probability $1/2$, giving an expected reward of $E[D, D] = V/2$. If a Hawk meets a Dove, the Hawk will escalate, the Dove will retreat (without injury) and so the Hawk will gain the reward V, and the Dove will receive 0. Hence, $E[H, D] = V$ and $E[D, H] = 0$. If two Hawks meet, they will escalate until one receives an injury, which is a cost $C > 0$ (the equivalent of a reward of $-C$). The injured animal retreats, leaving the other with the reward. The expected reward for each individual is thus $E[H, H] = (V - C)/2$. We have a matrix game with two pure strategies, where the payoff matrix is

$$
\begin{array}{cc}
 & \begin{array}{cc} \text{Hawk} & \text{Dove} \end{array} \\
\begin{array}{c} \text{Hawk} \\ \text{Dove} \end{array} & \left(\begin{array}{cc} \dfrac{V-C}{2} & V \\ 0 & \dfrac{V}{2} \end{array} \right).
\end{array}
\tag{4.1}
$$

We denote a mixed strategy $\mathbf{p} = (p, 1 - p)$ to mean to play Hawk with probability p and to play Dove otherwise.

4.1.3   Mathematical analysis

Since $E[H, D] > E[D, D]$, Dove is never an ESS. Hawk (or the strategy $(1, 0)$) is a pure ESS if $E[H, H] > E[D, H]$ i.e. if $V > C$. In fact, Hawk is also an ESS when $V = C$ since in this case $E[H, H] = E[D, H]$ and $E[H, D] > E[D, D]$. This is an example of the non-generic case that we usually ignore, which is discussed in Chapter 2. Thus Hawk is a pure ESS if $V \geq C$.

A general method for finding all, including the mixed, ESSs of matrix games is shown in Section 6.2. It follows that $\mathbf{p} = (V/C, 1 - V/C)$ is the unique ESS when $V < C$ (this can also be proved directly, see Exercise 4.9). This means that there is a unique ESS in the Hawk-Dove game, irrespective of the values of V and C.

4.1.4 An adjusted Hawk-Dove game

It should be noted here that the above Hawk-Dove game model has been criticised for its description of the Dove versus Dove contests. If there is a protracted series of displaying between the two contestants, there will be a cost to each competitor in terms of the time and energy wasted, so that the payoff would be less than $V/2$. In fact, if we model the Dove versus Dove contest as the war of attrition (see Section 4.3 below) we get that $E[D, D] = 0$, so that the adjusted payoff matrix is now

$$
\begin{array}{cc}
& \begin{array}{cc} \text{Hawk} & \text{Dove} \end{array} \\
\begin{array}{c} \text{Hawk} \\ \text{Dove} \end{array} &
\left(\begin{array}{cc} \dfrac{V-C}{2} & V \\ 0 & 0 \end{array} \right).
\end{array}
\tag{4.2}
$$

If $V \geq C$ Hawk is still the unique ESS, and if $V < C$ there is still a unique mixed ESS, now with the Hawk frequency $p = 2V/(V + C)$. Thus the fundamentals of the game are not affected by this change, and so we shall not consider it further.

4.1.5 Replicator dynamics in the Hawk-Dove game

Now letting $\mathbf{p} = (p, 1-p)$ describe the mixed population of pure strategists with proportion of Hawks being p and proportion of Doves being $1 - p$, the replicator dynamics described in Section 3.1.1 gives the following equation for the evolution of p

$$
\begin{aligned}
\frac{\mathrm{d}}{\mathrm{d}t} p &= p((A\mathbf{p}^T)_1 - \mathbf{p}A\mathbf{p}^T) \\
&= p\left\{ V - p\frac{V+C}{2} - \left[p\left(V - p\frac{V+C}{2}\right) + (1-p)^2\frac{V}{2} \right] \right\} \\
&= p(1-p)\frac{V - pC}{2}.
\end{aligned}
\tag{4.3}
$$

Thus for $V < C$, p is driven towards the ESS value $p = V/C$ in all cases. Moreover, small deviations in the value of p away from the ESS will always see the population return to the ESS; see Figure 4.1.

4.1.6 Polymorphic mixture versus mixed strategy

We have already discussed in Section 2.2.1 that for matrix games, it does not make a significant difference if the focal individual is in a monomorphic population $\delta_{\mathbf{p}}$ or a polymorphic population $\mathbf{p}^{\mathbf{T}} = \sum_i \mathbf{p}_i \delta_i$ because a random opponent will always play strategy S_i with probability p_i. Let us, in this part, suppose $V < C$ and investigate what happens in a population consisting of three different types of individuals: Hawks, Doves and individuals playing

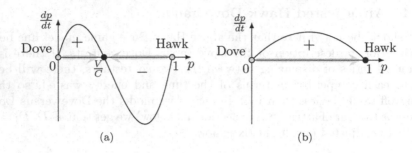

FIGURE 4.1: The replicator dynamics of the Hawk-Dove game (4.3); (a) $V < C$, (b) $V > C$. Arrows indicate the direction of evolution of p, grey discs are unstable and black disks are stable rest points of the dynamics.

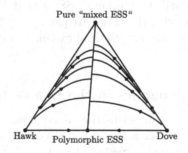

FIGURE 4.2: Continuous replicator dynamics for the Hawk-Dove game given by the matrix (4.4); $V = 5, C = 9$.

mixed ESS strategy $\mathbf{p}_{\text{ESS}} = (V/C, 1 - V/C)$. The payoff matrix is thus

$$
\begin{array}{cc}
 & \begin{array}{ccc} \text{Hawk} & \text{Dove} & \mathbf{p}_{\text{ESS}} \end{array} \\
\begin{array}{c} \text{Hawk} \\ \text{Dove} \\ \mathbf{p}_{\text{ESS}} \end{array} &
\left(\begin{array}{ccc}
\frac{V-C}{2} & V & \frac{V(C-V)}{2C} \\
0 & \frac{V}{2} & \frac{V(C-V)}{2C} \\
\frac{V(V-C)}{2C} & \frac{V(V+C)}{2C} & \frac{V(C-V)}{2C}
\end{array} \right).
\end{array}
\tag{4.4}
$$

It should be noted that the matrix (4.4) is non-generic, as all entries in the third column are identical. This is caused by adding an ESS strategy to the original game; all payoffs against an internal ESS must be equal in a matrix game as we see in Chapter 6 (see Lemma 6.7). This addition of the strategy $\mathbf{p}_{\text{ESS}}$ makes the ESS analysis uninteresting in this case; see Exercise 6.17.

It is more illuminating to analyze this game using replicator dynamics as done in Hofbauer and Sigmund (1998, p.75), see Figure 4.2. Once the population starts inside of the triangle, it evolves towards a point that lies on the line connecting $\delta_{\mathbf{p}_{\text{ESS}}}$ and $\frac{V}{C}\delta_H + (1 - \frac{V}{C})\delta_D$ (which is in fact $\mathbf{p}_{\text{ESS}}^{\mathbf{T}}$). This line is an example of an ESSet (see Section 3.2.3). If the population is on that line, the random opponent plays Hawk with probability V/C exactly as it should in the ESS of the Hawk-Dove game (4.1). Also, any point on the line can be reached from some initial values and once on the line, the population reaches an equilibrium and does not evolve further under the replicator dynamics. Non-generic games often generate such problems.

This is not the end of the story, however, as by considering the "trembling hand" (Section 3.2.4) we come to a different conclusion. If the population deviates from this line, then the mixed strategy is always fitter than the mean of this perturbed population, and so should increase in density; see Exercise 4.8. Since the continuous supply of some imperfections or newcomers (or stochastic effects due to the finiteness of the population in any real scenario) causes small deviations from equilibria and thus from the line, the population eventually evolves towards $\mathbf{p}_{\mathrm{ESS}}$ strategists only, provided that these deviations do not systematically involve reduction in the proportion of $\mathbf{p}_{\mathrm{ESS}}$ strategists (possible for example if these were due to errors, at the points where $\mathbf{p}_{\mathrm{ESS}}$ is in the majority). In any case, with precise description of the nature of the error mechanism, the equilibrium line will, in general, become a single point.

4.2   The Prisoner's Dilemma

The Prisoner's Dilemma is arguably the most famous game in all of game theory, its application straddling the disciplines of biology, economics, psychology and more. The game dates back to 1950 to M. Dresher and M. Flood at RAND corporation and was formalised by Tucker (Straffin Jr, 1980; Tucker and Straffin Jr, 1983); see also Axelrod (1984), Poundstone (1992) and Walker (2008).

The apparent paradox at the heart of the game (to non-game theorists) has been and continues to be the subject of much debate about how cooperation between individuals can evolve, when such cooperation is mutually beneficial when compared to non-cooperating, but when some individuals can cheat and become "free-riders" on others. We explore this issue more deeply in Chapter 15, but here we shall introduce the fundamentals of the game. A closely related scenario, with the same fundamental problem at its heart, is the tragedy of the commons (Hardin, 1998).

Although it is most commonly applied to modelling cooperation amongst humans, it can also be applied to many situations involving social animals. A number of examples are discussed by Dugatkin (2009) including social grooming in primates, the transfer of information on predators in fish shoals and reciprocity of various kinds, including food sharing, in vampire bats; see also Exercise 4.1.

4.2.1   The underlying conflict situation

Although the terminology used is now often in terms of competing companies, or more general social interactions, the original story behind the game, from whence it got its name, is an engaging one of the type rarely found in mathematical texts. It is thus worth telling, and we give a version below. For a biological motivation, see Exercise 4.1.

Two people, Walter and Monica, successfully commit an armed robbery. They hide the money, put their guns safely under their beds, and go to sleep satisfied with their work. A third person who bears a grudge against them knows of the crime, and informs the police, who raid their homes in the early morning. They find no money or other evidence of the crime, but they do find the guns each criminal possesses. Walter and Monica are put in separate cells and the police try to make each confess to the crime implicating the other, with the promise of immunity if their confession leads to the conviction of the other prisoner.

Both prisoners are intelligent and experienced and know both their options, and the consequences of their potential actions. Their options are to tell on their colleague to the police, to *Defect*, or to say nothing, to *Cooperate*. The consequences are as follows:

- If one Defects and the other Cooperates, the Defector's confession will be used to convict the Cooperator, with the result that the Cooperator gets 10 years in jail and the Defector gets a suspended sentence, i.e. goes free.

- If both Cooperate, the police have no evidence against them, but can convict them of possessing a gun, so each receives one year in jail.

- If both Defect, then both have confessed, so there is no immunity, as the confession was not needed to convict the other. Judges like people to plead guilty, however, so they each receive 8 years rather than 10.

The sole aim of each player is to minimise his or her own time in jail, and they are neither concerned to help the other nor to "beat" the other.

4.2.2   The mathematical model

The above description leads to a two-player game where the payoffs are given by the following payoff matrix

$$
\begin{array}{c}
\quad\quad\quad \text{Cooperate} \quad \text{Defect} \\
\begin{array}{c} \text{Cooperate} \\ \text{Defect} \end{array}
\left(\begin{array}{cc} -1 & -10 \\ 0 & -8 \end{array} \right).
\end{array}
\tag{4.5}
$$

In general, the rewards for the Prisoner's Dilemma are usually written as in the payoff matrix

$$
\begin{array}{cc}
 & \text{Cooperate} \quad \text{Defect} \\
\begin{array}{c}\text{Cooperate}\\\text{Defect}\end{array} &
\left(\begin{array}{cc} R & S \\ T & P \end{array}\right),
\end{array}
\qquad (4.6)
$$

where R, S, T and P are termed the *Reward, Sucker's payoff, Temptation* and *Punishment*, respectively. Whilst the individual numbers are not important, for the classical dilemma we need $T > R > P > S$. It turns out that the additional condition $2R > S + T$ is also necessary for the evolution of cooperation, as we see below in Section 4.2.5.

4.2.3 Mathematical analysis

Rather than presenting analysis in formulae, we will present the thinking process that leads to the solution. Walter sits in his cell and thinks about what to do. If Monica cooperates then the reward for him to play Defect is bigger than that if he plays Cooperate, so if he thinks that she will play Cooperate, he should Defect. But if she plays Defect it is also the case that Defect is best. Thus irrespective of what he thinks that Monica should do, Walter should defect. Monica goes through exactly the same logical process, and so also decides to play Defect. Thus both players defect, and receive a payoff of -8.

The same thinking process is valid for the general payoff matrix (4.6) as long as $T > R$ and $P > S$. In this case, both players conclude that they should defect. The outcome seems paradoxical because $R > P$, i.e. they would both be better off if they cooperated.

4.2.4 Interpretation of the results

Each player following his own self-interest in a rational manner has led to the worst possible collective result, with a total payoff of -16. This problem raised two significant issues for economists. Firstly, the free market is supposed to work so that everyone pursuing his own interest should lead, at least in some sense, to the common good. Yet here is a game which shows that the opposite can happen. Secondly, even though such problems clearly do happen in real-life situations, e.g. in the over-exploitation of resources like fishing stocks but also the problem of global warming (related to the "tragedy of the commons"), there is still a lot of cooperation in the world. How could it occur? This question has been addressed in many ways, the first and most well-known development is the idea of repeated plays and the Iterated Prisoner's Dilemma.

4.2.5   The IPD, computer tournaments and Tit for Tat

For the above analysis, as in the case of the Hawk-Dove game or matrix games, in general, an individual may play a number of games, but either any individual is played only once, or there is no memory of previous interactions. One reason why a player may be tempted to play Defect, is that there is only this single game played, so that there is no "relationship" between the players. Suppose that the players will now play each other in a sequence of n games, where each can make his choice based upon what has gone before. In this case a strategy is a specification of what to do in every possible situation, based upon all of the possible choices that both players have made in earlier rounds. Thus potential strategies could be very complex.

The Iterated Prisoners Dilemma (IPD) was popularised in Axelrod (1984); see also Axelrod (1981), which describes a tournament organised by the author. In total 14 participants, who were fellow academics from a variety of disciplines, submitted a computer programme which played their chosen strategy in an n-step Iterated Prisoner's Dilemma, where n was fixed to be equal to 200. Every programme played a round robin tournament against every other programme, as well as against a copy of themselves and a random strategy. There was great variety in the type of programme entered in terms of their willingness to defect, and in the general complexity of the programmes.

The best strategy with the highest average score was the one submitted by Anatol Rapoport, which he called *Tit for Tat*. The surprising thing about this was that this was the simplest of the programmes in the tournament, with only four lines of code. The play of Tit for Tat in rounds $j = 1, \ldots, n$ can be described as follows:

(1) If $j = 1$, play Cooperate.

(2) If $j > 1$, copy your opponent's strategy from round $j - 1$.

The top scoring strategies had three key properties in common, and we can see clearly that they apply to the winner, Tit for Tat. A strategy should be *nice*, and not be the first to defect. It should *retaliate*, so be able to defect against an opponent who defects. Tit for Tat defects immediately after an opponent does (but e.g. *Tit for two Tats* only defects against two successive defections by opponents). It should *forgive*, so that if an opponent, having once defected, starts to cooperate again, so will our strategy. Tit for Tat forgives immediately it observes its opponent cooperate.

A second tournament was held after the results of this one were known. This time there were 63 entries, but Tit for Tat was again the winner. This ability to beat a variety of other strategies was coined "robustness" by Axelrod. It should be noted that although Tit for Tat won the tournaments, it is impossible for it to score more highly than any opponent in their individual contest (compare to Exercise 4.10). Moreover, other strategies (that were not present at the tournaments) would have beaten it; see Axelrod (1984).

There was a slight modification in the second tournament in that the contests were not for a fixed number of rounds anymore, but rather after every play another round was held with probability $w = 0.99654$. Thus the number of rounds in each game followed a geometrical distribution with the median round length of $n = 200$. Note that for fixed n rounds, Tit for Tat can never be best, as it will be invaded by a strategy that plays as Tit for Tat on moves 1 to $n-1$, but then defects on move n. Such a strategy would have won the first computer tournament above, but this development was not thought of at the time. In general, any ESS strategy for the IPD with a fixed number of rounds should play Defect on the last round; and by backwards induction, it should play Defect on any round. Thus, *All Defect*, the strategy of defecting on every round, is the only ESS in this case.

In the IPD game the optimal strategy depends upon the strategies of likely opponents. We investigate below when Tit for Tat can be an ESS in a population playing an IPD. We note, however, that in the setting of IPD, one should use the term ESS cautiously and always keep in mind what are the potential invading strategies. Many different strategies (such as TFT and ALLC) can become practically indistinguishable and thus one strategy can invade the other by drift; in the discussion below we ignore invasion by such indistinguishable strategies.

The IPD for a random number of rounds was considered analytically in Axelrod and Hamilton (1981). They showed that All Defect is always an ESS, and Tit for Tat is an ESS if and only if it could resist invasion by All Defect and Alternate, where Alternate plays $DCDCDC\ldots$ irrespective of its opponent. The contests between Tit for Tat (TFT), All Defect (ALLD) and Alternate (ALT) against Tit for Tat give the following plays:

$$
\begin{array}{c|l}
TFT & CCCCCC\ldots \\
TFT & CCCCCC\ldots \\
\hline
ALLD & DDDDDD\ldots \\
TFT & CDDDDD\ldots \\
\hline
ALT & DCDCDC\ldots \\
TFT & CDCDCD\ldots
\end{array}
\tag{4.7}
$$

The payoffs are thus given by

$$
E[TFT, TFT] = R + Rw + Rw^2 + \ldots = \frac{R}{1-w}, \tag{4.8}
$$

$$
E[ALLD, TFT] = T + Pw + Pw^2 + \ldots = T + \frac{Pw}{1-w}, \tag{4.9}
$$

$$
E[ALT, TFT] = T + Sw + Tw^2 + Sw^3 + \ldots = \frac{T+Sw}{1-w^2}. \tag{4.10}
$$

Thus Tit for Tat is an ESS if and only if

$$
\frac{R}{1-w} > \max\left(T + \frac{Pw}{1-w}, \frac{T+Sw}{1-w^2}\right) \tag{4.11}
$$

i.e. if and only if

$$w > \max\left(\frac{T-R}{T-P}, \frac{T-R}{R-S}\right). \tag{4.12}$$

Tit for Tat is thus an ESS if there is a sufficiently large chance of a repeat interaction. It is clear here why we also need the condition $2R > (S+T)$, as otherwise Alternate will invade for all w. We will explore the idea of the evolution of cooperation and more modern developments in the theory in Chapter 15, where we see that some refinements of Tit for Tat are needed to help individuals perform well in more realistic scenarios.

Note that the Hawk-Dove game and the Prisoner's Dilemma are two distinct types of two-strategy matrix games. In the generic case there are $4! = 24$ possible orderings of the four entries in the payoff matrix, and each of the Prisoner's Dilemma and the Hawk-Dove game (with $V < C$) can occur for two of these orderings. When considering a single interaction, there are essentially only three distinct cases as we see in Chapter 6: those with a single pure ESS like the Prisoner's Dilemma, those with a mixed ESS like the Hawk-Dove game with $V < C$, and those with two pure ESSs, sometimes called the *coordination game*. However when more complex situations occur, as above, there are (sometimes many) more distinctions.

4.3   The war of attrition

The two games above are classic examples of a matrix game, where there are a finite number of pure strategies. We now introduce a game with an infinite (and uncountable) number of pure strategies; see for example Bishop and Cannings (1978), and Parker and Thompson (1980) who studied struggles between male dung flies (*Scatophaga stercoraria*).

4.3.1   The underlying conflict situation

Consider a situation that arises in the Hawk-Dove game type conflict, namely two individuals compete for a reward of value V and both choose to display. Both individuals keep displaying for some time, and the first to leave does not receive anything, the other gaining the whole reward. For simplicity, we assume that individuals have no active strategy, i.e. the time they are prepared to wait is determined before the conflict begins and cannot be adjusted during the contest. Although, unlike in the Hawk-Hawk contest, there is no real harm done to the individuals during their displays, each individual pays a cost related to the length of the contest. This can be interpreted as an opportunity cost for being in the contest, and thus not being able to engage is some other useful activity.

4.3.2 The mathematical model

A pure strategy of an individual is to be prepared to wait for a predetermined time $t \geq 0$. We will denote such a strategy by S_t. It is clear that there are uncountably many pure strategies. A mixed strategy of an individual is a measure $\mathbf{p}$ on $[0, \infty)$. The measure $\mathbf{p}$ determines that an individual chooses a strategy from a set A with probability $\mathbf{p}(A)$. One can see the mixed strategy $\mathbf{p}$ given by a density function $p(x)$, that is the probability that an individual leaves between x and $x + \mathrm{d}x$ is $p(x)\mathrm{d}x$. A pure strategy S_t can be seen as a Dirac measure at the point t, i.e. for any function f that is continuous on $[0, \infty)$ we have

$$\int_0^\infty f(x)\mathrm{d}S_t(x) = f(t). \tag{4.13}$$

Let us now determine the payoffs. A payoff is the reward minus the cost. The reward for the winner of the contest is given by V. The one that is prepared to display the longer gets the reward; we assume that in the event that both individuals leave at exactly the same time, the reward will be won by one of them at random. We also assume that the cost is proportional to the length of the contest, i.e. the cost of the contest of length x is $C(x) = cx$ for some given constant c. For pure strategies S_x and S_y we get

$$E[S_x, S_y] = \begin{cases} V - cy & x > y, \\ V/2 - cx & x = y, \\ -cx & x < y. \end{cases} \tag{4.14}$$

For mixed strategies $\mathbf{p}, \mathbf{q}$ we get

$$E[\mathbf{p}, \mathbf{q}] = \iint_{(x,y) \in [0,\infty)^2} E[S_x, S_y]\mathrm{d}\mathbf{p}(x)\mathrm{d}\mathbf{q}(y); \tag{4.15}$$

see Exercise 4.11.

4.3.3 Mathematical analysis

It is immediately clear that there is no pure ESS, as for any $\tau > 0$

$$E[S_{t+\tau}, S_t] = V - ct > \frac{V}{2} - ct = E[S_t, S_t] \tag{4.16}$$

and thus for any small $\varepsilon > 0$

$$\mathcal{E}[S_{t+\tau}; (1 - \varepsilon)\delta_{S_t} + \varepsilon\delta_{S_{t+\tau}}] > \mathcal{E}[S_t; (1 - \varepsilon)\delta_{S_t} + \varepsilon\delta_{S_{t+\tau}}]. \tag{4.17}$$

To find candidates for a mixed ESS, we use a version of an argument that we develop in Chapter 6, Lemma 6.7, modified appropriately for our particular situation of uncountably many strategies. If $\mathbf{p}$ with a density function p is an ESS, then for almost all $t \geq 0$ for which $p(t) > 0$ we have

$$E[S_t, \mathbf{p}] = E[\mathbf{p}, \mathbf{p}], \tag{4.18}$$

which, by (4.15) and using (4.13), yields

$$\int_0^t (V - cx)p(x)\mathrm{d}x + \int_t^\infty (-ct)p(x)\mathrm{d}x = E[\mathbf{p}, \mathbf{p}]. \tag{4.19}$$

Differentiating (4.19) with respect to t (assuming that such a derivative exists) gives

$$(V - ct)p(t) - c \int_t^\infty p(x)\mathrm{d}x + ctp(t) = 0. \tag{4.20}$$

If $p(x) = P'(x)$ for some function P and all $x \geq 0$ such that $p(x) > 0$, then (4.20) becomes

$$VP'(t) + cP(t) = 0 \tag{4.21}$$

which yields $P(t) = K \exp(-ct/V)$ for some constant K and hence

$$p(t) = \frac{-Kc}{V} \exp(-ct/V); \quad \text{for almost all } t \text{ with } p(t) \neq 0, \tag{4.22}$$

where the constant K is chosen so that $\mathbf{p}$ is a probability measure, i.e. so that $\int_0^\infty p(x)\mathrm{d}x = 1$. Note that there are many solutions of (4.22). The most notable one is

$$p_{\mathrm{ESS}}(t) = \frac{c}{V} \exp\left(-\frac{ct}{V}\right) \tag{4.23}$$

but others include

$$p(t) = \begin{cases} 0 & t \in [0, 1), \\ p_{\mathrm{ESS}}(t - 1) & t > 1, \end{cases} \tag{4.24}$$

and more generally any (appropriately normalised) restriction of p_{ESS} on any subset of $[0, \infty)$. In other words, we have many candidates for an ESS (and in fact we could have missed some by the way we arrived at (4.22)).

To find out which (if any) of the candidates for ESSs are indeed ESSs, we use the inequality

$$E[\mathbf{r}, \mathbf{r}] - E[\mathbf{q}, \mathbf{r}] - E[\mathbf{r}, \mathbf{q}] + E[\mathbf{q}, \mathbf{q}] < 0, \quad \text{for all } \mathbf{q} \neq \mathbf{r}, \tag{4.25}$$

derived in Bishop and Cannings (1976). Here $\mathbf{q} \neq \mathbf{r}$ means two distributions with different distribution functions, i.e. does not include a pair of strategies with trivial differences, only occurring on a set of measure zero. It follows from (4.25) that there cannot be two ESSs, since for any pair of strategies $\mathbf{q}$ and $\mathbf{r}$, if $\mathbf{q}$ cannot invade $\mathbf{r}$, then $\mathbf{r}$ can invade $\mathbf{q}$. It follows from (4.18) that

$$E[\mathbf{p}_{\mathrm{ESS}}, \mathbf{p}_{\mathrm{ESS}}] - E[\mathbf{q}, \mathbf{p}_{\mathrm{ESS}}] = 0 \tag{4.26}$$

for all pure strategies $\mathbf{q} = S_t$ and any $t \geq 0$, and thus for all mixed strategies $\mathbf{q}$, since $p_{\mathrm{ESS}}(t) > 0$ for all $t \geq 0$. This means that by (4.25) we have that $E[\mathbf{p}_{\mathrm{ESS}}, \mathbf{q}] > E[\mathbf{q}, \mathbf{q}]$ and so $\mathbf{p}_{\mathrm{ESS}}$ is the unique ESS.

4.3.4 Some remarks on the above analysis and results

The above analysis should be compared with Theorem 6.9 stating that an internal ESS of a matrix game is the unique ESS of the game. Also, the inequality (4.25) should be compared to the Haigh criteria (see Section 6.3 and Haigh, 1975).

Note that the expected payoff for $\mathbf{p}_{ESS}$ in a population of $\mathbf{p}_{ESS}$-players, and so the average gain in resources from entering into such a contest, is zero. This can be calculated directly from (4.15) and (4.23); or one can realize that the duration of a contest is the minimum of two exponential distributions with mean V/c, and so has expectation $V/(2c)$. Hence if we assume that two Doves from a Hawk-Dove game play a war of attrition, their reward would become 0, as we describe above in Section 4.1.4. It is perhaps surprising that the expected reward for entering a contest over a valuable resource is zero. Why then should an individual enter such a contest? Thinking of the cost as relating to missed opportunities, the energy gain from other activities will be at a rate c per unit time, meaning the reward for entering the contest is not zero, it is just that the expected gain over doing other activities instead is zero. It is probably more accurate to say that there is a precise exchange of energy for reproductive reward, and if there is no way to achieve such a reward without entering a contest of this type, the saved energy from effectively not playing (playing 0) can only be invested in another such contest.

Also, note that the strategy $\mathbf{p}_{ESS}$ follows an exponential distribution, i.e. a process "without memory", so that the probability of waiting for an additional time t conditional on having stayed until now is always the same. This is equivalent to the individuals leaving the contest as a Poisson process with (constant) rate c/V. Thus contests can be very long. In particular, real animals of course do not have an infinite life span, and an unfortunate feature of this interpretation is that there is a theoretical possibility that the individuals wait for a longer time than this life span.

Finally, realize that $\mathbf{p}_{ESS}$ is truly a mixed strategy and we cannot interpret it as a mixed population as this would require uncountably many types of individuals in the population. In particular, if for whatever reason, the individuals cannot play a mixed strategy, the mixed population can never be in an equilibrium.

The last two potential shortcomings will be addressed in the sections below.

4.3.5 A war of attrition game with limited contest duration

Bishop and Cannings (1978) generalise the war of attrition model in a number of ways, allowing cost and reward values to depend upon the length of the contest, and also allowing for a maximum time in the game. In particular, suppose that a game can only last for a maximum time of T, e.g. until nightfall. This is equivalent to restricting the strategy space so that any strategy $\mathbf{p} = (p(x)\mathrm{d}x)$ must satisfy $p(t) = 0$ for all $t > T$.

The analysis of this modified game is effectively similar to the original game. The ESS would have to follow exactly the same distribution as given in (4.22). It means we just have to identify the support of the ESS and also the set (of measure zero) of points where the strategy would not follow (4.22). It turns out that it does not pay to wait for time $t \in [T - V/(2c), T)$ because heuristically, if one waited until $T - V/(2c)$ with the contest continuing, then waiting for an additional time $V/(2c)$ increases the cost by at most $V/2$ but also guarantees a win of at least $V/2$. Thus waiting the additional time to the end of the contest is better than stopping at any intermediate time.

Also, as in Section 4.3.3, the uniqueness of the internal ESS means that the ESS strategy will be supported on $[0, T - V/(2c)]$, following the standard exponential distribution with mean V, truncated at $T - V/2$, with an atom of probability at T (this has the property of having a constant hazard function for all allowable time values apart from T, and a mean game duration of $V/2$), as in Section 4.3.3.

4.3.6   A war of attrition with finite strategies

Suppose individuals are forced to be pure strategists and suppose even further that there are only a finite number of available strategies $S_{t_1}, S_{t_2}, \ldots, S_{t_n}$, where the leaving time of strategy S_{t_i} is t_i and the times are sorted that $t_i < t_j$ if $i < j$. The payoff to S_{t_i} if playing against S_{t_j}, assuming a cost of $c = 1$, is thus

$$E[S_{t_i}, S_{t_j}] = \begin{cases} a_{ij} = V - t_j & i > j, \\ a_{ij} = V/2 - t_i & i = j, \\ a_{ij} = -t_i & i < j. \end{cases} \tag{4.27}$$

It is shown in Bishop and Cannings (1976) that there is a unique ESS to this game. Alternatively, the same follows from the fact that the full payoff matrix satisfies the negative-definiteness condition of Haigh (1975); see also Section 6.3. In this case, there is a simple iterative procedure which determines whether a pure strategy is in the support of the ESS or not, working from the longest waiting time t_n downwards; see for example Cressman (2003, Section 7.4.2). This largest value must always be in the support of the ESS, since it can invade any other combination. In addition, no values within a time $V/2$ of t_n can be included, as for S_{t_i} to invade a population of pure S_{t_n} we need $a_{in} > a_{nn}$ which is equivalent to $t_n - t_i > \frac{V}{2}$, and similarly any strategy S_{t_i} within $V/2$ of S_{t_n} does no better than S_{t_n} in an individual contest against any other strategy.

Despite this it is not simple to say, in general, terms which strategies will or will not be included in the ESS; see for example Cressman (2003, Figure 7.4.2, p. 213). There is also no simple way to assess the effect of introducing a new strategy into the population.

4.3.7 The asymmetric war of attrition

We consider asymmetric contests in general, including the asymmetric war of attrition, in Chapter 8. However, it is worth briefly looking at it here, as we see how the introduction of distinct roles can recover pure strategy solutions for the individuals involved. Suppose that we have a population where individuals can occupy one of two roles, *owner* and *intruder*, and suppose further that the reward is worth more to the owner V than the intruder $v < V$. A strategy in this game determines how the individual should behave as an owner and how it should behave as an intruder (the behaviour can be different in different roles).

Maynard Smith and Parker (1976) showed that there are two alternative types of pure ESS.

ESS$_1$:   as owner, play T (for some sufficiently large T),
            as intruder, play 0,
ESS$_2$:   as owner, play 0,
            as intruder, play T (for some sufficiently large T).

However, out of the above two strategies, only ESS$_1$ could invade the strategy which ignores roles (and which, assuming an individual is equally likely to enter the contests as owner or as the intruder, plays an exponential distribution with mean $(v + V)/2$).

This result, that in asymmetric contests ESSs are pure, is a general one due to Selten (1980). One problem with the above result of Maynard Smith and Parker (1976) is that there are many possible values of T that can be chosen, and all do equally well, as long as they are sufficiently large to resist invasion. The "trembling hand" (Section 3.2.4) comes to our rescue again (see Parker and Rubenstein, 1981; Hammerstein and Parker, 1982). As soon as there is some small probability of mistaking the role an individual is in, there is a unique equilibrium solution, and T in ESS$_1$ follows a probability distribution, with an exponential distribution of mean V.

We see this trembling hand result and discuss the asymmetric war of attrition more in Chapter 8.

4.4 The sex ratio game

The underlying problem of the sex ratio was the first problem to receive a treatment which can be recognised as game-theoretical in character. This problem, and the essential nature of the solution, was realised by Darwin (1871) (as noted in Bulmer (1994), it was omitted in Darwin (1874)), and was expressed more clearly and popularised in Fisher (1930); an earlier mathematical treatment was given in Düsing (1884); see also Edwards (2000). It was not until Hamilton (1967) that a clear game-theoretical argument for the explanation was developed, and this remains one of the most well-known, important and elegant results of evolutionary game theory.

4.4.1 The underlying conflict situation

Why is the sex ratio in most animals close to a half? A naive look at this problem would contend that for the good of the species, there should be far fewer males than females, since while females usually must make a significant investment in bringing up offspring, often the only involvement of the male is during mating and consequently males can have many children. Indeed it is often the case that most offspring are sired by a relatively small number of males and most males make no contribution at all. What is the explanation for this?

To make the problem more specific, let us suppose that in a given population, any individual will have a fixed number of offspring, but that it can choose the proportion which are male and female. We also assume that all females (males) are equally likely to be the mother (father) of any particular child in the next generation. Our task is to provide a reasonable argument to answer the question of what is the best choice of the sex ratio.

4.4.2 The mathematical model

A strategy of an individual will be a choice of the proportion of male offspring. We note that in this model we explicitly discuss males and females, and it is important how the male and female strategies influence the sex ratio in the offspring. Thus this is a model where genetics plays a role, see Argasinski (2012) and also Karlin and Lessard (1986). We consider a small proportion of the population (fraction ε) of individuals playing a (potentially) different strategy to the rest. Let p denote the strategy of our invading group, and m denote the strategy played by the rest of the population.

Since every individual has the same number of offspring, we cannot use this number as the payoff. Instead we have to consider the number of grandchildren produced by each strategy. Let us assume that the total number of individuals in the next generation is N_1, and the total number in the following generation, the generation of grandchildren, is N_2. In fact, as we shall see, these numbers are irrelevant, but it is convenient to consider them at this stage.

When the grandchildren are taken as a measure of success, we also need to know the proportion of males in the generation of offspring, which is given by a combination of the strategies p and m and is thus $m_1 = (1 - \varepsilon)m + \varepsilon p$.

The question we have to answer is: given m, what is the best reply p to m? A strategy p then will be an ESS if p is the best reply to itself and uninvadable by any other strategy.

We will not investigate singular cases of $m = p = 0$ or $m = p = 1$ since if every individual adopts this strategy, there would be no grandchildren; and

$p = 1$ is thus a better reply to $m = 0$ than $p = 0$. So $p = 0$ is not an ESS, and similarly $p = 1$ cannot be an ESS.

4.4.3 Mathematical analysis

For large N_1, the total number of males in the next generation will be $m_1 N_1$ and the total number of females $(1 - m_1)N_1$. Since all females (males) are equally likely to be the mother (father) of any particular grandchild, a female will on average be the mother of $N_2/((1 - m_1)N_1)$ offspring and a male will be the father of $N_2/(m_1 N_1)$ offspring. Hence, the expected number of offspring of one of our focal individual's offspring is given by

$$\mathcal{E}[p; \delta_m] = p \times \frac{1}{m_1 N_1} \times N_2 + (1 - p) \times \frac{1}{(1 - m_1)N_1} \times N_2$$
$$= \frac{N_2}{N_1}\left(\frac{p}{m_1} + \frac{1 - p}{1 - m_1}\right) \approx \frac{N_2}{N_1}\left(\frac{p}{m} + \frac{1 - p}{1 - m}\right). \tag{4.28}$$

To find the best choice of p, we simply maximise this function $\mathcal{E}[p; \delta_m]$; see also Exercise 4.13. This gives $p = 1$ if $m < 1/2$ (although any $p > m$ can invade), and $p = 0$ if $m > 1/2$ (similarly, any $p < m$ can invade). If $m = 1/2$ we have to consider the case where the invading group fraction is non-negligible i.e. $\varepsilon > 0$, and then $m = 1/2$ performs strictly better than the mutant. Note that $p = 1/2$ invades any other strategy (or combination of strategies, which can be seen by replacing m by the appropriate weighted average in the above), so that it completely replaces it.

We have seen that $p = 1/2$ is an ESS. This does not necessarily mean that at equilibrium all individuals play $1/2$. If the average population strategy is $m = 1/2$ then all individuals do equally well. In general, if we have a restricted strategy set of pure strategies, there will be a stable mixture if and only if there is at least one pure strategy with $p_i > 1/2$, and at least one with $p_i < 1/2$; see Exercise 6.16. For a set of three or more strategies satisfying this property there are an infinite number of mixtures of this type.

4.5 Python code

In this section we show how to use Python to run a version of Axelrod's tournament of IPD from Section 4.2.5. There are two files. The first file contains a generic definition of a Strategy class and a single IPD game.

```
1   """  Iterated Prisonners Dilemma game
2   This script provides a basic class for the IPD strategies
8   It also provides the function running the IPD game once
```

```
4   """
5
6   # Import rand to generate random numbers
7   from numpy.random import rand
8
9   # Set the moves
10  Defect = 0        # Defect is indexed by 0 in the payoff matrix
11  Cooperate = 1     # Cooperate is indexed by 1 in the payoff matrix
12
13
14  # Define the strategy class, parent class to all strategies
15  class Strategy:
16      def __init__(self):
17          self.History = []   # Initially no history of my moves
18          self.Payoff = []    # Initially no payoff history
19
20      def Intended(self, opponent):
21          " Returns the intended move against given opponent "
22          # Empty now, will be defined for specific strategies
23          pass
24
25      def Move(self, opponent, ErrorProb):
26          " The `actual' move. May modify the intended move "
27
28          # Get the intended move
29          move = self.Intended(opponent)
30
31          # If error occured, do the opposite
32          if rand(1) < ErrorProb:
33              move = 1 - move
34          return move
35
36      def Update(self, myMove, oppMove, PD_payoff):
37          " Adds the myMove and payoff to the histories "
38          self.History.append(myMove)
39          self.Payoff.append(PD_payoff[myMove][oppMove])
40
41      name = ""   # Empty name so far
42
43  # Define a game of IPD between two players
44  def IPDgame(Player1, Player2, PD_payoff, next_rnd, ErrorProb):
45      """ Single IPD game between two players
46          PD_payoff is the 2x2 payoff matrix
47          ErrorProb is the probability of error in the move
48
49          next_rnd determines the duration of the game
50          If next_rnd in (0,1), the game has a random
51          number of rounds. The next round happen next_rnd
52
53          If next_rnd>1, the game will have round(next_rns) rounds
54
55          The function returns mean payoffs to the players
56      """
57      # The first round will be played
58      prob = 0
59
60      # Play the rounds with probability p_next_round
```

```
61    while prob<next_rnd:
62            # Get the move of player 1, this updates p1 history
63            p1Move = Player1.Move(Player2, ErrorProb)
64
65            # Get the move of player 2, this updates p2 history
66            p2Move = Player2.Move(Player1, ErrorProb)
67
68            # Update the histories and payoffs
69            Player1.Update(p1Move, p2Move, PD_payoff)
70            Player2.Update(p2Move, p1Move, PD_payoff)
71
72            # See if the next round being played
73            if next_rnd<1:
74                # The game has random number of rounds
75                prob = rand(1)
76            else:
77                # The game has fixed number of rounds
78                # Increase the count of rounds played
79                prob +=1
80
81        # Game is done, return the payoffs
82        return Player1.Payoff, Player2.Payoff
```

The second file contains definitions of several specific strategies and actually codes the tournament.

```
1    """ Round robin tournament of the Iterated Prisoner's Dilemma
2
3    Specify PD payoff matrix, probability of next round
4    and probability of an error (in execution) of the move
5    Specify the list of strategies and their definitions
6
7    If p_next_round is integer N, the game will have exactly N rounds
8
9    The script will run the tournament with the given Strategies
10   """
11
12   ## Define parameters
13   PD_payoff =[[1,5],   # Payoff to D against D, C
14               [0,3]]   # Payoff to C against D, C
15   ErrorProb = 0.01     # Probability of an error in the move
16   p_next_round = 0.99654 # prob. of a subsequent round
17
18   ## Import basic packages
19   from numpy.random import seed, randint, rand
20   import numpy as np
21   import copy
22
23   # Import class Strategy and function IPDgame
24   from IPD import Strategy, IPDgame, Defect, Cooperate
25
26   # Seed random number generator for reproducibility
27   seed(1)
28
29   ## Actual definitions of specific strategies
```

```
30
31   # A strategy inherits everything from the parent class Strategy
32   # We just have to define the actual "intended move" and name
33
34   class AllC(Strategy):
35       def Intended(self, opponent):
36           return Cooperate # Always cooperate
37       name = "AllC"
38
39   class AllD(Strategy):
40       def Intended(self, opponent):
41           return Defect # Always defect
42       name = "AllD"
43
44   class TFT(Strategy):
45       # Tit for tat
46       def Intended(self, opponent):
47           # Cooperate on the first move
48           if len(self.History)==0:
49               return Cooperate
50           else:
51               # If not first move, repeat opponent's last move
52               return opponent.History[-1]
53       name = "TFT"
54
55   class Grimm(Strategy):
56       def Intended(self, opponent):
57           # Defect if opponent ever defected
58           if Defect in opponent.History:
59               return Defect;
60           else:
61               # If opponent always cooperated, cooperate
62               return Cooperate;
63       name = "Grimm"
64
65   class Random(Strategy):
66       def Intended(self, opponent):
67           # Return random integer 0 or 1
68           return randint(0,2)
69       name = "Random"
70
71
72   ## Insert/import additional strategy definitions above if needed
73
74
75   # Create a list of participating strategies
76   Strategies = [AllC(), AllD(), TFT(), Grimm(), Random()]
77
78   # Initialize the tournament matrix
79   Results = np.zeros((len(Strategies), len(Strategies)))
80
81   # Run the tournament
82   # Every strategy will play against every strategy
83   for s1Count in range(len(Strategies)):
84       s1 = Strategies[s1Count]
85       print(s1.name)
86       for s2Count in range(len(Strategies)):
```

```
87      s2 = Strategies[s2Count]
88      # Create "copies" to prevent overrides and issues
89      # of playing against oneself
90      p1 = copy.deepcopy(s1)
91      p2 = copy.deepcopy(s2)
92
93      # Play the actual game and get the payoffs
94      p1Payoff, p2Payoff = IPDgame(p1, p2, PD_payoff,
95                             p_next_round, ErrorProb)
96
97      # Get the average payoffs
98      p1Avg = np.mean(p1Payoff)
99      p2Avg = np.mean(p2Payoff)
100
101     Results[s1Count, s2Count] = p1Avg
102     Results[s2Count, s1Count] = p2Avg
103
104 print(Results)
```

4.6 Further reading

The four main games involved in this chapter are the most well-known classical games and so are discussed in most books that consider evolutionary games. A particularly good investigation into all four can be found in Maynard Smith (1982). Other good books include Hofbauer and Sigmund (1998), Mesterton-Gibbons (2000) and Gintis (2000, Chapters 4, 5, and 6).

The Hawk-Dove game was introduced in Maynard Smith and Price (1973). It is perhaps still best described in Maynard Smith (1982). Essentially the same game also appears under other names (chicken, the snowdrift game); see for instance Doebeli and Hauert (2005). There is a lot of work which deals with the Prisoner's Dilemma and the evolution of cooperation. The classical work is Axelrod (1984), a more modern book is Nowak (2006a). See also McNamara et al. (2004) on variation in strategies promoting cooperation. The Hawk-dove game and especially the Prisoner's dilemma feature as part of more complex models in a variety of settings, for instance in the structured population models of Chapter 13 or the time constraint games of Section 7.4. Important mathematical results on the war of attrition are shown in Bishop and Cannings (1976, 1978). See Iyer and Killingback (2016) for more recent work on war of attrition evolutionary dynamics. A variant on the war of attrition (Parker's model), where the losing player pays the cost indicated by their own strategy, was considered in Cannings (2015). Another good book is Charnov (1982) which considers the sex ratio problem in more detail; see also Hardy (2002). For more on the sex ratio game see Pen et al. (1999); Pen and Weissing (2000) and also Argasinski (2012, 2013, 2018).

4.7   Exercises

Exercise 4.1 (Prisoner's Dilemma among fish, Gintis, 2000, p.119). Many species of fish, for example sunfish, suffer from a parasite that attaches to fish gills. Such fish can form a symbiotic relationship with smaller species (cleaner fish) that can eat the parasite. Big fish have two available strategies, to defect (eat the small fish) or to cooperate (do not attack or eat the small fish). The small fish has also two options, to cooperate (working hard by picking the tiny parasite) or defect (taking a whole chunk out of the big fish). Set up and analyse the game. Explain how cooperation can occur between the fish.

Exercise 4.2 (Blotto game or resource allocation, Roberson, 2006). Two males are courting N females. They can each court any (and all) females for an integer multiple of minutes, but the total length of courting of all of them cannot be more than T minutes. A female will mate with the male that courted longer (or chooses randomly, if there is a draw). The males do not know how long the opponents have been courting any female. How should the males divide their courting? Solve the specific case of $N = 3$ and $T = 6$.

Exercise 4.3 (Traveller's dilemma, Basu, 2007). Directly after being released from prison, the unlucky robbers Walter and Monica we met in Section 4.2 committed another armed robbery, were caught and are again facing a punishment of 0-10 years. They are again put in separate cells, but this time the investigator plays another trick on them. They can tell him what punishment they deserve, any integer between 2 to 8. If they give the same number, they will serve that many years. If their numbers differ, the one that offered more will serve that number minus 2 years and the one that offered less will serve the choice of other player plus an additional 2 years (i.e. 4 more years than the player that offered more). Set up the game and decide what numbers they should offer.

Hint. 2 seems to be the best answer, but isn't 3 then a better one?

Exercise 4.4 (Stag hunt, Skyrms, 2004). Two hunters wait for a large stag. The stag has not come by yet, but each of them see a (separate) hare. They can either choose to hunt a hare, or wait for the stag. If a hunter chooses the hare, he will eat. However, if either kills a hare, it disturbs the stag who will not come, and so if the other hunter chooses to wait for the stag, he will starve. If both wait the stag will eventually arrive, and either can kill it and both will eat. What should the hunters do?

Exercise 4.5 (Ideal free distribution, Fretwell and Lucas, 1969). Consider m individuals that can feed at n distinct but otherwise equivalent places. The quality of food diminishes with the number of individuals present at the feeding area. Where should an individual go to eat? Set up and analyse the game for $m = n = 2$.

Exercise 4.6 (El Farol bar problem, Arthur, 1994). Consider m individuals; each one can either get water from its own small puddle or go to drink at a pond. Water in the pond is much better, but if more than cm individuals (for some $c \in (0,1)$) come at the same time, the water quality gets low and it can even attract some predators. Set up and analyse the game for $m = 2$ and $c = 1/2$.

Exercise 4.7 (Diner's dilemma, Glance and Huberman, 1994). Consider m players that go to eat at a restaurant. They agree to split the bill. They can each order from several differently priced options. Assume the value for player $i \in \{1, 2, \ldots, m\}$ of any menu is a function $f_i(c)$ of the cost of the item. For $m = 2$, and some simple functions f_i, write down the payoffs of the game and try to analyse the game.

Exercise 4.8. Consider once more the matrix game given by (4.4). Show that once the population is away from the equilibrium line (see Figure 4.2), the fitness of $\mathbf{p}_{\text{ESS}}$ is higher than the mean fitness of the population.

Exercise 4.9. For the Hawk-Dove game of Section 4.1, show that $\mathbf{p} = (V/C, 1 - V/C)$ is the unique ESS if $V < C$ and that Hawk is the unique ESS otherwise.

Exercise 4.10. Show that if S is any strategy in the Iterated Prisoner's Dilemma and TFT is the Tit for Tat strategy, then $\mathcal{E}[S; \delta_{\text{TFT}}] \geq \mathcal{E}[\text{TFT}; \delta_S]$. In other words, TFT will never do strictly better against any strategy in a single contest (yet it can win a round robin tournament).

Exercise 4.11. Derive the war of attrition payoff (4.15) for mixed strategies $\mathbf{p}$ and $\mathbf{q}$ that have density functions p and q.

Exercise 4.12. Consider the war of attrition from Section 4.3 where $V = 6$ and $c = 2$.
(a) Find all of the best responses in a population playing the pure strategy S_t for general t.
(b) What is the unique ESS for the game? How would this change if (i) the maximum allowable playing time is 10? (ii) the only allowable strategies are S_0, S_5, S_8 and S_{10}?

Exercise 4.13. The results of Section 4.4.3 were based on the optimisation of the approximated $\mathcal{E}[p; \delta_m] \approx \frac{N_2}{N_1} \left(\frac{p}{m} + \frac{1-p}{1-m} \right)$ from (4.28). Show that the results do not change if we optimise the actual payoff $\mathcal{E}[p; m] = \frac{N_2}{N_1} \left(\frac{p}{m_1} + \frac{1-p}{1-m_1} \right)$ except that now $m = 1/2$ strictly outperforms any mutant.

Exercise 4.14. For the sex ratio game, assume a population that consists only of a finite number of strategies. Find the ESSs in the population under the following sets of allowable male offspring proportions:
(i) $p_1 = 0.2, p_2 = 0.6$;
(ii) $p_1 = 0.2, p_2 = 0.3$;
(iii) $p_1 = 0.2, p_2 = 0.3, p_3 = 0.6$.

Chapter 5

The underlying biology

So far we have considered evolution in a population in a very idealised mathematical context, and we shall continue to do so. First we will pause to consider some of the assumptions that we make, and discuss the underlying rationale for considering the evolution of biological populations in the way that we do. The foundations of the study of evolution involve a vast amount of empirical observation, and it is possible to understand much of how evolution works without considering any mathematics at all. Similarly, the level of complexity of observations about the world's natural history means that there is much that is beyond mathematical explanation, and circumstances where game-theoretical analysis is not appropriate. Nevertheless, game theory has a very important role to play. We shall consider the meaning of the types of games that we discuss in the context of evolution, discuss limitations of the models, especially in relation to genetics, and finally briefly discuss the merits of simple mathematical models in general.

5.1 Darwin and natural selection

The starting point of game-theoretical modelling in biology is evolutionary theory, originated by Darwin (1859). His treatise *On the Origin of Species* is arguably the greatest of all biological books, and one of the most important in the history of science. The central argument is that any trait possessed by organisms which positively affects the reproductive success of that organism will be favoured over alternative traits, and will spread through the population. Similarly, individuals who are better adapted to their environment will have more offspring than others, and so their characteristics will spread. In fact, the last two sentences put the idea in two different ways, and the meanings are not identical, as we discuss below. The term *fitness* was coined as a measure of how successful any trait or organism is likely to be. Natural selection favours the individuals/traits with the greatest fitness, relative to the rest of the population. Fitness is a term with a number of potential meanings, and we discuss the concept in general later in this chapter. Fitness for us will be determined by the games that the animals play (our individuals generally being animals). In terms of the game-theoretical terminology invoked earlier

DOI: 10.1201/9781003024682-5

73

an individual's fitness is closely linked to the payoff of the game. In its simplest form the two are identical (or the fitness is a linear function of the payoff), although in more complex situations the relationship between the two can be less direct. For now it is enough to think of the fitness and the payoff as the same thing.

A trait can be morphological (horns, scales) or behavioural (food stealing, aggression in contests). The term that corresponds to a trait in the notation of game theory is strategy, in particular a *pure strategy*. Although the term strategy invokes the idea of a behavioural choice, it equally applies to a physical form, although in that case some of the implications are different (in particular, the meaning of a mixed strategy; for example an individual cannot have horns with probability $1/2$). The central feature of game theory is that the strategies of others affect an individual's payoff, and so potentially its optimal strategy. Thus the environment in which natural selection acts is not a static one, or one that changes only under outside influence; it changes as a result of the composition of the individuals within it. They constitute part of the environment under which natural selection acts, and influence which individuals are favoured and which are not. This, in turn, alters the environment, so that which are the most successful strategies can continually change only due to the population composition. It is upon strategies (or alternatively the individuals who play those strategies) that natural selection acts, and we seek the strategies that are favoured by evolution. We note that this game-theoretical idea of fitness explicitly depending upon the composition of the population, although more formally defined in the 1960s (e.g. Hamilton, 1964, 1970), is present in some of Darwin's work. In particular, in the first edition of *The Descent of Man, and Selection in Relation to Sex* (Darwin, 1871) he discusses the sex-ratio problem in these terms (see our Chapter 4).

When we analyse evolutionary games, the first thing we need to do is correctly identify the players and their potential strategies. Identifying all of the possible strategies for real populations is very difficult. We can see which strategies are there, but not those which are possible but currently absent; thus we are likely to only see the strategies that make up one ESS of the population. We can surmise other strategies by a mixture of a scientific understanding of reasonable possibilities and the observation of similar populations, but we cannot with certainty identify all potential strategies. It is important not to include unrealistic strategies which would be unable to be formed by plausible mutations, e.g. the trait of "fire-breathing" would likely give great benefits to individuals, and it would be wrong to conclude that it is not observed because it cannot invade a population.

In this book, we generally think of the individuals in our populations, the players of the games, as single animals. These are the units of selection. An individual plays a strategy (possesses a trait) and that individual will produce a greater number of offspring than average, and so its strategy/trait will spread in the population, if its payoff (fitness) is larger than the average. The idea of the individual animal as the unit of selection is the classical one, and the most convenient from the purposes of game-theoretical modelling, but nature is not as simple as that. How do traits or strategies transfer themselves to the next generation? Whilst many species reproduce asexually, typically single cell organisms such as bacteria, most multi-cellular species do not, and individuals are not mere copies of a single parent. It should be noted that even for asexual organisms there are mechanisms for gene transfer which mean that individuals are not just direct copies of distant ancestors. Thus in the rest of this chapter, we will sometimes look beyond the individual to the gene as the unit of selection (Dawkins, 1999). We recall from Chapter 1 that in some earlier models of evolution the group was initially taken as the unit of selection. Whilst this idea as originally postulated was effectively refuted by Maynard Smith (1964), a more sophisticated version of this idea has experienced a revival with work by e.g. David Sloan Wilson (Wilson, 1975), see also Hertler et al. (2020) and Section 15.7. To explore some of the above, we must consider some basic genetics.

5.2 Genetics

Individual humans and other mammals are not just direct copies of those in previous generations. Reproduction is sexual, and individuals inherit their traits from their parents, with some contribution from each. Such populations are known as diploid populations, and the traits are controlled by genes, each of which are, in turn, comprised of a single allele from each parent. The collection of all genes in an organism is called a *genome* which is stored in a number of *chromosomes* (46 in the case of humans). The genes are located throughout these chromosomes (although much of the genome has no obvious function). Traits are sometimes controlled by multiple genes at varied locations (loci) on the genome, although some are just controlled by a single gene. This is the case in the classic example of the colour of pea plants observed by Mendel (1866) (see also Weiling, 1986; Fairbanks and Rytting, 2001, for some criticism and defence of the Mendel experiments, respectively), and which led him to the first explanation of classical genetics.

The evolution of a population is affected by a number of important processes. We are already familiar with *selection* where each genotype will have a certain fitness which will influence how likely it is to spread through the population. We have also mentioned *mutation* in terms of individuals being replaced by others of different types; more properly mutation occurs at the genetic level, and alleles can mutate into others, which then can affect the behaviour or traits possessed by the individual. *Drift* is a property of finite populations, where the random selection of individuals, or here alleles, albeit with probabilities determined by their fitness, can cause fluctuations in the composition of populations due to chance. In particular, drift will cause many advantageous mutations never to fixate in the population (see Chapter 12).

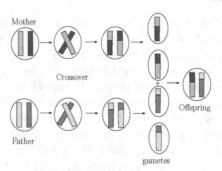

FIGURE 5.1: Schematic description of meiosis and recombination.

The above three processes occur both in asexual haploid populations where genes are transferred from generation to generation almost unaltered (except for mutations) as well as sexual diploid populations. In the sexual population, there are further processes that affect evolution, namely *recombination* and *genetic draft*. In order to understand these processes, let us briefly introduce sexual reproduction. Both parents are diploid, i.e. each bear two sets of chromosomes (one from their own mother and one from their father). A process called *meiosis* causes these two sets to separate and produces a *gamete* (a sperm cell from a father and an egg cell from a mother) that bears only one set of chromosomes. When a sperm cell merges with an egg cell, the resulting cell will have two sets of chromosomes again, one coming from the mother and one coming from the father. If meiosis was a "simple" process equivalent to separating the two sets of chromosomes and the passing of them to the gamete, the offspring would have one set of chromosomes coming from either his mother's mother or mother's father, and another set from one of the father's parents. However, during meiosis the pairs of chromosomes are broken at one or more points and then joined together again, so that the set of chromosomes in a gamete is not like any of the chromosomes from a parent. This process is known as *chromosomal crossover*; see Figure 5.1.

The choice of which alleles are selected from each parent for loci on different chromosomes can be assumed to be independent because recombination takes place over each specific chromosome only. However, the situation is different for two alleles at two loci on the same chromosome. Assume that one of the alleles becomes part of the gamete. As long as there is no crossover (or if there is an even number of them) between the two loci, the other allele will become part of the gamete as well. The closer the position of the loci, the higher the correlation between the two alleles transferred, since the chance of a break point (strictly an odd number of break points) occurring between the alleles is smaller. This can lead to certain traits being correlated such as mimicry in African swallowtail butterflies *Papilio memnon* or heterostyly in *Primula plants* (see Maynard Smith, 1989, Chapter 5) and is known as *linkage*. Thus the process of inheritance is significantly more complicated than we have assumed. It is possible for genes which are neutral or even mildly deleterious (i.e. have a negative effect on fitness) to spread through the population because they happen to be strongly linked to an advantageous mutant, and this happens for no other reason than that the mutation occurred in an individual with that gene. This is called *genetic hitchhiking*, which we discuss in Section 5.5.1. Consequently, a neutral or slightly disadvantageous allele may fixate in a population; this is called a *genetic draft*.

So far there has been an implicit assumption that if a chromosome bears two alleles on the same loci, say A and a, then during meiosis either A or a will make it to the gamete, each one with probability 1/2 and unless we specifically say otherwise, we will assume this in the future. However, there have been documented cases of *segregation distorter* or *meiotic drive* genes that are able to influence meiosis in such a way that they will end up in a gamete with probability larger than 1/2; see Section 5.5.2. Any allele that has a greater probability than 1/2 of being transmitted can spread in the population even if it is slightly deleterious.

5.2.1   Hardy-Weinberg equilibrium

Let us consider a simple example. Assume that at a given locus there are two possible alleles which we will call A and a. This leads to three distinct combinations, or genotypes, AA, Aa and aa. If a gene has two copies of the same allele it is referred to as a *homozygote*; if it does not then it is a *heterozygote*. These genotypes, in turn, control traits, or phenotypes. Each of the three genotypes could lead to distinct phenotypes, or two could lead to the same type, so that for instance AA and Aa could both lead to one phenotype, and aa to the other. In this case, A is said to be a *dominant* allele and a a *recessive* allele, as the presence of a single copy of A leads to the phenotype associated with AA, which we may call the type A phenotype. This was the case for Mendel's peas, when a number of traits exhibited this property. For instance seeds were either pure yellow or pure green, with yellow associated with a dominant allele, and green a recessive one.

If two parents are of genotype Aa, then the genotype of their offspring is determined by a random selection of an allele from each pair. Thus the father has probability of 1/2 of contributing each of A and a, as has the mother. Thus the genotypes AA, Aa and aa appear with expected frequencies 0.25, 0.5 and 0.25, respectively. In particular, if A is dominant, then both parents will have trait A, but a quarter of their children on average will have trait a. Recessive alleles are responsible for the persistence of a number of human diseases, for example sickle cell anaemia, see Neel (1949), where both parents can carry a single copy of the recessive gene and be perfectly healthy (heterozygotes can even be at an advantage due to extra resistance to malaria), but the transmission of the recessive gene by both leads to the disease in their offspring.

TABLE 5.1: Mating table.

Mother	Father	Pair frequency	Offspring genotype frequencies		
			AA	Aa	aa
AA	AA	p_{AA}^2	p_{AA}^2	0	0
AA	Aa	$p_{AA}p_{Aa}$	$\frac{1}{2}p_{AA}p_{Aa}$	$\frac{1}{2}p_{AA}p_{Aa}$	0
AA	aa	$p_{AA}p_{aa}$	0	$p_{AA}p_{aa}$	0
Aa	AA	$p_{AA}p_{Aa}$	$\frac{1}{2}p_{AA}p_{Aa}$	$\frac{1}{2}p_{AA}p_{Aa}$	0
Aa	Aa	p_{Aa}^2	$\frac{1}{4}p_{Aa}^2$	$\frac{1}{2}p_{Aa}^2$	$\frac{1}{4}p_{Aa}^2$
Aa	aa	$p_{Aa}p_{aa}$	0	$\frac{1}{2}p_{Aa}p_{aa}$	$\frac{1}{2}p_{Aa}p_{aa}$
aa	AA	$p_{AA}p_{aa}$	0	$p_{AA}p_{aa}$	0
aa	Aa	$p_{Aa}p_{aa}$	0	$\frac{1}{2}p_{Aa}p_{aa}$	$\frac{1}{2}p_{Aa}p_{aa}$
aa	aa	p_{aa}^2	0	0	p_{aa}^2

If we know the genotype frequencies p_{AA}, p_{Aa}, p_{aa} of genotypes AA, Aa, aa in the population, we can easily calculate the frequencies P_A, P_a of the alleles A and a in the population by

$$P_A = p_{AA} + \frac{1}{2}p_{Aa}, \tag{5.1}$$

$$P_a = p_{aa} + \frac{1}{2}p_{Aa}. \tag{5.2}$$

As far as evolution is concerned, we are interested in how the frequencies P_A (and $P_a = 1 - P_A$) will evolve over time. The next example shows that under certain assumptions, evolution can reach an equilibrium very fast.

Example 5.1 (Hardy-Weinberg equilibrium, Hardy, 1908; Weinberg, 1908). Assume that the population is diploid, sexually reproducing and randomly mating. In addition the population does not suffer drift, selection, mutation or migration. Study the evolution of the allele frequencies P_A (and $P_a = 1 - P_A$).

According to (5.1)–(5.2), it is enough to study the evolution of frequencies of p_{AA}, p_{Aa} and p_{aa}. Under the assumptions, we can construct Table 5.1 to show that already after one generation of mating, the population will reach an equilibrium with frequencies

$$p_{AA} = P_A^2, \tag{5.3}$$

$$p_{aa} = P_a^2, \tag{5.4}$$

$$p_{Aa} = 2P_A P_a. \tag{5.5}$$

Such a distribution is called the *Hardy-Weinberg distribution* and the population is said to be in *Hardy-Weinberg equilibrium*.

Using a slightly different approach, the principle can easily be generalised to the case with multiple alleles as seen below.

Example 5.2 (Hardy-Weinberg equilibrium, multiple alleles). Suppose that there are n alleles $A_1, A_2, \ldots, A_n$ at a locus and these have population frequencies of $P_{A_1}, \ldots, P_{A_n}$, respectively. Under the same assumptions on the population as in Example 5.1, study the evolution of P_{A_i}.

Under the assumptions a mother's gamete with allele $M \in \{A_1, \ldots, A_n\}$ meets a father's gamete with allele $F \in \{A_1, \ldots, A_n\}$ with probability $P_M P_F$, then under Hardy-Weinberg equilibrium the population frequencies are

$$p_{A_i A_i} = P_{A_i}^2 \qquad i = 1, \ldots, n; \text{ for homozygotes,} \qquad (5.6)$$

$$p_{A_i A_j} = 2 P_{A_i} P_{A_j} \qquad 1 \le i < j \le n; \text{ for heterozygotes.} \qquad (5.7)$$

In general, a low level of selection, as with a low level of mutation, migration or drift, will not greatly disturb the population from the equilibrium. However, such factors need to be carefully considered before assuming Hardy-Weinberg equilibrium in any model, since they can disrupt the equilibrium by disproportionately removing or introducing individuals of a given genotype when levels are not low.

There is also the so-called *Wahlund effect*, the departure from Hardy-Weinberg equilibrium due to mixing of individuals from groups with different allelic frequencies; see Exercise 5.1.

5.2.2 Genotypes with different fitnesses

In Section 5.2.1 we saw that with no selection or other advantages/disadvantages among genotypes the population will settle to a Hardy-Weinberg equilibrium. Now, assume that genotypes differ in viabilities (probability of survival) but otherwise are equally fertile. Let W_{xy} denote the viability of a genotype xy where $x, y \in \{A, a\}$. The viability may be frequency dependent as we shall see in Section 5.3.1 and as is common in other game-theoretical models. Note that, if it is not frequency dependent, it is customary to normalise, for example to have

$$W_{aa} = 1, \qquad (5.8)$$

$$W_{AA} = 1 + s, \qquad (5.9)$$

$$W_{Aa} = 1 + hs, \qquad (5.10)$$

where s represents the selection for (or against) AA relative to aa and choice of the parameter h gives the fitness of the heterozygote, including the cases of dominant ($h = 1$), recessive ($h = 0$) or additive ($h = 1/2$) genes.

Example 5.3. Under the above assumptions and notation, find the dynamics and the associated rest points of P_A.

Let $P_A(t)$ be the frequency of allele A in generation t and suppose that the total number of offspring born in generation $t + 1$ is K. For generation

$t+1$ we would thus have $KP_A^2(t)$ offspring of type AA, but only $KW_{AA}P_A^2(t)$ such individuals will survive to the reproductive stage. Similarly, there will be $KW_{aa}(1 - P_A(t))^2$ individuals of type aa and $2KW_{Aa}P_A(t)(1 - P_A(t))$ individuals of type Aa. Using ideas similar to those used in (5.1)–(5.2) we get the following discrete dynamics

$$P_A(t+1) = \frac{W_{AA}P_A(t)^2 + W_{Aa}P_A(t)(1 - P_A(t))}{W_{AA}P_A(t)^2 + 2W_{Aa}P_A(t)(1 - P_A(t)) + W_{aa}(1 - P_A(t))^2}. \quad (5.11)$$

We remark that we would get formula (5.11) even if different genotypes had the same viabilities but W_{xy} represented the fertility of genotype xy.

How $P_A(t)$ behaves with t is investigated in Exercise 5.3. For now, we will be concerned with the value of P_A at its rest points. At a rest point, we should have $P_A(t) = P_A(t+1)$ and thus, using (5.11), we get that the frequency has to solve

$$P_A(1 - P_A)\{(W_{aa} - W_{Aa}) - P_A(W_{AA} - 2W_{Aa} + W_{aa})\} = 0, \quad (5.12)$$

which has infinitely many roots if the main bracketed term is identically zero, has two roots 0 and 1 if $W_{AA} - 2W_{Aa} + W_{aa} = 0$ but not all coefficients are 0, and otherwise has three roots 0, 1 and

$$P_A = \frac{W_{aa} - W_{Aa}}{W_{AA} - 2W_{Aa} + W_{aa}}. \quad (5.13)$$

Since we must have $P_A \in [0,1]$, we get the following:

- If $W_{aa} = W_{Aa} = W_{AA}$, then any $P_A \in [0,1]$ is a rest point;

- if $W_{aa} = W_{Aa} \neq W_{AA}$, then the only rest points are $P_A = 0$ or $P_A = 1$;

- if the sign of $W_{AA} - W_{Aa}$ is the opposite as the sign of $W_{aa} - W_{Aa}$, then the only rest points are $P_A = 0$ or $P_A = 1$;

- if the sign of $W_{AA} - W_{Aa}$ is the same as the sign of $W_{aa} - W_{Aa}$, then $\frac{W_{aa} - W_{Aa}}{W_{AA} - 2W_{Aa} + W_{aa}}$ lies in the interval (0,1) and $P_A = 0$, $P_A = 1$ and $P_A = \frac{W_{aa} - W_{Aa}}{W_{AA} - 2W_{Aa} + W_{aa}}$ are rest points.

We note that when $W_{AA} - W_{Aa} < 0$ and $W_{aa} - W_{Aa} < 0$ the third rest point involving both alleles is stable and the population converges to it, but that when $W_{AA} - W_{Aa} > 0$ and $W_{aa} - W_{Aa} > 0$ this is unstable, and the population converges to either $P_A = 0$ or $P_A = 1$, depending upon the initial conditions. See Figure 5.2 for an illustration of the evolution of P_A and also the evolution of p_{AA}, p_{Aa}, p_{aa}.

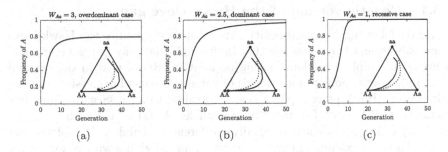

FIGURE 5.2: Evolution of allele (main graphs) and genotype (inserted triangles) frequencies based on equation (5.11). In all cases $W_{AA} = 2.5, W_{aa} = 1$. In triangles, the full curve is the actual genotype frequency, the dotted curve is the frequency representing Hardy-Weinberg equilibrium.

5.3 Games involving genetics

Under what circumstances do we need to take genetics into consideration in our models? Genetics clearly places some restrictions on which strategies are possible. We may carry out an ESS analysis, but genetic restrictions may prevent the strategies which emerge from our analyses from appearing; a strategy may not be in the *repertoire* of the available genes. Thus the fire-breathing strategy discussed above is unlikely to be feasible, but also perhaps some otherwise plausible strategies from the behavioural point of view are simply unrealisable. If an ESS strategy can be exactly represented by a homozygote, then we know that it will still be an ESS. If not, can the appropriate strategy be generated as a mixture of genotypes (in the correct proportions) and if so, will it be stable? As the following example shows, this is not always the case.

Example 5.4 (Maynard Smith, 1982, p. 40). If three strategies are equally common in the ESS with frequencies $1/3$, but are governed by genotypes AA, Aa and aa, respectively, then no proportion P_A of As can produce the ESS mixture (in a population satisfying conditions for Hardy-Weinberg equilibrium; see Example 5.1).

Indeed, if the mixture is an ESS, then the fitness of all the genotypes must be the same (see Lemma 6.7), in which case the frequencies of the genotypes would have to be in Hardy-Weinberg equilibrium; but there is no P_A such that

$$P_A^2 : 2P_A(1 - P_A) : (1 - P_A)^2 = 1/3 : 1/3 : 1/3. \tag{5.14}$$

We will see below that considering genetics may place some additional restrictions on the ESS.

5.3.1   Genetic version of the Hawk-Dove game

Again following Maynard Smith (1982), here we modify the Hawk-Dove game of Section 4.1 by assuming that individuals that play the game are part of an infinite diploid random mating population. Recall that in the original game individuals compete for rewards of value V, losers of aggressive contests between two Hawk players pay a cost C, leading to an ESS with Hawk frequency V/C if $V < C$. Let there be two alleles A and a, and each genotype $xy \in \{AA, Aa, aa\}$ has a (potentially) different probability H_{xy} of playing Hawk. Thus, for example, in any interaction, an individual with genotype Aa plays the mixed strategy that selects Hawk with probability H_{Aa}. Writing the payoffs of Hawk and Dove as E_H and E_D, the payoff W_{xy} to the genotype xy is given by

$$W_{xy} = H_{xy}E_H + (1 - H_{xy})E_D. \qquad (5.15)$$

Looking for an equilibrium frequency P_A, we use results from Section 5.2.2. We distinguish two cases

(i) either $E_H = E_D$,

(ii) or $E_H \neq E_D$.

In the first case, Hawks and Doves are in their ESS mixture, and also, by (5.15) $W_{AA} = W_{Aa} = W_{aa}$. Thus frequencies of the genotypes must be in Hardy-Weinberg equilibrium. Hence, the frequency of Hawks is

$$H_{AA}P_A^2 + H_{Aa}2P_A(1 - P_A) + H_{aa}(1 - P_A)^2, \qquad (5.16)$$

which must (because of the ESS) be equal to V/C. This gives a solution for P_A.

In the second case (see Exercise 5.4), a polymorphism where both alleles feature in the population in positive proportions, can result only if

$$P_A = \frac{W_{aa} - W_{Aa}}{W_{AA} - 2W_{Aa} + W_{aa}} = \frac{H_{aa} - H_{Aa}}{H_{AA} - 2H_{Aa} + H_{aa}}. \qquad (5.17)$$

When this is the case, the relative frequencies of the two alleles are the same among individuals that play Hawk and those that play Dove (so each benefits equally from the fitness contribution of the superior strategy). Also, the signs of $H_{aa} - H_{Aa}$ and $H_{AA} - H_{Aa}$ must be the same, i.e. there is overdominance (where the heterozygote lies outside the fitness range of the homozygotes). Consequently, if there is no overdominance (for example in the case of additive alleles) either a mixed ESS is reached as in case (i), or one of the alleles is eliminated.

5.3.2   A rationale for symmetric games

Suppose, as we described above, we can treat alleles as strategies, so that our strategy set is $\{A_1, A_2, \ldots, A_n\}$ say. The payoff in our game then corresponds to the fitness of the genotype, so that $a_{ij} = w_{ij}$ is the fitness of genotype A_iA_j. Clearly as genotype A_iA_j can also be written as A_jA_i then we must have $a_{ij} = a_{ji}$ and we have a class of games which have symmetric matrices, which we call *symmetric games*. Thus under random mating genetic games can be thought of as a special class of matrix games, with viability matrix $W = (w_{ij})$ identical to the payoff matrix $A = (a_{ij})$ (see Edwards, 2000; Hofbauer and Sigmund, 1998). Such games have some distinct properties as compared to standard matrix games, due to the extra restriction of symmetry. They were discussed in Cannings et al. (1993) who considered possible evolutionary pathways to multi-allelic systems using the sequential introduction of alleles. For any stable polymorphism (ESS) there is always at least one route to the ESS using such sequential introduction (this is not the case for non-symmetric matrices) and they showed that there may be many or few such routes, depending upon the payoff matrix involved.

5.3.3 Restricted repertoire and the streetcar theory

Recall that above we discussed the problem of restricted repertoire, so that potential ESSs could not be reached because they did not have a genetic representation from among the currently available alleles. How much is this a problem in reality, i.e. to what extent is the scope for phenotypic adaptation restricted by genetic factors such as recombination in sexually reproducing diploid populations? The above idea of the sequential introduction of mutants can be used to help explain this as well, as is shown by Hammerstein (1996); see also Weissing (1996). Hammerstein (1996, p. 512) argues, using what he terms the *streetcar theory*, that successive mutations can destabilise current populations which have reached temporarily stable positions due to current genetic restrictions and move in a series of steps towards an ESS, provided that there is sufficient flexibility of possible alleles. He summarises the argument as follows.

> A streetcar comes to a halt at various temporary stops before it reaches a final stop where it stays for much longer. ... an evolving population resembles a streetcar in the sense that it may reach several "temporary stops" that depend strongly on genetic detail before it reaches a "final stop" which has higher stability properties and is mainly determined by selective forces at the phenotypic level.

There are some new mathematical intricacies based upon the arguments raised, but the central conclusions of the paper are that perhaps these genetical worries are not so problematic after all. Thus while it is important to have consideration of genetic factors at the back of our minds when modelling using game theory, they should not overly dominate our approach.

5.4   Fitness, strategies and players

In this book, we use the term *fitness* as a measure to compare different individuals in a population. For static analysis, a strategy is an ESS if it is fitter than any potential invader in a population comprising almost all individuals playing that strategy, and a small number playing the invader. We will generally think of fitness as the expected number of offspring of an individual which survive to breeding age. This is made more explicit in the consideration of the replicator dynamics where the composition of the population changes, with a strategy's frequency changing at a rate proportional to the difference between its fitness and the mean population fitness. We saw in Chapter 4 in the consideration of the sex-ratio problem that this measure is not always sufficient (in this case the number of grandchildren needs to be considered).

Indeed the whole of this book (with the exception of this chapter and parts of Chapters 21 and 22) essentially works on the assumption of the individual as the unit of selection, which has a property called fitness (albeit a variable one depending upon its surroundings, including the composition of the population). Thus the players of our games are individuals. Dawkins (1999) makes it clear that this is a very simplified assumption and also that the fitness in question is an impossible thing to measure, being the property of a single individual who will never reappear (as reproduction is not asexual). Thus whilst it is ideal for modelling, there are real practical problems. Dawkins (1999, Chapter 10) describes five notions of fitness, and we follow his arguments below.

5.4.1   Fitness 1

This is the original usage of the term, as described in Spencer (1864, pp. 444-445) and used by Darwin and Wallace. Without a firm definition, it was the properties which a priori allowed individuals to do well in their environment, and thus made them likely to prosper e.g. for Darwin's finches, large beaks where the food supply was tough nuts or long thin beaks when it was insects.

5.4.2   Fitness 2

This is the fitness of a particular class of individuals defined by its genotype, often at a single locus. The fitness is defined as the expected number of offspring that a typical individual of that type will bring up to reproductive age. It is usually expressed as a relative fitness, compared to one particular genotype (see Chapter 12 for an example of this). Thus this fitness can be measured by considering the whole population of individuals and averaging over them. It is said that there is selection for or against a particular genotype at a given locus. As we have seen above, we can consider the evolution of a population of alleles in this way as a game with a symmetric matrix.

5.4.3 Fitness 3

This is the lifetime reproductive success of an individual and is the measure we described at the start of this section, and, as stated above, is the most common usage of fitness throughout the book. Whilst fitness 2 is very localised, it is easy to measure since the genotype continually recurs. However, each individual only occurs once. Just because an individual has a large number of offspring that survive to reproductive age, it does not mean that its genes are more likely to survive long term, since its children may be less fit (see the sex ratio problem in Section 4.4–although considering grandchildren only puts the problem off for one generation). Thus this definition, and the non-genetic methodology of our book, works in an idealised world of direct asexual copies, in the hope that this is a good approximation to reality.

5.4.4 Fitness 4

This is the idea of inclusive fitness developed in Hamilton (1964). Individuals in a population are not completely different to others, but have properties in common. No offspring will be a direct copy of its parents, but it will carry forward some genes from one and some from the other. Similarly, since our relatives have some genes in common with us, the offspring of our relatives will contain some of our genes (though generally less than our own offspring). Thus it may be beneficial for us to help our relatives at some cost to ourselves, and this is one of the central explanations for the evolution of altruism (see Chapter 15). There are some difficulties with using this as an absolute measure, which Dawkins discusses, and the concept is more properly used to analyse the effects of different actions (e.g. will helping my brother at my own expense increase or decrease my inclusive fitness). This, of course, is an ideal approach for game theory, and it is made use of in this book in a number of places (see Chapters 15, 16 and 18).

5.4.5 Fitness 5

This is termed personal fitness and is essentially fitness 4 thought of in the reverse direction, in that it is the number of an individual's own offspring (so just as in fitness 3) but bearing in mind that this includes the contributions of relatives (so a change in strategy in the population against helping relatives may reduce this number). It has the merit that each individual is only counted once in any analysis, but consequences of the analysis should be the same as for fitness 4 in terms of finding the "right" strategy.

5.4.6 Further considerations

Another complication with determining fitness is the *Fisher process* (Fisher, 1915; Kirkpatrick, 1982; Lande, 1981) which was used to explain seemingly paradoxical traits such as the large ornamental (but essentially functionally useless) peacock's tail (but see Chapter 17 for an alternative, and perhaps more plausible, explanation for the peacock's tail). If for whatever reason females prefer a certain trait in males this will be selected for in the short term and evolution generates a process of reinforcement (if other females prefer males with a certain trait then all should, as this will ensure that their offspring have the desired trait; this is also known as the "sexy son" hypothesis). Thus traits can enhance fitness in a more complex way than originally envisaged when considering fitness 1. Given that such preferences may originate at random, long term success may depend upon such fluctuations in the future.

We should also note at this point a further problem that we encounter with the concept of fitness from the modelling perspective. When populations are considered infinite then evolution follows dynamics such as the replicator dynamics, and it is possible to predict the number of descendants of any strategy arbitrarily far into the future. For finite populations even simple dynamics will lead to the number of individuals several generations into the future following a complex multivariate distribution. This is illustrated particularly for games on graphs, where there are a number of forms of finite population dynamics (each effectively equivalent on an infinite population), where a straightforward adaptation of the classical definition of fitness for an infinite population can lead to non-intuitive results, see Chapter 12.

5.5 Selfish genes: How can non-beneficial genes propagate?

In Section 5.4 we have discussed the fitness of the members of our populations, and have seen that the simple view of payoffs and fitnesses that we have discussed in Chapter 2 and use in the rest of the book is, necessarily, an idealised approach. However, as mentioned in Section 5.2, the underlying reality is more complex still. We think of individuals being self-enclosed units which propagate or become extinct based upon how well they perform against the natural environment, including other players. However, within these individuals there is also a battle for survival. Why does a particular part of an individual's genome survive and spread in the population?

The usual assumption, from Section 5.2, is that with probability 1/2 a gene found in a particular parent will be found in a particular offspring, and so if this gene contributes to a fit genotype then the offspring will, in turn, have more offspring in the next generation, and so the gene will spread. We are effectively making (at least) three assumptions here. Firstly, that all genetic material contributes to the fitness of the individuals of which it is part, secondly that the fitnesses of individuals are effectively comprised of the genetic contributions of many genes independently, and thirdly that the probability that each gene will be found in the offspring really is $1/2$. We will briefly look at these assumptions.

5.5.1 Genetic hitchhiking

We saw in Section 5.2 that the genes in a population occupy positions on 23 pairs of chromosomes for humans, and that if genes are closely located on the same chromosome, there is a strong correlation between whether one of the genes is selected and whether the other is also selected. Thus if a particular gene is advantageous, then its neighbours will benefit by its presence, even if they are not. Maynard Smith and Haigh (1974) considered the following model to demonstrate this effect; see also Exercise 5.9.

Example 5.5 (Hitchhiking, Maynard Smith and Haigh, 1974). Consider a haploid model where there are two closely linked loci, each with a pair of alleles A and a, and B and b, respectively. Thus there are four types of individual, AB, Ab, aB and ab. We assume that the alleles A and a are neutral, but that B has a fitness advantage over b. Let the recombination fraction between the two, the probability that the two are separated during meiosis, be c. Suppose that B is introduced into the population as a single individual, and that this happens to be an individual containing a. Assume a deterministic updating of the population, so that our B individuals are not eliminated by chance (see Section 12.1). Determine the evolution of frequencies of A and a.

We may assume that the fitness of b is 1 and assume that the fitness of B is $1 + s$ for some $s > 0$. Let β_t be the frequency of B in generation t and let $\alpha_{B,t}$ and $\alpha_{b,t}$, respectively be the frequency of A in chromosomes containing B and b, respectively in generation t. Since genotypes AB and aB have fitness $1 + s$, and genotypes Ab and ab have fitness 1, assuming random mating and considering all ten possible matings, Maynard Smith and Haigh (1974) found that expressions for the frequencies $\beta_{t+1}, \alpha_{B,t+1}$ and $\alpha_{b,t+1}$ can be determined as follows (see Exercise 5.7).

$$(1 + \beta_t s)\beta_{t+1} = \beta_t(1 + s), \tag{5.18}$$

$$(1 + \beta_t s)\alpha_{B,t+1} = (1 + \beta_t s)\alpha_{B,t} + c(1 - \beta_t)(\alpha_{b,t} - \alpha_{B,t}), \tag{5.19}$$

$$(1 + \beta_t s)\alpha_{b,t+1} = (1 + \beta_t s)\alpha_{b,t} + c(1 + s)\beta_t(\alpha_{B,t} - \alpha_{b,t}). \tag{5.20}$$

Note that the mean population growth rate $(1 + \beta_t s)$ appears on the left-hand side of all terms and the population fractions at time $t + 1$ needs to be scaled by this. It follows from (5.18) that β_t evolves independently of the values of

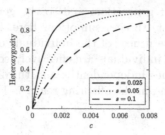

FIGURE 5.3: Effects of genetic hitch-hiking on heterozygosity, $h = 4\alpha_{B,\infty}(1 - \alpha_{B,\infty})$, at fixation of B where $\alpha_{B,\infty}$ given by (5.24) for $\beta_0 = 10^{-6}$, $\alpha_{b,0} = 1/2$.

$\alpha_{B,t}$ and $\alpha_{b,t}$ and

$$\beta_t = \frac{\beta_0(1 + s)^t}{1 - \beta_0 + \beta_0(1 + s)^t}. \tag{5.21}$$

Since

$$\lim_{t \to \infty} \beta_t = \beta_\infty = 1, \tag{5.22}$$

B eventually fixates, and the eventual proportion of A will be given by the limit of $\alpha_{B,t}$. Solving the system (5.18)–(5.20), see Exercise 5.8, yields

$$\alpha_{B,t+1} = \alpha_{B,t} + c\alpha_{b,0} \frac{(1 - \beta_0)(1 - c)^t}{1 - \beta_0 + \beta_0(1 + s)^{t+1}} \tag{5.23}$$

and thus

$$\alpha_{B,\infty} = c\alpha_{b,0}(1 - \beta_0) \sum_{t=0}^{\infty} \frac{(1 - c)^t}{1 - \beta_0 + \beta_0(1 + s)^{t+1}}. \tag{5.24}$$

We shall briefly consider a simple numerical example. The heterozygosity at a locus in a system with two alleles is defined as 4 times the product of the two allele frequencies. Hence the eventual heterozygosity at the first locus when B has fixated is $4\alpha_{B,\infty}(1 - \alpha_{B,\infty})$. For $\alpha_{b,0} = 1/2$ (giving maximum initial heterozygosity $4\alpha_{b,0}(1 - \alpha_{b,0}) = 1$) and population size $N = 10^6$ (i.e. $\beta_0 = 1/N = 10^{-6}$) we obtain the eventual heterozygosity for various c and s as seen in Figure 5.3. As we see, if the loci are very close together then a gets close to fixation due to the hitchhiking effect of initial linkage to B. Also, for fixed c, the heterozygosity decreases with increasing s.

5.5.2   Selfish genes

In Section 5.5.1 we considered how the frequencies of alleles A and a might change, even though they were neutral, and the allele that an individual possessed had no effect on its fitness. In general, much of the DNA of the genomes of different species seems to have no function, in terms of encoding traits. Thus what is it for? This was discussed in detail in Dawkins (1976, 1999). Maybe it performs some function of which we are currently unaware, and it is likely that this is true for much of it. However, as Dawkins (1999) points out, why do different species have such widely varying lengths of genome? He cites the example of salamanders, which have a genome 20 times the length of that of humans, but also have great variation between very similar species. The explanation that he puts forward is that the function of all this extra DNA is simply to survive. A good way for DNA to ensure its survival is to per-form functions which make the individual in which it resides fitter than other individuals, but this is not strictly necessary. Doing nothing, in the sense of neither helping nor hindering the organism, and simply acting as a passenger, can also enable it to survive.

Some genes, even though they do nothing to help the organism, can never-theless enable their survival by more active means (see Crow, 1979; Dawkins, 1999, Chapter 8). So-called segregational distorter genes propagate by means of increasing the probability that they are passed on to offspring with proba-bility above $1/2$. The exact mechanism how they do this is often unclear, but for example in Drosophila such genes (as with many things) are well-known. It appears that while still paired during meiosis, a distorter gene might damage its partner, to make the partner less able to be passed on (e.g. to cause the associated sperm to be damaged). However, if the probability of transmission can be increased sufficiently above $1/2$, even some reduction in the fitness of the organism would still allow the gene to prosper.

5.5.3 Memes and cultural evolution

We also briefly note that the transmission of behaviour by cultural means, rather than direct genetics, can also be thought of in evolutionary terms. Cultural transmission occurs through the spread of memes (ideas or pieces of information) which are transferred from the brains of one individual to another, through different but related mechanisms to the transfer of genes (Dawkins, 1976). Thus behaviours are transferred by learning from parents and group-mates, rather than being an instinctive behaviour controlled by the genes (Stephens and Clements, 1998). In the theory of memes (Cloak, 1975; Dawkins, 1976), as in genetics with genotypes and phenotypes, there is a distinction between the information itself which physically resides in the individual in some sense, and the expression of the idea in the population.

In many cases the link between genetics and cultural learning can be com-plicated, and most complex social behaviour will have elements of both. In particular, for humans cultural evolution can happen quickly, and it is pos-sible that any genetic adaptation to new environments has been preceded by much faster cultural adaptation. For an in-depth look at the theory of cultural evolution see Boyd and Richerson (1988).

Whilst we have aimed to give some of the biology in this section, the au-thors are both mathematicians by background, and we consider this deep and rather contentious area beyond our expertise. Thus this important subject will not be discussed in our book separately, although it is possible to argue that the game-theoretical framework that we use applies equally well to cultural transmission as to genetic transmission. As an example, see Fogarty and Kandler (2020) which develops a stochastic model of cultural evolution using processes of the same type as the finite population evolutionary models of Chapter 12.

5.5.4   Selection at the level of the cell

As can be seen from the above discussions, selection is a complex process, and the simple assumptions of our mathematical models cannot hope to take into account all of the different factors without becoming very complex. In general, selection can be thought to occur at a number of different levels. As well as selection at the level of the individual and of the gene, we have discussed selection at the level of the group in Chapter 1, and we will briefly discuss it further in Chapter 15. Between the gene and the individual, there is a further important unit of selection, the cell. Buss (1987) considers the evolution of animal life within the context of evolutionary processes at the cellular level, and in particular, contends that these processes play a central role in how organisms develop. The most striking examples of cellular evolution occur when evolution at the cellular level is in conflict with that at the individual level. A clear example is that of cancer cells, and we consider this in some detail in Chapter 22. Within an individual organism, cancer occurs when individual cells mutate and then multiply in an unconstrained manner (see Gatenby et al., 2009, 2010, for an evolutionary modelling approach to cancer). This is against the interests of the animal, very different from how cells within the body usually behave. Mostly the cells that make up an individual can be thought to contribute to its fitness in some way, e.g. in the reproductive or immune system. However, our point is that we cannot simply assume that this is the case, just as we cannot assume that all aspects of the genome exist to enhance individual fitness. Evolution is a complex interplay between these different levels.

5.6   The role of simple mathematical models

There is not a simple answer to the question of what constitutes a good mathematical model; we discuss this further in Chapter 23. It is also not easy to answer the question of how much biological reality should be used in game-theoretical models.

Typical genetic models focus on the genetic mechanism of inheritance and (generally) avoid the description of complex strategic interactions. In contrast to this, game-theoretical models generally ignore the genetic mechanisms completely and focus on strategic interactions. An ideal model, it seems, would include both aspects of the problem. Moreover, as we saw in Section 5.3.1, inclusion of genetics may significantly alter the result of the ESS analysis and this may be another reason why genetic considerations should be included in the model.

On the other hand, including genetics may (and in many cases will) unnecessarily complicate the model, potentially making the model predictions and conclusions less clear and consequently making the whole model less useful. The final judgement on what mechanism and degree of reality to include in the model should thus depend on the main objective of the model itself.

The explicit use of genetics in game-theoretical models is not widespread, although there are some important examples where the genetics plays a key role. One of these is in the case of the modelling of brood parasitism carried out by Takasu (see for example Takasu, 1998a; Takasu et al., 1993; Takasu, 1998b) and also Yamauchi (1995); we consider models of brood parasitism in Chapter 20. A distinctive feature of these models is the presence of dominant egg rejector and recessive egg acceptor alleles. The inclusion of such dominant and recessive alleles which can significantly affect the model, whilst not grossly complicating the mathematics, as in this case, is a common reason to include an explicit genetic element.

We cannot forget that most models (and game-theoretic models in particular) are very simplified representations of a more complex reality which contains important phenomena that need to be explained. Whilst such models do not closely approximate reality, they focus on important features and give insights into the nature of real systems, without pretending to explain them entirely. The models of Chapter 4 are particularly good examples of this. The Hawk-Dove game does not explain real interactions between stags in any detail, but does give a strong insight into why aggressive contests are so rare. Similarly, the sex ratio game is a wonderfully simple but believable (and mainly correct) explanation of why the sex ratio in many species is so close to 1/2. It is, in general, not the case that most of the models considered in this book can be fitted to data, although many such models do exist for specific situations; see for example the reproductive skew model in Section 16.2.1, the interference model of Section 19.6 or the virus model in Section 21.3. Thus we are not generally testing a statistical hypothesis to see whether it is plausible or not, but trying to gain insights, which experimentalists can use to formulate explicit tests.

5.7 Python code

In this section we show how to use Python to generate Figure 5.3.

```
1   """
2   Hitch-hiking game from Maynard Smith and Haigh 1974
3   The script calculates eventual heterozygosity for the
4   fitness neutral locus due to its linkage to another locus
5   """
6
7   ## Import basic packages
8   import numpy as np
9   import matplotlib.pyplot as plt
10
11  ## Define inputs/variables
12  # Frequency of B in generation 0 (1/population size)
13  beta0 = 1/10**6
14
15  # Frequency of A in chromozones containing b in generation 0
16  alpha_b0 = 0.5
17
18  # Fitness advantage. Fitness of b is 1, fitness of B is 1+s
19  s=0.025
20
21  # Maximal linkage strength
22  MaxLinkStr = 0.008
23
24  ## Auxiliary variables
25  # Range of linkage strength from 0 to MaxLinkStr
26  # c is the probability of separation during meiosis
27  c=np.linspace(0, MaxLinkStr, 101)
28  # Make it a vertical array
29  c = c[:,np.newaxis]
30
31  infinity = 1000 # A large number, used for sum from 1 to infinity
32  t=np.linspace(0,infinity, infinity+1) # Time 0 to large number
33
34  ## Main calculation
35  # Get the summands in the main formula
36  summands = (1-c)**t/(1-beta0+beta0*(1+s)**(t+1))
37  # summands is 2D array summands[a,b] = value for c[a] and t[b]
38
39  # Evaluate the main formula
40  alphaBinfty = c[:,0]*alpha_b0*(1-beta0)*np.sum(summands, axis =1)
41  # Since c is vertical, have to use c[:,0] for its first column
42  # sum over `axis = 1' means to sum over t
43
44  # Get the heterozygozity
45  h = 4*alphaBinfty*(1-alphaBinfty)
46
47  ## Plot
48  plt.plot(c[:,0],h)
49  plt.xlabel('Probability of separation during meiosis, c');
50  plt.ylabel('Heterozygozity');
51  plt.show()
```

5.8 Further reading

For mathematics related to Darwin's work in general see Chalub and Rodrigues (2011), and for a specific discussion of the Darwinian relevance of evolutionary games see Brown (2016). For genetics and the sex-ratio see Karlin and Lessard (1986), and also Argasinski (2012, 2013, 2018) for a game theoretical analysis involving genetics. The issue of the attainability of ESSs using specific genetics is discussed in more detail in Lloyd (1977); Eshel and Cavalli-Sforza (1982); Hines and Bishop (1984); Hines (1994). For some other examples on dynamical approaches to genetic models of game theory, see for example Cressman (1988); Cressman et al. (1996); Yi et al. (1999).

Some recent work developing a genetic approach to evolutionary games has been carried out in Fishman (2016, 2018) and Fishman (2020), particularly considering classical games including the Hawk-Dove game and Prisoner's dilemma. Barreto et al. (2017) considers a genetic model of the rock-sissors-paper game. A model of epigenetic effects is considered in Wang et al. (2017a).

The evolution of a genetic system with a single locus and two alleles was modelled by both Wright (1930) and Fisher (1930). The classical Wright-Fisher model originally modelled neutral populations, and so was a model primarily of random drift. It has been developed in many ways since incorporating selection and multiple alleles (see Edwards, 2000; Waxman, 2011). For a large but finite population this process can be approximated by a diffusion process, which was first done for the original Wright-Fisher model by Kimura (1964); see also Chapter 12. This approximation has also been developed in a significant number of ways for example to include recombination (Ohta and Kimura, 1969b) and mutations (Ohta and Kimura, 1969a) and to include cases where fixation of one or other allele may occur (McKane and Waxman, 2007). Wang et al. (2019b) consider a model where strategies have different selection intensities.

For much more on important aspects of genetics see Dawkins (1999), especially chapter 5 for units of selection, and chapter 8 for meotic drive genes. For more on fitness, see Grafen (2009). For selfish genes including segregational distorter genes see Dawkins (1976, 1999); Crow (1979); van Boven et al. (1996); van Boven and Weissing (1998). For genetic hitchhiking see for example Maynard Smith and Haigh (1974); Barton (2000); Charlesworth (2007); Stephan and Langley (1989); Stephan et al. (1992). For evolution at the level of the cell see Buss (1987).

5.9 Exercises

Exercise 5.1. Consider two different and separated populations of a diploid, sexually reproducing species without any drift, selection, mutation or migration. Assume that individuals mate randomly within each population, with frequency of an allele A in one population $P_A = 1/10$ and in another $P_A = 1/2$. We take a single sample, with 50% of the captured individuals from each population. Calculate the frequencies of different genotypes in the sample of captured individuals. Based on those frequencies, calculate the overall frequency of allele A, and finally, calculate the frequencies of genotypes as if the samples came from one randomly mating population.

Hint. You should find that the sample contained less heterozygotes than the number predicted by Hardy-Weinberg equilibrium. This is called the *Wahlund effect*.

Exercise 5.2. Consider a population of N individuals, N_A of them of type A and $N_a = N - N_A$ of type a. At each time step, one individual, selected at random, makes an exact copy of itself, which replaces another randomly selected individual. Calculate the average time to when all individuals will be of the same type.

Hint. Consider the relationship of this process to the simple random walk.

Exercise 5.3 (Maynard Smith, 1989, pp. 40-43). Let $P_A(t)$ denote the proportion of allele A in a two allele system under random mating with different viabilities, as in Section 5.2.2. Show that (5.11) is a version of the discrete replicator dynamics of Section 3.1.1.1. Study the dependence of $P_A(t)$ on t.

Exercise 5.4. For the genetic Hawk-Dove game described in Section 5.3.1, show that there can be a genetic equilibrium in which Hawks have different payoffs to Doves.

Exercise 5.5. In a similar way to the genetic version of the Hawk-Dove game from Section 5.3.1, set up and analyse a genetic version of the Prisoner's Dilemma game in Section 4.2.

Exercise 5.6. Set up and analyse genetic versions of other games introduced in the book. In particular, we suggest those from the exercises from Chapter 4, such as the Stag Hunt game (Exercise 4.4) and Blotto's game (Exercise 4.2).

Exercise 5.7. Obtain the expressions (5.18)–(5.21) for successive gene frequencies from Example 5.5.

Exercise 5.8. Using the equations (5.18)–(5.21) from Example 5.5 show that (5.24) holds. Hence discuss the relationship in the eventual heterozygosity at the locus associated with A/a and the parameters c, s, β_0 and $\alpha_{b,0}$.

Exercise 5.9. Show that the haploid model from Example 5.5 can be considered effectively equivalent to a diploid model with additive genes.

Exercise 5.10 (Van Boven and Weissing, 1998). Consider alleles A_1 and A_2, where A_2 is a segregational distorter, so that in an $A_1 A_2$ individual, the probability of A_2 being passed on is $S > 1/2$. Find an analogous formula to (5.11) for the dynamics, and find the rest points of the dynamics.

Chapter 6

Matrix games

We introduced the concept of a matrix game in Example 2.5 in Chapter 2. Recall that contests involve two players, each with n available pure strategies $\{1, 2, \ldots n\}$ where the payoff to an individual playing i against one playing j is a_{ij}, so all payoffs can be summarised by the matrix

$$A = (a_{ij})_{i,j=1,\ldots,n}. \tag{6.1}$$

The key property is that the payoffs are given by the function

$$E[\mathbf{x}, \mathbf{y}] = \mathbf{x} A \mathbf{y}^{\mathbf{T}}, \tag{6.2}$$

which is continuous and linear in each variable. The players are matched randomly and the payoff to an individual using strategy σ in a population $\sum_j \alpha_j \delta_{\mathbf{p}_j}$ is given by

$$\mathcal{E}\left[\sigma; \sum_j \alpha_j \delta_{\mathbf{p}_j}\right] = \sum_j \alpha_j E[\sigma, \mathbf{p}_j]. \tag{6.3}$$

In this chapter we show how to find all of the ESSs of any given matrix in a systematic way. We then look at the possible combinations of ESSs that can occur, through the study of patterns of ESSs. Finally we look at some specific examples of matrix games which are developments of the Hawk-Dove game that we looked at in Chapter 4.

6.1 Properties of ESSs

6.1.1 An equivalent definition of an ESS

Recall that an ESS was defined as follows.

Definition 6.1. *A mixed strategy* $\mathbf{p}$ *is an ESS if for every* $\mathbf{q} \neq \mathbf{p}$ *there is an* $\varepsilon_{\mathbf{q}}$ *such that for all* ε, $0 < \varepsilon < \varepsilon_{\mathbf{q}}$

$$\mathcal{E}[\mathbf{p}; (1-\varepsilon)\delta_{\mathbf{p}} + \varepsilon\delta_{\mathbf{q}}] > \mathcal{E}[\mathbf{q}; (1-\varepsilon)\delta_{\mathbf{p}} + \varepsilon\delta_{\mathbf{q}}]. \tag{6.4}$$

DOI: 10.1201/9781003024682-6

We note that, by (6.3) and (6.2) (see also Exercise 2.6), (6.4) is equivalent to

$$E[\mathbf{p}, (1 - \varepsilon)\mathbf{p} + \varepsilon\mathbf{q}] > E[\mathbf{q}, (1 - \varepsilon)\mathbf{p} + \varepsilon\mathbf{q}]. \tag{6.5}$$

Moreover, for matrix games, the next theorem provides another equivalent definition. The main reason why Theorem 6.2 holds is because the payoff function $E[\mathbf{x}, \mathbf{y}]$ is, by (6.2), linear in the second variable.

Theorem 6.2. *For a matrix game, $\mathbf{p}$ is an ESS if and only if, for all $\mathbf{q} \neq \mathbf{p}$,*

$$(i) E[\mathbf{p}, \mathbf{p}] \geq E[\mathbf{q}, \mathbf{p}], \tag{6.6}$$

$$(ii) \text{ If } E[\mathbf{p}, \mathbf{p}] = E[\mathbf{q}, \mathbf{p}], \text{ then } E[\mathbf{p}, \mathbf{q}] > E[\mathbf{q}, \mathbf{q}]. \tag{6.7}$$

Proof. By (6.2), (6.5) is equivalent to

$$(1 - \varepsilon)E[\mathbf{p}, \mathbf{p}] + \varepsilon E[\mathbf{p}, \mathbf{q}] > (1 - \varepsilon)E[\mathbf{q}, \mathbf{p}] + \varepsilon E[\mathbf{q}, \mathbf{q}]. \tag{6.8}$$

If (6.6) and (6.7) hold, then (6.8) holds for all $\varepsilon > 0$ small enough and thus $\mathbf{p}$ is an ESS. Conversely, if (6.7) does not hold, i.e. if $E[\mathbf{p}, \mathbf{p}] = E[\mathbf{q}, \mathbf{p}]$ but $E[\mathbf{p}, \mathbf{q}] \leq E[\mathbf{q}, \mathbf{q}]$, then (6.8) cannot hold for any $\varepsilon > 0$. If (6.6) does not hold, then (6.8) cannot hold for small enough $\varepsilon > 0$. $\square$

The condition (6.6) is often referred to as the *equilibrium condition*. It means that the individuals using a mutant strategy $\mathbf{q}$ cannot do better than residents using $\mathbf{p}$ in the most common contests, i.e. those against individuals using $\mathbf{p}$. The condition (6.7) is often called the *stability condition*. It means that if $\mathbf{q}$ does as well against $\mathbf{p}$ as $\mathbf{p}$ does, then $\mathbf{q}$ must do worse than $\mathbf{p}$ in the rare contests against $\mathbf{q}$.

Note that Theorem 6.2 does not hold, in general, for non-matrix games; see Exercise 6.7.

6.1.2   A uniform invasion barrier

An ESS was defined in a way that no strategy could invade it; more precisely, any other strategy could not invade if its frequency was below a certain threshold. In general, the threshold could be different for different strategies. We shall see that in matrix games there is a threshold which works for all invading strategies.

Definition 6.3. *A strategy $\mathbf{p}$ is called* uniformly uninvadable *if there is $\varepsilon_{\mathbf{p}} > 0$ so that for every other strategy $\mathbf{q} \neq \mathbf{p}$ and for all $0 < \varepsilon < \varepsilon_{\mathbf{p}}$ we have*

$$\mathcal{E}[\mathbf{p}; (1 - \varepsilon)\delta_{\mathbf{p}} + \varepsilon\delta_{\mathbf{q}}] > \mathcal{E}[\mathbf{q}; (1 - \varepsilon)\delta_{\mathbf{p}} + \varepsilon\delta_{\mathbf{q}}]. \tag{6.9}$$

Thus when $\mathbf{p}$ is uniformly uninvadable, there is a uniform invasion barrier $\varepsilon_{\mathbf{p}}$.

Theorem 6.4. *For a matrix game, $\mathbf{p}$ is an ESS if and only if it is uniformly uninvadable, i.e. there is an $\varepsilon_{\mathbf{p}}$ such that (6.4) holds for all $\mathbf{q}$ and all $\varepsilon \in (0, \varepsilon_{\mathbf{p}})$.*

Proof. We only need to prove that if $\mathbf{p}$ is an ESS, then there is an $\varepsilon_{\mathbf{p}}$ that works for all $\mathbf{q} \neq \mathbf{p}$. We will use the facts that the strategy simplex is compact, the payoff function $E[\mathbf{x}, \mathbf{y}]$ is continuous and that lower semi-continuous functions attain minima on compact sets.

For any $\mathbf{q} \neq \mathbf{p}$, let $\varepsilon_{\mathbf{p},\mathbf{q}} \leq 1$ be the maximal value such that (6.4) holds for that $\mathbf{q}$ and all $\varepsilon \in (0, \varepsilon_{\mathbf{p},\mathbf{q}})$. We now show that the function $\mathbf{q} \to \varepsilon_{\mathbf{p},\mathbf{q}}$ is lower semi-continuous. Let

$$
\begin{aligned}
h_{\mathbf{p},\mathbf{q},\varepsilon} &= \mathcal{E}[\mathbf{p}; (1-\varepsilon)\delta_{\mathbf{p}} + \varepsilon\delta_{\mathbf{q}}] - \mathcal{E}[\mathbf{q}; (1-\varepsilon)\delta_{\mathbf{p}} + \varepsilon\delta_{\mathbf{q}}] \\
&= E[\mathbf{p}, (1-\varepsilon)\mathbf{p} + \varepsilon\mathbf{q}] - E[\mathbf{q}, (1-\varepsilon)\mathbf{p} + \varepsilon\mathbf{q}] \\
&= \sum_i (p_i - q_i) f_i((1-\varepsilon)\mathbf{p} + \varepsilon\mathbf{q})
\end{aligned}
\tag{6.10}
$$

be the incentive function, where

$$
f_i((1-\varepsilon)\mathbf{p} + \varepsilon\mathbf{q}) = E[S_i, (1-\varepsilon)\mathbf{p} + \varepsilon\mathbf{q}]. \tag{6.11}
$$

The functions f_i are linear and thus uniformly continuous in $\mathbf{q}$. Hence, $h_{\mathbf{p},\mathbf{q}',\varepsilon} > 0$ whenever $h_{\mathbf{p},\mathbf{q},\varepsilon} > 0$ and $\mathbf{q}'$ is close to $\mathbf{q}$. In other words, if $\varepsilon_{\mathbf{p},\mathbf{q}} > \alpha$ for some $\alpha > 0$, then $\varepsilon_{\mathbf{p},\mathbf{q}'} > \alpha$ for all $\mathbf{q}'$ close enough to $\mathbf{q}$. Thus the function $\mathbf{q} \to \varepsilon_{\mathbf{p},\mathbf{q}}$ is lower semi-continuous. Since the strategy simplex is compact, there is a $\mathbf{q}_0$ such that $\varepsilon_{\mathbf{p},\mathbf{q}_0} = \min_{\mathbf{q}}\{\varepsilon_{\mathbf{p},\mathbf{q}}\} > 0$. Now we just have to set $\varepsilon_{\mathbf{p}} = \varepsilon_{\mathbf{p},\mathbf{q}_0}$. $\square$

Example 6.5. Find the uniform invasion barrier of the ESS $(1/2, 1/2, 0)$ for the matrix game

$$
\begin{pmatrix} 0 & 1 & 0 \\ 1 & 0 & 0 \\ 0 & 0 & 1 \end{pmatrix}. \tag{6.12}
$$

It is easy to see that $(\mathbf{p} = 1/2, 1/2, 0)$ is an ESS of the above game (as is $(0, 0, 1)$). For $\mathbf{q} = (q_1, q_2, q_3)$ we have

$$
E[\mathbf{q}, (1-\varepsilon)\mathbf{p} + \varepsilon\mathbf{q}] = \frac{1}{2}(q_1 + q_2) + \varepsilon\left(q_3^2 + 2q_1 q_2 - \frac{q_1}{2} - \frac{q_2}{2}\right), \tag{6.13}
$$

$$
E[\mathbf{p}, (1-\varepsilon)\mathbf{p} + \varepsilon\mathbf{q}] = \frac{1}{2} + \frac{1}{2}\varepsilon(q_1 + q_2 - 1), \tag{6.14}
$$

and thus

$$
h_{\mathbf{p},\mathbf{q},\varepsilon} = \frac{1}{2}q_3 - \varepsilon\left(\frac{1}{2} + q_3^2 + 2q_1 q_2 - q_1 - q_2\right). \tag{6.15}
$$

The strategy $\mathbf{q}$ can invade if $h_{\mathbf{p},\mathbf{q},\varepsilon} < 0$, which is easiest to achieve if $q_1 = q_2$. Substituting into (6.15), we see that invasion is easiest to achieve at $q_3 = 1$, which gives $h_{\mathbf{p},\mathbf{q},\varepsilon} = (1 - 3\varepsilon)/2$. Hence the uniform invasion barrier is $\varepsilon_{\mathbf{p}} = 1/3$.

It is not the case that all ESSs in all games have the property of uniform uninvadability; see Exercise 6.6.

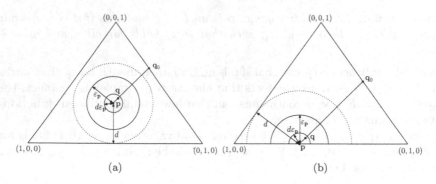

FIGURE 6.1: Illustrating the proof of the local superiority of an ESS. (a) A case with **p** an internal strategy, (b) a case with **p** on a face of the strategy simplex.

6.1.3   Local superiority of an ESS

The definition of the ESS is such that in a population of ESS players with only a small number of mutants, playing the ESS is better than playing any mutant strategy. The next theorem shows that playing the ESS is better than playing the resulting mean strategy in the population.

Theorem 6.6 (Local superiority condition, Hofbauer et al., 1979; Hofbauer and Sigmund, 1988; Weibull, 1995; Křivan, 2009). *For a matrix game, a mixed strategy* **p** *is an ESS if and only if*

$$E[\mathbf{p}, \mathbf{q}] > E[\mathbf{q}, \mathbf{q}] \tag{6.16}$$

for all $\mathbf{q} \neq \mathbf{p}$ *sufficiently close to* **p**.

Proof. Let **p** be an ESS and $\varepsilon_{\mathbf{p}}$ be given by Theorem 6.4. Let d be the minimal distance of **p** to any face of a strategy simplex that does not contain **p** and let **q** be in the $d\varepsilon_{\mathbf{p}}$ neighbourhood of **p**. Let $\mathbf{q}_0$ be on a face of the strategy simplex and also on the half line going from **p** through **q**; see Figure 6.1. Without loss of generality we can assume that **q** is not on the same "face" as **p**, since in such a case we can consider a reduced space ignoring any strategies where both **p** and **q** have zero entries. Then, the distance from **p** to $\mathbf{q}_0$ is at least d and thus

$$\mathbf{q} = (1 - \varepsilon)\mathbf{p} + \varepsilon\mathbf{q}_0 \tag{6.17}$$

for some $\varepsilon \in (0, \varepsilon_{\mathbf{p}})$. Since **p** is an ESS, we get

$$E[\mathbf{p}, (1 - \varepsilon)\mathbf{p} + \varepsilon\mathbf{q}_0] > E[\mathbf{q}_0, (1 - \varepsilon)\mathbf{p} + \varepsilon\mathbf{q}_0] \tag{6.18}$$

which is equivalent to

$$E[\mathbf{p}, \mathbf{q}] > E[\mathbf{q}_0, \mathbf{q}]. \tag{6.19}$$

Thus

$$E[\mathbf{p}, \mathbf{q}] - E[\mathbf{q}, \mathbf{q}] = \varepsilon \Big(E[\mathbf{p}, \mathbf{q}] - E[\mathbf{q}_0, \mathbf{q}] \Big)$$
$$+ (1 - \varepsilon)\Big(E[\mathbf{p}, \mathbf{q}] - E[\mathbf{p}, \mathbf{q}] \Big) > 0. \tag{6.20}$$

Conversely, if $E[\mathbf{p}, \mathbf{q}] > E[\mathbf{q}, \mathbf{q}]$ holds for all $\mathbf{q}$ whose distance to $\mathbf{p}$ is smaller than $\varepsilon_{\mathbf{p}}$, pick any $\mathbf{q}_0 \neq \mathbf{p}$ and $\varepsilon \in (0, \varepsilon_{\mathbf{p}}/2)$ and set $\mathbf{q} = (1 - \varepsilon)\mathbf{p} + \varepsilon\mathbf{q}_0$. We will have

$$E[\mathbf{p}, \mathbf{q}] > E[\mathbf{q}, \mathbf{q}] = (1 - \varepsilon)E[\mathbf{p}, \mathbf{q}] + \varepsilon E[\mathbf{q}_0, \mathbf{q}] \tag{6.21}$$

and thus

$$E[\mathbf{p}, (1 - \varepsilon)\mathbf{p} + \varepsilon\mathbf{q}_0] = E[\mathbf{p}, \mathbf{q}] > E[\mathbf{q}_0, \mathbf{q}] = E[\mathbf{q}_0, (1 - \varepsilon)\mathbf{p} + \varepsilon\mathbf{q}_0], \tag{6.22}$$

which means that $\mathbf{p}$ is an ESS. □

6.1.4 ESS supports and the Bishop-Cannings theorem

Recall that for a strategy $\mathbf{p}$, the support of $\mathbf{p}$ is $S(\mathbf{p}) = \{i : p_i > 0\}$. The next lemma shows that any pure strategy in the support of an ESS strategy $\mathbf{p}$ does equally as well against $\mathbf{p}$ as $\mathbf{p}$ itself. Thus the payoffs to all pure strategies within the support of an ESS must be identical. This is one of the three key conditions for $\mathbf{p}$ to be an ESS identified in Haigh (1975).

Lemma 6.7. *Let* $\mathbf{p} = (p_i)$ *be an ESS. Then* $E[S_i, \mathbf{p}] = E[\mathbf{p}, \mathbf{p}]$ *for any* $i \in S(\mathbf{p})$.

Proof. As $\mathbf{p}$ is an ESS, we have $E[\mathbf{p}, \mathbf{p}] \geq E[S_k, \mathbf{p}]$ for any pure strategy S_k. If we had $E[\mathbf{p}, \mathbf{p}] > E[S_i, \mathbf{p}]$ for $i \in S(\mathbf{p})$, we would have

$$E[\mathbf{p}, \mathbf{p}] = \mathbf{p}A\mathbf{p}^{\mathbf{T}} = \sum_{k,j} p_k a_{kj} p_j = \sum_k p_k \sum_j a_{k,j} p_j$$

$$= \sum_k p_k E[S_k, \mathbf{p}] = p_i E[S_i, \mathbf{p}] + \sum_{k \neq i} p_k E[S_k, \mathbf{p}] \tag{6.23}$$

$$< p_i E[\mathbf{p}, \mathbf{p}] + \sum_{k \neq i} p_k E[\mathbf{p}, \mathbf{p}] = E[\mathbf{p}, \mathbf{p}],$$

which is a contradiction. Hence,

$$E[S_i, \mathbf{p}] = E[\mathbf{p}, \mathbf{p}], \quad \text{for all } i \in S(\mathbf{p}). \tag{6.24}$$

□

Note that we did not really use the full strength of the linearity of payoffs in matrix games. Thus Lemma 6.7 can be generalised; see for example Theorem 7.5 in Chapter 7.

We also note that Lemma 6.7 is intuitively obvious if we think of replicator dynamics in polymorphic populations rather than ESSs. If $E[\mathbf{p},\mathbf{p}] \neq E[S_i,\mathbf{p}]$ for some $i \in S(\mathbf{p})$, then strategy S_i would have to do better or worse than some other strategy S_j with $j \in S(\mathbf{p})$, and the population would not be in a dynamic equilibrium.

Now, we define $T(\mathbf{p})$ to be the set of indices of pure strategies which have equal payoffs against the strategy $\mathbf{p}$, i.e.

$$T(\mathbf{p}) = \{i : E[S_i,\mathbf{p}] = E[\mathbf{p},\mathbf{p}]\}. \tag{6.25}$$

Lemma 6.7 shows that

$$S(\mathbf{p}) \subseteq T(\mathbf{p}). \tag{6.26}$$

It is possible that $S(\mathbf{p}) \neq T(\mathbf{p})$; however, for generic games (see Section 2.1.3.3), $S(\mathbf{p}) = T(\mathbf{p})$; see Exercise 6.11 and Exercise 6.12. The next Lemma shows that $T(\mathbf{p})$ is of special importance to an ESS.

Lemma 6.8. *Let $\mathbf{p}$ be an ESS. Then $E[\mathbf{q},\mathbf{p}] = E[\mathbf{p},\mathbf{p}]$ if and only if $S(\mathbf{q}) \subseteq T(\mathbf{p})$.*

Proof. Clearly, if $S(\mathbf{q}) \subseteq T(\mathbf{p})$, then

$$\begin{aligned}
E[\mathbf{q},\mathbf{p}] &= \sum_i q_i E[S_i,\mathbf{p}] = \sum_{i \in S(\mathbf{q})} q_i E[S_i,\mathbf{p}] \\
&= \sum_{i \in S(\mathbf{q})} q_i E[\mathbf{p},\mathbf{p}] = E[\mathbf{p},\mathbf{p}].
\end{aligned} \tag{6.27}$$

On the other hand, if there is $i \in S(\mathbf{q})$ such that $E[S_i,\mathbf{p}] < E[\mathbf{p},\mathbf{p}]$, then because $\mathbf{p}$ is an ESS and thus $E[S_j,\mathbf{p}] \leq E[\mathbf{p},\mathbf{p}]$ we get, similarly to the proof of Lemma 6.7,

$$E[\mathbf{q},\mathbf{p}] = \sum_{j \in S(\mathbf{q})} q_j E[S_j,\mathbf{p}] < \sum_{j \in S(\mathbf{q})} q_j E[\mathbf{p},\mathbf{p}] = E[\mathbf{p},\mathbf{p}]. \tag{6.28}$$

$\square$

Theorem 6.9 (Bishop and Cannings, 1976). *If $\mathbf{p}$ is an ESS of the matrix game A and $\mathbf{q} \neq \mathbf{p}$ is such that $S(\mathbf{q}) \subseteq T(\mathbf{p})$, then $\mathbf{q}$ is not an ESS of matrix game A.*

Proof. If $S(\mathbf{q}) \subseteq T(\mathbf{p})$ then $E[\mathbf{q},\mathbf{p}] = E[\mathbf{p},\mathbf{p}]$ by Lemma 6.8. Thus for $\mathbf{p}$ to be an ESS we need $E[\mathbf{p},\mathbf{q}] > E[\mathbf{q},\mathbf{q}]$ which clearly means that $\mathbf{q}$ cannot be an ESS. $\square$

Corollary 6.10. *If $\mathbf{p}$ is an internal ESS (i.e. all $p_i > 0$) of a matrix game, then it is the only ESS.*

Proof. If $\mathbf{p}$ is an internal ESS, then for any $\mathbf{q}$, $S(\mathbf{q}) \subseteq \{1,\dots,n\} = S(\mathbf{p}) = T(\mathbf{p})$. $\square$

Since $S(\mathbf{p}) \subseteq T(\mathbf{p})$, $S(\mathbf{p_i})$ is not a subset of $S(\mathbf{p_j})$ if both $\mathbf{p_i}$ and $\mathbf{p_j}$ are ESSs of A. This is a slightly weaker result than the full theorem, but recall that removing non-generic cases implies that $S(\mathbf{p}) = T(\mathbf{p})$, which makes the two statements equivalent.

6.2   ESSs in a 2×2 matrix game

Let us first consider the simplest matrix game where there are only two strategies so that the payoff matrix is

$$\begin{pmatrix} a & b \\ c & d \end{pmatrix}. \tag{6.29}$$

Any strategy $\mathbf{x}$ can be written as $(x, 1 - x)$ for $x \in [0, 1]$. If pure strategy 1 is an ESS, then strategy 2 does not invade it. Thus, following Theorem 6.2, a necessary condition for 1 to be an ESS is $a \geq c$ (equilibrium) and if $a = c$ then $b > d$ (stability). Since the payoff function E is linear in the first variable, the necessary condition is also sufficient. Removing the non-generic case gives that strategy 1 is a pure ESS if and only if $a > c$. Similarly, strategy 2 is a pure ESS if and only if $d \geq b$ and if $d = b$ then $c > a$, while removing the non-generic case gives $d > b$. The natural extension of the necessary condition above for games with $n > 2$ pure strategies to consider invasion by each pure strategy separately holds for generic games, but not non-generic ones; see Exercise 6.13.

To see whether and when a mixed strategy $\mathbf{p} = (p, 1 - p)$ for $0 < p < 1$ is an ESS, recall that, by Lemma 6.8, a necessary condition for $\mathbf{p}$ to be an ESS is that

$$
\begin{aligned}
E[S_1, \mathbf{p}] &= E[S_2, \mathbf{p}] \Rightarrow & (6.30) \\
ap + b(1 - p) &= cp + d(1 - p) & (6.31)
\end{aligned}
$$

which yields

$$\mathbf{p} = \frac{1}{b + c - a - d}(b - d, c - a). \tag{6.32}$$

Since we require $p \in (0, 1)$, (6.32) makes sense only if $b > d$ and $c > a$ or $d > b$ and $a > c$ (excluding the cases of equality since these would yield $p = 0$ or $p = 1$). Hence $\mathbf{p}$ is the only candidate for an internal ESS. By calculations as in (6.27), it follows from (6.30) that, for any $\mathbf{q}$,

$$E[\mathbf{q}, \mathbf{p}] = E[\mathbf{p}, \mathbf{p}] \tag{6.33}$$

and thus for $\mathbf{p}$ to be an ESS we need to check that $E[\mathbf{q}, \mathbf{q}] < E[\mathbf{p}, \mathbf{q}]$. We get, see Exercise 6.2,

$$E[\mathbf{p}, \mathbf{q}] - E[\mathbf{q}, \mathbf{q}] = -(a - b - c + d)(p - q)^2. \tag{6.34}$$

TABLE 6.1:   ESS of a 2 × 2 game with payoff matrix (6.29).

Parameters	ESSs
$a > c, b > d$	$(1,0)$
$a > c, b < d$	$(1,0)$, $(0,1)$
$a < c, b > d$	$\left(\dfrac{b-d}{b+c-a-d}, \dfrac{c-a}{b+c-a-d}\right)$
$a < c, b < d$	$(0,1)$

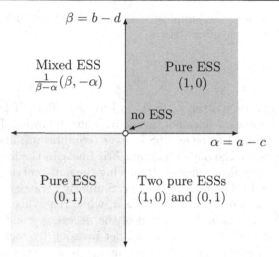

FIGURE 6.2: Patterns of ESS in a 2 × 2 game with payoff matrix $\left(\begin{smallmatrix} a & b \\ c & d \end{smallmatrix}\right)$.

Thus $E[\mathbf{p},\mathbf{q}] - E[\mathbf{q},\mathbf{q}] > 0$ if and only if $a - b - c + d < 0$. Thus $\mathbf{p}$ is an ESS if $b > d$ and $c > a$ and it is an unstable equilibrium if $b < d$ and $c < a$. Without non-generic cases, we have four possible ESS combinations in all, shown in Table 6.1. The situation is summarised in Figure 6.2 where even non-generic cases are considered.

As seen in Exercises 3.3 and 6.2 the dynamics and static analysis for a 2 × 2 game yield the same outcome. In particular, as seen above, the incentive function

$$h(p) = \mathcal{E}[S_1, \delta_{(p,1-p)}] - \mathcal{E}[S_2, \delta_{(p,1-p)}]$$
$$= p(a - b - c + d) + b - d \tag{6.35}$$

plays an important role in both approaches. See also Section 9.1.4.

A more in-depth investigation of ESS combinations is pursued in Section 6.5.

6.3   Haigh's procedure to locate all ESSs

The following procedure to find all of the ESSs of a matrix game is due to Haigh (1975). It was subsequently shown by Abakuks (1980) that in some non-generic cases this method will not always identify the set of ESSs correctly. Nevertheless, this does not devalue the procedure to any significant extent.

So far we have seen a necessary condition (6.26) for $\mathbf{p}$ to be an ESS; in particular, if $\mathbf{p}$ is an ESS, then

$$E[S_i, \mathbf{p}] = E[S_j, \mathbf{p}], \quad \text{for all } i, j \in S(\mathbf{p}) = \{k; p_k > 0\}. \qquad (6.36)$$

Letting B be the submatrix of A restricted to the elements of $S(\mathbf{p})$ and $\mathbf{p}^*$ being the similar restriction of $\mathbf{p}$, $B\mathbf{p}^{*T}$ is equal to a constant times the unit vector $\mathbf{1}$, and consequently $\mathbf{p}$ satisfies (6.36) if and only if

$$\mathbf{p}^* = \frac{B^{-1}\mathbf{1}^T}{\mathbf{1}B^{-1}\mathbf{1}^T}. \qquad (6.37)$$

Note that B^{-1} exists since A is generic. Thus such a vector $\mathbf{p}$ exists if and only if the right-hand term of (6.37) is a non-negative vector. Also, as $\mathbf{p}$ is an ESS, we get that

$$E[S_i, \mathbf{p}] < E[\mathbf{p}, \mathbf{p}], \quad \text{for all } i \notin T(\mathbf{p}) = \{k; E[S_k, \mathbf{p}] = E[\mathbf{p}, \mathbf{p}]\}. \qquad (6.38)$$

Conversely, if (6.36) and (6.38) hold, then, for any $\mathbf{q}$, we have

$$E[\mathbf{q}, \mathbf{p}] = \sum_i q_i E[S_i, \mathbf{p}] \leq \sum_i q_i E[\mathbf{p}, \mathbf{p}] = E[\mathbf{p}, \mathbf{p}] \qquad (6.39)$$

which means that $\mathbf{p}$ is a good candidate for an ESS. Moreover, the inequality (6.39) is strict if $S(\mathbf{q}) \not\subseteq T(\mathbf{p})$; we thus only have to worry about $\mathbf{q}$ with $S(\mathbf{q}) \subseteq T(\mathbf{p})$. From the example of a 2×2 matrix game (6.29), e.g. with $a = d = 0, b = c = -1$, we have also seen that conditions (6.36) and (6.38) are not enough. We find below what further condition needs to hold for $\mathbf{p}$ to be an ESS.

If $\mathbf{q}$ is such that $S(\mathbf{q}) \subseteq T(\mathbf{p})$, then

$$E[\mathbf{q}, \mathbf{p}] = \sum_i q_i E[S_i, \mathbf{p}] = \sum_i q_i E[\mathbf{p}, \mathbf{p}] = E[\mathbf{p}, \mathbf{p}] \qquad (6.40)$$

and thus $(\mathbf{p} - \mathbf{q})A\mathbf{p}^{\mathbf{T}} = E[\mathbf{p}, \mathbf{p}] - E[\mathbf{q}, \mathbf{p}] = 0$. Thus

$$\begin{aligned}
E[\mathbf{p}, \mathbf{q}] - E[\mathbf{q}, \mathbf{q}] &= \mathbf{p}A\mathbf{q}^{\mathbf{T}} - \mathbf{q}A\mathbf{q}^{\mathbf{T}} \\
&= (\mathbf{p} - \mathbf{q})A(\mathbf{q} - \mathbf{p} + \mathbf{p})^{\mathbf{T}} \qquad (6.41) \\
&= -(\mathbf{p} - \mathbf{q})A(\mathbf{p} - \mathbf{q})^{\mathbf{T}}.
\end{aligned}$$

For **p** to be an ESS, we thus need

$$(\mathbf{p} - \mathbf{q})A(\mathbf{p} - \mathbf{q})^{\mathbf{T}} < 0 \quad \text{for all } \mathbf{q} \text{ such that } S(\mathbf{q}) \subseteq T(\mathbf{p}). \tag{6.42}$$

If we deal with the generic case only, then $T(\mathbf{p}) = S(\mathbf{p})$; see Exercise 6.12. In this case, (6.42) is equivalent to

$$\mathbf{z}A\mathbf{z}^{\mathbf{T}} < 0 \quad \text{for all } \mathbf{z} \neq \mathbf{0} \text{ such that } S(\mathbf{z}) \subseteq S(\mathbf{p}) \text{ and } \sum_i z_i = 0. \tag{6.43}$$

It should be noted (6.42) and (6.43) are not equivalent in the non-generic case; see Abakuks (1980) and also Exercise 6.4. For any $\mathbf{z}$ as in (6.43) and any $k \in T(\mathbf{p})$, we have

$$\begin{aligned}
\mathbf{z}A\mathbf{z}^{\mathbf{T}} &= \sum_{i,j \in T(\mathbf{p})} z_i a_{ij} z_j \\
&= \sum_{i,j \neq k} z_i a_{ij} z_j + \left(\sum_{i \neq k} z_i a_{ik} + \sum_{j \neq k} a_{kj} z_j \right) z_k + a_{kk} z_k^2 \\
&= \sum_{i \neq k} z_i \sum_{j \neq k} (a_{ij} - a_{ik} - a_{kj} + a_{kk}) z_j,
\end{aligned} \tag{6.44}$$

where we use the fact that $z_k = -\sum_{i \neq k} z_i$. Consequently, the next condition for **p** to be ESS is that

$$(a_{ij} - a_{ik} - a_{kj} + a_{kk})_{i,j \in T(\mathbf{p}) \backslash \{k\}} \quad \text{is negative definite.} \tag{6.45}$$

Note that this is true for any choice of k. We have arrived at the following theorem.

Theorem 6.11 (Haigh, 1975). **p** *is an ESS of a generic matrix game A if and only if (6.36), (6.38) and (6.45) hold.*

Proof. We have seen that (6.36), (6.38) and (6.45) are necessary conditions. It remains to show that **p** is an ESS if (6.36), (6.38) and (6.45) hold. If (6.36) and (6.38) hold, we have already seen that $E[\mathbf{q}, \mathbf{p}] \leq E[\mathbf{p}, \mathbf{p}]$. If $E[\mathbf{q}, \mathbf{p}] = E[\mathbf{p}, \mathbf{p}]$, then reversing the process by which we arrived at (6.45) yields that $E[\mathbf{q}, \mathbf{q}] < E[\mathbf{p}, \mathbf{q}]$, which means that **p** is an ESS. $\qquad \square$

We have thus arrived at an algorithmic procedure that finds (all) ESS(s) of a given generic $n \times n$ matrix game A.

We note that from a practical point of view if investigating the ESSs of a matrix "by hand", ESSs with the smallest supports are easiest to check first, and subsequent application of the Bishop-Cannings Theorem 6.9 may mean that the larger supports do not need to be checked at all.

Example 6.12. Find the ESSs of the matrix game

$$\begin{pmatrix} 0 & 1 & 2 & -2 \\ -1 & 0 & 3 & -1 \\ -3 & 2 & 0 & -1 \\ -2 & 1 & 1 & 0 \end{pmatrix}. \tag{6.46}$$

Let us first check for pure ESSs. Strategy S_1 is a pure ESS as the leading diagonal entry is the largest in its column, and due to similar reasoning strategy S_4 is also a pure ESS and neither strategy S_2 nor S_3 is a pure ESS. Now we check for pair ESSs, which have support composed of two pure strategies. There cannot be a pair with support $(1,2), (1,3), (1,4), (2,4)$ or $(3,4)$ due to Bishop-Cannings since S_1 and S_4 are pure ESSs. Thus the only pair ESS could be with support $(2,3)$. This has a valid equilibrium $(0,3/5,2/5,0)$, but this is invaded by pure strategy S_1, so is not an ESS. Finally the Bishop-Cannings theorem also rules out any triple or internal ESS, as clearly strategy S_1 and/or S_4 would have to be involved. Hence, the ESSs are pure S_1 and pure S_4 only.

6.4 ESSs in a 3×3 matrix game

We will illustrate a general method of finding ESSs in a 3×3 matrix game. There are in total 9 entries of the payoff matrix. However, by Exercise 6.3, we can assume that our matrix has 0s down the leading diagonal and thus there are only 6 independent entries, and so we consider the following matrix

$$\begin{pmatrix} 0 & a & b \\ c & 0 & d \\ e & f & 0 \end{pmatrix}. \tag{6.47}$$

This simplification makes the analysis, as well as the mathematical expressions for the conditions for each ESS, neater. We will concentrate on the generic case only.

6.4.1 Pure strategies

A pure strategy is an ESS (in the generic case, see Exercise 6.13) if it does better than any other pure strategy invader, and so:

- $(1,0,0)$ is an ESS if $c < 0, e < 0$;

- $(0,1,0)$ is an ESS if $a < 0, f < 0$;

- $(0,0,1)$ is an ESS if $b < 0, d < 0$.

6.4.2   A mixture of two strategies

For a pair ESS, the pair needs to be both an ESS within its own strategy set, plus resist invasion from the third strategy. By the results from Section 6.2 (Table 6.1), there will be an ESS involving strategies S_1 and S_2 only if $a > 0, c > 0$ and the potential ESS will be given by

$$\mathbf{p} = \left(\frac{a}{a+c}, \frac{c}{a+c}, 0 \right). \tag{6.48}$$

This strategy resists invasion by any strategy $\mathbf{q}$ such that $S(\mathbf{q}) \subset \{1, 2\}$. Hence $\mathbf{p}$ resists invasion by any strategy if it resists invasion by pure strategy S_3. By Lemma 6.7

$$E[\mathbf{p}, \mathbf{p}] = E[S_1, \mathbf{p}] = \frac{ac}{a+c} \tag{6.49}$$

and thus, recalling that $S(\mathbf{p}) = T(\mathbf{p})$ in the generic case, $\mathbf{p}$ resists invasion by pure strategy S_3 if

$$\frac{ac}{a+c} = E[\mathbf{p}, \mathbf{p}] > E[S_3, \mathbf{p}] = e\frac{a}{a+c} + f\frac{c}{a+c} = \frac{ae+cf}{a+c}, \tag{6.50}$$

which is equivalent to

$$ae + cf - ac < 0. \tag{6.51}$$

Conditions for mixed strategies $(p_1, 0, p_3)$ or $(0, p_2, p_3)$ to be ESSs can be derived in a similar way. If we set

$$\begin{align} \alpha &= ad + bf - df, \tag{6.52}\\ \beta &= bc + de - be, \tag{6.53}\\ \gamma &= ae + cf - ac, \tag{6.54} \end{align}$$

we get:

- $\frac{1}{a+c}(a, c, 0)$ is an ESS if and only if $a > 0, c > 0, \gamma < 0$;

- $\frac{1}{b+e}(b, 0, e)$ is an ESS if and only if $b > 0, e > 0, \beta < 0$;

- $\frac{1}{d+f}(0, d, f)$ is an ESS if and only if $d > 0, f > 0, \alpha < 0$.

6.4.3   Internal ESSs

To find candidates for an internal ESS, we need to solve for the probability vector $\mathbf{p}$ satisfying $E[S_1, \mathbf{p}] = E[S_2, \mathbf{p}] = E[S_3, \mathbf{p}]$, i.e. we need to find a solution of

$$\begin{align} ap_2 + bp_3 &= cp_1 + dp_3, \tag{6.55}\\ ap_2 + bp_3 &= ep_1 + fp_2, \tag{6.56}\\ p_1 + p_2 + p_3 &= 1. \tag{6.57} \end{align}$$

The solution is given by, see Exercise 6.5,

$$\mathbf{p} = \frac{1}{\alpha + \beta + \gamma}(\alpha, \beta, \gamma). \tag{6.58}$$

For $\mathbf{p}$ to be an internal strategy, we need $\alpha > 0, \beta > 0, \gamma > 0$, which rules out (not surprisingly) the existence of any pair ESS. For the stability or negative-definiteness condition, we need the matrix

$$C = \begin{pmatrix} -b-e & a-b-f \\ c-d-e & -d-f \end{pmatrix} \tag{6.59}$$

to be negative definite. To find an equivalent condition, we note that since $\mathbf{z}C\mathbf{z}^{\mathbf{T}} = \mathbf{z}C^T\mathbf{z}^{\mathbf{T}}$, C is negative definite if and only if

$$C + C^{\mathbf{T}} = \begin{pmatrix} -2(b+e) & (a+c)-(b+e)-(d+f) \\ (a+c)-(b+e)-(d+f) & -2(d+f) \end{pmatrix} \tag{6.60}$$

is negative definite. For the latter, we only need to check sub-determinants (Sylvester's criterion, Meyer, 2001). Specifically, we need

$$0 > -2(b+e), \tag{6.61}$$

$$0 < 4(b+e)(d+f) - \big((a+c)-(b+e)-(d+f)\big)^2. \tag{6.62}$$

The condition (6.62) is equivalent to

$$(a+c)^2+(b+e)^2+(d+f)^2-2(a+c)(b+e)-2(a+c)(d+f)-2(b+e)(d+f)<0 \tag{6.63}$$

which can be treated as a quadratic inequality for $(a+c)$. Since the coefficient of $(a+c)^2$ is 1, we find that (6.63) is satisfied if and only if

$$(b+e)+(d+f)-2\sqrt{(b+e)(d+f)} < (a+c) < (b+e)+(d+f)+2\sqrt{(b+e)(d+f)} \tag{6.64}$$

where clearly for $\sqrt{(b+e)(d+f)}$ to exist with $(b+e) > 0$ we also need $(d+f) > 0$ which consequently gives $(a+c) > 0$.

Thus the conditions for an internal ESS are

- $a+c > 0, b+e > 0, d+f > 0$, and

- $|\sqrt{b+e} - \sqrt{d+f}| < \sqrt{a+c} < \sqrt{b+e} + \sqrt{d+f}$, or equivalently that $\sqrt{a+c}, \sqrt{b+e}$ and $\sqrt{d+f}$ form a triangle.

6.4.4 No ESS

There is also the possibility of there being no ESS. The conditions for this are that no pure, pair or internal ESS can exist. Recall from Chapter 2 we discussed the ESSs of the Rock-Scissors-Paper game with payoffs

$$\begin{pmatrix} 0 & a_3 & -b_2 \\ -b_3 & 0 & a_1 \\ a_2 & -b_1 & 0 \end{pmatrix}. \tag{6.65}$$

It is clear from the above that there are no pure or pair ESSs. The equilibrium condition for an internal ESS is satisfied for any set of positive a_i and b_i, $1 \leq i \leq 3$. The key condition is the stability condition, which is satisfied if and only if $a_i > b_i$ for $i = 1, 2, 3$ and $(a_1 - b_1)^{1/2}, (a_2 - b_2)^{1/2}, (a_3 - b_3)^{1/2}$ form a triangle. In particular, if $a_i - b_i < 0$ for any $i \in \{1, 2, 3\}$, then there is no ESS.

6.5　Patterns of ESSs

An interesting question is what ESSs can occur for a particular game, and what restrictions upon ESSs there are. Why might we be interested in investigating this? One reason is that often, observed differences in behaviour lead to inferences of differences between two systems. The following example was used in Vickers and Cannings (1988b). It may be, for example, that frogs in different pools behave in different ways. Is this due to inherent differences in their two circumstances (different payoffs to the strategies in the games that they play) or could it simply be that there is more than one ESS and different initial conditions lead to one solution in one pool and another in the other? If we know that some combinations of ESSs are not possible, then we might be able to conclude that in some cases behaviour is inconsistent with common payoffs and so there must be real differences and in other circumstances that behaviour is consistent with a common matrix, so that the payoffs may be the same (of course this does not show that they, in fact, are). The following results make it clear that there are some significant restrictions on what is possible.

We now consider what ESSs can and cannot co-exist. To try to do this for a combination of exact probability vectors is impractical, and so we simplify this by representing any given ESS by its support. A *pattern of an $n \times n$ matrix game* is a collection $\mathcal{P}$ of subsets of $\{1, 2, \ldots, n\}$ such that if $\mathbf{p}_1, \mathbf{p}_2, \ldots, \mathbf{p}_N$ is the list of all ESS of the matrix game A, then

$$\mathcal{P} = \{S(\mathbf{p}_1), S(\mathbf{p}_2) \ldots, S(\mathbf{p}_N)\}. \tag{6.66}$$

We say that a collection of finite subsets of natural numbers $\mathcal{P}$ is an *attainable pattern*, if there is a matrix game A such that (6.66) holds. If the matrix game A could be chosen to be an $n \times n$ matrix, we say that $\mathcal{P}$ is *attainable on n pure strategies*. The numbers associated with the strategies are essentially arbitrary, so any pattern is (or is not) attainable if and only if any pattern with a renumbering is (or is not) attainable, and we will thus not list such equivalent patterns. A pattern $\mathcal{P}$ of $\{1, 2, \ldots, n\}$ is called a *maximal pattern* if $\mathcal{P}$ is attainable and $\mathcal{P} \cup \{Q\}$ is not an attainable pattern for any $Q \subset \{1, 2, \ldots, n\}, Q \neq \emptyset$.

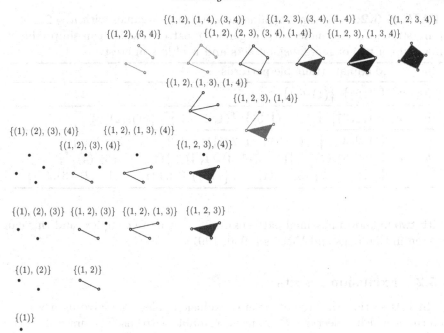

FIGURE 6.3: All attainable patterns for matrix games with $n = 2, 3, 4$ strategies. The grey patterns are attainable but not maximal. Supports of ESSs are drawn as sub-simplexes of the strategy simplex; in particular, a pure ESS is drawn as a full dot, a pair ESS as a line connecting two hollow dots, an ESS with support of 3 strategies is drawn as a solid triangle and an ESS with a support of 4 strategies as a tetrahedron.

6.5.1 Attainable patterns

All attainable patterns on $n = 1, 2, 3, 4$ strategies are shown in Figure 6.3. One can easily find an $n \times n$ game for various n with the same patterns of ESSs and for small n, one can verify the following still unproven conjecture.

Conjecture (Maximal Pattern Conjecture, Cannings and Vickers, 1990). *If $\mathcal{P}^* \subset \mathcal{P}$ and $\mathcal{P}$ is attainable on n pure strategies, then $\mathcal{P}^*$ is attainable on n pure strategies.*

The weaker result, that if $\mathcal{P}^* \subset \mathcal{P}$ and $\mathcal{P}$ is attainable on n pure strategies then $\mathcal{P}^*$ is attainable on $n + k$ pure strategies for all $k \geq K$ for some K, was proved in Broom (2000b). On the assumption that the conjecture is true, finding a full list of maximal patterns for a given number of strategies n is equivalent to finding all of the attainable patterns for n. The set of maximal patterns for the cases $n = 2, 3, 4$ (up to renumbering) are shown in Table 6.2 and also illustrated in Figure 6.3. The case $n = 5$ is almost complete as well

TABLE 6.2:   Maximal attainable patterns for games with $n = 2, 3, 4$ pure strategies. A $*$ indicates a degenerate pattern, which can simply be made up of two or more disjoint sets and is of less interest.

n	Maximal attainable patterns
$n = 2$	$\{(1,2)\}, \{(1),(2)\}^*$
$n = 3$	$\{(1,2,3)\}, \{(1,2),(1,3)\}, \{(1,2),(3)\}^*, \{(1),(2),(3)\}^*$
$n = 4$	$\{(1,2,3,4)\}, \{(1,2,3),(1,2,4)\}, \{(1,2,3),(2,4),(3,4)\},$ $\{(1,2,3),(4)\}^*, \{(1,2),(1,3),(1,4)\}, \{(1,2),(1,3),(4)\}^*,$ $\{(1,2),(2,3),(3,4),(1,4)\}, \{(1,2),(3),(4)\}^*, \{(1),(2),(3),(4)\}^*$

(with two unknown maximal patterns), but a lot more complex, and this can be seen in Cannings and Vickers (1990, 1991).

6.5.2   Exclusion results

In this section we present several exclusion rules that give us a way to determine which collections $\mathcal{P}$ are not attainable patterns. The first rule is the Bishop-Cannings Theorem 6.9; the rest, together with others, can be found in Vickers and Cannings (1988b).

Rule 1 (Anti-chain). If $T_1 \subseteq T_2$ then no pattern can contain both T_1 and T_2.

Thus any attainable pattern must form an anti-chain, and $\{(1,\dots,n)\}$, representing a single internal ESS, is a maximal pattern.

Rule 2 (No triangles). If $Q \subseteq \{4, 5, \dots, n\}$ and $T_1 = \{1,2\} \cup Q, T_2 = \{1,3\} \cup Q, T_3 = \{2,3\} \cup Q$, then no pattern can contain $\{T_1, T_2, T_3\}$.

A special case of Rule 2 (for $Q = \emptyset$) is that no pattern can contain $\{(1,2),(1,3),(2,3)\}$, and so, renumbering, no pattern can contain $\{(i,j),(i,k),(j,k)\}$ for any triple $(i,j,k) \in \{1,2,\dots,n\}$. Another special case (for $Q = \{4,5,\dots,n\}$) is that no pattern can contain three or more sets of $n-1$ elements. For example when $n = 4$, $\{(1,2,3),(1,2,4),(1,3,4)\}$ is not an attainable pattern.

Rule 3 (No general simplexes). If $Q \subseteq \{l+1, l+2, \dots, n\}$, $T_i = \{i\} \cup Q$, for $i = 1, \dots, l$ and $T_{l+1} = \{1,2,3,\dots,l\}$, then no pattern can contain $\{T_1, T_2 \dots, T_l, T_{l+1}\}$.

A special case of Rule 3 is for $l = n-1$ and $Q = \{n\}$ which yields that $\{(1,2,\dots,n-1),(1,n),(2,n),\dots,(n-1,n)\}$ is not an attainable pattern. For example when $n = 4$, $\{(1,2,3),(1,4),(2,4),(3,4)\}$ is not attainable.

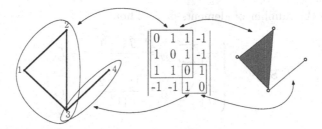

FIGURE 6.4: Graph of a clique matrix and the cliques. A line between vertices i, j means that $a_{ij} = a_{ji} = 1$. The cliques of this matrix are $(1, 2, 3)$ and $(3, 4)$. $(1, 2)$ is not a clique since 3 is connected to both 1 and 2. By Theorem 6.13 the pattern of the matrix is $\{(1, 2, 3), (3, 4)\}$. The two ESSs are $(1/3, 1/3, 1/3, 0)$ and $(0, 0, 1/2, 1/2)$. The pattern itself is also drawn using the convention from Figure 6.3.

6.5.3 Construction methods

To resolve whether a collection $\mathcal{P}$ is an attainable pattern, we need to do one of two things. Firstly, we can show that it is impossible either using one of the known rules, or by a direct proof that it cannot occur. Alternatively, we can prove that it is possible by finding a matrix with that pattern, or show that it can be constructed using an iterative construction method (a number of such methods are given in Broom et al., 1994).

Here we present a method from Cannings and Vickers (1988). We say that a matrix A is a *clique matrix*, if

- $a_{ii} = 0$ for all i, and

- $a_{ij} = a_{ji} = \pm 1$ if $i \neq j$.

For a matrix game A we define the graph $G(A)$ as follows. Every vertex represents a strategy, and vertices i and j are connected by an edge if and only if $a_{ij} = 1$. A *clique* of $G(A)$ is a maximal set of pairwise connected vertices, i.e. a set T is a clique if

- $a_{ij} = 1$ for all $i, j \in T$, $i \neq j$, and

- there is no $k \notin T$ for which $a_{kj} = 1$ for all $j \in T$.

Figure 6.4 shows an example of a graph of a particular clique matrix and its cliques.

Theorem 6.13. *For any clique matrix A, there is an ESS with support T if and only if T is a clique of $G(A)$.*

Proof. Suppose that T is a clique of $G(A)$. We define **p** by

$$p_i = \begin{cases} |T|^{-1} & \text{if } i \in T, \\ 0 & \text{otherwise,} \end{cases} \tag{6.67}$$

where $|T|$ is the number of elements of T. Then

$$E[S_i, \mathbf{p}] = \begin{cases} \dfrac{1}{|T|} \sum_{j \in T} a_{ij} = \dfrac{|T| - 1}{|T|} & i \in T, \\[4mm] \dfrac{1}{|T|} \sum_{j \in T} a_{ij} \leq \dfrac{|T| - 2}{|T|} & i \notin T, \end{cases} \tag{6.68}$$

because if $i \in T$, then there are $|T| - 1$ 1s and one 0 in the sum of a_{ij}, whereas for $i \notin T$ there are at most $|T| - 1$ 1s and at least one -1 in the sum. Thus we have $E[\mathbf{p}, \mathbf{p}] \geq E[\mathbf{q}, \mathbf{p}]$ for all $\mathbf{q} \neq \mathbf{p}$, with the inequality being strict if $S(\mathbf{q}) \not\subseteq T$.

We must now consider $\mathbf{q}$ such that $S(\mathbf{q}) \subseteq T$. Letting $\mathbf{z} = \mathbf{p} - \mathbf{q}$,

$$E[\mathbf{q}, \mathbf{q}] - E[\mathbf{p}, \mathbf{q}] = E[\mathbf{q}, \mathbf{q}] - E[\mathbf{q}, \mathbf{p}] - E[\mathbf{p}, \mathbf{q}] + E[\mathbf{p}, \mathbf{p}]$$
$$= \mathbf{z} A \mathbf{z}^{\mathbf{T}} = \sum_{i, j \in T} z_i z_j = - \sum_{i \in T} z_i^2 \leq 0, \tag{6.69}$$

with strict inequality unless $\mathbf{z} = 0$, i.e. unless $\mathbf{p} = \mathbf{q}$. Thus $\mathbf{p}$ is an ESS of A. $\qquad\qquad\square$

It should be noted that all of the attainable patterns for $n = 3$ could be found using clique matrices; see Exercise 6.10. This is not true for general n. For the $n = 4$ case all attainable maximal patterns except $\{(1, 2, 3), (2, 4), (3, 4)\}$ can be found using cliques; compare to Exercise 6.15. As n gets larger, proportionally fewer and fewer patterns are attainable through cliques.

6.5.4   How many ESSs can there be?

A question that follows from the consideration of patterns of ESSs is how many ESSs is it possible to have for a given number of strategies? Let M_n denote the maximum number of ESSs on n pure strategies. It was proved in Broom et al. (1993) that M_n increases exponentially, in particular that

$$M_n M_m \leq M_{n+m}, \tag{6.70}$$

and so the maximum number M_n increases so that

$$\lim_{n \to \infty} (M_n)^{1/n} = \gamma; \text{ for some } \gamma \leq 2. \tag{6.71}$$

In fact, $(M_n)^{1/n} \leq \gamma$ for any n, (Broom et al. (1993)).

We can find upper and lower bounds on γ as follows. Since the number of subsets of $\{1, 2, \ldots, n\}$ is 2^n, and since no different ESSs can have the same support (Theorem 6.9), $\gamma \leq 2$. It has not been possible to improve on this bound, and it may indeed be that $\gamma = 2$.

For a lower bound, we use a construction method using clique matrices as in Theorem 6.13. In Cannings and Vickers (1988) it is shown that $\gamma \geq 3^{1/3} \approx 1.442$. The outline of the procedure is given below; refer to Cannings and Vickers (1988) for the omitted details.

For $n = 3r + s$ where $s = 2, 3, 4$ it is possible to construct a matrix where a set T is a clique if and only if T contains exactly one element from each column of the following table with $r+1$ columns (the entries in $\{\}$ are excluded if greater than n):

$$
\begin{array}{llll}
1 & 4 & 7 & \ldots \quad 3\left\lfloor \frac{n+1}{3} \right\rfloor - 2 \\
2 & 5 & 8 & \ldots \quad 3\left\lfloor \frac{n+1}{3} \right\rfloor - 1 \\
3 & 6 & 9 & \ldots \quad \left\{ 3\left\lfloor \frac{n+1}{3} \right\rfloor \right\} \\
& & & \quad\quad \left\{ 3\left\lfloor \frac{n+1}{3} \right\rfloor + 1 \right\}.
\end{array}
\tag{6.72}
$$

It follows that $M_n \geq s3^r$ and thus $\gamma \geq 3^{1/3}$.

In fact, it follows from (6.70) that if we can find a matrix with K ESSs on n strategies, then $\gamma \geq K^{1/n}$. A matrix with 9 strategies and 30 ESSs was found in Broom et al. (1993) and so $\gamma \geq 30^{1/9} \approx 1.459$ which was the best lower bound found until Bomze et al. (2018), who found a matrix with 24 strategies and 15120 ESSs ($15120^{1/24} \approx 1.493$). This paper had built on earlier work (Bomze, 1992, 2002) to address this and related problems (as in Broom et al. (1993) these papers considered the standard quadratic optimisation problem relevant to other areas beyond matrix games) and this paper contains ideas for finding matrices with a large number of ESSs for larger n more generally, see also Bomze and Schachinger (2020) for a number of interesting related analyses. We note that 1.493 and 2 are far apart, and so there is still much uncertainly about the true value of γ.

6.6 Extensions to the Hawk-Dove game

Finally, we shall return to a specific matrix game, the classical Hawk-Dove game that we discussed in Section 4.1. This game can be extended by adding in two new strategies which combine aspects of the two original strategies. Recall that a *Dove* begins the contest by displaying, but retreats if its opponent escalates, and a *Hawk* escalates and never retreats unless it receives an injury. The strategy *Retaliator* begins by displaying but if an opponent escalates it escalates in response, so that it begins by playing like a Dove but will then change to Hawk behaviour when challenged. The strategy *Bully* begins by escalating, but if its opponent escalates in return, then it retreats. Thus it begins to play like a Hawk, but will then change to Dove-like behaviour when challenged. The behaviour in pairwise contests is detailed in Table 6.3. This

TABLE 6.3: Pairwise contests in the extended Hawk-Dove game with strategies Hawk, Dove, Retaliator and Bully.

Opponents	Behaviour
D vs D	both display, each winning with probability 1/2
H vs H	both fight until one is injured
R vs R	both display as Doves
B vs B	both escalate, one retreats immediately with no injury
H vs D	H escalates and D retreats
H vs B	Both escalate but B then retreats
H vs R	H escalates, R responds, and they fight until one is injured
B vs D	B escalates and D retreats
D vs R	both display as Ds
B vs R	B escalates, R responds, B retreats

leads to the payoff matrix (6.73) where, for convenience, we set $v = V/2$ and $c = C/2$:

$$
\begin{array}{c}
\begin{array}{cccc}
\quad\text{Hawk} & \text{Dove} & \text{Retaliator} & \text{Bully}
\end{array} \\
\begin{array}{c}
\text{Hawk} \\
\text{Dove} \\
\text{Retaliator} \\
\text{Bully}
\end{array}
\left(
\begin{array}{cccc}
v-c & 2v & v-c & 2v \\
0 & v & v & 0 \\
v-c & v & v & 2v \\
0 & 2v & 0 & v
\end{array}
\right).
\end{array}
\qquad (6.73)
$$

6.6.1 The extended Hawk-Dove game with generic payoffs

Maynard Smith (1982, Appendix E) pointed out a number of problems with the model as it stands above, both in terms of biological plausibility and mathematically. The key biological problem was in how non-escalated contests between Doves and/or Retaliators resolved themselves, and this is the same as we mention in Chapter 4, in terms of such a contest becoming a war of attrition in terms of its time-related costs. We did not let this problem affect our analysis then, as qualitatively the results were unchanged. This is still true for the current game as well, so we will not consider this further.

From a mathematical viewpoint the biggest problem relates to the fact that a number of rewards are identical, or exactly twice other rewards (in particular, when Bully and Hawk are absent, the rewards to Retaliator and Dove are identical). This is just the problem of non-generic payoffs that we discussed in Section 2.1.3.3 and above, and it is best to avoid these unlikely parameter coincidences if we can. We can avoid these non-generic payoffs as done in Zeeman (1981) by assuming that when Retaliator meets Dove, there is a small chance that Retaliator will become aware that Dove will never escalate, when it will escalate itself and win. Thus Retaliator wins slightly over half the Dove v Retaliator contests. Also, when Retaliator meets Hawk, Hawk escalates first, so that it has a slightly higher chance of winning, and so

the payoff to Hawk is slightly larger than that to Retaliator. This leads to a
revised payoff matrix:

$$
\begin{array}{c}
\begin{array}{cccc}
\text{Hawk} & \text{Dove} & \text{Retaliator} & \text{Bully}
\end{array}\\
\begin{array}{c}
\text{Hawk}\\
\text{Dove}\\
\text{Retaliator}\\
\text{Bully}
\end{array}
\left(
\begin{array}{cccc}
v-c & 2v & v-c+\varepsilon & 2v\\
0 & v & v-\varepsilon & 0\\
v-c-\varepsilon & v+\varepsilon & v & 2v\\
0 & 2v & 0 & v
\end{array}
\right).
\end{array}
\qquad (6.74)
$$

The ESSs of the matrix game (6.74) depend on the sign of $v - c$. If $v > c$
then both Hawk and Retaliator are pure ESSs. If $v < c$ then Retaliator is still
a pure ESS, and there is a mixed ESS of Hawk and Bully, with probability
of playing Hawk $v/c = V/C$, as in the original frequency of the Hawk-Dove
mixture. See Exercise 6.18.

6.6.2 ESSs on restricted strategy sets

It should be noted that the conclusions of Section 6.6.1 are valid when all
four strategies are allowed. When there was just Hawk and Dove as in Section
4.1, then for $v > c$ we had pure Hawk, and for $v < c$ a Hawk-Dove mixture.
The Hawk-Dove mixture (i.e. a strategy that plays Bully with probability 0)
is not an ESS of (6.74) because it can be invaded by Bully; Bully dominates
Dove since it does equally well against Hawk as Dove, but scores better against
Dove than Dove does.

On the other hand, by eliminating Bully from the strategy set and considering only Hawk, Dove and Retaliator, we get the payoff matrix:

$$
\begin{array}{c}
\begin{array}{ccc}
\text{Hawk} & \text{Dove} & \text{Retaliator}
\end{array}\\
\begin{array}{c}
\text{Hawk}\\
\text{Dove}\\
\text{Retaliator}
\end{array}
\left(
\begin{array}{ccc}
v-c & 2v & v-c+\varepsilon\\
0 & v & v-\varepsilon\\
v-c-\varepsilon & v+\varepsilon & v
\end{array}
\right).
\end{array}
\qquad (6.75)
$$

Retaliator is still a pure ESS of (6.75) irrespective of the relative size of v and
c. If $v > c$ then Hawk is also an ESS, otherwise a mixture of Hawk and Dove
is an ESS.

The full set of ESSs under all possible different available combinations of
the four strategies is shown in Table 6.4 (clearly all are ESSs if they are the
only strategy).

6.6.3 Sequential introduction of strategies

We have seen that what ESSs can be seen in a Hawk-Dove game depends
on what strategies are originally available in the population. For now, suppose
that all of the strategies H, D, B, R are potentially available, but each strategy
is introduced to the population one at a time. If the time between successive
introductions is sufficiently large, there is time for the population to settle

TABLE 6.4: Patterns of ESS under different available strategies in the extended Hawk-Dove game. If only one strategy is available, it is clearly an ESS.

Available strategies				ESSs if $v < c$	ESSs if $v > c$
H	D	B	R	R, (H,B)	R, H
H	D	B		(H,B)	H
H	D		R	R, (H,D)	R, H
H	D			(H,D)	H
H		B	R	R, (H,B)	R, H
H		B		(H,B)	H
H			R	H, R	R, H
	D	B	R	R	R
	D	B		B	B
	D		R	R	R
		B	R	R	R

to an ESS. Which ESS eventually persists will depend upon the order of introduction. It can be directly inferred from Table 6.4 that for $v < c$ only R or an (H,B) mixture persists, each in 12 out of 24 possible orders of sequential introduction. For instance, once R is an ESS, nothing can further invade it, so it will stay an ESS. The same is true for an (H,B) mixture. The other potential ESSs such as H, B, (H,D) which are ESSs on some restricted strategy set can be invaded by some of the other strategies and thus do not persist. In fact, Retaliator is an ESS whenever it comes before Hawk and the (H,B) mixture is an ESS whenever Hawk comes before Retaliator. Compare this to Exercise 6.19.

6.7   Python code

In this section we show how to use Python to find the ESSs of matrix games.

```python
""" ESSs of the matrix games
Identifies all ESS for a game given by a payoff matrix
For generic matrix games, it follows Haigh procedure
For non-generic matrix games, it tests against random
invasions by random strategies (if needed)
"""

## Import basic packages
import numpy as np
from numpy.random import seed, rand
```

```
11  import matplotlib.pyplot as plt
12
13  # Seed the random number generator for reproducibility
14  seed(1)
15
16  ## Give the payoff matrix, must be square
17
18  A = np.array([[ 1, -1,  2, -3,  2],
19                [ 2,  0, -1,  3, -1],
20                [-1,  2,  0,  2, -1],
21                [-8,  7,  1,  0,  1],
22                [-1,  2, -1,  2,  0]])
23  """
24
25  A = np.array([[0, 2, 2],
26                [0, 1, 0],
27                [0, 0, 1]])
28  """
29  """
30
31
32  A = np.array([[0, 2, 2, 0],
33                [0, 1, 0, 0],
34                [0, 0, 1, 0],
35                [0, 0, 0, 1]])
36  """
37  """
38  A = np.array([[0, 0, 2, 2, 0],
39                [0, 0, 1, 0, 0],
40                [0, 0, 0, 1, 0],
41                [0, 0, 0, 0, 1],
42                [0, 1, 0, 0, 0]])
43
44  """
45  """
46  A = np.array([[ 0,  1,  2, -2],
47                [-1,  0,  3, -1],
48                [-3,  2,  0, -1],
49                [-2,  1,  1,  0]])
50  """
51
52  ## Basic init for the algorithm
53  # Number of strategies
54  N = A.shape[1]
55
56
57  # All supports
58  all_s = np.arange(0,N)
59
60  # Identity matrix used for getting pure strategies
61  # S[i-1,:] will be pure ith strategy
62  S = np.identity(N)
63
64  # Initialize the list of ESSs as empty for now
65  ESSs = []
66
67  ## Actual algorithm
```

```
68
69   # Go over all 2^N subsets of {0, 1, 2, ..., N-1}
70   # Each subset will represent the support of potential ESS
71   # Start from 1 because at least one strategy must be involved
72   for ind in range(1,2**N):
73
74       # Get the binary representation of ind
75       # Keep it at length N and reverse the order for readibility
76       support = np.binary_repr(ind, N)[::-1]
77
78       # For indexing, change the characters to true/false
79       # Make it np array for quick negations
80       support = np.asarray([int(char)>0 for char in support])
81
82       # List involved strategies
83       involved_strategies = all_s[support]
84
85       # Number of strategies in the support
86       n=len(involved_strategies);
87
88       ## Check whether payoff of Si against p is the same as
89       ## payoff of Sj against p for all i, j in the support of p
90       ## i.e. solve S_i*A*p.T=S_j*A*p.T for p
91       if n==1:
92           # If the support has only 1 element
93           # the condition is trivially satisfied
94
95           # So far, it can be an ESS
96           CanBeESS = 1;
97
98           # Get the potential ESS strategy p
99           p = S[support,:]
100
101      else:
102          # If the support has several elements, we need to find p
103          # that solves S_i*A*p.T=S_j*A*p.T
104          # for all i, j in the support.
105
106          # Pick i to be the first strategy in the support
107          first_involved = involved_strategies[0];
108
109          # Define the system of linear equations that will give p
110          # as a solution to (S_i-S_j)*A*p.T=0
111          E = -S
112
113          # The above put -1 on the diagonal
114          # -1 will stay there in rows not in the support of p,
115          # which in turn will force p to be 0 outside the support
116          # Now we need to redefine E for rows in support of p
117          for j in involved_strategies:
118              # j is in the support of p
119              # this will be for (Si-Sj)*A*p'=0
120              E[j,:] = np.dot((E[j,:]+S[first_involved,:]),A)
121
122          # Elements of p need to add to 1
123          E[first_involved,:] = np.ones(N)
124
```

```
125        # The system is now defined. Need to solve it
126        if np.linalg.matrix_rank(E)<N:
127            # E is singular, we can't solve E*p'=B and so
128            # we do not have ESS
129            CanBeESS = 0
130        else:
131            # Set the right hand side of the equation E*p'=B
132            B = np.zeros(N)
133            B[first_involved] = 1
134
135            # Solve  E*p'=B
136            p = np.linalg.solve(E,B)
137
138            # Due to rounding, p may be a slightly off zero
139            # outside of support
140            # Force it to live within support and be nonnegative
141            p = abs(p*support)
142
143            # ESS must have positive coordinates on the support
144            CanBeESS = all(p[support]>0)
145
146    ## If we still believe that p can be an ESS,
147    ## check that S_i*A*p.T < p*A*p.T for every i not in T(p)
148    ## Here T(p) = {k, S_k*A*p.T = p*A*p.T}
149    if CanBeESS == 1:
150
151        # For now, set T(p) = support(p)
152        # This is true if A is generic.
153        #If it turns out not true we will update Tp later
154        Tp = support == True
155        # We cannot just say Tp = support because
156        # modifying Tp would modify the support too
157
158        # for all i not in the support of p
159        for i in all_s[¬support]:
160
161            # Calculate (S_i-p)*A*p.T
162            Si_against_p = (np.dot(S[i,:]-p,A)).dot(p.T)
163
164            if Si_against_p > 0:
165                # If Si does better than p against p,
166                # i.e. if Si can invade
167                # p cannot be an ESS
168                CanBeESS = 0
169            elif Si_against_p == 0:
170                # For non-generic A, i can be in T(p) even if
171                # not in Supp(p). p can still be ESS but support
172                # of p differs from T(p)
173
174                # Update T(p) for later testing
175                Tp[i]=True
176
177    ## If we still believe that p is an ESS,
178    ## perform the negative definite test
179    if CanBeESS == 1:
180        if all(support == Tp):
181            # if p satisfies S(p)=T(p), do Haigh procedure
```

```
182            for k, sk in enumerate(involved_strategies):
183                # Go through all involved_strategies
184
185                # Initialize the aij-aik-akj+akk matrix
186                H = np.zeros((n-1, n-1))
187
188                for i, si in enumerate(involved_strategies):
189                    # for all si not equal sk
190                    if si == sk:
191                        continue  # this skips sk
192
193                    for j, sj in enumerate(involved_strategies):
194                        # for all sj not equal sk
195                        if sj == sk:
196                            continue  # this skips k
197
198                        # We need to skip k. Note that i-(i>k)
199                        # is i if i<k and  it is i-1 if i>k
200                        H[i-(i>k), j-(j>k)] = A[si,sj] - ...
                              A[si,sk] - A[sk,sj] + A[sk,sk]
201
202            # Now, the matrix H is defined
203            # Check whether H+H.T is negative definite
204            # Cholesky method raises an exception when the
205            # matrix not positive definite
206            try:
207                np.linalg.cholesky(-(H+H.T))
208            except np.linalg.linalg.LinAlgError as err:
209                # Exception raised, not negative definite
210                CanBeESS=0
211     else:
212            # S(p) is not equal T(p). Do random invasions
213            # by random strategies with support in T(p)
214
215            # Set initial count of invasions to 0
216            invasion = 0
217
218            while invasion < N*N*10 and CanBeESS == 1:
219                # While we did not try too many times
220                # and while we still believe we can have an ESS
221
222                # Try an invasion by a strategy supported on T(p)
223                for ind_q in range(1,2**N):
224                    # Generate every subset of {0,..., N-1}
225
226                    # Get the binary representation as before
227                    supp_q = np.binary_repr(ind_q, N)[::-1]
228                    # For indexing, change the characters ...
                          true/false
229                    supp_q = [int(char)>0 for char in supp_q]
230
231                    # Restrict support of q on Tp
232                    invading_supp = supp_q and Tp
233
234                    # Get the size of the support of q
235                    n = sum(invading_supp)
236
```

```
237          # Initialize q
238          q = np.zeros(N)
239
240          if n==1:
241              # If pure strategy
242              q[invading_supp] = 1
243          elif n>1:
244              ## Generate random q by generating
245              ## q1, q1+q2, q1+q2+q3 etc first
246
247              # Generate random points in 0, 1
248              points=rand(n-1)
249
250              # Add 0 and 1 to the points
251              points = np.append(points, [0,1])
252
253              # Sort the points
254              points.sort()
255
256              # Get differences to get the strategy
257              q[Tp] = np.diff(points);
258
259          # We already know q does equally well
260          # against p as p itself does
261          # We need to check that p does better
262          # against q than q does
263          if (np.dot((p-q),A).dot(q.T)≤0) and not ...
                 (np.all(q == p)):
264              CanBeESS = 0;
265              # No need to test any further,
266              # break from the ind_q for cycle
267              break
268
269          # Increase the count of invasions
270          invasion +=1
271
272      # If even after all three tests we believe we have an ESS,
273      # add the strategy to the list of ESSs
274      if CanBeESS == 1:
275          ESSs.append(p)
276
277  # We went through all possible strategies. Print out the ESSs
278  print('ESSs is (are): \n', ESSs)
```

6.8 Further reading

Good books which explore important properties of matrix games include Hofbauer and Sigmund (1988, 1998); Weibull (1995). A systematic method to find all of the ESSs of a particular (non-generic) matrix game was given in Haigh (1975). The Bishop-Cannings theorem which established the most important restriction on which ESSs could occur for the same matrix, and which underpins the concept of patterns of ESSs, was shown in Bishop and Cannings (1976).

More recent work considering the ESSs of a variant of the clique matrix defined above, but with $a_{ij} = a_{ji} = 1$ or 0 has been considered in Wang et al. (2019a) and Hao and Wu (2018). This was chosen for application reasons, leading to more complex conditions less useful for constructing ESS patterns.

Broom (2000a) considers a question of how large the number of ESSs of a given mean support size could be. Haigh (1988) asked how many ESSs you would actually expect to see, assuming that all payoffs a_{ij} were generated from a uniform distribution, for small n. Cannings and Vickers (1988) considers the simpler problem of the distribution of the number of ESSs using clique matrices, for general n. Haigh (1989) considered how large the support of any ESS could be expected to be, and found in general that ESSs with large supports are very unlikely to occur.

Another way to avoid non-generic payoffs in the extended Hawk-Dove game is presented in Maynard Smith and Price (1973)). The dynamics of the game (6.74) are completely analysed in Zeeman (1981); see also Maynard Smith (1982, Appendix E).

In general, the idea of sequential introduction as in Section 6.6.3, especially in relation to ESSs with large supports, is discussed in more detail in Cannings and Vickers (1988) and Cannings et al. (1993).

6.9   Exercises

Exercise 6.1. Find the ESSs of the following matrices.

$$A = \begin{pmatrix} 0 & -1 & 3 \\ 1 & 0 & 1 \\ 2 & -2 & 0 \end{pmatrix}, B = \begin{pmatrix} 0 & 9 & -2 \\ 4 & 0 & 3 \\ 1 & 6 & 0 \end{pmatrix}, C = \begin{pmatrix} 0 & 3 & 4 \\ 4 & 0 & 1 \\ 4 & 7 & 0 \end{pmatrix}, D = \begin{pmatrix} 0 & 2.1 & 4 \\ -2 & 0 & 7 \\ 4 & 3 & 0 \end{pmatrix}.$$

Exercise 6.2. Show that for a 2×2 game with payoff matrix $\begin{pmatrix} a & b \\ c & d \end{pmatrix}$ and any strategies $\mathbf{p} = (p, 1-p), \mathbf{q} = (q, 1-q)$ where $\mathbf{p} = 1/(b+c-a-d)(b-d, c-a)$, we get $E[\mathbf{p}, \mathbf{q}] - E[\mathbf{q}, \mathbf{q}] = -(a-b-c+d)(p-q)^2$. Compare with Haigh's condition (6.43).

Exercise 6.3 (Zeeman, 1980). Show that if a matrix B is derived from matrix A by adding a constant to any column of matrix A, then the games A and B have exactly the same collection of ESSs.

Exercise 6.4 (Abakuks, 1980). Show that (6.42) and (6.43) are not equivalent in a non-generic case.

Exercise 6.5. Solve the system (6.55)–(6.57) for an internal equilibrium in a three strategy matrix game.

Exercise 6.6 (Vickers and Cannings, 1987). Consider the game with an infinite number of strategies $i = 1, 2, \ldots$ and payoffs given by $a_{ii} = 0$, for all $i > 0$, $a_{1j} = -1, a_{j1} = -2^{-j}$ for all $j > 1$, and $a_{ij} = 2$ for all $i, j \geq 2, i \neq j$. Prove that pure strategy 1 is an ESS but not uniformly uninvadable.

Exercise 6.7 (Bomze and Pötscher, 1989, p. 28). Show that Theorem 6.2 does not hold in general for non-matrix games.

Exercise 6.8. Show that if $\mathbf{p}$ is an internal ESS of a matrix game, then (6.16) holds for all $\mathbf{q} \neq \mathbf{p}$.

Exercise 6.9 (Vickers and Cannings, 1988b). Prove directly, without the use of exclusion Rule 2, that $\{(1, 2), (1, 3), (2, 3)\}$ is not an attainable pattern for a 3×3 game.

Exercise 6.10. Identify all attainable ESS patterns for a 3×3 game.

Exercise 6.11. Show that for the payoff matrix

$$A = \begin{pmatrix} 0 & 1 & 3 - x \\ 1 & 0 & 3 - y \\ x & y & 0 \end{pmatrix}, \tag{6.76}$$

there is an ESS $\mathbf{p}$ with support $S(\mathbf{p}) = \{1, 2\}$, if and only if $x + y \leq 1$. Show that $S(\mathbf{p}) \neq T(\mathbf{p})$, when $x + y = 1$. Explain why this is a non-generic case.

Exercise 6.12. Show that for a generic payoff matrix, and an ESS $\mathbf{p}$, we have $S(\mathbf{p}) = T(\mathbf{p})$.

Exercise 6.13. Show that for a generic matrix game, a pure strategy is an ESS if and only if no other pure strategy can invade. Show that the same is true for any (even non-generic) 2×2 matrix game. Give an example of a non-generic matrix game where non-invadability by other pure strategies is not enough.

Exercise 6.14. Let A be a clique matrix with a pure ESS S_i. Change one entry of A to get a matrix B with exactly the same ESSs as A except the one pure ESS.

Exercise 6.15 (Broom et al., 1994). Show that the matrices below have patterns $\{(1, 2, 3), (2, 4), (3, 4)\}$ and $\{(1, 2, 3), (2, 4), (3, 4), (1, 2, 5), (4, 5)\}$, respectively:

$$A = \begin{pmatrix} 0 & -1 & 2 & -3 \\ 2 & 0 & -1 & 3 \\ -1 & 2 & 0 & 2 \\ -8 & 7 & 1 & 0 \end{pmatrix}, \quad B = \begin{pmatrix} 0 & -1 & 2 & -3 & 2 \\ 2 & 0 & -1 & 3 & -1 \\ -1 & 2 & 0 & 2 & -1 \\ -8 & 7 & 1 & 0 & 1 \\ -1 & 2 & -1 & 2 & 0 \end{pmatrix}. \tag{6.77}$$

Exercise 6.16 (Discrete variant of sex-ratio game, Maynard Smith, 1982, p. 25). Consider a population playing the sex ratio game (4.4) but for which only the strategies $p_1 = 0.1$ and $p_2 = 0.6$ of male proportions are available. Find the "payoff matrix" with entries a_{ij}, meaning the payoff to an individual using strategy p_i in a population where everybody else uses strategy p_j. Find an ESS of this game. Show that the method of finding ESSs for matrix games in Chapter 6 yields a different result. Explain why and decide which method for finding the ESS should be used.

Exercise 6.17. Show that a Hawk-Dove game with payoffs given by (4.4) in Section 4.1 has no pure ESS. Identify potential mixed ESSs supported by two or all three strategies and show that the game has no ESS.

Hint. $\mathbf{p}_{\mathrm{ESS}}$ is not an ESS as it can be invaded by $\mathbf{p} = (V/C, 1 - V/C, 0)$.

Exercise 6.18. Show that the only ESSs of the matrix (6.74) are: Retaliator, together with Hawk if $v \geq c$ or with a mixture of Hawk and Bully if $v < c$.

Hint. You must also show that there are no other ESSs, but the Bishop-Cannings theorem can reduce the work that needs to be done.

Exercise 6.19. Consider an extended Hawk-Dove game with $c \geq v$ when strategies are introduced in the population sequentially, but repeatedly and randomly; i.e. every once in a while, H,D,B or R is introduced in the population, each with probability $1/4$. If the time intervals between introductions are long enough for a potentially new ESS to settle after the invasion, describe the long-term evolution of the system.

Chapter 7

Nonlinear games

7.1 Overview and general theory

In Chapter 6 we considered matrix games, where individuals play one of a number of pure strategies according to a probability vector $\mathbf{p}$ and the opposing pure strategy is chosen according to the frequencies chosen by their opponent (or the mixture of opponents that occur in the population). Thus as in (6.2)

$$\mathcal{E}[\mathbf{p}; \mathbf{q^T}] = E[\mathbf{p}, \mathbf{q}] = \mathbf{p}A\mathbf{q^T}$$
$$= \sum_i p_i(A\mathbf{q^T})_i = \sum_j (\mathbf{p}A)_j q_j \qquad (7.1)$$
$$= \sum_{i,j} a_{ij} p_i q_j.$$

We can see from the above that payoffs are linear in both the strategy of the focal individual and the strategy of the population, yielding a quadratic form as the payoff function $\mathcal{E}[\mathbf{p}; \mathbf{q^T}]$. This has many nice static properties that were explored in Chapter 6, but also useful dynamical properties as well (see Chapter 3, Section 3.1 and Hofbauer and Sigmund, 1998).

Definition 7.1. *We say that $\mathcal{E}$ is* linear in the focal player strategy *(or* linear on the left*) if*

$$\mathcal{E}\left[\sum_i \alpha_i \mathbf{p}_i; \Pi\right] = \sum_i \alpha_i \mathcal{E}[\mathbf{p}_i; \Pi] \qquad (7.2)$$

for every population Π, every $m-$tuple of individual strategies $\mathbf{p}_1, \ldots, \mathbf{p}_m$ and every collection of constants $\alpha_i \geq 0$ such that $\sum_i \alpha_i = 1$. Also, we say that $\mathcal{E}$ is linear in the population strategy *(or* linear on the right*) if*

$$\mathcal{E}\left[\mathbf{p}; \sum_i \alpha_i \delta_{\mathbf{q}_i}\right] = \sum_i \alpha_i \mathcal{E}[\mathbf{p}; \delta_{\mathbf{q}_i}] \qquad (7.3)$$

for every individual strategy $\mathbf{p}$, every $m-$tuple $\mathbf{q}_1, \ldots, \mathbf{q}_m$ and every collection of α_i's from $[0, 1]$ such that $\sum_i \alpha_i = 1$.

We noted in Chapter 2 that for matrix games, the payoff to an individual is the same whether it faces opponents playing a polymorphic mixture of pure

DOI: 10.1201/9781003024682-7

strategies or a monomorphic population playing the equivalent mixed strategy. We saw from Example 2.6 that such equivalence does not always hold.

Definition 7.2. *We say that a game has* polymorphic-monomorphic equivalence *if for every strategy* **p***, any finite collection of strategies* $\{\mathbf{q}_i\}_{i=1}^{m}$ *and any corresponding collection of m constants* $\alpha_i \geq 0$ *such that* $\sum_i^m \alpha_i = 1$ *we have*

$$\mathcal{E}\left[\mathbf{p}; \sum_i \alpha_i \delta_{\mathbf{q}_i}\right] = \mathcal{E}\left[\mathbf{p}; \delta_{\sum_i \alpha_i \mathbf{q}_i}\right]. \tag{7.4}$$

We note that polymorphic-monomorphic equivalence holds only in respect of the static notion of ESSs, and there is no such equivalence in terms of dynamics.

Quite often, the payoff is linear in the focal player strategy because by its very definition $\mathcal{E}[\mathbf{p}; \Pi]$ is set to equal the expected payoff of the focal individual playing a pure strategy S_i with probability p_i for all i. This is the case for matrix games but also for example in the sex ratio game (see Section 4.4 and also 7.2.2). It is common, however, that the payoff is nonlinear in the population strategy. This occurs whenever the game does not involve pairwise contests against opponents playing pure strategies (or equivalent mixed combinations). Alternatively, the payoff will also be nonlinear in the population strategy if it does involve such pairwise contests, but that these are not independent contests against randomly chosen opponents. A third situation occurs when a strategy is a pure strategy drawn from a continuum, such as a level of defence or a volume of sperm, as we see in Section 7.5.1, but that the payoff is nonlinear as a function of this pure strategy. This will lead to games which are nonlinear in the focal player strategy.

It can be argued that nearly all real situations feature nonlinearity of at least one of the types described above, and so nonlinear payoffs should occur for most models. As we see in later chapters, when models of real behaviours are developed, the payoffs involved are indeed almost always nonlinear in some way.

There is not as yet a significantly developed general theory of nonlinear games. However, some results for linear games can be generalised and reformulated even for nonlinear games. We do not attempt a detailed analysis of all of the important properties here. For the reader interested in such an analysis, we recommend Bomze and Pötscher (1989). We note that the theory of adaptive dynamics, for which a significant body of work has been produced, as we mention later in Section 7.5 (see also Chapter 14), is closely related.

For example, Theorem 6.2 can be generalised as follows

Theorem 7.3. *For games with generic payoffs, if the incentive function*

$$h_{\mathbf{p},\mathbf{q},u} = \mathcal{E}[\mathbf{p}; (1-u)\delta_{\mathbf{p}} + u\delta_{\mathbf{q}}] - \mathcal{E}[\mathbf{q}; (1-u)\delta_{\mathbf{p}} + u\delta_{\mathbf{q}}] \tag{7.5}$$

is differentiable (from the right) at $u = 0$ *for every* **p** *and* **q***, then* **p** *is an ESS if and only if for every* $\mathbf{q} \neq \mathbf{p}$;

1. $\mathcal{E}[\mathbf{p}; \delta_{\mathbf{p}}] \geq \mathcal{E}[\mathbf{q}; \delta_{\mathbf{p}}]$ and

2. if $\mathcal{E}[\mathbf{p}; \delta_{\mathbf{p}}] = \mathcal{E}[\mathbf{q}; \delta_{\mathbf{p}}]$, then $\frac{\partial}{\partial u} h_{\mathbf{p},\mathbf{q},u}\big|_{u=0} > 0$.

Proof. Recall that, by Definition 3.4, a strategy $\mathbf{p}$ is an ESS if and only if for every $\mathbf{q}$ there is $u_{\mathbf{q}} > 0$ such that for every $u \in (0, u_{\mathbf{q}})$

$$h_{\mathbf{p},\mathbf{q},u} > 0. \qquad (7.6)$$

So, if $\mathbf{p}$ is an ESS, then by the continuity of $h_{\mathbf{p},\mathbf{q},u}$ in u (it is differentiable so is clearly continuous) as we take a limit of (7.6) for $u \to 0+$, we get that $\mathcal{E}[\mathbf{p}; \delta_{\mathbf{p}}] \geq \mathcal{E}[\mathbf{q}; \delta_{\mathbf{p}}]$. If $\mathcal{E}[\mathbf{p}; \delta_{\mathbf{p}}] = \mathcal{E}[\mathbf{q}; \delta_{\mathbf{p}}]$, then differentiating (7.6) at $u = 0$ yields $\frac{\partial}{\partial u} h_{\mathbf{p},\mathbf{q},u}\big|_{u=0} \geq 0$. Since we assume the payoffs are generic, we cannot have $\frac{\partial}{\partial u} h_{\mathbf{p},\mathbf{q},u}\big|_{u=0} = 0$ and the statement of Theorem 7.3 thus follows. The proof of the reverse implication is left as Exercise 7.1; note that we do not need the generic payoff requirement in this part. $\qquad \square$

Example 7.4. Consider a two strategy game which satisfies polymorphic-monomorphic equivalence, where the payoff to an individual playing pure strategy S_1 with probability p in a population where the probability of an individual playing S_1 is r is given by

$$\mathcal{E}[\mathbf{p}; \delta_{\mathbf{r}}] = p(a_1 r^2 + a_2 r + a_3). \qquad (7.7)$$

Find conditions on the ESSs of the game in terms of the parameters a_1, a_2 and a_3.

The condition $\mathcal{E}[\mathbf{p}; \delta_{\mathbf{p}}] \geq \mathcal{E}[\mathbf{q}; \delta_{\mathbf{p}}]$ (for all $\mathbf{q}$) means that we need

$$(p - q)(a_1 p^2 + a_2 p + a_3) \geq 0, \qquad (7.8)$$

for all $q \in [0, 1]$. Hence, the pure strategy S_2 (or $p = 0$) is an ESS if $a_3 < 0$ ($a_3 = 0$ is a non-generic case) and similarly the pure strategy S_1 (or $p = 1$) is an ESS if $a_1 + a_2 + a_3 > 0$. An internal strategy $\mathbf{p}$ (or $0 < p < 1$) can be an ESS only if $a_1 p^2 + a_2 p + a_3 = 0$. Depending upon the values of a_1, a_2 and a_3 there may be 0, 1 or 2 values of p which satisfy this condition, the values of which can be easily found. It is clear that (7.8) is always satisfied with equality, so that we need

$$\frac{\partial}{\partial u} h_{\mathbf{p},\mathbf{q},u}\bigg|_{u=0} > 0, \qquad (7.9)$$

where

$$h_{\mathbf{p},\mathbf{q},u} = (p - q)\left(a_1 \left(p + u(q - p)\right)^2 + a_2 \left(p + u(q - p)\right) + a_3\right), \qquad (7.10)$$

giving the required condition as $2a_1 \hat{p} + a_2 < 0$ at the root $\hat{p}$. We note that if $2a_1 \hat{p} + a_2 = 0$ we would again have a non-generic example.

Under most (if not all) biologically reasonable conditions, the function

$$u \mapsto \mathcal{E}[\mathbf{r}; (1-u)\delta_{\mathbf{p}} + u\delta_{\mathbf{q}}] \qquad (7.11)$$

is differentiable for every $u \in [0,1]$ and thus Theorem 7.3 can still be used. In Section 7.2.2.1 we will see one of the rare instances where the function $h_{\mathbf{p},\mathbf{q},u}$ is not differentiable for all $\mathbf{p}$; yet it is still differentiable for all important $\mathbf{p}$s.

We will see later in Section 9.1.3 that the Bishop-Cannings theorem does not hold in general for non-matrix games. Yet, as seen in Section 7.2, many properties of games with linear payoffs carry over to games with payoffs that are linear in the focal player strategy.

In Sections 7.2, 7.3 and 7.5 we investigate a number of scenarios where the payoffs are nonlinear, divided into the three types that we outlined above. Firstly we look at games which are nonlinear in the strategy of the opponent, focusing on playing the field games, which directly incorporates the frequencies of the different strategies in the population into a fitness function.

7.2   Linearity in the focal player strategy and playing the field

As we commented above, it is quite common that games are linear in the focal player strategy, and this leads to some results that are similar to those of matrix games. We again look at some key results only rather than carry out a more extensive analysis. We also investigate the implications of polymorphic-monomorphic equivalence (7.4).

7.2.1   A generalisation of results for linear games

The following theorem is analogous to Lemma 6.7.

Theorem 7.5. *Let $\mathcal{E}$ be linear in the focal player strategy, i.e. (7.2) holds, and let the function $h_{\mathbf{p},\mathbf{q},u}$ be differentiable w.r.t u at $u = 0$. Let $\mathbf{p} = (p_i)$ be an ESS. Then $\mathcal{E}[\mathbf{p}; \delta_{\mathbf{p}}] = \mathcal{E}[S_i; \delta_{\mathbf{p}}]$ for any pure strategy S_i such that $i \in S(\mathbf{p}) = \{j; p_j > 0\}$.*

We note that it is enough to assume $h_{\mathbf{p},\mathbf{q},u}$ to be continuous; see Exercise 7.7.

Proof. By Theorem 7.3, $\mathbf{p}$ is a best response to itself and thus

$$\max_{j \in S(\mathbf{p})} \left(\mathcal{E}[S_j; \delta_{\mathbf{p}}] \right) = \sum_{i \in S(\mathbf{p})} p_i \max_{j \in S(\mathbf{p})} \left(\mathcal{E}[S_j; \delta_{\mathbf{p}}] \right) \geq \sum_{i \in S(\mathbf{p})} p_i \mathcal{E}[S_i; \delta_{\mathbf{p}}] \qquad (7.12)$$

$$= \mathcal{E}[\mathbf{p}; \delta_{\mathbf{p}}] \geq \max_{j \in S(\mathbf{p})} \left(\mathcal{E}[S_j; \delta_{\mathbf{p}}] \right). \qquad (7.13)$$

Hence, we must have equalities everywhere in the above and the statement
follows. □

Note that if the payoff is not linear but strictly convex, i.e.

$$\sum_i p_i \mathcal{E}[S_i; \delta_{\mathbf{q}}] > \mathcal{E}[\mathbf{p}; \delta_{\mathbf{q}}] \tag{7.14}$$

for all $\mathbf{q}$ and all $\mathbf{p}$ with at least two elements in $S(\mathbf{p})$, then the inequality in
(7.13) would be strict for any ESS $\mathbf{p}$ that is not pure. Thus strict convexity
of payoffs in the focal player strategy forces the ESSs to be pure.

Lemma 7.6 below shows that the payoffs of games that are linear in the focal player strategy and satisfy polymorphic monomorphic equivalence (7.4)
must be of a special form (Sandholm (2010) calls such games *population
games*).

Lemma 7.6. *If the payoffs of the game are linear in the focal player strategy
(i.e. satisfy (7.2)) and satisfy polymorphic monomorphic equivalence (7.4),
then for every* $\mathbf{x}, \mathbf{y}, \mathbf{z}$ *and every* $\varepsilon \in [0,1]$

$$\mathcal{E}[\mathbf{x}; (1-\varepsilon)\delta_{\mathbf{y}} + \varepsilon\delta_{\mathbf{z}}] = \sum_i x_i f_i((1-\varepsilon)\mathbf{y} + \varepsilon\mathbf{z}) \tag{7.15}$$

where $f_i(\mathbf{q}) = \mathcal{E}[S_i; \delta_{\mathbf{q}}]$.

Proof. From the assumptions of Lemma 7.6, it follows that for every $\mathbf{x}, \mathbf{y}, \mathbf{z}$
and every $\varepsilon \in [0,1]$

$$\mathcal{E}[\mathbf{x}; (1-\varepsilon)\delta_{\mathbf{y}} + \varepsilon\delta_{\mathbf{z}}] = \sum_i x_i \mathcal{E}[S_i; (1-\varepsilon)\delta_{\mathbf{y}} + \varepsilon\delta_{\mathbf{z}}] \tag{7.16}$$

$$= \sum_i x_i \mathcal{E}[S_i; \delta_{(1-\varepsilon)\mathbf{y}+\varepsilon\mathbf{z}}] \tag{7.17}$$

$$= \sum_i x_i f_i((1-\varepsilon)\mathbf{y} + \varepsilon\mathbf{z}). \tag{7.18}$$

□

Hence, as in Theorem 7.7 below, we can sometimes write that payoffs are
such that $\mathcal{E}[\mathbf{p}; \delta_{\mathbf{q}}] = \sum_i p_i f_i(\mathbf{q})$ for some functions f_i and in accordance with
Lemma 7.6 mean that the payoffs to the game are linear in the focal player
strategy (i.e. satisfy (7.2)) and satisfy polymorphic monomorphic equivalence
(7.4).

Theorem 7.7 (Uniform invasion barrier, Crawford, 1990a,b)**.** *Let the payoffs
be such that* $\mathcal{E}[\mathbf{p}; \delta_{\mathbf{q}}] = \sum_i p_i f_i(\mathbf{q})$ *for some continuous functions* f_i. *Then* $\mathbf{p}$
is an ESS if and only if there exists $\varepsilon_{\mathbf{p}} > 0$ *such that for all* $\mathbf{q} \neq \mathbf{p}$ *and all*
$\varepsilon \in (0, \varepsilon_{\mathbf{p}})$ *we have*

$$\mathcal{E}[\mathbf{p}; (1-\varepsilon)\delta_{\mathbf{p}} + \varepsilon\delta_{\mathbf{q}}] > \mathcal{E}[\mathbf{q}; (1-\varepsilon)\delta_{\mathbf{p}} + \varepsilon\delta_{\mathbf{q}}]. \tag{7.19}$$

Proof. We have

$$h_{\mathbf{p},\mathbf{q},\varepsilon} = \mathcal{E}[\mathbf{p}; (1-\varepsilon)\delta_{\mathbf{p}} + \varepsilon\delta_{\mathbf{q}}] - \mathcal{E}[\mathbf{q}; (1-\varepsilon)\delta_{\mathbf{p}} + \varepsilon\delta_{\mathbf{q}}] \qquad (7.20)$$

$$= \sum_i (p_i - q_i) f_i\big((1-\varepsilon)\mathbf{p} + \varepsilon\mathbf{q}\big). \qquad (7.21)$$

Thus if $\mathbf{p}$ is an ESS and $\mathbf{q}$ is any strategy, there is $\varepsilon_{\mathbf{p},\mathbf{q}} \in (0,1]$ such that $h_{\mathbf{p},\mathbf{q},\varepsilon} > 0$ for all $\varepsilon \in (0, \varepsilon_{\mathbf{p},\mathbf{q}})$. We can pick $\varepsilon_{\mathbf{p},\mathbf{q}}$ to be the maximum possible. Since the strategy simplex is compact and the f_i's are continuous, it follows that the f_i's are uniformly continuous. From uniform continuity of the f_i's it follows that $h_{\mathbf{p},\mathbf{q}',\varepsilon} > 0$ for all $\mathbf{q}'$ close enough to $\mathbf{q}$. Hence, for every strategy $\mathbf{q}$, there exists a neighbourhood $U(\mathbf{q})$ of $\mathbf{q}$ such that for every $\mathbf{q}' \in U(\mathbf{q})$, $\varepsilon_{\mathbf{p},\mathbf{q}'} \geq \varepsilon_{\mathbf{p},\mathbf{q}}$. Since the strategy simplex is compact, it follows that there exists $\mathbf{q}_0$ such that $\varepsilon_{\mathbf{p},\mathbf{q}_o} \leq \varepsilon_{\mathbf{p},\mathbf{q}}$ for every $\mathbf{q}$. $\qquad\square$

Theorem 7.8 (Local superiority of an ESS, Palm, 1984). *Let the payoffs be such that* $\mathcal{E}[\mathbf{p}; \delta_{\mathbf{q}}] = \sum_i p_i f_i(\mathbf{q})$ *for some continuous functions* f_i. *Then the strategy* $\mathbf{p}$ *is an ESS if and only if it is locally superior, i.e. there is* $U(\mathbf{p})$ *a neighbourhood of* $\mathbf{p}$ *such that*

$$\mathcal{E}[\mathbf{p}; \delta_{\mathbf{q}}] > \mathcal{E}[\mathbf{q}; \delta_{\mathbf{q}}], \text{ for all } \mathbf{q}(\neq \mathbf{p}) \in U(\mathbf{p}). \qquad (7.22)$$

Proof. Let $\mathbf{p}$ be a strategy and let d denote the shortest distance from $\mathbf{p}$ to a face of a strategy simplex not containing $\mathbf{p}$.

If $\mathbf{p}$ is an ESS, then by Theorem 7.7 there exists a uniform invasion barrier $\varepsilon_{\mathbf{p}}$. Now, if $\mathbf{x}$ is such that $|\mathbf{x} - \mathbf{p}| < d\varepsilon_{\mathbf{p}}$; $\mathbf{x} \neq \mathbf{p}$, there is $\mathbf{q}$ such that $\mathbf{x} = (1-\varepsilon)\mathbf{p} + \varepsilon\mathbf{q}$, for some $\varepsilon \in (0, \varepsilon_{\mathbf{p}})$; compare with Figure 6.1. Since the following series of inequalities are equivalent

$$\mathcal{E}[\mathbf{p}; (1-\varepsilon)\delta_{\mathbf{p}} + \varepsilon\delta_{\mathbf{q}}] > \mathcal{E}[\mathbf{q}; (1-\varepsilon)\delta_{\mathbf{p}} + \varepsilon\delta_{\mathbf{q}}], \qquad (7.23)$$

$$\varepsilon \sum_i p_i f_i(\mathbf{x}) > \varepsilon \sum_i q_i f_i(\mathbf{x}), \qquad (7.24)$$

$$\sum_i p_i f_i(\mathbf{x}) > \sum_i \big((1-\varepsilon)p_i + \varepsilon q_i\big) f_i(\mathbf{x}), \qquad (7.25)$$

$$\mathcal{E}[\mathbf{p}; \delta_{\mathbf{x}}] > \mathcal{E}[\mathbf{x}; \delta_{\mathbf{x}}], \qquad (7.26)$$

we get that $\mathbf{p}$ is locally superior.

Conversely, if $\mathbf{p}$ is locally superior and $\mathbf{q}$ is an invading strategy, take ε small enough so that $\mathbf{x} = (1-\varepsilon)\mathbf{p} + \varepsilon\mathbf{q} \in U(\mathbf{p})$ and thus (7.26) holds. Thus (7.23) holds and so $\mathbf{p}$ is an ESS. $\qquad\square$

7.2.2 Playing the field

There are some biological situations which require game-theoretical modelling but, unlike in matrix games, do not involve individuals interacting in direct pairwise contests. Thus the term $E[\mathbf{p}, \mathbf{q}]$ is undefined for such games, and we deal with $\mathcal{E}[\mathbf{p}; \delta_\mathbf{q}]$ directly.

In this section we consider payoff functions of the form

$$\mathcal{E}[\mathbf{p}; \Pi] = \sum p_i f_i(\Pi) \tag{7.27}$$

where the f_i's are in general nonlinear functions, and Π represents the strategy played by the population. In accordance with the discussion after Lemma 7.6, in general, here and elsewhere we shall assume that polymorphic monomorphic equivalence (7.4) holds by definition (i.e. that it does not matter whether an individual is in a population that is polymorphic or one that plays a mixed strategy equal to the mean population strategy). These games will be called *playing the field*.

Playing the field games are perhaps the most straightforward way of incorporating nonlinearity into a game model, as the fitness function of the individuals involved automatically includes the population frequencies of the different strategies. An example is the sex ratio game of Section 4.4, where a strategy was effectively mixed, with two pure strategies "male" and "female". The fitness of an individual with strategy p was given by (4.28) as

$$\mathcal{E}[p; \delta_m] = \frac{p}{m} + \frac{1-p}{1-m} \tag{7.28}$$

so that in the notation of equation (7.27) we have

$$f_1(m) = \frac{1}{m}, \tag{7.29}$$

$$f_2(m) = \frac{1}{1-m}. \tag{7.30}$$

We saw that the strategy $\mathbf{p}_{\text{ESS}} = 0.5$ was not just an ESS, but a strategy which beats any other strategy in any mixed population, in the sense that it can invade any other, and will always be the rest point of the replicator dynamics.

7.2.2.1 Parker's matching principle

Parker (1978) considers a single species foraging on N patches, with resources $r_i > 0$ for $i = 1, \ldots, N$ which are shared equally by all who choose the patch. There are N pure strategies for this game, each corresponding to foraging on one patch only, a mixed strategy $\mathbf{x} = (x_i)$ meaning to forage at patch i with probability x_i. The general payoff to an individual using strategy $\mathbf{x} = (x_i)$ against a population playing $\mathbf{y} = (y_i)$ is

$$\mathcal{E}[\mathbf{x}; \delta_\mathbf{y}] = \begin{cases} \infty, & \text{if } x_i > 0 \text{ for some } i \text{ such that } y_i = 0, \\ \sum_{i; x_i > 0}^{N} r_i \dfrac{r_i}{y_i} & \text{otherwise,} \end{cases} \tag{7.31}$$

where r_i is a constant corresponding to the quality of a patch i. It is obvious from (7.31) that if there is an ESS $\mathbf{p}$, it must be internal, i.e. we must have $p_i > 0$ for all $i = 1, \ldots, N$. Thus any problematic cases with potentially infinite payoffs do not arise in any analysis. In particular, (7.7) still holds even though the fitness functions are not continuous everywhere, since they are continuous in the vicinity of the ESS. Note that the sex ratio game is a special case with $N = 2$ and $r_1 = r_2$. Now, let us show that $\mathbf{p} = (p_i)$ given by

$$p_i = \frac{r_i}{\sum\limits_{i=1}^{N} r_i} \tag{7.32}$$

is an ESS. Clearly, $\mathcal{E}[\mathbf{q}; \delta_{\mathbf{p}}] = \mathcal{E}[\mathbf{p}; \delta_{\mathbf{p}}]$ for all $\mathbf{q}$. Moreover, since this game satisfies polymorphic monomorphic equivalence (7.4) then

$$\mathcal{E}[\mathbf{x}; (1 - u)\delta_{\mathbf{y}} + u\delta_{\mathbf{z}}] = \mathcal{E}[\mathbf{x}; \delta_{(1-u)\mathbf{y}+u\mathbf{z}}] \tag{7.33}$$

and thus

$$h_{\mathbf{p},\mathbf{q},u} = \mathcal{E}[\mathbf{p}; (1 - u)\delta_{\mathbf{p}} + u\delta_{\mathbf{q}}] - \mathcal{E}[\mathbf{q}; (1 - u)\delta_{\mathbf{p}} + u\delta_{\mathbf{q}}] \tag{7.34}$$

$$= \sum_{i=1}^{N} (p_i - q_i) \frac{r_i}{p_i + u(q_i - p_i)} \tag{7.35}$$

$$= \sum_{i=1}^{N} \frac{p_i - q_i}{p_i} r_i \left(1 - u \frac{q_i - p_i}{p_i} + \ldots \right), \tag{7.36}$$

which implies that

$$\frac{\partial}{\partial u} h_{\mathbf{p},\mathbf{q},u} \Big|_{u=0} = \sum_{i=1}^{N} r_i \left(\frac{p_i - q_i}{p_i} \right)^2 > 0. \tag{7.37}$$

Thus $\mathbf{p}$ is an ESS, using Theorem 7.3. Also, note (and this is called *Parker's matching principle*), that at the ESS strategy $\mathbf{p}$ given by (7.32), we have

$$\frac{p_i}{p_j} = \frac{r_i}{r_j}. \tag{7.38}$$

7.3   Nonlinearity due to non-constant interaction rates

The second scenario where games can be nonlinear is where the strategies employed by the players affect the frequency of their interactions. The actual pairwise interactions themselves can be very simple, but if the strategy affects the interaction rate, then the overall payoff function can be complicated. We note that the idea of nonlinear interactions occurs in other cases as well. In Chapter 9, we look at multi-player games comprised of pairwise interactions in a structure where selection of opponents depends on the results of previous interactions so that, for example, a single multi-player game of four players is constructed from three non-independent pairwise games (Broom et al., 2000c).

In this section, we look at examples from the modelling of kleptoparasitism using compartmental games to illustrate the effect of nonlinear interaction rates, but starting with some more general results.

7.3.1 Nonlinearity in pairwise games

The simplest non-trivial scenario to consider where interaction rates are not constant is a two-player contest with two pure strategies S_1 and S_2, with payoffs given by a standard payoff matrix

$$\begin{pmatrix} a & b \\ c & d \end{pmatrix}, \tag{7.39}$$

but where the three types of interaction happen with probabilities not simply proportional to their frequencies. This is the scenario in Taylor and Nowak (2006), where it is assumed that each pair of S_1 individuals meet at rate r_{11}, each pair of S_1 and S_2 individuals meet at rate r_{12} and each pair of S_2 individuals meet at rate r_{22}. Thus the frequency of interactions of an S_1 individual with other S_1 individuals is $r_{11}(N-1)p_1$ and the frequency with S_2 individuals is $r_{12}Np_2$, where N is the population size and p_i is the proportion of S_i-individuals in the population. For N large enough, we get $(N-1)/N \approx 1$ and thus this yields the following nonlinear payoff function

$$\mathcal{E}[S_1; \mathbf{p^T}] = \frac{ar_{11}p_1 + br_{12}p_2}{r_{11}p_1 + r_{12}p_2}, \tag{7.40}$$

$$\mathcal{E}[S_2; \mathbf{p^T}] = \frac{cr_{12}p_1 + dr_{22}p_2}{r_{12}p_1 + r_{22}p_2}. \tag{7.41}$$

This reduces to the standard payoffs for a matrix game when $r_{11} = r_{12} = r_{22}$, but otherwise does not.

How do these non-uniform interaction rates affect the game? In particular, when are there differences between this case and the simple two-player matrix game?

In the simple game (uniform interaction rates) if $a < c$ and $b > d$ there is a mixed ESS, and this is not altered by the use of non-uniform interaction rates, although the ESS proportions of the strategies do change. If $a > c$ and $b < d$ then there are two ESSs in the simple case, and this is also always true for non-uniform interactions, although the location of the unstable equilibrium between the pure strategies changes, which affects the dynamics. See Exercise 7.11 and also Figure 7.1.

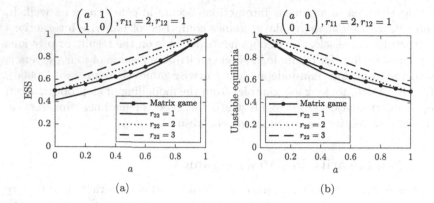

FIGURE 7.1: ESS (a) and unstable equilibria (b) for a game (7.39) with non-uniform interaction rates.

In the simple game, if $a < c$ and $b < d$, then pure strategy 2 is a unique ESS. It turns out that this is the most interesting case, as in the non-uniform interaction rates, it yields a solution that does not occur for matrix games. Under some circumstances there are two ESSs rather than just one, a pure $(0, 1)$ ESS, but also a mixed ESS. Setting $r_{12} = 1$ without loss of generality, this occurs if either $c > a > d > b$ and

$$r_{11}r_{22} > \left(\frac{\sqrt{(a-b)(c-d)} + \sqrt{(a-c)(b-d)}}{d-a} \right)^2, \qquad (7.42)$$

or $d > b > c > a$ and

$$r_{11}r_{22} < \left(\frac{\sqrt{(a-b)(c-d)} - \sqrt{(a-c)(b-d)}}{d-a} \right)^2. \qquad (7.43)$$

Considering the Prisoner's Dilemma (see Chapter 4 Section 4.2), which is an example where $c > a > d > b$, Taylor and Nowak (2006) show that cooperation is possible if interaction rates are non-uniform, there being a mixed ESS if

$$r_{11}r_{22} > \frac{1}{(R-P)^2} \left(\sqrt{(T-R)(P-S)} + \sqrt{(R-S)(T-P)} \right)^2. \qquad (7.44)$$

In particular, setting $r_{11} = r_{22}(= r)$ and with payoffs given by the matrix

$$\begin{pmatrix} 3 & 0 \\ 5 & 1 \end{pmatrix} \qquad (7.45)$$

from Axelrod and Hamilton (1981), they show that there is a stable mixture of cooperators and defectors when $r > 2.44$. As $r \to \infty$ the proportion of cooperators in the mixture tends to 1, and the basin of attraction of the proportion of cooperators p in the replicator dynamics increases, tending to $p \in (0,1]$. Thus in extreme cases, the eventual outcome of the game can be essentially the opposite to that implied by the game with uniform interaction rates.

7.3.2 Other games with nonlinear interaction rates

In matrix games such as the Hawk-Dove game there are rewards and costs, with costs given as a negative reward. Often real costs can be a loss of time, which does not directly reduce an individual's fitness, but does so indirectly by removing the opportunity of taking other actions (see Section 7.4). The size of this cost will thus often depend upon the opportunities missed, and the strategies of both our focal individual and others. The classical game of the cost as lost time is the war of attrition, where this is generally expressed just as a penalty proportional to the length of a contest. This is not completely realistic for a breeding season of finite duration, where lost time may depend upon the remaining season duration, and this was explored by Cannings and Whittaker (1995).

Such non-uniform interaction rates occur naturally in settings where individuals move between different states, or compartments, which have different options associated with them. Strategies affect which compartment an individual will be in at any one time, and thus will also affect the rate of interactions between individuals. This means that polymorphic-monomorphic equivalence does not hold. We discuss examples of these games in more depth in Chapters 19 and 21.

7.4 Nonlinearity due to games with time constraints

In almost all of the games that we have considered up until now, the times of interactions have not been considered. Interactions have been instantaneous, and irrespective of strategy each individual will play the same number of games (whether one or many, depending upon the scenario). This is the assumption for matrix games, for example. The one exception is the war of attrition, where time is central, but (as mentioned above) here time directly affects payoffs, but does not (perhaps rather unrealistically) affect the frequency of any further interactions. This leads to linearity of payoffs in some classes of game such as matrix games and the war of attrition.

Imagine, however a population involved in simple pairwise interactions, but where as well as leading to some reward these also take some time, which depends upon the strategies selected by the players. Further individuals can be repaired after games finish, and can be engaged only in at most one pairwise interaction at a time, so long contests explicitly reduce further opportunities within the game. This is the subject of research by Křivan and Cressman (2017), where repairing was instantaneous and so all individuals had exactly one interaction at any time, Garay et al. (2017), where there was a non-zero searching time and so individuals had at most one interaction at any time, and more recently, by Garay et al. (2018), Cressman and Křivan (2019), Bodnar et al. (2020), Varga et al. (2020b).

7.4.1　The model

We shall, in fact, consider two closely-related models mentioned above, Křivan and Cressman (2017) and Garay et al. (2017).

We shall first consider the most straightforward way of introducing time constraints, as in Křivan and Cressman (2017). A large (effectively infinite) population of size N consists of individuals who play pairwise games selecting one of two available pure strategies (although the methodology naturally extends to any number of pure strategies). The available payoffs are as in the standard payoff matrix:

$$A = \begin{pmatrix} a_{11} & a_{12} \\ a_{21} & a_{22} \end{pmatrix}. \tag{7.46}$$

As well as providing payoffs, each interaction causes the participants to lose some time, depending upon their payoffs. This is denoted by the time constraint matrix:

$$T = \begin{pmatrix} \tau_{11} & \tau_{12} \\ \tau_{21} & \tau_{22} \end{pmatrix}. \tag{7.47}$$

We note that the time spent is not necessarily the same for both participants of a contest; for example the recovery time can be different, depending upon the individual strategies played. Thus τ_{12} and τ_{21} are not necessarily the same (though in Křivan and Cressman (2017) they are).

As fights take time, there will at any point in time be some members of the population occupied in fights. In Křivan and Cressman (2017) it is assumed that when individuals are free, they instantly re-pair, so consequently all individuals are competing at essentially all times. Contests are assumed to finish following a Poisson process with constant rate, which must be one divided by the mean contest time. Thus those involved in shorter contests get to repair more quickly, and there is thus a continuous time Markov process following interacting pairs, with no free single individuals. It was shown that the number of pairs involving an i versus j contest, n_{ij}, satisfied the following differential equations:

$$\frac{dn_{11}}{dt} = -\frac{n_{11}}{\tau_{11}} + \frac{\left(\frac{2n_{11}}{\tau_{11}} + \frac{n_{12}}{\tau_{12}}\right)^2}{4\left(\frac{n_{11}}{\tau_{11}} + \frac{n_{12}}{\tau_{12}} + \frac{n_{22}}{\tau_{22}}\right)}, \tag{7.48}$$

$$\frac{dn_{12}}{dt} = -\frac{n_{12}}{\tau_{12}} + \frac{\left(\frac{2n_{11}}{\tau_{11}} + \frac{n_{12}}{\tau_{12}}\right)\left(\frac{2n_{22}}{\tau_{22}} + \frac{n_{12}}{\tau_{12}}\right)}{4\left(\frac{n_{11}}{\tau_{11}} + \frac{n_{12}}{\tau_{12}} + \frac{n_{22}}{\tau_{22}}\right)}, \tag{7.49}$$

$$\frac{dn_{22}}{dt} = -\frac{n_{22}}{\tau_{22}} + \frac{\left(\frac{2n_{22}}{\tau_{22}} + \frac{n_{12}}{\tau_{12}}\right)^2}{4\left(\frac{n_{11}}{\tau_{11}} + \frac{n_{12}}{\tau_{12}} + \frac{n_{22}}{\tau_{22}}\right)}. \tag{7.50}$$

Setting the right hand side of these equations equal to 0 (see Exercise 7.48), we note that we have the Hardy-Weinberg equilibrium

$$\left(\frac{n_{12}}{\tau_{12}}\right)^2 = 4\frac{n_{11}}{\tau_{11}}\frac{n_{22}}{\tau_{22}}. \tag{7.51}$$

The sizes of the two sub-populations of strategies then must satisfy

$$n_1 = 2n_{11} + n_{12}, \tag{7.52}$$

$$n_2 = 2n_{22} + n_{12}, \tag{7.53}$$

since every individual is in precisely one fighting pair at any time. Noting that $n_{11} + n_{12} + n_{22} = 2N$ and $n_2 = N - n_1$, we can thus find the number of each type of pair in terms of n_1 and n_2, and we obtain n_{ij} in terms of the sizes of the two populations as follows. If $D = \tau_{12}^2 - \tau_{11}\tau_{22} = 0$, we get

$$n_{11} = \frac{n_1^2}{2N}, \tag{7.54}$$

$$n_{12} = \frac{n_1 \cdot (N - n_1)}{N}, \tag{7.55}$$

$$n_{22} = \frac{(N - n_1)^2}{2N}, \tag{7.56}$$

and otherwise we get

$$n_{11} = \frac{n_1 D - \tau_{12}^2 \frac{N}{2} + \tau_{12}\sqrt{\frac{N^2}{4}\tau_{12}^2 - n_1(N - n_1)D}}{2D}, \tag{7.57}$$

$$n_{12} = \frac{\tau_{12}^2 \frac{N}{2} - \tau_{12}\sqrt{\frac{N^2}{4}\tau_{12}^2 - n_1(N - n_1)D}}{D}, \tag{7.58}$$

$$n_{22} = \frac{N}{2} - n_{11} - n_{12}. \tag{7.59}$$

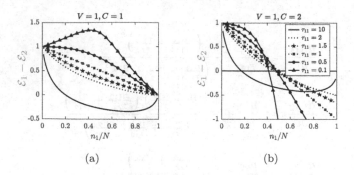

(a)   (b)

FIGURE 7.2: A Hawk-Dove game with time constraints as in Křivan and Cressman (2017) where the duration of the fight between two Hawks is τ_{11} and any interaction involving a Dove takes $\tau = 1$. Note that the Hawk-Dove game here was formulated slightly differently, leading to the usual payoff matrix as described in Section 4.1.2 being multiplied by 2.

The payoffs are then

$$\mathcal{E}_1 = \frac{2n_{11}}{2n_{11} + n_{12}} \frac{a_{11}}{\tau_{11}} + \frac{n_{12}}{2n_{11} + n_{12}} \frac{a_{12}}{\tau_{12}}, \tag{7.60}$$

$$\mathcal{E}_2 = \frac{2n_{22}}{2n_{22} + n_{12}} \frac{a_{22}}{\tau_{22}} + \frac{n_{12}}{2n_{22} + n_{12}} \frac{a_{12}}{\tau_{12}}. \tag{7.61}$$

We thus have payoffs as a function of the strategy composition, and as previously discussed in this chapter, for a game with two pure strategies the ESSs of such a game can easily be found from this information by considering the incentive function $\mathcal{E}_1 - \mathcal{E}_2$ (see Theorem 7.3 and example 7.4, and also Section 9.1.3).

The work of Garay et al. (2017) started from the same basic idea, that there is a population which interacts in pairwise games governed by the payoff matrix (7.46) and where contests take times following the time constraint matrix (7.47). They considered mixed strategy players rather than pure strategy ones, and there were also other differences; in particular, there was a non-zero search time τ_S to find an opponent, and so individuals transitioned between searching and fighting states. Their payoffs are written as the expected reward per cycle of behaviour divided by the expected time spent per cycle. Here a cycle runs from the start of a search for an opponent, to the conclusion of the interaction after that opponent is found. We note that the reward obtained and the time taken in an interaction will be different for different opponents and strategies selected. Indeed when a potential opponent is found it may already be involved in a contest, and then both reward and interaction time are considered to be 0. The payoffs to a **p**-strategist and to a **p***-strategist in a mixed population of $\mathbf{p}$ and $\mathbf{p}^*$-strategists are then, respectively,

$$\mathcal{E}_p = \frac{r\mathbf{p}A\mathbf{p^T} + r^*\mathbf{p}A\mathbf{p^{*T}}}{\tau_S + r\mathbf{p}T\mathbf{p^T} + r^*\mathbf{p}T\mathbf{p^{*T}}}, \tag{7.62}$$

$$\mathcal{E}_{p^*} = \frac{r\mathbf{p^*}A\mathbf{p^T} + r^*\mathbf{p^*}A\mathbf{p^{*T}}}{\tau_S + r\mathbf{p^*}T\mathbf{p^T} + r^*\mathbf{p^*}T\mathbf{p^{*T}}}, \tag{7.63}$$

where r (r^*) are the probabilities that a focal individual meets an active $\mathbf{p}$ or $\mathbf{p}^*$ -strategist, respectively. Individuals are only active if they are not involved in a contest already, and whether an individual is likely to be active depends upon its strategy. Thus r and r^*, whilst depending upon the actual fractions of each type in the population, are not proportional to them.

At first sight the form of the payoffs functions (7.60) and (7.62) seem different, the first being an average over the payoffs per unit time of the different types of interaction, and the second the ratio of the total payoff to total time. The casual belief of the equivalence of the two concepts is even a famous fallacy as debated in Templeton and Lawlor (1981); Stephens and Krebs (1986). However, when properly interpreted, this is not the case. The fact that these two approaches to defining the payoff function are identical was proved in Broom et al. (2019a). We can see that the expression (7.60) can be rearranged as

$$\mathcal{E}_1^* = \frac{T\frac{2n_{11}}{(2n_{11}+n_{12})\tau_{11}}a_{11} + T\frac{n_{12}}{(2n_{11}+n_{12})\tau_{12}}a_{12}}{T}. \tag{7.64}$$

Here the numerator is the expected number of interactions with individuals of the first type multiplied by the payoff for each interaction in time T added to the similar product for interactions with individuals of the second type, and so giving the expected payoff in time T, and the denominator is the time T. We see that this is of the same form as in (7.62) when noticing that the payoff associated with the searching event is 0. The key thing to observe is that the distributions are different in the two methods; the former gives the probability distribution of the activity being carried out at a random time point, the latter the probability distribution in terms of the count of events. These two are different if the times for events are different (mistaking the former for the latter is known as length-biased sampling, see Qin (2017)).

We see that in either the case of Křivan and Cressman (2017) or of Garay et al. (2017) (which we have of course said are equivalent for models with the same assumptions) this leads to non-linear payoff functions, very different to those in matrix games. In fact, the simplest time constraint game has a very close connection to the games with non-constant interaction rates from Section 7.3.1, based upon the work of Taylor and Nowak (2006). The model in Section 7.3.1 considered two strategies where individuals encountered others at different rates, leading to three possible pairs formed at the three different rates r_{AA}, r_{AB} and r_{BB}. This is actually the practical consequence of the Křivan and Cressman (2017) time constraints. There is a 1-1 mapping between the sets of time constraints here and the set of interaction rates from Section 7.3.1. Note that the resulting payoffs are not equivalent, however. In Section 7.3.1 the payoffs are simply an average of the payoffs over all of the games played, whereas here (Křivan and Cressman, 2017) they explicitly depend upon the interaction times beyond just through their effect on the pairing probabilities.

We also note that there is earlier work that involves time constraints explicitly in a similar way, modelling kleptoparasitic behaviour, with similar non-linear results. This is discussed in Section 19.5. What distinguishes the current work is the fact that it focuses on a general theory closely related to classical matrix games, which has such games as a special case.

7.5 Nonlinearity in the strategy of the focal player

In this section we consider the third case, involving games where the strategy of an individual is described by a single number (or a vector) but where this does not describe the probability of playing a given pure strategy, but effectively describes a unique behaviour such as the intensity of a signal. We note that this is also the scenario generally considered in adaptive dynamics, see Chapter 14, and the most general treatment of adaptive dynamics can be thought to include the games that we consider in this section (though in practice stronger assumptions are generally made than we use here).

For such a strategy set, some regularity properties of the payoffs as a function of the strategy are generally assumed. It is often assumed that the function is continuous and even differentiable everywhere, or almost everywhere. In practice, almost everywhere will usually mean everywhere but a small number of critical points. There may be good biological reasons to suppose other properties, e.g. that the function is convex, as in the case of the handicap principle in Chapter 17. In fact, many of the above assumptions are implicit, and it is often the case that specific functional forms are used, and even that investigations are conducted entirely numerically.

An example is the war of attrition, where the payoff function is piecewise linear (see Section 4.3). In general, such payoff functions will not be linear. Often they will yield a unique local maximum given the strategies of the opponent as in the example concerning tree height in Section 7.5.2 below. It should be noted that this does not occur for matrix games or indeed for games with payoffs nonlinear in the population strategy but linear in the strategy of the focal player, where the linear nature of the function means that either maxima occur on the boundaries or that payoffs are equivalent for many strategies. Thus in the Hawk-Dove game with $C > V$, the payoff to mixed strategy q is constant (equal to $(C - V)V/2C$) when the population plays the ESS proportion $p = V/C$, so that all values of q perform equally well. Nonlinearity on the left, whilst generally complicating the analysis, can thus also be beneficial in one respect, as it often means that the ESS condition reduces to $\mathcal{E}[\mathbf{p}; \delta_{\mathbf{p}}] > \mathcal{E}[\mathbf{q}; \delta_{\mathbf{p}}]$.

It should be noted that two-strategy matrix games can be viewed as a special case of games described above. In this case, the set of possible playable (mixed) strategies is described by p, the probability of playing strategy 1. However, the payoffs are still linear in the focal player strategy, which is not generally the case for the type of game we consider here.

7.5.1 A sperm allocation game

Ball and Parker (2007) discuss a game of sperm competition (see Chapter 17). A male has to allocate sperm at different matings, and the probability of fathering offspring depends upon the amount of sperm contributed, as well as the amount contributed by other males in previous or subsequent matings with females. There is not a fixed amount of sperm available, but there are energetic costs to producing sperm, and there is a maximum amount of energy usage allowed. In the model the male can be in one of the following three situations

(1) he is the only mating partner of a female,

(2) he is the first of two mating partners,

(3) he is the second of two mating partners.

When mating, the male has to decide how much sperm he should contribute; a (pure) strategy is thus given by a triple (s_0, s_1, s_2) where s_0, s_1 and s_2 are non-negative real numbers, corresponding to the above cases (1), (2) and (3), respectively. When two males mate with a female the probability of a male being the one to fertilise the female is proportional to how much sperm he contributes.

Let us assume that the values s_0 and s_2 are fixed at s_0^* and s_2^*, respectively, so we only consider variation in s_1. Ball and Parker (2007) show that the payoff function to a male playing (s_0^*, s_1, s_2^*) in a population playing (s_0^*, s_1^*, s_2^*) is

$$\frac{k}{1+q} \left(\frac{s_0^*(1-q)}{s_0^* + \varepsilon} + \frac{qs_1}{s_1 + rs_2^* + \varepsilon} + \frac{qrs_2^*}{s_1^* + rs_2^* + \varepsilon} \right), \qquad (7.65)$$

where q is the proportion of females that mate twice, r is the discount factor for being the second male to mate and k is the number of matings per male. $\varepsilon > 0$ indicates that there is some probability that there will be no fertilisation from a particular mating.

We can see that the payoff function is nonlinear in the strategy of the focal player (in s_1) and in the population strategy (in s_1^*). We investigate this and related models in greater detail in Chapter 17.

7.5.2 A tree height competition game

Kokko (2007, Section 6.1) considered the following game-theoretical model of tree growth. We assume that a tree has to grow large enough in order to get sunlight and not get overshadowed by neighbours; yet the more the tree grows the more of its energy has to be devoted to "standing" rather than photosynthesis. Let $h \in [0,1]$ be the normalised height of the tree; here 1 means the maximum possible height of a tree (Koch et al., 2004). We define the fitness of a tree of height h in the forest where all other trees are of height H by

$$\mathcal{E}[h; \delta_H] = (1 - h^3) \cdot (1 + \exp(H - h))^{-1}, \qquad (7.66)$$

where $f(h) = 1 - h^3$ represents the proportion of leaf tissue of a tree of height h and $g(h - H) = (1 + \exp(H - h))^{-1}$ represents the advantage or disadvantage of being bigger/smaller than one's neighbour; see Kokko (2007, Section 6.1) for some justification of this function and Exercise 7.14 for a generalisation.

As in the example in Section 7.5.1, this game is nonlinear in the focal player strategy.

Now we will determine the ESSs for the tree, i.e. what are the evolutionarily stable heights. First, in accordance with Theorem 7.3 we will determine the best response h to a general tree height H. We get

$$\frac{\partial}{\partial h} \mathcal{E}[h; \delta_H] = \frac{-3h^2(1 + \exp(H - h)) + (1 - h^3)\exp(H - h)}{(1 + \exp(H - h))^2}. \qquad (7.67)$$

Thus the derivative is positive for $h = 0$ and negative for $h = 1$ and it is decreasing in h. So, there is only one root of the derivative and the function $h \mapsto \mathcal{E}[h; \delta_H]$ attains a maximum at that root, i.e. the root is the best response.

Since the ESS must be a best response to itself, we get that the ESS must thus solve an equation

$$0 = \frac{-3H^2(1 + \exp(H - H)) + (1 - H^3)\exp(H - H)}{(1 + \exp(H - H))^2} \qquad (7.68)$$

$$= \frac{1}{4}(-6H^2 + (1 - H^3)). \qquad (7.69)$$

The above equation has only one root and the crossing of the x axis happens with negative derivative, so that the root is the unique ESS; see Figure 7.3.

7.6   Linear versus nonlinear theory

When games are both linear on the left and the right we have matrix games and strong results hold. For example not only can there not be two

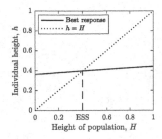

FIGURE 7.3: The best response and ESS for a tree height competition game with payoff function given by (7.66).

ESSs with the same support, but the stronger Bishop-Cannings Theorem 6.9 holds. All ESSs are also uniformly uninvadable. For games which are linear on the left but not on the right, uniform uninvadability still holds provided that payoffs are differentiable, however there can be more than one ESS (in fact, an arbitrarily large number of ESSs) with the same support. The important condition of polymorphic-monomorphic equivalence also holds when games are linear on the right and on the left, but not necessarily otherwise. If this condition does not hold we must be careful about the precise make up of the population (are the players pure strategists or mixed strategists?). As long as a game is linear on the left, then all strategies with support that is a subset of the support of an ESS perform equally well, and this gives a natural set of simultaneous equations to find ESS candidates. Games that are not linear on the left in contrast usually have a unique maximum best reply to any strategy in the population, leading to the simplified ESS condition $\mathcal{E}[\mathbf{p}; \delta_{\mathbf{p}}] > \mathcal{E}[\mathbf{q}; \delta_{\mathbf{p}}]$ as described in the previous section.

7.7 Python code

In this section we show how to use Python to identify an ESS.

```
1    """ ESSs in nonlinear games
2    Computes the best response  for a 2 player 2 strategy game.
3    Identifies candidates for ESS as strategies that are best
4    responses to itself. Naively performs local superiority test.
5
6    It prints out the strategies that are best responses to itself,
7    and strategies that are locally superior.
8    It plots best response strategies to any population strategy
9
```

```
10   One can tweak the payoff function, discretization of the strategy
11   space and tolerance level
12   """
13
14   ## Define parameters
15   # Tolerance level. If the payoff is within the tolerance level
16   # from the maximal value, it is considered maximal.
17   tol = 0.0001     # One may need to test several values of tol
18                    # too low or too high is not good
19
20   num = 101;   # Number of points for strategy space
21                    # If a suspected ESS is around 1/3,
22                    # use num = (large multiple of 3) + 1
23
24   def payoff(s1,s2): # Payoff functions
25       """ Returns payoff to a player using s1 if everybody else
26            uses s2 """
27       return s1*s2+(1-s1)*(1-s2)
28
29   ## Import basic packages
30   import numpy as np
31   import matplotlib.pyplot as plt
32
33   ## Basic set up
34   # Strategy space
35   strategy_Space = np.linspace(0, 1, num)
36
37   # Make a vertical array for the individual's strategy
38   ind_Strategy = strategy_Space[:, np.newaxis]
39
40   # Make a horizontal array for strategy in the population
41   pop_Strategy = strategy_Space
42
43   ## Actual calculations
44   # Payoff[a,b] = payoff to individual using ind_Strategy[a]
45   # against pop_Strategy[b]
46   Payoff = payoff(ind_Strategy, pop_Strategy)
47
48   # Identify maximal payoff to p1 given pop_Strategy
49   P_max = np.amax(Payoff, axis = 0);
50
51   # Identify best responses
52   # Best Response[a, b] is true if Payoff[a, b]>P_max[b] - tol,
53   # When true, the individual should use ind_Strategy[a]
54   # in the population playing pop_Strategy[b]
55   BestResponse = Payoff > P_max - tol
56
57   ## Find best responses to itself
58   print('The following strategies are best responses to itself')
59   for n in range(num):
60       if BestResponse[n, n]:
61           print(strategy_Space[n])
62
63   ## Test for local superiority
64   # p is locally superior if
65   # Payoff(p,q)>Payoff(q,q) for all q close to p
66   # We will test only the closest neighbours
```

```
67  print('The following strategies may be locally superior')
68  for n in range(num):
69      if  ((n == 0) and (Payoff[0, 1] > Payoff[1, 1])) or \
70          ((n > 0) and (n < num - 1)
71                   and (Payoff[n, n-1] > Payoff[n-1, n-1]) \
72                   and (Payoff[n, n+1] > Payoff[n+1, n+1])) or \
73          ((n == num - 1) and  (Payoff[n, n-1]>Payoff[n-1, n-1])):
74              print(strategy_space[n])
75
76  ## Plot best responses
77
78  # Get the min and max values for the axis
79  xMin = min(strategy_space)
80  xMax = max(strategy_space)
81
82  # Plot the boolean array
83  # Use negation so that true is plotted as black
84  plt.imshow(np.logical_not(BestResponse), aspect='auto',
85          cmap=plt.cm.gray, origin='lower',
86          extent = [xMin, xMax, xMin, xMax])
87
88  plt.xlabel('Population strategy')
89  plt.ylabel('Best response')
90  plt.show()
```

7.8 Further reading

For general games which are linear in the focal player, which incorporates playing the field games, Bomze and Pötscher (1989) give extensive analysis of different general conditions on $h_{\mathbf{p},\mathbf{q},u}$. They also study the concept of the uniform invasion barrier in great detail. For more on such games see Pohley and Thomas (1983), Crawford (1990a,b). The games that we shall meet in Section 9.1 also fall within this class. Other games of this type are the Ideal Free Distribution games of Section 19.2; see Fretwell and Lucas (1969), Cressman et al. (2004) and Cressman and Křivan (2006). For work on population games, see Sandholm (2015) and Hofbauer and Sandholm (2009).

For pairwise games with non-uniform interaction rates, as well as Taylor and Nowak (2006) which we have already looked at, Cannings and Whittaker (1995) consider a war of attrition model where individuals continuously play, so those who play shorter times get to play in more contests. See also Pacheco et al. (2006a), Dong et al. (2017), Peña et al. (2011) and Tang et al. (2012) for other games which are pairwise but not in well-mixed populations.

The models of kleptoparasitism explored in Chapter 19 involve such pairwise contests and are closely aligned to the time constraints work described in Section 7.4; see for example Broom et al. (2008a). For further work on time constraints see Garay et al. (2020, 2017, 2018); Varga et al. (2020a); for time constraint games on a spatial structure, see Broom and Křivan (2020). For related models see Wang et al. (2017b) and Alboszta and Miękisz (2004).

Games which are not linear on the left include most interesting applications of adaptive dynamics; see Chapter 14, Metz et al. (1996), and Geritz et al. (1998). A more complex model of the "tree height competition game" appears in Givnish (1982); see also Falster and Westoby (2003). The sperm competition game is looked at in more detail in Chapter 17; see also Ball and Parker (2000).

For non-linear public goods games, see for example Archetti (2009b,a); Archetti and Scheuring (2012); Archetti (2018)

In fact, most games of real situations are nonlinear, so there are examples of such games throughout the remaining chapters of this book.

7.9　Exercises

Exercise 7.1. Prove the "if" implication of Theorem 7.3.

Exercise 7.2. Consider the game from Example 7.4. Find the ESSs of this game when (a) $a_1 = 9, a_2 = -9$ and $a_3 = 2$, (b) $a_1 = 4, a_2 = -6$ and $a_3 = 3$.

Exercise 7.3. Verify that the results from Theorems 7.5, 7.7 and 7.8 hold for Example 7.4.

Exercise 7.4. Explain why if the incentive function $h_{\mathbf{p},\mathbf{q},u}$ defined by (7.5) satisfies $h_{\mathbf{p},\mathbf{q},0} = 0$ and $\frac{\partial}{\partial u} h_{\mathbf{p},\mathbf{q},u}\big|_{u=0} = 0$, then the payoffs are not generic.

Exercise 7.5. Give an example of (non-generic) payoffs in a game and an ESS strategy $\mathbf{p}$ such that for $h_{\mathbf{p},\mathbf{q},u}$ defined by (7.5) we have $h_{\mathbf{p},\mathbf{q},0} = 0$ and $\frac{\partial}{\partial u} h_{\mathbf{p},\mathbf{q},u}\big|_{u=0} = 0$.

Hint. For $\mathbf{p}$ to be an ESS, one needs $h_{\mathbf{p},\mathbf{q},u} > 0$ for all $\mathbf{q} \neq \mathbf{p}$ and all u small enough. Even with the requirement on h, this can be achieved if $\frac{\partial^2}{\partial u^2} h_{\mathbf{p},\mathbf{q},u}\big|_{u=0} = 0$ and $\frac{\partial^3}{\partial u^3} h_{\mathbf{p},\mathbf{q},u}\big|_{u=0} > 0$.

Exercise 7.6. Consider a matrix game with a payoff matrix A and let $h_{\mathbf{p},\mathbf{q},u}$ be defined as in (7.5). Evaluate $\frac{\partial}{\partial u} h_{\mathbf{p},\mathbf{q},u}\big|_{u=0}$ and use it to compare Theorem 6.2 and Theorem 7.3.

Exercise 7.7. Show that the conclusions of Theorem 7.5 hold even if we assume only that the incentive function $h_{\mathbf{p},\mathbf{q},u}$ is continuous at $u = 0$.

Hint. See the proof of Theorem 7.5 and check what can be proved without assuming that h is differentiable.

Exercise 7.8. Show that in the patch foraging game of Section 7.2.2.1, the strategy $\mathbf{p}$ given by (7.32) is the only candidate for an ESS in this game.

Hint. Use Theorem 7.5.

Exercise 7.9. Consider a two-strategy game with

$$\mathcal{E}[\mathbf{x}; \mathbf{y}^{\mathbf{T}}] = x_1(y_1 + 2y_2)^\alpha + x_2(2y_1 + y_2)^\alpha. \tag{7.70}$$

(i) Show that if $\alpha = 1$, this is a matrix game with payoff matrix $\left(\begin{smallmatrix} 1 & 2 \\ 2 & 1 \end{smallmatrix}\right)$ which has a mixed ESS $\mathbf{p}_{\mathrm{ESS}} = (0.5, 0.5)$.

(ii) For a given population $\mathbf{y}^{\mathbf{T}}$, identify the best response strategy $\mathbf{x}$.

(iii) Find the only ESS for $\alpha > 0$.

(iv) Show that there are two pure ESSs for $\alpha < 0$.

Exercise 7.10 (Taylor and Nowak, 2006). Show that a strategy $\mathbf{p}$ is an ESS of the game described in Section 7.3.1 if and only if it is a stable rest point of the replicator dynamics.

Hint. The replicator dynamics is $\frac{d}{dt}p = p(1 - p)\left(\mathcal{E}[S_1; \mathbf{p}^{\mathbf{T}}] - \mathcal{E}[S_2; \mathbf{p}^{\mathbf{T}}]\right)$.

Exercise 7.11 (Taylor and Nowak, 2006). Identify the ESS(s) of the game described in Section 7.3.1 if $b = c = 1, d = 0$ and a is a parameter between $[0, 1]$. Assume $r_{12} = 1$ and keep r_{11} and r_{22} general.

Hint. One needs $\mathcal{E}[S_1; \mathbf{p}^{\mathbf{T}}] = \mathcal{E}[S_2; \mathbf{p}^{\mathbf{T}}]$ for a mixed ESS.

Exercise 7.12. Specify a game payoff function $\mathcal{E}[\mathbf{p}; \Pi]$ so that $h_{\mathbf{p},\mathbf{q},u}$ is not continuous at 0 and neither assumptions nor conclusions of Theorem 7.3 hold.

Exercise 7.13. Consider a game with two strategies where the payoffs are given by $\mathcal{E}[\mathbf{x}; \mathbf{y}^{\mathbf{T}}] = x_1^\alpha(y_1 + 2y_2) + x_2^\alpha(2y_1 + y_2)$. Identify the ESSs in this game for various values of the parameter $\alpha > 0$.

Hint. If $\alpha = 1$, then we have a linear matrix game as in Exercise 7.9. If $\alpha > 1$, then the payoffs are strictly convex in the focal player strategy and thus ESSs must be pure by the discussion after (7.14).

Exercise 7.14 (Tree competition game in general, Kokko, 2007). With the notation as in Section 7.5.2, consider that the fitness of the tree of height h in the forest where other trees have height H is given by $\mathcal{E}[h; \delta_H] = f(h)g(h - H)$ for some general functions f and g. Investigate the ESSs of the game.

Exercise 7.15 (Křivan and Cressman, 2017). Using the expressions (7.48)-(7.50), show that the Hardy-Weinberg formula for the numbers of interacting pairs (7.51) holds for the time constraint model from Section 7.4.

Chapter 8

Asymmetric games

Up until now, when we have considered pairwise contests, we have assumed symmetric games where players are indistinguishable in every way, except perhaps their strategy. In this chapter we investigate games where this is no longer the case. We will classify such games into different types, and look at examples including from the classic paper by Maynard Smith and Parker (1976). They identified three types of asymmetry in relation to a contest.

(i) Asymmetries in payoff, where the rewards to one of the players in a contest may be different to the other. A territory owner may have invested significant time in exploring a territory and is able to make better use of it than the intruder, so that it will be more valuable to the owner. Note it is not enough that the owner has expended resources in the territory if there is not some future advantage (the Concorde fallacy, Dawkins and Brockmann (1980); Weatherhead (1979)).

(ii) Asymmetry in fighting ability, alternatively called *Resource Holding Potential*, which will affect the chances of victory in a contest (we look at this idea, in particular, in Chapter 16). For example, one individual may be larger than another.

(iii) Asymmetries which affect neither the payoffs nor success probabilities, so called *uncorrelated asymmetries*. At first sight such asymmetries sound irrelevant, but as we shall see, they can be highly important.

In mathematical terms, the three types are not different and can be modelled by bimatrix games given by a pair of matrices (A, B), A representing the payoffs to a player in role 1, B representing the payoff to a player in role 2. See Section 8.2 for more details.

Case (iii) of uncorrelated asymmetries is especially interesting biologically, as it involves a bimatrix game where $B = A$, and thus there is no real difference except perceived role. If animals can distinguish their roles during the conflict (e.g. dominance relating to past performance), it leads to this bimatrix game, and if they cannot, it leads to the standard matrix game situation.

Cases (i) and (ii) are both examples of situations where the payoffs are distinct, and both can be expressed in terms of bimatrix games where $B \neq A$. We thus do not distinguish the two here, but instead discuss *correlated asymmetries* in general.

DOI: 10.1201/9781003024682-8

151

8.1   Selten's theorem for games with two roles

In this section we will consider games where individuals will always be in role 1 or in role 2. An individual in role 1 can play the game only against an individual in role 2. Selten (1980) called such games, where individuals could never face opponents in the same role, *truly asymmetric games*. When the individual is in role r, it has a strategy set $\mathbf{S}_r = \{S_{r1}, S_{r2}, \ldots, S_{rn_r}\}$ available to it. We assume that in role r, an individual can play mixed strategies $\mathbf{p}_r = (p_{r,i})_{i=1}^{n_r}$ where $p_{r,i}$ is the probability of playing strategy S_{ri} given an individual is already in role r. The strategy in the game is thus determined by a pair $(\mathbf{p}_1, \mathbf{p}_2)$.

A population is, similarly to Section 2.2.1, represented by a measure on the strategy space. We can view a given population Π as a pair (Π_1, Π_2) where Π_r represents the population structure of individuals in role r.

Definition 8.1. *A strategy $(\mathbf{p}_1, \mathbf{p}_2)$ is an ESS of the game if for every other strategy $(\mathbf{q}_1, \mathbf{q}_2)$ there exists $\varepsilon_0 > 0$ such that for all $\varepsilon \in (0, \varepsilon_0)$*

$$\mathcal{E}[(\mathbf{p}_1, \mathbf{p}_2); \Pi] > \mathcal{E}[(\mathbf{q}_1, \mathbf{q}_2); \Pi], \tag{8.1}$$

where $\Pi = (1 - \varepsilon)\delta_{(\mathbf{p}_1, \mathbf{p}_2)} + \varepsilon\delta_{(\mathbf{q}_1, \mathbf{q}_2)}$.

We define $\mathcal{E}[(\mathbf{p}_1, \mathbf{p}_2); \Pi]$, the payoff to a strategy $(\mathbf{p}_1, \mathbf{p}_2)$ in a population $\Pi = (\Pi_1, \Pi_2)$, by

$$\mathcal{E}[(\mathbf{p}_1, \mathbf{p}_2); \Pi] = \rho_1(\mathbf{p}_1, \mathbf{p}_2)\mathcal{E}_1[\mathbf{p}_1; \Pi_2] + \rho_2(\mathbf{p}_1, \mathbf{p}_2)\mathcal{E}_2[\mathbf{p}_2; \Pi_1], \tag{8.2}$$

where, for example, $\mathcal{E}_1[\mathbf{p}_1; \Pi_2]$ is the payoff to the focal individual in role 1 when using strategy $\mathbf{p}_1$ when it is effectively playing in a population of "role 2 players" described by Π_2, which happens with probability $\rho_1(\mathbf{p}_1, \mathbf{p}_2)$. Our player must be in one of the two roles, and so $\rho_2(\mathbf{p}_1, \mathbf{p}_2) = 1 - \rho_1(\mathbf{p}_1, \mathbf{p}_2)$.

Definition 8.2. *We will say that a game is* strategy-role independent *if the probability of an individual occupying a particular role does not depend upon their chosen strategy, i.e. if the functions $\rho_1(\mathbf{p}_1, \mathbf{p}_2)$ and $\rho_2(\mathbf{p}_1, \mathbf{p}_2)$ from (8.2) are in fact constants ρ_1 and $\rho_2 = 1 - \rho_1$ not depending on an individual's strategy $(\mathbf{p}_1, \mathbf{p}_2)$.*

This means that the payoff to a strategy $(\mathbf{p}_1, \mathbf{p}_2)$ in a population $\Pi = (\Pi_1, \Pi_2)$, becomes

$$\mathcal{E}[(\mathbf{p}_1, \mathbf{p}_2); \Pi] = \rho_1\mathcal{E}_1[\mathbf{p}_1; \Pi_2] + (1 - \rho_1)\mathcal{E}_2[\mathbf{p}_2; \Pi_1]. \tag{8.3}$$

We note that often ρ will be equal to $1/2$, e.g. in bimatrix games; see Exercise 8.1.

In Chapter 7 we defined the term polymorphic-monomorphic equivalence in symmetric games (7.4). We define an extension of this concept as follows.

Definition 8.3. *An asymmetric game is* polymorphic-monomorphic equivalent *if*

$$\mathcal{E}\left[(\mathbf{p}_1, \mathbf{p}_2); \sum_i \alpha_i \delta_{(\mathbf{q}_{i1}, \mathbf{q}_{i2})}\right] = \mathcal{E}\left[(\mathbf{p}_1, \mathbf{p}_2); \delta_{(\sum_i \alpha_i \mathbf{q}_{i1}, \sum_i \alpha_i \mathbf{q}_{i2})}\right]. \qquad (8.4)$$

For example, if the game is polymorphic-monomorphic equivalent and the population is given by $\Pi = (1 - \varepsilon)\delta_{(\mathbf{p}_1, \mathbf{p}_2)} + \varepsilon\delta_{(\mathbf{q}_1, \mathbf{q}_2)}$, then an individual in role 1 will play against a population described by $\Pi_2 = (1 - \varepsilon)\delta_{\mathbf{p}_2} + \varepsilon\delta_{\mathbf{q}_2}$.

In the rest of this section we will assume that the game is both polymorphic-monomorphic equivalent and strategy-role independent.

Theorem 8.4 (Selten, 1980). *Assume that the payoff function $\mathcal{E}_1$ is linear in the focal player strategy, i.e. that there are functions $(f_{1i})_{i=1}^{n_1}$ such that*

$$\mathcal{E}_1[\mathbf{p}_1; \Pi_2] = \sum_{i=1}^{n_1} p_{1i} f_{1i}(\Pi_2). \qquad (8.5)$$

Then $(\mathbf{p}_1, \mathbf{p}_2)$ is an ESS only if $\mathbf{p}_1$ is a pure strategy. Consequently, if both $\mathcal{E}_1$ and $\mathcal{E}_2$ are linear in the focal player strategy, then $(\mathbf{p}_1, \mathbf{p}_2)$ can be an ESS only if $\mathbf{p}_1$ and $\mathbf{p}_2$ are pure strategies.

Proof. Let $(\mathbf{p}_1, \mathbf{p}_2)$ be given and let i_0 be such that

$$f_{1i_0}(\delta_{\mathbf{p}_2}) = \max_i\{f_{1i}(\delta_{\mathbf{p}_2})\}. \qquad (8.6)$$

Now, assume that $\mathbf{p}_1$ is not a pure strategy, i.e. $\mathbf{p}_1 \neq S_{i_0}$. Since $(\mathbf{p}_1, \mathbf{p}_2)$ is an ESS, it has to resist invasion by $(S_{i_0}, \mathbf{p}_2)$ which by (8.1) means that

$$\mathcal{E}[(\mathbf{p}_1, \mathbf{p}_2); \Pi] > \mathcal{E}[(S_{i_0}, \mathbf{p}_2); \Pi], \qquad (8.7)$$

where $\Pi = (1 - \varepsilon)\delta_{(\mathbf{p}_1, \mathbf{p}_2)} + \varepsilon\delta_{(S_{i_0}, \mathbf{p}_2)}$. (8.7) is, by (8.3), equivalent to

$$\mathcal{E}_1[\mathbf{p}_1; \delta_{\mathbf{p}_2}] > \mathcal{E}_1[S_{i_0}; \delta_{\mathbf{p}_2}]. \qquad (8.8)$$

However, we know that

$$\mathcal{E}_1[S_{i_0}; \delta_{\mathbf{p}_2}] = f_{1i_0}(\delta_{\mathbf{p}_2}) = \sum_i p_{1i} f_{1i_0}(\delta_{\mathbf{p}_2}) \qquad (8.9)$$

$$\geq \sum_i p_{1i} f_{1i}(\delta_{\mathbf{p}_2}) = \mathcal{E}_1[\mathbf{p}_1; \delta_{\mathbf{p}_2}], \qquad (8.10)$$

which contradicts inequality (8.8), so that $\mathbf{p}_1$ must be a pure strategy. □

We should note here that sometimes individuals will not be able to change role; for instance roles may be "male" or "female". In such circumstances Theorem 8.4 still holds; the key assumption is that the strategy employed by individuals does not affect which role they will be in, and we can think

TABLE 8.1: Three game properties – strategy-role independence (SRI), polymorphic-monomorphic equivalence (PME) and polymorphic-monomorphic equivalence within roles (PMEWR) – lead to eight potential combinations. Three combinations are not possible. The fifth combination labelled * was only demonstrated by a non-generic example (Broom and Rychtář, 2014).

SRI	PME	PMEWR	Is the combination possible?
Y	Y	Y	Yes
Y	Y	N	No
Y	N	Y	No
Y	N	N	Yes
N	Y	Y	Yes*
N	Y	N	No
N	N	Y	Yes
N	N	N	Yes

of our population comprising half males and half females as equivalent to a population comprising individuals which are equally likely to play either role, i.e. that we have polymorphic-monomorphic equivalence. For example, the use of the replicator dynamics given in Section 8.2.1 implies two distinct types of individual, each occupying a different role.

Let us also note that there are asymmetric games, such as the Owner-Intruder game which we consider in Section 8.3, a game of brood care and desertion from Section 8.4.2.2, the asymmetric war of attrition from Section 8.4.3 or a signalling game from Section 18.2 where the assumptions of Selten's Theorem 8.4 do not hold and the ESS strategies can be mixed.

Games with two roles were further considered in Broom and Rychtář (2014). There, a game had *polymorphic-monomorphic equivalence within roles* if individuals that were already in role 1 (or 2) could not tell the difference between polymorphic and monomorphic populations, i.e. whenever $\Pi = \sum_i \alpha_i \delta_{(\mathbf{q}_{i1}, \mathbf{q}_{i2})}$ and $\Pi' = \delta_{\sum_i \alpha_i (\mathbf{q}_{i1}, \mathbf{q}_{i2})}$, then

$$\mathcal{E}_1 \left[\mathbf{p}_1; \Pi_2 \right] = \mathcal{E}_1 \left[\mathbf{p}_1; \Pi_2' \right], \tag{8.11}$$

$$\mathcal{E}_2 \left[\mathbf{p}_2; \Pi_1 \right] = \mathcal{E}_2 \left[\mathbf{p}_2; \Pi_1' \right]. \tag{8.12}$$

They then investigated the relationship between this property and strategy-role independence and polymorphic-monomorphic equivalence as described earlier in this section. The eight possible combinations were considered, and either proved incompatible or an example of them was found. Results are summarised in Table 8.1.

Finally, we have restricted ourselves to games with two roles only, however we can consider a more general setting than games with two roles. Assume that there are any finite number of distinct roles, $r \in R$. The strategy is determined by $(\mathbf{p}_r)_{r \in R}$ and the payoff to this strategy in a population described by Π is given by

$$\mathcal{E}[(\mathbf{p}_r)_{r\in R}; \Pi] = \sum_{r\in R}\sum_{s\in R} w_{rs}\mathcal{E}_{rs}[\mathbf{p}_r; \Pi_s], \tag{8.13}$$

where w_{rs} are interaction probabilities (the probability of player 1 being in role r and player 2 being in role s), Π_s describes a population structure of individuals in role s and $\mathcal{E}_{rs}$ describes the payoff to a focal individual in role r when in the contest with an individual in role s.

Selten (1980) showed that Theorem 8.4 still holds in this more general setting as long as

1. the players unambiguously know their role and that of their opponent;

2. it is not possible for two players to have the same role in a given interaction ($w_{rr} = 0$);

3. the probability of any particular interaction occurring does not depend on the players' strategies, i.e. we have strategy-role independence (this was not explicitly stated, but was implicit in the way w_{rs} was defined).

We shall see examples of where condition 3 is violated in Sections 8.3 and 8.4.2, and an example of where conditions 1 and 2 are violated in Section 8.4.3.

8.2 Bimatrix games

Recall that a bimatrix game is given by a pair of matrices (A, B) where

$$A = (a_{ij})_{i=1,\dots,n;j=1,\dots,m}, B = (b_{ij})_{i=1,\dots,m;j=1,\dots,n}. \tag{8.14}$$

This is a game between two players where player 1 (alternatively the player in role 1) has strategy set $\mathbf{S} = \{S_1, \dots, S_n\}$, player 2 (the player in role 2) has strategy set $\mathbf{T} = \{T_1, \dots, T_m\}$ and a_{ij} and b_{ji} represent rewards to player 1 and player 2 after player 1 chooses a pure strategy S_i and player 2 chooses pure strategy T_j.

The usual setting of bimatrix games is similar to matrix games (Example 2.5) and is as follows. A population of individuals is engaged in pairwise contests. Every single contest can be represented as a game (the same one for every contest) whose payoffs are given by a bimatrix as discussed above (so that the set of pure strategies is finite). Every game is completely independent of each other. Any particular individual can play one or several such games against randomly chosen opponents. The total payoff to the individual is taken as an average payoff of all the games it plays. Usually, especially in more simple games, we assume an infinite population of absolutely identical individuals in every way except, possibly, their chosen strategy.

In other words, the payoffs are given by

$$\mathcal{E}_1[\mathbf{p}; \delta_\mathbf{q}] = \mathbf{p}A\mathbf{q}^T, \tag{8.15}$$

$$\mathcal{E}_2[\mathbf{q}; \delta_\mathbf{p}] = \mathbf{q}B\mathbf{p}^T. \tag{8.16}$$

This means that Selten's Theorem 8.4 applies and that there can be only pure ESSs in bimatrix games.

A point worth making about bimatrix games is that they can be converted into standard matrix games by denoting all of the combinations of pure strategy pairs (S_i, T_j) as the pure strategies of the new matrix game. Thus we can define the new pure strategies W_k by $W_{(i-1)m+j} = (S_i, T_j)$ (see below in Section 8.3 for an example of this approach). This is generally fine if all of the solutions of interest are pure, although "non-generic" payoffs will now appear automatically, but not with mixed strategies. For instance, if an individual is equally likely to be player 1 or player 2, then playing (S_1, T_1) and (S_2, T_2) each with probability 0.5 is indistinguishable from playing (S_1, T_2) and (S_2, T_1) each with probability 0.5. Thus mixed strategies involving completely different pure strategy components are essentially identical, and this can lead to significant problems with the analysis of the game. More generally this is one disadvantage of habitually expressing games in normal form, where as above it is possible that a number of mixed strategies can, in fact, be identical from the point of view of observable behaviour (see Selten, 1975; Selten, 1980; Kuhn, 1953).

8.2.1   Dynamics in bimatrix games

Note that the replicator dynamics can be extended to the bimatrix case in a natural way, so that

$$\frac{d}{dt}p_{1i}(t) = p_{1i}\left((A\mathbf{p_2}^T)_i - \mathbf{p_1}A\mathbf{p_2}^T\right) \quad i = 1, \ldots, n; \tag{8.17}$$

$$\frac{d}{dt}p_{2j}(t) = p_{2j}\left((B\mathbf{p_1}^T)_j - \mathbf{p_2}B\mathbf{p_1}^T\right) \quad j = 1, \ldots, m; \tag{8.18}$$

where $\mathbf{p_k} = (p_{ki})$ is the population mixture of individuals in role k, for $k = 1, 2$.

This corresponds to a polymorphic population where individuals in role 1 meet individuals in role 2 at a rate proportional to their frequency within the population of players in that role. It is generally assumed that the population sizes are constant, or at least that the relative size of the two populations is constant. We discuss the reasonableness of this assumption at the end of this section.

We follow Hofbauer and Sigmund (1998), assuming that we start with a population where there are at least two non-zero elements of both vectors $\mathbf{p}_1$ and $\mathbf{p}_2$ (if we do not it is easy to see that the above equations reduce to a single set of differential equations and the population tends to a pure strategy in the generic case).

To be an equilibrium point, the strategy $(\mathbf{p}_1, \mathbf{p}_2)$ must satisfy the following

$$(A\mathbf{p}_2^T)_i = c_1 \text{ for } i = 1, \ldots, n_1 \text{ for some constant } c_1 \text{ and} \qquad (8.19)$$

$$(B\mathbf{p}_1^T)_j = c_2 \text{ for } j = 1, \ldots, n_2 \text{ for some constant } c_2, \qquad (8.20)$$

where there are n_1 and n_2 elements in the supports of $\mathbf{p}_1$ and $\mathbf{p}_2$, respectively, and without loss of generality we number these $i = 1, \ldots, n_1$ and $j = 1, \ldots, n_2$. We note that c_1 and c_2 are the mean payoffs to the two sub-populations. Since $\mathbf{p}_1$ is a vector with n_1 elements (and $\sum_i p_{1i} = 1$) and $\mathbf{p}_2$ is a vector with n_2 elements (and $\sum_j p_{2j} = 1$) if $n_1 > n_2$ then (8.19) gives $n_1 - 1$ equations in $n_2 - 1$ unknowns which in the generic case yields no solutions.

Thus there can only be an isolated rest point in the case where $n_1 = n_2$, and there will be exactly one solution of (8.19)–(8.20) in the generic case, which to exist as a state of the population must satisfy the obvious additional conditions that all elements of the probability vectors are non-negative. Thus there is either one or no such rest point(s). It is shown in Hofbauer and Sigmund (1998) that the rest point, if it exists, is not an Evolutionarily Stable State (ESState). Recall from Section 3.2.3 that this was termed an ESState in Hofbauer and Sigmund (1998), to distinguish monomorphic populations (strategy) from polymorphic populations (state). Thus there are no non-pure solutions to the dynamics, in a similar way to the static case above.

Hofbauer and Sigmund (1998) also consider the following example of a two-player, two-strategy game, with payoff matrices

$$A = \begin{pmatrix} 0 & a_{12} \\ a_{21} & 0 \end{pmatrix}, \quad B = \begin{pmatrix} 0 & b_{12} \\ b_{21} & 0 \end{pmatrix}. \qquad (8.21)$$

Note that this is generic, since taking a constant off any column does not affect the game. Setting $x = p_{11}$ and $y = p_{21}$, this yields the replicator equations

$$\frac{dx}{dt} = x(1 - x)(a_{12} - (a_{12} + a_{21})y), \qquad (8.22)$$

$$\frac{dy}{dt} = y(1 - y)(b_{12} - (b_{12} + b_{21})x). \qquad (8.23)$$

They show that orbits converge to the boundary in all cases except if $a_{12}a_{21} > 0, b_{12}b_{21} > 0$ and $a_{12}b_{12} < 0$, which yield closed periodic orbits around the internal equilibrium

$$\left(\frac{b_{12}}{b_{12} + b_{21}}, \frac{a_{12}}{a_{12} + a_{21}} \right). \qquad (8.24)$$

This situation where an equilibrium point is stable (but not asymptotically stable) is reminiscent of non-generic games such as the Rock-Scissors-Paper game from Chapter 2, where $a_{12} = a_{23} = a_{31} = 1$ and $a_{21} = a_{32} = a_{13} = -1$, which also featured closed orbits. Yet this comes from a game with apparently generic payoffs. We note that the differential equations for the frequencies of

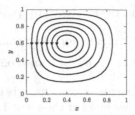

FIGURE 8.1: Dynamics (8.22)–(8.23) of the bimatrix game with the matrices $A = \left(\begin{smallmatrix} 0 & 3 \\ 2 & 0 \end{smallmatrix}\right)$, $B = \left(\begin{smallmatrix} 0 & -2 \\ -3 & 0 \end{smallmatrix}\right)$.

one role only contain the payoffs associated with that role, and involve the other role only through the frequencies of the strategies.

Recall the origin of the replicator equation from Section 3.1.1.2. We started with a population size N_i, considered $\frac{d}{dt}N_i = N_i f_i(\mathbf{p}(t))$ and eventually obtained a differential equation (3.5) for $\mathbf{p}$,

$$\frac{d}{dt}p_i = p_i\Big(f_i(\mathbf{p}(t)) - \bar{f}(\mathbf{p}(t))\Big). \tag{8.25}$$

In the procedure, we made no assumptions about the total population size. It can change, but the frequencies p_i vary with time irrespective of the size of N. The differential equations (8.17)–(8.18) only depend upon the proportions of the two types of individuals and not on the total number of each. However suppose, for example, that there were large numbers of role 1 players and small numbers of role 2 players. Each role 1 player would face few interactions and would be far less affected than role 2 players; in particular, the effect on role 1 players would be increased if the number of role 2 players was large. Thus there is an important population dynamics aspect that is being neglected in this situation, which can be more readily ignored for matrix games, namely the influence of the size/densities of the populations (for a discussion of this issue see Argasinski, 2006). In effect for these replicator equations to hold, the two populations must increase at precisely the same rate, so that the relative size of the two populations is constant. For this to occur we would need $\mathbf{p_1}A\mathbf{p_2}^T = \mathbf{p_2}B\mathbf{p_1}^T$ or $c_1 = c_2$ if in the equilibrium from (8.19)–(8.20). This is a specific assumption that is unlikely to hold in practice. Indeed when strategies oscillate as in Figure 8.1, population sizes are likely to as well.

8.3   Uncorrelated asymmetry—The Owner-Intruder game

We first look at the case of uncorrelated asymmetries where there is an asymmetry between the two types, but this does not affect the elements of the game. The key function of the asymmetry is that the individuals can distinguish between the types, and can choose strategies based upon their type. Thus suppose we have a contest over a territory, between an owner and an intruder. The strategy of an individual will consist of what to play when the owner and what to play when the intruder. Maynard Smith (1982, Chapter 8) considered such a game where the contest was a Hawk-Dove game. There are four pure strategies:

Hawk	—	play Hawk when both owner and intruder,
Dove	—	play Dove when both owner and intruder,
Bourgeois	—	play Hawk when owner and Dove when intruder,
Marauder	—	play Dove when owner and Hawk when intruder.

"Marauder" was coined for this type of behaviour in a model of kleptoparasitism by Broom et al. (2004b). Maynard Smith (1982) simply called this strategy "Strategy X".

The payoffs to this game can be summarised in the following payoff matrix:

$$
\begin{array}{c}
\\
\text{Hawk} \\
\text{Dove} \\
\text{Bourgeois} \\
\text{Marauder}
\end{array}
\begin{array}{cccc}
\text{Hawk} & \text{Dove} & \text{Bourgeois} & \text{Marauder} \\
\begin{pmatrix}
(V-C)/2 & V & (3V-C)/4 & (3V-C)/4 \\
0 & V/2 & V/4 & V/4 \\
(V-C)/4 & 3V/4 & V/2 & (2V-C)/4 \\
(V-C)/4 & 3V/4 & (2V-C)/4 & V/2
\end{pmatrix}
\end{array}
\qquad (8.26)
$$

For example when a Hawk faces a Bourgeois, half of the time the Hawk will be the owner; the Hawk then plays Hawk in that encounter, the Bourgeois plays Dove, so the rewards are respectively V and 0. Half of the time the Bourgeois will be the owner, and it then plays Hawk in that encounter. "Hawk" of course still plays Hawk, and each receives $(V-C)/2$. Thus averaging the payoffs gives $(3V - C)/4$ for Hawk and $(V - C)/4$ for Bourgeois.

If $V \geq C$ then Hawk is the unique ESS. Assuming that $V < C$, then we can see that both Bourgeois and Marauder are pure ESSs, and that Hawk and Dove are not. By the Bishop-Cannings Theorem 6.9, the only other possible ESS would be a Hawk-Dove mixture.

This is an example where non-generic payoffs have appeared (see Section 2.1.3.3, and see Section 6.6.1 for a way to avoid non-generic payoffs in this kind of situation), as either Bourgeois or Marauder have equal payoff to the equilibrium mixture of Hawks and Doves, and so this mixture cannot be an ESS (Exercise 8.3).

Thus we have two ESSs for the game when $V < C$, when intruders or owners always concede. The former is the more natural, and nature is full of examples where when roughly evenly matched individuals meet over a territory, the intruder gives ground without a fight. In the game above, this has nothing to do with the territory being more valuable to the owner. It should be noted that often territories really will be more valuable to owners, but what this game makes clear is that concession of intruders to owners does not imply that this is the case. This conclusion is not unique to Hawk-Dove type games, and as Maynard Smith and Parker (1976) observe, uncorrelated asymmetries generally yield pairs of solutions like this (they give the war of attrition as an example). Maynard Smith and Parker (1976) cite the swallowtail butterfly *Papillo zelicaon* as an example where uncorrelated asymmetries settle contests over territories (hilltop sites) in the mating season.

Interestingly, although it is common for owners to win equal contests, the counterintuitive strategy that we have called Marauder has been shown to occur in nature. The social spider *Oecibus civitas* gives way to intruders and flees from its hole, often leading to a sequence of displacements as, in turn, displaced spiders become intruders and displace others (Burgess, 1976; Maynard Smith, 1982).

Note that the assumptions behind this game are clearly oversimplified—if a Hawk is more likely to win than a Bourgeois, then in reality a Hawk will have a probability greater than $1/2$ to be the territory owner. In general, if the strategy that you choose affects the probability of occupying a given role, then the assumption of no correlation between the strategy played and the role of the individual required for Selten's Theorem 8.4 will not hold. Let us suppose that a population of Hawks, strategy $(1,1)$, and Bourgeois, strategy $(1,0)$, have occupied some territories for some time, with repeated fights. It is clear that, if the proportion of Hawks is p, then if $p > 1/2$ then all Bourgeois will be intruders, and the Hawks will be divided between owners and intruders. In this case the proportion of Hawks that are owners will be $1/2p$. Similarly, if $p < 1/2$ all Hawks will be owners, and the Bourgeois individuals will be divided between owners and intruders. Thus

$$\rho(1,1) = \begin{cases} 1 & p < 1/2, \\ \frac{1}{2p} & p > 1/2; \end{cases} \tag{8.27}$$

$$\rho(1,0) = \begin{cases} \frac{1/2-p}{1-p} & p < 1/2, \\ 0 & p > 1/2. \end{cases} \tag{8.28}$$

It is easy to see that Hawks always do better when $p < 1/2$, since they always win any contest with no cost, so we shall concentrate on the case $p > 1/2$. Π comprises a population where all owners are Hawks, and the proportion of intruders that are Hawks (Bourgeois) is just the proportion of individuals that are intruders and Hawks (Bourgeois) divided by $1/2$ and is thus $2p - 1$

$(2-2p)$. Here (8.2) becomes

$$\mathcal{E}[(1,1);\Pi] = \frac{1}{2p}\left(\frac{V-C}{2}(2p-1) + V(2-2p)\right) + \left(1 - \frac{1}{2p}\right)\frac{V-C}{2} \quad (8.29)$$

$$= \frac{V-C}{2} + \frac{1-p}{p}\frac{V+C}{2}, \quad (8.30)$$

and $\mathcal{E}[(1,0);\Pi] = 0$ since all Bourgeois individuals play Dove as intruders against Hawk owners. Equating (8.30) to zero, we see that there is an equilibrium at $p = (V+C)/(2C)$, which can be shown to be an ESS (see Exercise 8.4).

Thus we have a mixed ESS, which happens because the assumptions of Selten's Theorem 8.4 are violated.

8.4 Correlated asymmetry

We consider three examples of asymmetries in payoff caused by different components of the payoff function. As already stated, although each of these has different features, the method that we use to approach the analysis is the same in each case.

8.4.1 Asymmetry in the probability of victory

We now consider a Hawk-Dove game where the contestants have differing resource holding potentials. We assume that this affects the probability of victory in a Hawk versus Hawk contest only. Thus suppose that individual A is larger than individual B, and that both contestants are aware of their sizes. All games feature two individuals of known different sizes, and hence the game is truly asymmetric. Each can play Hawk or Dove. If both play Dove the reward is divided equally between them; if one plays Hawk and the other Dove, then the Hawk player gains the reward V and the Dove player gains 0; and finally if they both play Hawk, the larger individual wins with probability $\alpha \geq 1/2$, the winner gaining the reward V and the loser receiving reward $-C$. The payoff bimatrix, with the larger individual in role 1, becomes

$$\begin{array}{cc} & \text{small Hawk} \qquad\qquad \text{small Dove} \\ \begin{array}{c} \text{large Hawk} \\ \text{large Dove} \end{array} & \left(\begin{array}{cc} \alpha V - (1-\alpha)C, (1-\alpha)V - \alpha C & V, 0 \\ 0, V & V/2, V/2 \end{array}\right). \end{array} \quad (8.31)$$

A strategy in the above game can be written as (p_1, p_2) where p_1 is the probability of playing Hawk in role 1, (i.e. when large) and p_2 is the probability of playing Hawk in role 2 (i.e. when small). More generally a strategy would involve $p(\alpha)$, the probability of playing Hawk for every value of α, but for perfect information (see Chapter 10) the choice that you make when your winning chance is α will only be dependent on the strategies of others with winning chance $1 - \alpha$, so considering such a pair is sufficient. In terms of Section 8.1, we have $w_{rs} = 0$ unless $r + s = 3$.

By Theorem 8.4, it is clear that there can be no mixed strategy solutions. Thus there are four potential solutions for (p_1, p_2), namely $(0, 0)$, $(1, 0)$, $(0, 1)$ and $(1, 1)$, and for any candidate ESS we need consider invasion by other pure strategies which differ in one role only. It is clear that $(0, 0)$ where both play Dove can never be an ESS, since

$$\mathcal{E}_1[D; \delta_D] = V/2 < V = \mathcal{E}_1[H; \delta_D]. \tag{8.32}$$

For $(1, 1)$ where individuals play Hawk in both roles to be an ESS we need

$$\mathcal{E}_1[H; \delta_H] = \alpha V - (1 - \alpha)C > \mathcal{E}_1[D; \delta_H] = 0, \text{ and} \tag{8.33}$$

$$\mathcal{E}_2[H; \delta_H] = (1 - \alpha)V - \alpha C > \mathcal{E}_2[D; \delta_H] = 0. \tag{8.34}$$

Inequality (8.34) is the harder condition to meet since $\alpha \geq 1/2$. Thus if

$$\frac{V}{C} > \frac{\alpha}{1 - \alpha}, \tag{8.35}$$

then $(1, 1)$ is an ESS i.e. if the reward is relatively valuable compared to the cost, and the chance of the weaker individual to win is not too small, then both play Hawk and there is a fight.

For $(1,0)$ where individuals play Hawk when large and Dove when small to be an ESS we need

$$\mathcal{E}_1[H; \delta_D] = V > \mathcal{E}_1[D; \delta_D] = V/2, \qquad \text{and} \tag{8.36}$$

$$\mathcal{E}_2[D; \delta_H] = 0 > \mathcal{E}_2[H; \delta_H] = (1 - \alpha)V - \alpha C. \tag{8.37}$$

Thus strategy $(1, 0)$ is an ESS precisely when (8.35) is not satisfied. There is a third possible ESS, $(0, 1)$, where the stronger individual gives ground to the weaker one, which occurs, analogously to the above, if

$$\frac{V}{C} < \frac{1 - \alpha}{\alpha} \left(< \frac{\alpha}{1 - \alpha} \right). \tag{8.38}$$

This "paradoxical ESS" could result from earlier conditions when the strength advantage was not apparent. For instance, an older individual may dominate a younger when it has an advantage in strength and then this dominance persists long after the strength ordering has been reversed, as the subordinate has got into the habit of defeat. The results are summarised in Figure 8.2.

8.4.2 A game of brood care and desertion

8.4.2.1 Linear version

Webb et al. (1999) consider a two-stage game of brood care and desertion which reduces to a bimatrix game as described above. At the start of a breeding

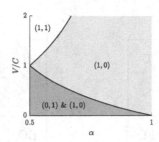

FIGURE 8.2: ESSs for the asymmetric Hawk-Dove game with payoffs (8.31).

TABLE 8.2: The ESSs in the brood care and desertion game (8.39).

ESS	Male	Female	Conditions	
Biparental care	Care	Care	$r_m < \frac{V_2-V_1}{V_2}$	$r_f < \frac{V_2-V_1}{V_2}$
Male uniparental care	Care	Desert	$r_m < \frac{V_1}{V_2}$	$r_f > \frac{V_2-V_1}{V_2}$
Female uniparental care	Desert	Care	$r_m > \frac{V_2-V_1}{V_2}$	$r_f < \frac{V_1}{V_2}$
Biparental desertion	Desert	Desert	$r_m > \frac{V_1}{V_2}$	$r_f > \frac{V_1}{V_2}$

season all individuals have found a mate. After their first breeding attempt there is the opportunity for a second, but to take this they will have to desert their offspring. If both parents desert, then the offspring will die and so they receive reward 0 from that mating. If both stay they receive reward V_2. If one deserts and one stays, they both receive $V_1 < V_2$, and the deserter has the chance to mate again. If the male deserts, the probability that he can mate again is r_m, and if the female deserts the probability that she can mate again is r_f. Since $V_2 > V_1$ and there is no opportunity for a third mating, it is clear that the best strategy in any second mating is for both individuals not to desert, so gaining reward V_2 for that mating.

The game thus reduces to whether to "Desert" or "Care" at the first mating. With the male in role 1 and the female in role 2, this gives us the bimatrix of payoffs as

$$
\begin{array}{cc}
 & \text{Female Care} \quad\quad \text{Female Desert} \\
\begin{array}{c}\text{Male Care}\\ \text{Male Desert}\end{array} &
\left(\begin{array}{cc}
V_2, V_2 & V_1, V_1 + r_f V_2 \\
V_1 + r_m V_2, V_1 & r_m V_2, r_f V_2
\end{array}\right).
\end{array}
\tag{8.39}
$$

By Selten's Theorem 8.4 we know that there will not be any mixed ESSs for such a game. Any of the pure strategy pairs can be ESSs for this game, as summarized in Table 8.2 and Figure 8.3 (see also Exercise 8.6). This game is similar to the battle of the sexes that we consider in Section 17.4 (see also Mylius, 1999).

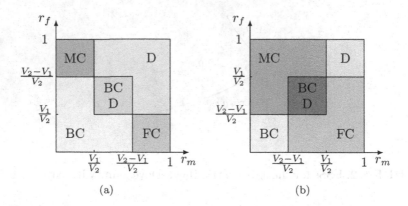

(a)   (b)

FIGURE 8.3: ESS for the brood care and desertion game (8.39). (a) $V_1 < V_2/2$, (b) $V_2/2 < V_1 < V_2$. BC = biparental care, MC = male uniparental care, FC = female uniparental care, D = desertion.

8.4.2.2 Nonlinear version

Now we consider a more complicated version of the brood care and desertion game from 8.4.2, in which the re-mating probabilities depend upon the choices of all individuals in the first stage, and so the payoffs become nonlinear. This, in common with the work in the current chapter, assumes that both individuals have to make their decision whether to desert or care simultaneously. In fact, there may be more than one opportunity to desert, and in particular, it could be the case that one parent (often the male) has the opportunity to choose to desert before the other can, leading to sequential decisions. We look at such games in Chapter 11 (see also Example 10.1 where we consider such a game in extensive form).

Assume, as would be expected in reality, that after desertion the potential number of available partners will depend upon whether other males or females desert, i.e. they will depend upon the strategies of the players in the current game. Webb et al. (1999) assume that remating probabilities are now functions of x and y, the proportion of males and females respectively which desert their first broods, using the remating probabilities

$$r_f(x,y) = k\frac{x}{\sqrt{x+y}}, \tag{8.40}$$

$$r_m(x,y) = k\frac{y}{\sqrt{x+y}} \tag{8.41}$$

in the payoff matrix (8.39). For instance we see that a male who cares at the first mating with probability p receives the expected reward

$$\mathcal{E}_1[\mathbf{p_1};\Pi_2] = p\big((1-y)V_2 + yV_1\big) + (1-p)\left((1-y)V_1 + V_2k\frac{y}{\sqrt{x+y}}\right) \tag{8.42}$$

which is a function of x, i.e. the population strategy of players in role 1. This violates the assumptions of Selten's Theorem 8.4 and results from Selten (1980).

Indeed, whilst there is a pure ESS (as biparental care is always an ESS), this game can also have a mixed ESS. With $k = 1$ and the same two rewards, V_1 and $V_2 > V_1$, it was shown that a mixed ESS occurs for sufficiently large V_1/V_2. If $3/4 < V_1/V_2$, then $x^* = y^* = 1/2$ is an ESS, and if $1/\sqrt{2} < V_1/V_2 < 3/4$, then the mixed ESS is

$$x^* = y^* = \frac{2(V_2 - V_1)^2}{(2V_1 - V_2)^2}. \tag{8.43}$$

Otherwise for $V_1/V_2 < 1/\sqrt{2}$ biparental care is still an ESS, but now so (perhaps implausibly) is biparental desertion.

8.4.3 Asymmetries in rewards and costs: the asymmetric war of attrition

We have already seen in Section 4.3.7 that when individuals play a war of attrition where all games involve a pair with distinct roles, and each can distinguish their roles unambiguously, there are ESSs where each player plays a pure strategy, and the pure strategy for one of the roles must be 0.

Hammerstein and Parker (1982) considered a war of attrition where costs and rewards are different for the two players, which we describe as being in role A or role B. This involves a different effect of obtaining the reward and different costs associated with waiting.

The roles of the players are selected at random. The probability that player 1 has role A (B) and player 2 has role B (A) is w_{AB} (w_{BA}), with the probability that they both have role A (B) being w_{AA} (w_{BB}). It is assumed that $w_{AB} = w_{BA} > 0$ and also that both $w_{AA} > 0$ and $w_{BB} > 0$. A player knows its own role in any given conflict, but not that of its opponent. We note that these values violate the assumptions of Selten's Theorem 8.4, that games are truly asymmetric, i.e. games between individuals in the same role were not allowed, and the roles occupied by both players were known to both.

The reward function to the individual in role A playing x against the individual in role B playing y is defined as

$$E_{AB}(x, y) = \begin{cases} V_{AB} - C_{AB}y & x > y, \\ V_{AB}/2 - C_{AB}x & x = y, \\ -C_{AB}x & x < y. \end{cases} \tag{8.44}$$

The role A is said to be the "favoured with respect to payoffs" role if

$$\frac{V_{AB}}{C_{AB}} > \frac{V_{BA}}{C_{BA}}, \tag{8.45}$$

and of course role B is "favoured" if the reverse inequality holds. It is assumed in Hammerstein and Parker (1982) that the following extra inequalities hold when (8.45) does:

$$\frac{V_{AB}}{C_{AB}} > \frac{V_{BB}}{C_{BB}}, \tag{8.46}$$

$$\frac{V_{AA}}{C_{AA}} > \frac{V_{BA}}{C_{BA}}. \tag{8.47}$$

There are a wide range of circumstances where this will be true, in particular, if rewards and costs to an individual depend upon its properties but not that of its opponent (e.g. $V_{AB} = V_{AA}(= V_A)$) where all three conditions (8.45)–(8.47) are identical.

Hammerstein and Parker (1982) proved that the ESS results in the unfavoured individual always playing a shorter time than the favoured individual, with both following a mixed strategy. Thus if A is the favoured individual, in A versus B contests, A will always receive the reward, and the variability in the chosen times only influences contests involving individuals in the same role. The strategies are as follows:

$$p_A(x) = \begin{cases} 0 & \text{if } 0 \le x < s, \\ \dfrac{C_{AA}}{V_{AA}} \exp\left(\dfrac{C_{AA}}{V_{AA}}(s-x)\right) & \text{if } x \ge s; \end{cases} \tag{8.48}$$

$$p_B(x) = \begin{cases} \dfrac{1}{V_{BB}} \dfrac{w_{BA}C_{BA} + w_{BB}C_{BB}}{w_{BB}} \exp\left(-\dfrac{C_{BB}}{V_{BB}}x\right) & \text{if } 0 \le x < s, \\ 0 & \text{if } x \ge s; \end{cases} \tag{8.49}$$

where

$$s = \frac{V_{BB}}{C_{BB}} \ln\left(\frac{w_{BA}C_{BA} + w_{BB}C_{BB}}{w_{BA}C_{BA}}\right). \tag{8.50}$$

It is straightforward to verify that the above are proper probability density functions, which is Exercise 8.8. For example, in the case where $C_{AA} = C_{AB} = C_{BA} = C_{BB} = 1$, $V_{AA} = V_{AB} = V$, $V_{BA} = V_{BB} = v < V$, $w_{AA} = w_{AB} = w_{BA} = w_{BB}$ (see Exercise 8.9) we obtain $s = v \ln 2$, and

$$p_A(x) = \begin{cases} 0 & \text{if } 0 \le x < v \ln 2, \\ \dfrac{1}{V} \exp\left(\dfrac{1}{V}(v \ln 2 - x)\right) & \text{if } x \ge v \ln 2; \end{cases} \tag{8.51}$$

$$p_B(x) = \begin{cases} \dfrac{2}{v} \exp\left(-\dfrac{1}{v}x\right) & \text{if } 0 \le x < v \ln 2, \\ 0 & \text{if } x \ge s. \end{cases} \tag{8.52}$$

We see that individuals in role B play the standard exponential distribution with mean v, but truncated at time $v \ln 2$. Individuals in role A play at least the time $v \ln 2$, and play an additional time taken from their standard exponential distribution with mean V.

Note that to obtain a situation where (almost) all games involve individuals one in each of the two roles, we just let w_{AA} and w_{BB} tend to zero (see also Section 4.3.7). In this case $s \to 0$ and we obtain the limiting result that

$$p_A(x) = \frac{C_{AA}}{V_{AA}} \exp\left(\frac{C_{AA}}{V_{AA}}\right) \qquad \text{if } x > 0 \qquad (8.53)$$

and B plays 0 with probability 1. Thus in this case the unfavoured individual immediately leaves. Note that here the favoured individual still plays a mixed strategy, even though the probability of meeting another A is vanishingly small. As almost all contests are against Bs it will wait no time at all, but the minute probability of playing an A is enough to provide a mixed, and pleasingly unique, solution.

An observer would see a game played as if the solution was a pure strategy, with B conceding to A immediately. This solution, of course, is not a pure strategy, as would seem to be required by the result of Selten (1980). This situation does not quite correspond, since Selten (1980) required that the probability of both individuals occupying the same role was 0 (i.e. $w_{rr=0}$) and even though the above situation gets as close as we like to a probability of zero, there is a minuscule probability of meeting an individual playing your own role. This is, in fact, a good example of Selten's trembling hand (Selten, 1975); see Section 3.2.4.

8.5 Python code

In this section we show how to use Python to identify an ESS in asymmetric games such as the Owner-Intruder game from Section 8.3.

```python
""" Owner-Intruder game
The script finds all possible ESSs in an asymetric game with
a user-defined payoff matrix PM. It identifies responses that are
"best to itself". Works for a GENERIC payoff matrix.

It may include non-ESSs for non-generic:
there may be more than one best response and thus
being best response to itself does not have to mean an ESS.
Tyr, for example, V = 2, C = 2 below for a non-generic matrix
"""

## Import basic packages
import numpy as np

## Define a payoff matrix
# An owner-intruder game, define resource value and fight cost
V = 1  # Reward for having the resource
C = 2  # Cost of the fight
```

```
19
20  # Give names to strategies
21  sNames = ['Hawk', 'Dove', 'Bourgeois', 'Marauder'];
22
23  PM = [[(V-C)/2,        V,     (3*V-C)/4,   (3*V-C)/4 ],   # Hawk
24        [0,             V/2,      V/4,          V/4 ],      # Dove
25        [(V-C)/4,       3*V/4,    V/2,       (2*V-C)/4],    # Bourgeois
26        [(V-C)/4,       3*V/4,  (2*V-C)/4,     V/2  ]]      # Marauder
27  # PM[a][b] is a payoff to a strategy sNames[a]  against  sNames[b]
28
29
30  ## Actual calculations
31
32  # Get the number of strategies = size of the payoff matrix
33  n=np.size(PM,0)
34
35  # Identify the maximal payoff for a given strategy
36  # max_P[b] is the maximal payoff one can get against sNames[b]
37  max_P = np.amax(PM,axis=0)
38
39  # Find out the best responses, BR[a,b] = true if a is BR to b
40  BR = PM == max_P
41
42  # Get best responses to itself
43  # BR_to_itself[a] is true if a is BR to a
44  BR_to_itself = np.diag(BR)
45
46  ## Printout results
47  # for every strategy s and its index i
48  for i, s in enumerate(sNames):
49      # If it is the best response to itself
50      if BR_to_itself[i]:
51          # If it is the only best response
52          if sum(BR)[i] == 1:
53              print(s + ' is an ESS')
54          else:
55              # otherwise it is not the only best response
56              print(s + ' is best response to itself, but')
57              print([t for (t, br) in zip(sNames, BR[i][:])
58                     if (br and t!=s)])
59              print('are also best responses to ' + s + '\n')
```

8.6 Further reading

For the classical mathematical treatment of asymmetric games see Selten (1980). For more mathematical results on two-population games see Fishman (2008). Whether mixed strategies can be stable is investigated in Eshel and Sansone (1995); Binmore and Samuelson (2001). Hofbauer and Sigmund (1998) give a good account of the dynamics of asymmetric games; see also

Hofbauer (1996); Hofbauer and Hopkins (2005); Gaunersdorfer et al. (1991); Samuelson and Zhang (1992).

The classical work that introduces asymmetric contests from a biological perspective is Maynard Smith and Parker (1976). Hammerstein (1981) provides an overview on the role of asymmetries in animal contests. The concept of the operational sex ratio, which can be useful when polymorphic-monomorphic equivalence does not occur, is discussed in Kokko and Johnstone (2002). Leimar and Enquist (1984) consider asymmetries in Owner-Intruder conflicts; see also Yee (2003) and Kokko et al. (2006b) for a related model of territorial behaviour.

Matsumura and Kobayashi (1998) consider an asymmetric Hawk-Dove game as a model of dominance relations in animal groups. Bishop and Cannings (1978) consider a generalised war of attrition where the reward and cost functions take a general form, and an asymmetric version of this was considered for example in Hammerstein and Parker (1982); Haccou et al. (2003); Haccou and van Alphen (2008).

Mesterton-Gibbons, Sherratt and colleagues have produced a series of papers further developing the owner-intruder model, for example in Mesterton-Gibbons et al. (2014) and Mesterton-Gibbons et al. (2016). Mesterton-Gibbons and Sherratt (2014) considers the idea of *infinite regress* originally raised by John Maynard Smith, where when an owner is displaced by an intruder there is a role reversal, so that Marauder might lead to an infinite contest. For a review of this and other work see Sherratt and Mesterton-Gibbons (2015).

The replicator dynamics for multiple populations (Weibull, 1995) under mutation was considered in Bauer et al. (2019). In particular, mutation often stabilised equilibria, and the resulting system is a potentially good model of learning behaviour related to reinforcement learning.

8.7 Exercises

Exercise 8.1. Consider a game with two roles as described in Section 8.1. Hawk-Dove games are played between territory owners and invaders. After each contest, one of the following scenarios happens. For each of the scenarios, say whether the game is strategy-role independent, and if it is, find the value of ρ_1.

1. Both individuals have to vacate the territory, and the owner in any subsequent game is the first individual to return.

2. The territory owner remains as the owner irrespective of the result of the contest

3. Games are played as in case (i) between females and males. The females do not move so widely about the habitat, and in any contest there is a probability $p > 1/2$ that the female takes the role of the owner.

4. The winner of the contest becomes the territory owner.

Exercise 8.2 (Hofbauer and Sigmund, 1998, Section 10.4). By considering the derivatives of the function $H(x, y) = a_{12} \log y + a_{21} \log(1 - y) - b_{12} \log x - b_{21} \log(1 - x)$, show that provided that $a_{12} a_{21} > 0, b_{12} b_{21} > 0$ and $a_{12} b_{12} < 0$, the dynamics from equations (8.22)–(8.23) yield closed orbits.

Exercise 8.3. Consider an Owner-Intruder game with payoffs given in (8.26). Show that a Hawk-Dove mixture cannot be an ESS because it will always be invaded by either Marauder or Bourgeois.

Exercise 8.4. For the Owner-Intruder game with payoffs given in (8.26) and role probabilities given by (8.27) and (8.28), show that a Hawk-Bourgeois mixture is an ESS when the equilibrium proportion of Hawks is $p = (V + C)/2C$.

Exercise 8.5. Show that in the Hawk-Dove game with asymmetric probability of victory given by the bimatrix in (8.31), $(0, 0)$ is never an ESS.

Exercise 8.6. Use results from Table 8.2 to show the following.

1. For this game of brood care and desertion there is always at least one ESS.

2. There can be two ESSs, with either biparental care and biparental desertion or male uniparental care and female uniparental care as ESSs simultaneously.

3. No other combination can occur.

Exercise 8.7. Show that the game of brood care and desertion described in Section 8.4.2.2 can be viewed as a playing the field game, i.e. expressed in the form of (7.27) for each sex.

Exercise 8.8. Verify that the general expressions (8.48)–(8.49) for the strategies for the individuals in roles A and B for the asymmetric war of attrition from Section 8.4.3 are proper probability density functions.

Exercise 8.9. For the asymmetric war of attrition in Section 8.4.3, verify the given formulae for the ESS in the case where $C_{AA} = C_{AB} = C_{BA} = C_{BB} = 1$, $V_{AA} = V_{AB} = V$, $V_{BA} = V_{BB} = v < V$, $w_{AA} = w_{AB} = w_{BA} = w_{BB}$. Find the ESS in the limiting cases $v \to 0$, $v \to V$ and when $v = V/2$.

Exercise 8.10. For the asymmetric war of attrition from Section 8.4.3, find the expected payoff to individuals from the two types. In particular, find the expected payoffs for the parameters defined in Exercise 8.9.

Chapter 9

Multi-player games

In previous chapters most of the games we have considered have either involved a pair of players only, albeit within a larger (infinite) population, or have been "playing the field" type games, where there are no direct conflicts, but where individuals play a strategy, the success of which is a function of the population strategy (see Chapter 7). We now consider situations where individual conflicts consist of a number of individuals greater than two.

Such games have, until fairly recently, only rarely been considered with regard to biological populations (although multi-player games are common in economics; see Chapter 1), with the first generalisation of the theory to the multi-player case due to Palm (1984). This lack of attention has been for two reasons. Firstly, real conflicts often comprise pairwise games, and a lot can be learnt from considering them. Secondly, the mathematics involved in the analysis of multi-player games is more complex, and it is harder to come up with generalisable results.

All the games that we consider are contests involving a randomly selected group of (at least three) players from a large population. As we shall see these games are linear in the strategy of the focal player but non-linear in the population strategy and so there is a similarity with many of the games described in Chapter 7. We could perhaps think of multi-player games as being a special class of nonlinear games, in the sense that all games which are not linear in both the focal player and the population strategy are nonlinear, but the games in Sections 9.2 and 9.3 have their own distinct character. The results from Chapter 7 are more directly relevant to the games in Section 9.1, but even here there is some benefit in discussing these separately, as games involve the sum of independent contests against specific opponents as in matrix games, and so have some features in common with matrix games.

We have already seen examples of multi-player games in the exercises in Chapter 4. Some games were directly formulated as m-player games, such as the El Farol bar problem or the Diner's dilemma. Some games, such as the stag hunt game or Blotto's game can be easily modified for m players.

Example 9.1. Killer whales *Orcinus orca* have been observed to engage in a collective hunting technique, so-called carousel feeding (Similä, 1997). A small group of whales releases bursts of bubbles to round prey into a tight defensive ball close to the surface and the whales then slap the ball with their tails, stunning or killing up to 10-15 (Domenici et al., 2000) fish with a successful

DOI: 10.1201/9781003024682-9

slap. To be successful, the technique requires good cooperation by a number of whales. This can be modelled as a multi-player stag hunt game since whales may be tempted not to cooperate but start feeding on their own.

We will now see how to set up and analyse these types of games.

9.1   Multi-player matrix games

Broom et al. (1997b) considered an infinite population, from which groups of m players were selected at random to play a game (see also Bukowski and Miękisz, 2004). The expected payoff to an individual is obtained by averaging over all of the possible rewards, weighted by their probabilities, just as in the case of matrix games (see Section 2.2.2 and Chapter 6). In its most general form where the ordering of individuals matter, and so effectively extending the bimatrix game case to m players, the payoff to an individual in position k is governed by an m-dimensional payoff matrix, and m such matrices are needed to summarise the game, each representing the possible payoffs to the player in one of the m distinct positions.

In Broom et al. (1997b) it was assumed that there was no significance to the ordering of the players, as a natural extension of matrix games (in contrast to bimatrix games). Thus the payoff to an individual depends only upon its strategy and the combination of the strategies of its opponents and only one such m-dimensional matrix is needed, with some of the entries being the same as others (if the same strategies are involved in a "different order").

We will call such games *symmetric* and write the payoffs for a three-player n-strategy game as $A = (A_1, A_2, \ldots, A_n)$ where A_j are the payoffs assuming the focal player plays pure strategy S_j, and

$$A_j = \begin{pmatrix} a_{j11} & a_{j12} & \cdots & a_{j1n} \\ a_{j21} & a_{j22} & \cdots & a_{j2n} \\ \vdots & \vdots & \ddots & \vdots \\ a_{jn1} & a_{jn2} & \cdots & a_{jnn} \end{pmatrix}. \tag{9.1}$$

To add an extra player, the number of matrices required in this formulation is multiplied by the number of strategies n. There are, in a full general case, n^m entries of the matrices. However, because we assume that the payoffs to the focal individual depend only on its strategy and on what other strategies are used in the contest, we have some symmetry conditions. For the three-player case, these are

$$a_{pqr} = a_{prq}, \quad \text{for all } p, q, r = 1, 2, \ldots, n. \tag{9.2}$$

In general, these are

$$a_{i_1 \ldots i_m} = a_{i_1 \sigma(i_2) \ldots \sigma(i_m)} \tag{9.3}$$

for any permutation σ of the indices $i_2, \ldots, i_m$.

Thus one needs to specify only $n\binom{n+m-2}{n-1}$ entries of the matrices (one for the combination of each of the n strategies an individual can use and for each of the $\binom{n+m-2}{n-1}$ combinations of the remaining individuals; see Exercise 9.1).

The payoff to an individual playing $\mathbf{p}$ in a contest with individuals playing $\mathbf{p}_1, \mathbf{p}_2, \ldots, \mathbf{p}_{m-1}$ respectively is written as $E[\mathbf{p}; \mathbf{p}_1, \mathbf{p}_2, \ldots, \mathbf{p}_{m-1}]$. As the ordering is irrelevant, when some strategies are identical a power notation is used, for example $E[\mathbf{p}; \mathbf{p}_1, \mathbf{p}_2, \mathbf{p}_3{}^{m-3}]$ stands for the payoff to a player using strategy $\mathbf{p} = (p_1, p_2, \ldots, p_n)$ in a group of m players where one other player plays $\mathbf{p}_1$, one plays $\mathbf{p}_2$ and $(m-3)$ players play $\mathbf{p}_3$. General payoffs are given as follows:

$$E[\mathbf{p}; \mathbf{p}_1, \mathbf{p}_2, \ldots, \mathbf{p}_{m-1}] = \sum_{i=1}^{n} p_i \sum_{i_1=1}^{n} \cdots \sum_{i_{m-1}=1}^{n} a_{i i_1 i_2 \ldots i_{(m-1)}} \prod_{j=1}^{m-1} p_{j, i_j}, \quad (9.4)$$

where $\mathbf{p}_j = (p_{j,1}, p_{j,2}, \ldots, p_{j,n})$.

As Gokhale and Traulsen (2010) point out, as long as groups are selected from the population completely at random, as is usually assumed, then there is no real difference between symmetric and non-symmetric games. For example in a population playing three-player games every individual is equally likely to occupy any of the ordered positions, and in particular, the term a_{ijk} will have identical weighting to a_{ikj} in the payoff to an i-player. Thus the sum of these two can be replaced by twice their average, and in such cases we are reduced to the payoffs satisfying (9.3).

Broom et al. (1997b) coined the term *super-symmetric* for the following class of games, where all players in a game get the same payoff, dependent on the profile of strategies (but not on which individuals play them). The formal definition of super-symmetric games was given in Bukowski and Miękisz (2004). In the context of matrix games, a two-player game is (super-) symmetric if $a_{ij} = a_{ji}$ for all i, j and a multi-player matrix is super-symmetric if

$$a_{i_1 \ldots i_m} = a_{\sigma(i_1) \ldots \sigma(i_m)} \quad (9.5)$$

for any permutation σ of the indices $i_1, \ldots, i_m$.

For example, for super-symmetric, three-player, three-strategy games, there are ten distinct payoffs. Without loss of generality we can define the three payoffs $a_{111} = a_{222} = a_{333} = 0$, and this leaves seven distinct payoffs to consider: $a_{112}, a_{113}, a_{221}, a_{223}, a_{331}, a_{332}$ and a_{123}.

Before we specify the payoff $\mathcal{E}[\mathbf{p}; \Pi]$ to an individual playing $\mathbf{p}$ in the population described by Π, let us consider the specific case of two-strategy games.

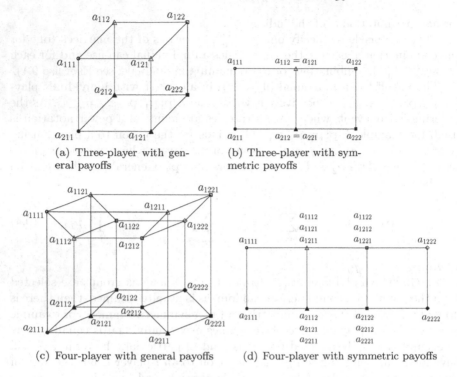

(a) Three-player with general payoffs

(b) Three-player with symmetric payoffs

(c) Four-player with general payoffs

(d) Four-player with symmetric payoffs

FIGURE 9.1: Visualisation of payoffs in two-strategy m-player games.

9.1.1 Two-strategy games

The complete payoffs to the three-player, two-strategy game can be written as

$$\begin{pmatrix} a_{111} & a_{112} \\ a_{121} & a_{122} \end{pmatrix} \begin{pmatrix} a_{211} & a_{212} \\ a_{221} & a_{222} \end{pmatrix} \tag{9.6}$$

where, as in (9.3) $a_{112} = a_{121}$ and $a_{212} = a_{221}$.

Similarly, for the four-player, two-strategy game we have

$$\begin{pmatrix} a_{1111} & a_{1112} \\ a_{1121} & a_{1122} \end{pmatrix} \quad \begin{pmatrix} a_{1211} & a_{1212} \\ a_{1221} & a_{1222} \end{pmatrix}$$
$$\begin{pmatrix} a_{2111} & a_{2112} \\ a_{2121} & a_{2122} \end{pmatrix} \quad \begin{pmatrix} a_{2211} & a_{2212} \\ a_{2221} & a_{2222} \end{pmatrix} \tag{9.7}$$

with symmetry conditions $a_{1112} = a_{1121} = a_{1211}$ etc; see Figure 9.1.

Thus the payoffs of the game are completely determined by specifying the payoffs to an individual playing pure strategy $i = 1, 2$ against $m - 1$ players, j of which play strategy S_1 (and the other $m - 1 - j$ play strategy S_2). Let us denote these payoffs by α_{ij}.

Example 9.2 (Bukowski and Miękisz, 2004). In Section 9.1.4 we look at the game with payoffs

$$
\begin{pmatrix} -3/32 & 0 \\ 0 & -13/96 \end{pmatrix} \quad \begin{pmatrix} 0 & -13/96 \\ -13/96 & 0 \end{pmatrix}
$$
$$
\begin{pmatrix} 0 & -13/96 \\ -13/96 & 0 \end{pmatrix} \quad \begin{pmatrix} -13/96 & 0 \\ 0 & -3/32 \end{pmatrix}.
$$

(9.8)

In this game if there are three players playing one strategy and one playing the other, the payoff to all players is 0; if the two strategies are played by two players each, all players receive $-13/96$; if all four players play the same strategy, then they receive $-3/32$.

Consider now what happens to an individual playing strategy **x** in a population where everybody else adopts a strategy **y**. A group of $m - 1$ opponents is chosen and each one of them chooses to play strategy S_1 with probability y_1 and strategy S_2 with probability y_2. Hence, with probability $\binom{m-1}{l}y_1^l y_2^{m-1-l}$, there will be l opponents playing strategy S_1 and $m-l-1$ opponents playing strategy S_2. On average, the payoff to the focal individual is thus

$$
\mathcal{E}[\mathbf{x}; \delta_\mathbf{y}] = \sum_{l=0}^{m-1} \binom{m-1}{l} y_1^l y_2^{m-1-l} E[\mathbf{x}; S_1^l S_2^{m-1-l}],
$$

(9.9)

where

$$
E[\mathbf{x}; S_1^l, S_2^{m-1-l}] = \sum_{i=1}^{2} x_i \alpha_{il}.
$$

(9.10)

A similar logic applies if the game is played in a population where a proportion y_1 of players plays pure strategy S_1 and a proportion $y_2 = 1 - y_1$ plays pure strategy S_2. Hence, it does not matter whether the population is polymorphic or monomorphic and playing the mean strategy, and so multi-player matrix games have the polymorphic-monomorphic equivalence property (see Chapter 7 and especially Definition 7.2).

With two strategies, as we see, the actual contest types will follow the binomial distribution. More generally with an arbitrarily large number of strategies the distribution will follow the appropriate multinomial distribution.

9.1.2 ESSs for multi-player games

The following definition of an ESS for an m-player game is a natural extension of the Definition 3.4 for a two-player game.

Definition 9.3. *A strategy* **p** *in an m-player game is called evolutionarily stable against a strategy* **q** *if there is an* $\varepsilon_\mathbf{q} \in (0,1]$ *such that for all* $\varepsilon \in (0, \varepsilon_\mathbf{q}]$

$$
\mathcal{E}[\mathbf{p}; (1-\varepsilon)\delta_\mathbf{p} + \varepsilon\delta_\mathbf{q}] > \mathcal{E}[\mathbf{q}; (1-\varepsilon)\delta_\mathbf{p} + \varepsilon\delta_\mathbf{q}],
$$

(9.11)

where

$$\mathcal{E}[\mathbf{x}; (1-\varepsilon)\delta_{\mathbf{y}} + \varepsilon\delta_{\mathbf{z}}] = \sum_{l=0}^{m-1} \binom{m-1}{l} (1-\varepsilon)^l \varepsilon^{m-1-l} E[\mathbf{x}; \mathbf{y}^l, \mathbf{z}^{m-1-l}]. \quad (9.12)$$

We say that $\mathbf{p}$ is an ESS *for the game if for every* $\mathbf{q} \neq \mathbf{p}$, *there is* $\varepsilon_{\mathbf{q}} > 0$ *such that* (9.11) *is satisfied for all* $\varepsilon \in (0, \varepsilon_{\mathbf{q}}]$.

Note that the payoff $\mathcal{E}[\mathbf{x}; (1-\varepsilon)\delta_{\mathbf{y}} + \varepsilon\delta_{\mathbf{z}}]$ in (9.12) was derived in a similar way as in Section 9.1.1.

Similarly, as in Theorem 6.2, we get the following theorem whose proof is left as Exercise 9.3.

Theorem 9.4. *For an m-player matrix game, the mixed strategy* $\mathbf{p}$ *is evolutionarily stable against* $\mathbf{q}$ *if and only if there is a* $j \in \{0, 1, \ldots, m-1\}$ *such that*

$$E[\mathbf{p}; \mathbf{p}^{m-1-j}, \mathbf{q}^j] > E[\mathbf{q}; \mathbf{p}^{m-1-j}, \mathbf{q}^j], \quad (9.13)$$

$$E[\mathbf{p}; \mathbf{p}^{m-1-i}, \mathbf{q}^i] = E[\mathbf{q}; \mathbf{p}^{m-1-i}, \mathbf{q}^i] \text{ for all } i < j. \quad (9.14)$$

A strategy $\mathbf{p}$ is called an *ESS at level J* if, for every $\mathbf{q} \neq \mathbf{p}$, the conditions (9.13)–(9.14) of Theorem 9.4 are satisfied for some $j \leq J$ and there is at least one $\mathbf{q} \neq \mathbf{p}$ for which the conditions are met for $j = J$ precisely.

If $\mathbf{p}$ is an ESS, then by Theorem 9.4, for all q,

$$E[\mathbf{p}; \mathbf{p}^{m-1}] \geq E[\mathbf{q}; \mathbf{p}^{m-1}]. \quad (9.15)$$

Since the payoffs are linear in the strategy of the focal player it follows from Theorem 7.5 that

$$E[\mathbf{p}; \mathbf{p}^{m-1}] = E[\mathbf{q}; \mathbf{p}^{m-1}], \text{ for all } \mathbf{q} \text{ with } S(\mathbf{q}) \subseteq S(\mathbf{p}). \quad (9.16)$$

Hence, in the generic case, any pure ESS is of level 0. We note that the generic case in this setting means that there is no non-trivial relationship between the payoffs apart from (9.3). A mixed ESS cannot be of level 0 because of (9.16); but in the generic case, any mixed ESS must be of level 1.

Also, similarly as in Theorem 6.6 and as a special case of Theorem 7.8, we get the following theorem whose proof is left as Exercise 9.6.

Theorem 9.5 (Local superiority for multi-player matrix games, Bukowski and Miȩkisz, 2004)**.** *In a multi-player matrix game, a strategy* $\mathbf{p}$ *is an ESS if and only if* $\mathbf{p}$ *is locally superior, i.e. there is a neighbourhood U of* $\mathbf{p}$ *such that*

$$\mathcal{E}[\mathbf{p}; \delta_{\mathbf{q}}] > \mathcal{E}[\mathbf{q}; \delta_{\mathbf{q}}], \text{ for all } \mathbf{q} \neq \mathbf{p}, \mathbf{q} \in U. \quad (9.17)$$

9.1.3   Patterns of ESSs

Analogues of the strong restrictions on possible combinations of ESSs we have seen in Chapter 6 do not hold for multi-player games. The Bishop-Cannings theorem fails already for $m = 3$; see Exercise 9.5. For $m > 3$, there can be more than one ESS with the same support as seen in Example 9.8. On the other hand, we still have the following for $m = 3$.

Theorem 9.6 (Broom et al., 1997b). *It is not possible to have two ESSs with the same support in a three-player matrix game.*

Proof. Suppose that **p** is an ESS of a three-player game. Then, by Theorem 9.4 exactly one of the following three conditions holds for any $\mathbf{q} \neq \mathbf{p}$,

(i) $E[\mathbf{p}; \mathbf{p}, \mathbf{p}] > E[\mathbf{q}; \mathbf{p}, \mathbf{p}]$,

(ii) $E[\mathbf{p}; \mathbf{p}, \mathbf{p}] = E[\mathbf{q}; \mathbf{p}, \mathbf{p}]$ and $E[\mathbf{p}; \mathbf{q}, \mathbf{p}] > E[\mathbf{q}; \mathbf{q}, \mathbf{p}]$,

(iii) $E[\mathbf{p}; \mathbf{p}, \mathbf{p}] = E[\mathbf{q}; \mathbf{p}, \mathbf{p}], E[\mathbf{p}; \mathbf{q}, \mathbf{p}] = E[\mathbf{q}; \mathbf{q}, \mathbf{p}]$ and
$E[\mathbf{p}; \mathbf{q}, \mathbf{q}] > E[\mathbf{q}; \mathbf{q}, \mathbf{q}]$.

Moreover, since **q** is also an ESS with $S(\mathbf{p}) = S(\mathbf{q})$ we have, by (9.16),

$$E[\mathbf{p}; \mathbf{p}, \mathbf{p}] = E[\mathbf{q}; \mathbf{p}, \mathbf{p}], \tag{9.18}$$

$$E[\mathbf{q}; \mathbf{q}, \mathbf{q}] = E[\mathbf{p}; \mathbf{q}, \mathbf{q}]. \tag{9.19}$$

Hence **p** must satisfy condition (ii) and thus

$$E[\mathbf{p}; \mathbf{q}, \mathbf{p}] > E[\mathbf{q}; \mathbf{q}, \mathbf{p}] = E[\mathbf{q}; \mathbf{p}, \mathbf{q}]. \tag{9.20}$$

However, by repeating the same process yet starting with **q** as an ESS, we get the reverse inequality in (9.20) which is a contradiction.   □

For super-symmetric three-player, three-strategy games, Broom et al. (1997b) found the complete set of attainable and unobtainable patterns of ESSs (except for one unresolved case), where a pattern is as defined in Section 6.5. Note that this is more complex than for matrix games since the Bishop-Cannings theorem does not hold. For example the pattern $\{(1), (2), (1, 2)\}$ is unobtainable, but the pattern $\{(3), (1, 3), (2, 3), (1, 2, 3)\}$ is obtained by the payoffs $a_{112} = 1, a_{113} = 1, a_{221} = -1, a_{223} = 1, a_{331} = -1, a_{332} = -1$ and $a_{123} = -1$.

9.1.4   More on two-strategy, m-player matrix games

Recall that as seen in Section 9.1.1, the payoffs of the m-player, two-strategy matrix game are given by α_{il} for $i = 1, 2$ and $l = 0, 1, \ldots, m - 1$. Let us define $\beta_l = \alpha_{1l} - \alpha_{2l}$ and consider the incentive function

$$h(p) = \mathcal{E}[S_1; \delta_{(p, 1-p)}] - \mathcal{E}[S_2; \delta_{(p, 1-p)}] \tag{9.21}$$

$$= \sum_{l=0}^{m-1} \binom{m-1}{l} \beta_l p^l (1-p)^{m-l-1}. \tag{9.22}$$

The function h quantifies the benefits of using strategy S_1 over strategy S_2 in a population where everybody else uses strategy $\mathbf{p} = (p, 1-p)$. In fact, the replicator dynamics now becomes (see Exercises 3.3 and 9.7)

$$\frac{dq}{dt} = q(1-q)h(q). \tag{9.23}$$

Also, note that h is differentiable. The following Theorem 9.7 should be compared to Theorem 7.3 as well as with Exercises 3.3 and 6.2 and results in Section 6.2.

Theorem 9.7 (Broom et al., 1997b; Bukowski and Miękisz, 2004). *In a generic two-strategy, m-player matrix game*

1. *pure strategy S_1 is an ESS (level 0) if and only if $\beta_{m-1} > 0$,*

2. *pure strategy S_2 is an ESS (level 0) if and only if $\beta_0 < 0$,*

3. *an internal strategy $\mathbf{p} = (p, 1-p)$ is an ESS, if and only if*

   (a) $h(p) = 0$, and

   (b) $h'(p) < 0$.

Proof. The result for pure strategies is a direct consequence of our discussion on generic cases and the fact that for S_1 to be an ESS, S_1 must do better in a population of S_1 strategists than S_2 does. Since in a population of S_1 strategists, most groups with a focal individual will contain $m-1$ players playing strategy S_1, we get that S_1 is an ESS if and only if $\alpha_{1(m-1)} > \alpha_{2(m-1)}$, i.e. $\beta_{m-1} > 0$.

By Theorem 9.5, an internal strategy $\mathbf{p} = (p, 1-p)$ is an ESS if and only if, for every $\mathbf{q} = (q, 1-q)$ close to $\mathbf{p}$ (with $q \neq p$) we have

$$0 < \mathcal{E}[\mathbf{p}; \delta_{\mathbf{q}}] - \mathcal{E}[\mathbf{q}; \delta_{\mathbf{q}}] = \sum_{i=1}^{2}(p_i - q_i)\mathcal{E}[S_i; \delta_{\mathbf{q}}] \tag{9.24}$$

$$= (p-q)\mathcal{E}[S_1; \delta_{\mathbf{q}}] + (q-p)\mathcal{E}[S_2; \delta_{\mathbf{q}}] = (p-q)h(q). \tag{9.25}$$

Thus $\mathbf{p} = (p, 1-p)$ is an ESS if and only if $h(q) > 0$ for all $q < p$ close enough to p and also $h(q) < 0$ for all $q > p$ close enough to p. Since payoffs are generic, it then follows that $h'(p) < 0$ (see also Section 7.1). The implication in the other direction follows in a similar manner (and generic payoffs are not required for it). $\square$

Note that the incentive function h given in (9.22) is a polynomial of degree $m-1$; thus it has at most $m-1$ distinct roots. If q_1 is a root of $h(q)$ with $h'(q_1) < 0$, we have that $h(q) > 0$ on some left neighbourhood of q_1 and $h(q) < 0$ on some right neighbourhood of q_1. Hence, if q_1 and q_2 are two consecutive roots of $h(q)$, $h'(q)$ cannot be negative at both of them. Consequently, exactly half of the roots of $h(q)$ between 0 and 1 will be ESSs if there is an even

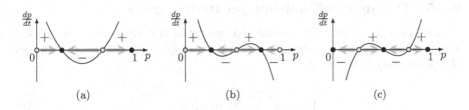

FIGURE 9.2: The incentive function and ESSs in multi-player games. The solid dots show the equilibrium points and the arrows show the direction of evolution under the replicator dynamics, see also Exercise 9.7.

number of such roots (and precisely one pure strategy will also be an ESS). If there is an odd number of roots between 0 and 1 then the number of ESSs will be either half of this number plus $1/2$ (with no pure ESSs) or half of this number minus $1/2$ (with two pure ESSs). This is illustrated in Figure 9.2.

Thus the possible sets of ESSs are one of the following:

1. 0 pure ESSs, and l internal ESSs with $l \leq \lfloor \frac{m}{2} \rfloor$;

2. 1 pure ESS, and l internal ESSs with $l \leq \lfloor \frac{m}{2} - 1 \rfloor$;

3. 2 pure ESSs, and l internal ESSs with $l \leq \lfloor \frac{m}{2} - 2 \rfloor$.

There can be more than one ESS with the same support in a four-player game as shown in the next example.

Example 9.8 (Bukowski and Miękisz, 2004). Consider the game from Example 9.2 given by

$$\begin{pmatrix} \alpha_{13} & 0 \\ 0 & \alpha_{22} \end{pmatrix} \quad \begin{pmatrix} 0 & \alpha_{11} \\ \alpha_{22} & 0 \end{pmatrix}$$
$$\begin{pmatrix} 0 & \alpha_{11} \\ \alpha_{22} & 0 \end{pmatrix} \quad \begin{pmatrix} \alpha_{11} & 0 \\ 0 & \alpha_{20} \end{pmatrix} \quad (9.26)$$

with $\alpha_{11} = \alpha_{22} = -\frac{13}{96}, \alpha_{13} = \alpha_{20} = -\frac{3}{32}$. Thus $\beta_0 = 3/32, \beta_1 = -13/96, \beta_2 = 13/96, \beta_3 = -3/32$ giving

$$h(p) = -\frac{3}{32}p^3 + \frac{13}{32}p^2(1-p) - \frac{13}{32}p(1-p)^2 + \frac{3}{32}(1-p)^3 \quad (9.27)$$

$$= -\left(p - \frac{1}{4}\right)\left(p - \frac{1}{2}\right)\left(p - \frac{3}{4}\right), \quad (9.28)$$

and thus the game has two internal ESSs at $\mathbf{p} = (1/4, 3/4)$ and $\mathbf{p} = (3/4, 1/4)$ (and no pure ESSs).

These two strategy games were further investigated by Peña et al. (2014), who considered both the general theory and some specific classes of games, such as public goods games and threshold games (see also Section 15.4).

9.1.5   Dynamics of multi-player matrix games

Broom et al. (1997b) considered evolutionary dynamics for super-symmetric three-player games with three strategies. The mean payoff over the population is given by

$$W = \sum_{i=1}^{3}\sum_{j=1}^{3}\sum_{k=1}^{3} a_{ijk}p_i p_j p_k. \tag{9.29}$$

The payoff to an individual playing pure strategy S_1 in such a population is

$$\sum_{j=1}^{3}\sum_{i=1}^{3} a_{1ij}p_i p_j = \frac{1}{3}\frac{\partial W}{\partial p_1}, \tag{9.30}$$

and thus using similar expressions for the payoffs to strategies S_2 and S_3, the continuous replicator equation is given by

$$\frac{dp_i}{dt} = p_i\left(\frac{1}{3}\frac{\partial W}{\partial p_i} - W\right) \qquad 1 \le i \le 3. \tag{9.31}$$

It should be noted that for super-symmetric games the discrete and continuous replicator dynamics lead to the same rest points, which are the local maxima of the function W (the result for recurrence relations, and thus for the discrete dynamics, comes from Baum and Eagon, 1967).

There are some results that occur for the dynamics on the three-player, super-symmetric games that cannot occur for two players (as perhaps we would expect). Two-player, super-symmetric games are important because they represent the genetic case where strategies represent alleles and a game a mating that leads to an offspring, where it can be reasonably assumed that the payoff to both players is the same, see Section 5.3.2 (e.g. see Edwards, 2000). Assuming that new strategies are allowed to enter a population sequentially and the population is allowed to converge to a new ESS, it is shown in Vickers and Cannings (1988a) that any new strategy that can invade the current ESS must subsequently feature in the support of the new ESS. This is not true in the multi-player case, as we see in Figure 9.3.

Cannings et al. (1993) show that in the two-player case any ESS can be reached by an appropriately ordered sequential introduction of strategies. This is again not true for the multi-player case, as we see in Figure 9.4 which has an unreachable ESS. This concept is not often discussed in static games or in replicator dynamics, where no new strategies are introduced but is of particular interest in adaptive dynamics, as we see in Chapter 14 (see also Exercise 14.9).

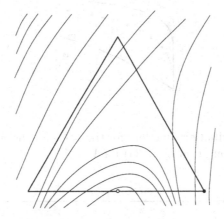

FIGURE 9.3: The game with payoffs $a_{111} = 0, a_{222} = 1.5, a_{333} = 0, a_{112} = 0.6, a_{113} = 0.05, a_{221} = 0, a_{223} = -1, a_{331} = 0, a_{332} = 0.5, a_{123} = 0.6$. The hollow dot on the base represents an equilibrium mixture of strategies 2 and 3, the solid dot at vertex 2 is the unique ESS and the solid lines represent contours of equal fitness. Here new strategy 1 can invade the (2, 3) mixture but the outcome is (2), i.e. 1 is not represented in the final outcome.

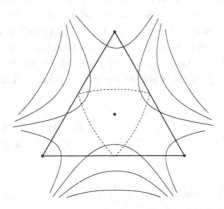

FIGURE 9.4: The game with payoffs $a_{111} = a_{222} = a_{333} = 0, a_{112} = a_{113} = a_{221} = a_{223} = a_{332} = -1, a_{123} = 1$. The solid dots in the centre and at the vertices are the ESSs, the solid lines are the contours and the dashed lines are the basins of attraction. Here we have ESSs with supports (1), (2), (3) and (1,2,3). With sequential introduction of strategies (1,2,3) can never be reached.

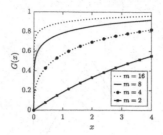

FIGURE 9.5: Distribution functions for the ESSs for the single reward multi-player war of attrition given by (9.34) for $V = 5$.

9.2   The multi-player war of attrition

The first explicit multi-player model in evolutionary games is that of the m-person war of attrition (Haigh and Cannings, 1989). They considered four related models, which we look at below.

9.2.1   The multi-player war of attrition without strategy adjustments

The first model of Haigh and Cannings (1989) is as follows. There are m players that compete for a single reward of value V. Individual i selects a random time X_i from some distribution (only) depending upon m and V. As in the standard war of attrition, a player receives the reward if the time that he selects is the largest, and pays a cost equivalent to the length of the time that he spends in the contest whether he wins or loses. The payoff to player i is thus given by

$$E[X_i; X_1, \ldots, X_{i-1}, X_{i+1}, \ldots, X_m] = \begin{cases} V - W_i & \text{if } X_i > W_i, \\ -X_i & \text{if } X_i < W_i, \end{cases} \tag{9.32}$$

where

$$W_i = \max(X_1, \ldots, X_{i-1}, X_{i+1}, \ldots, X_m). \tag{9.33}$$

In this game there is a unique ESS with all players choosing a value following a random distribution, with distribution function

$$G(x) = \left(1 - \exp(-x/V)\right)^{1/(m-1)} \quad x \ge 0, \tag{9.34}$$

illustrated in Figure 9.5.

How can we show this? Recall that a strategy $\mathbf{p}$ is given by a probability distribution $\left(p(t)\right)_{t \ge 0}$ and that the support of $\mathbf{p}$ is $S(\mathbf{p}) = \{t; p(t) > 0\}$. By a generalisation of Theorem 7.5 we must have that if $\mathbf{p}$ is an ESS, then for any $t \in S(\mathbf{p})$ and any pure strategy S_t (wait until time t and then give up if still in the game) we have

$$\mathcal{E}[S_t; \delta_{\mathbf{p}}] = \mathcal{E}[\mathbf{p}; \delta_{\mathbf{p}}]. \tag{9.35}$$

Suppose that $X_i : i = 2, \ldots, m$ are random variables with distributions given by $\mathbf{p}$. Since the X_is are independent, using (9.33) we get that

$$P(W_1 \leq t) = P(X_1 \leq t)^{m-1}. \tag{9.36}$$

Moreover, the first player is, in fact, effectively playing a two-person war of attrition against an opponent using strategy W_1 (ie. to wait for a time W_1 determined by (9.33) from $m-1$ randomly chosen times $X_j; j = 1, \ldots, m, j \neq i$). If $E[x, y]$ are the payoffs of the two-player war of attrition, (9.35) can be rewritten as

$$E[S_t, W_1] = K_1 \tag{9.37}$$

for some constant K_1. By closely following the analysis of a two-player war of attrition from Section 4.3.3, we obtain that w_1, the probability density function of W_1, must be of the form

$$w_1(t) = \frac{-K_2}{V} \exp(-t/V) \quad \text{for almost all } t \text{ with } w_1(t) \neq 0, \tag{9.38}$$

where the constant K_2 is chosen so that $\int_0^\infty w_1(x)\mathrm{d}x = 1$. As in Section 4.3.3, one candidate for an ESS stands out, namely

$$w(t) = \frac{1}{V} \exp(-t/V). \tag{9.39}$$

Let W be the strategy of a two-player war of attrition with density function w, and consider $W_1 \neq W$. Let $\mathbf{q}$ be the strategy of the m-player war of attrition so that when $m-1$ players follow that strategy, the last one gives up at the time with the same distribution as W_1. We get that $\mathbf{q}$ is not an ESS because by (4.25),

$$0 > E[W, W] - E[W_1, W] - E[W, W_1] + E[W_1, W_1] \tag{9.40}$$
$$= -E[W, W_1] + E[W_1, W_1], \tag{9.41}$$

and thus

$$\mathcal{E}[W; \delta_{\mathbf{q}}] = E[W, W_1] > E[W_1, W_1] \tag{9.42}$$
$$= E[\mathbf{q}, W_1] = \mathcal{E}[\mathbf{q}; \delta_{\mathbf{q}}]. \tag{9.43}$$

It follows that if there is an ESS strategy $\mathbf{p}$, it must correspond to w and from (9.39) it follows that $P(W \leq t) = 1 - \exp(-t/V)$ and thus by (9.36), the distribution function of $\mathbf{p}$ is $P(X_1 \leq t) = \left(1 - \exp(-t/V)\right)^{1/(m-1)}$. The proof that the ESS exists is more complicated and can be found in Haigh and Cannings (1989).

9.2.2   The multi-player war of attrition with strategy adjustments

The second model of Haigh and Cannings (1989) allowed players to adjust their strategy as others dropped out. Thus with m players all individuals select a time they are prepared to wait before becoming the first to leave; let us denote the time individual i selects as $X_i^{(m)}$. When the time

$$\min\{X_1^{(m)}, X_2^{(m)}, \dots, X_m^{(m)}\} \tag{9.44}$$

is reached one individual drops out and all remaining individuals select a new time $X_i^{(m-1)}$ that they are prepared to wait above and beyond the current time. We suppose that two or more individuals cannot leave simultaneously. If times are selected so that this will occur, then we select one of the individuals at random to leave, and all others are allowed to update following the described procedure.

Hence, the strategy in this case is an $(m-1)$−tuple $(\mathbf{p}_{m-1}, \mathbf{p}_{m-2}, \dots, \mathbf{p}_1)$ where $\mathbf{p}_i$ specifies how long the individual will wait when currently in the contest with i other individuals.

An intuitive argument based on induction over m can establish the only candidate for an ESS in this game. Suppose that there are two players remaining. We know that the optimal play is then to choose $\mathbf{p}_{\text{ESS}}$, an exponential distribution with mean V, given by (4.23). However, we also know that the expected reward from being in this position is zero. Thus for more than two players, the reward to get down to the final two is (at most) effectively zero. Thus it is optimal to quit immediately. Of course if everyone quits immediately then nobody wins the reward, but we did not allow simultaneous departures. Thus all choose 0, but are allowed to update if another is randomly chosen to leave first until there are two remaining players, when the usual exponential distribution is selected.

Above, we have identified that $P = (0, 0, \dots, 0, \mathbf{p}_{\text{ESS}})$ is the only candidate for an ESS. Now we will, however, illustrate that it is, in fact, not an ESS. Consider a three-player game. Then $P = (0, \mathbf{p}_{\text{ESS}})$. Let $Q = (q, 0)$ for small $q > 0$. We have

$$E[P; P^2] = 0 = E[Q; P^2], \tag{9.45}$$

$$E[P; PQ] = \frac{V}{2} = E[Q; PQ], \tag{9.46}$$

$$E[P, Q^2] = 0 < \frac{V}{3} - q = E[Q; Q^2]. \tag{9.47}$$

Thus, by Theorem 9.4, P is not an ESS and consequently, there is no ESS of the game. The game where all but one player receives the same reward (zero) can be thought of as non-generic. We now look at the case with multiple rewards.

ɪlti-player war of attrition with several rewards

remaining models are equivalent to the above models, except in
re is more than one reward available, and the ith individual to
collect reward V_{m+1-i}.

lividuals are allowed to update their strategy there is a unique
ase $V_m < V_{m-1} < \ldots < V_1$, where individuals play an exponential
ean $(i-1)(V_{i-1} - V_i)$ when the remaining number of individuals
e same tie-breaking rule as above, we can deal with ties at any
the existence of a unique candidate for an ESS extends to $V_m \leq$
$\leq V_1$. There may be no ESS as demonstrated in Section 9.2.2.
plex updating rule is required if the values of the V_is are not
this way.

the final two players with remaining rewards $V_2 > V_1$. Clearly
quit immediately, with expected reward $(V_1 + V_2)/2$ (as both
lo this). Thus in the position when there are still three players,
ch rewards V_3, V_2, V_1 has the same optimal strategy as that with
$/2, (V_1 + V_2)/2$. If $V_3 \geq (V_1 + V_2)/2$ then quitting immediately is
ise optimal play here is to use an exponential distribution with
$+ V_2)/2 - V_3) = V_1 + V_2 - 2V_3$. This method can be adapted
working from the contest with the final two players backwards
nduction) to give a general solution for the best play at all times
9.14).

case with more than one reward but no possibility to update the
is perhaps less interesting biologically and mathematically, and
scuss it.

.9. Find the candidate ESS for the multi-player war of attrition
$V_1 = 2, V_2 = 10, V_3 = 9, V_4 = 0$.

$< V_2$ so that the last two players quit immediately, receiving on
$< V_3$, so when there are three players the best strategy is still
ediately, with mean reward $(V_1 + V_2 + V_3)/3 = 7$. This gives the
vard to win the first contest as 7. Thus all individuals play an
time with mean $3 \times 7 = 21$, and then quit immediately after the
eaves.

ɪctures of dependent pairwise games

consider models which again involve the random selection of
ze m from a larger population, but within the m players a set
pendent pairwise contests are played. In particular, individuals

evious contests. These models were used to represent the forma-
nance hierarchies in Broom et al. (2000a,c) and we investigate this
el in Chapter 16. Here we introduce the mathematical structure
me observations only.

ockout contests

ut contest is a multi-player game comprising a number of pairwise
ssume initially that there are 2^M players which each play another
game where there is a "winner". The winners are then repaired
ubsequent round of games, and this process continues until there
erall winner. Rewards are allocated to the players based upon the
nd that they are eliminated. Thus an individual receives V_j if it is
n the round with 2^j players remaining (denoted round i, so that
occurs directly before round j), and the overall winner receives
$V_j \geq V_{j+1}$ for all j. Extra payoffs (rewards or more usually costs)
ned depending upon the strategy selected by the individual and
in any particular game. This process is modelled on real human
tests, for example tennis tournaments such as Wimbledon or the
though these involve some non-randomness by seeding the leading
Olympic boxing competitions. It is assumed in the mathematical
all pairings in a round are done at random, so each collection of
en with equal probability. The game is easily adapted to the case
are not exactly 2^M players by the introduction of byes, where
mprises only enough contests to ensure exactly 2^{M-1} individuals
-1 (again which individuals are involved in round M, and which
ith which, is decided randomly).

e many other structures of pairwise contests are possible (for ex-
e the defending champion is challenged by successive opponents
d as in professional boxing, which occurs in established dominance
where there is a dominant male which others seek to supplant, or
urnament (Broom and Cannings, 2002) commonly used in chess
). One advantage of the knockout structure from the perspective
tants is that it can give a final outcome with a unique winner and
ften the top positions only are important in real dominance hier-
lving relatively few contests (one less than the number of players,
s lose exactly one game except the overall winner).

ntage from the perspective of the modeller is that any pairwise
incorporated into such a structure, as long as in every possible
ither a clear winner can be identified or the winner can be selected
a given probability. An example is the Hawk-Dove game, as in
. (2000a). In this case the reward V for winning the contest is
progressing to the next round.

esting question concerns whether individuals are able to vary their
ing the sequence of pairwise interactions. Broom et al. (2000a)

ne extreme where individuals had to play a fixed pure strategy
nd. Broom et al. (2000c) considered the other extreme where
ould play distinct mixed strategies in every round (see also Broom
s, 2002). In this latter case, the term *strategy* is reserved for a
scription of the play of an individual in all rounds; the actual
ble in each round are referred to as *options*.

al. (2000c) developed an iterative backwards induction procedure
at from Haigh and Cannings (1989) to find the optimal strategy
nds based upon best play later on (i.e. rounds with lower index).
as the expected reward for winning in round k (with 2^k players
he expected rewards in round k could be organised into a payoff
re *Evolutionarily Stable Options* (ESOs) could be found in the
ESSs in a standard matrix game. Note that the concept of an
s similar to the choice of a single play in an iterated game (Section
oice from an extensive form game (Section 10.1).
; that individuals play $\mathbf{r}_j$ in round j for all $j < k$, and denoting
) an individual playing $\mathbf{p}_k$ against an individual playing $\mathbf{q}_k$ in
$\Xi[\mathbf{p}_k, \mathbf{q}_k; \mathbf{r}_{k-1}, \ldots, \mathbf{r}_1]$, then $\mathbf{p}_k$ is an evolutionarily stable option
onditional on $\mathbf{r}_{k-1}, \ldots, \mathbf{r}_1$ if, for all $\mathbf{q}_k \neq \mathbf{p}_k$,

$_k; \mathbf{r}_{k-1}, \ldots, \mathbf{r}_1] \geq E[\mathbf{q}_k, \mathbf{p}_k; \mathbf{r}_{k-1}, \ldots, \mathbf{r}_1]$ and

bove condition is satisfied with equality, then
$_k; \mathbf{r}_{k-1}, \ldots, \mathbf{r}_1] > E[\mathbf{q}_k, \mathbf{q}_k; \mathbf{r}_{k-1}, \ldots, \mathbf{r}_1].$

which is an ESS of the overall game must involve play of an
y stage of the game, conditional upon playing ESOs at all later
ered) stages. A strategy which plays an ESO at every stage of
ists invasion from any strategy which plays differently in a single
but this is not sufficient to ensure resistance from invasion by
hich differs in two or more rounds. Note the difference to the
trix games and their generalisations considered by Selten (1980)
8.1), where if there is any invading strategy, there is also one
n one place only. This difference occurs because if an individual
trategy in an earlier position it directly affects its probability of
ter contest.
wo-option case where S_1 beats S_2 with probability $1/2 + \Delta$, and
l playing S_i against S_j and losing pays a cost c_{ij}, the round k
x was given by

$$
\left(
\begin{array}{cc}
\frac{\tilde{V}_k - c_{11}}{2} & \frac{\tilde{V}_k - c_{12}}{2} + \Delta(\tilde{V}_k - 2V_k + c_{12}) \\
\frac{-c_{21}}{2} - \Delta(\tilde{V}_k - 2V_k + c_{21}) & \frac{\tilde{V}_k - c_{22}}{2}
\end{array}
\right), \quad (9.48)
$$

als V_k plus the expected reward for winning in round k (and thus eral, if

$$\frac{c_{11}}{2} - c_{12}\left(\frac{1}{2} - \Delta\right) - c_{21}\left(\frac{1}{2} + \Delta\right) + \frac{c_{22}}{2} > 0 \qquad (9.49)$$

ique ESO for each round, and thus a unique candidate ESS; see . For the Hawk-Dove game, the above parameters become $c_{11} = $ $= c_{22} = 0$ and $\Delta = 1/2$, and so this condition is automatically note that similar methods for Swiss tournaments are shown in ,3.

utionarily stable probability of playing Hawk in each round p_k is $= \min(1, z_k)$ where

$$z_1 = \frac{V_0 - V_1}{C}, \qquad (9.50)$$

$$z_{k+1} = \frac{z_k}{2} + \frac{V_k - V_{k+1}}{C} - \frac{p_k^2}{2}. \qquad (9.51)$$

vn in Broom et al. (2000c) that for the two-strategy, two-round •gy **p** which satisfies the above conditions with $0 < p_1, p_2 < 1$ is d only if

$$\left[2\Delta(V_0 - V_1) + \left(\frac{1}{2} + \Delta\right)c_{21} - \left(\frac{1}{2} - \Delta\right)c_{12}\right] <$$

$$\sqrt{2}\left(\frac{c_{11} + c_{22} - c_{12} - c_{21}}{2} + \Delta(c_{12} - c_{21})\right). \qquad (9.52)$$

-Dove case (9.52) reduces to

$$\frac{V_0 - V_1}{C} < \sqrt{2}, \qquad (9.53)$$

arly satisfied for $p_1 < 1$. An interesting feature of this condition es not include the payoff for losing in the first contest V_2, which s in the solution through the value of p_2.

s no ESS from the above system, then this can lead to interesting nd in certain cases at least, the candidate ESS can still play an ple, being the end point of the replicator dynamics from any initial hough these dynamics can on their trajectory visit a position very candidate ESS, irrespective of how close to it they start (this was al dynamics").

hon code

ction we show how to use Python to find the outcome of a multi-

```
i-player war of attrition

ers 'generate' the leaving times 'before' the game and
not change the times as others drop.

lation consists of m-1 residents and 1 mutant

 can define
of leaving times
d value(s) for the winner(s)
any players play the WA game

 calculates and outputs an average players' payoffs ...
er 1000 games).


t basic packages
py.random import seed, rand
umpy as np

he random number generator for reproducibility


e inputs
s to players, V[i] is reward for ith largest time
] is not explicitly defined, it is 0
natural but not required to have V[a]>V[b] for a<b
 8]

 of players in a contest together


e the player of WA game.
_Player:      # Defines the `resident' population
__init__(self):
self.Payoff = []   # Initially no payoffs
time_left = 0      # Time when left the contest

pdf(self, x):
""" Probability density function for leaving times
Does not have to integrate to 1 but should be pdf(x) < 1
Bulk of support must be between 0 and MaxLeavingTime """

return np.exp(-x/V[0])

getLeavingTime(self):
""" Generates random number between 0 and MaxLeavingTime
    based on the pdf
Tt uses rejection-sampling method:
 - Generates two random numbers, Time and u.
 - Returns Time only if u < pdf(Time)
 - Otherwise regenerates Time and u """

# Set "artificial" limit for max leaving time
MaxLeavingTime = 100
```

```
# Follow the rejection sampling method
Time = rand(1) * MaxLeavingTime
u    = rand(1)
while (u > self.pdf(Time)):
    Time = rand(1)*MaxLeavingTime
    u    = rand(1)
return float(Time)


tant(WA_Player):
tant is a WA player, with a different pdf
pdf(self,x):
""" Redefines pdf for leaving times.
This particular function is for uniformly random
distribution on [0 20] """
if (x<0) or (x>20):
    return  0
else:
    return  1/20


ion describing one round of WA game
round_wa(Players):
Simulates one round of WA game
fies Players Payoffs based on the outcomes of the game"""

etermine when Players want to leave
ingTime=[]        # Initiate a list storing leaving times
p in Players:
# Get when the player p wants to leave
p.time_left = p.getLeavingTime()
# Update the list of leaving times
LeavingTime.append(p.time_left)

rt Leaving times in descending order
ingTime.sort(reverse = True)

ssign payoffs to Players
p in Players:
# Find the order they left
order = LeavingTime.index(p.time_left)

if order == 0:
    # If this player won (wanted to wait max time)
    # it actually did not wait as long as intended
    # It left when the second to last player left
    p.time_left = LeavingTime[1]
    # Get the reward and subtract the waiting time
    p.Payoff.append(V[0] - p.time_left)

elif order<len(V):
    # Not the winner, but the reward is still specified
    # Get the reward and subtract the waiting time
    p.Payoff.append(V[order] - p.time_left)

else:
```

```
113                    # Player gets no reward and pays only the cost
114                    p.Payoff.append(0 - p.time_left)
115        return ()
116
117    # Create a list of players. m-1 regular players and one mutant
118    players = [WA_Player() for aux in range(m-1)]
119    players.append(Mutant())
120
121    # Play the WA game many times
122    for rep in range(1000):
123        one_round_wa(players)
124
125    # Calculate and print out mean payoffs
126    mp = [np.mean(p.Payoff) for p in players]
127    print(mp)
```

9.5 Further reading

Multi-player evolutionary games were first considered by Palm (1984). Multi-player matrix games were first considered in Broom et al. (1997b). Dynamical aspects of multi-player games are studied in the following papers: Bukowski and Miękisz (2004); Płatkowski (2004); Miękisz (2008); Pacheco et al. (2011). The differences between two-player and m-player games under the replicator dynamics was studied in Plank (1997). Wu et al. (2016) considered the emergence of multi-player games from pairwise ones.

Płatkowski and Stachowska-Pietka (2005) investigate the concept of an m-player mixed game where random subsets of the m players play games, and the total game is a probabilistically weighted combination of these, see also Płatkowski (2004); Płatkowski and Bujnowski (2009). In Płatkowski (2016) he introduced the idea of evolutionary coalitional games , incorporating classical coalition games into the evolutionary setting, with the population evolving following the replicator dynamics. Kamiński et al. (2005) consider stochastic games where evolution occurs in a finite population of m players, with games played between randomly selected groups of three players.

Miękisz and co-authors consider the stochastic stability of a finite population playing three player games with two strategies in Kamiński et al. (2005) for the well-mixed case, Miękisz (2004) for a spatially structured population and Miękisz et al. (2014); Bodnar et al. (2020) when there was an added time-delay in the games.

Gokhale and Traulsen (2010) develop the multi-player matrix game model of Section 9.1, including linking it to finite population games and the Moran process (see Chapter 12), see also Gokhale and Traulsen (2011); Wu et al. (2013); Han et al. (2012) and Duong and Han (2016). Gokhale and Traulsen

(2014) is a good review of work prior to its publication. The link to finite populations is further studied in Lessard (2011).

The evolution of cooperation in m-player games is considered in Bach et al. (2006) and Płatkowski and Bujnowski (2009) as well as in Kurokawa and Ihara (2009). Pacheco et al. (2009) study m-person stag hunt games, and Souza et al. (2009) study m-person Hawk-Dove games. We also consider m-person public goods games (Fehr and Gachter, 2002) in Section 15.4.1. The two-strategy game of Broom et al. (1997b) was also, recently, used in laboratory experiments to investigate human cooperative behaviour in Kuzmics and Rodenburger (2020).

The multi-player war of attrition was revisited in Helgesson and Wennberg (2015) who considered the model without strategy adjustments in the limit of infinitely many players.

Some examples of models involving games with a multi-player character within animal groups are van Doorn et al. (2003b), Hock and Huber (2006), Conradt and List (2009), and Broom et al. (2009).

Game theory, including evolutionary game theory, has been combined with quantum dynamics to give *quantum games*, (Eisert et al., 1999) including applications to evolution at the molecular level. Evolutionary games have particularly been considered by Iqbal and colleagues for both the two-player and multi-player case (Iqbal and Toor, 2001; Iqbal et al., 2008; Chappell et al., 2012; Iqbal and Cheon, 2008). An introduction to this area is given in Grabbe (2005).

9.6   Exercises

Exercise 9.1. Show that $m - 1$ individuals can be divided into k groups in $\binom{m-2+k}{k-1}$ ways, assuming that the size of each group matters, but that the identity of the individuals within each group does not.

Hint. There is a one-to-one correspondence between divisions and sequences of $m - 1$ I's and $k - 1$ D's, (the jth group will contain as many individuals as there are I's between the $j - 1$st and jth D in the sequence). The sequence is determined by the positions of $k - 1$ D's in it.

Exercise 9.2. How many payoffs have to be specified for a super-symmetric m-player game with k strategies?

Exercise 9.3. Prove Theorem 9.4.

Exercise 9.4. Prove (9.16).

Exercise 9.5. Give an example of a three-player matrix game where the Bishop-Cannings theorem fails.

Hint. One way is to start with a function h such as in (9.21) and derive the payoffs from there.

Exercise 9.6. Prove Theorem 9.5.

Exercise 9.7. Show that when $h(q)$ is a function defined by (9.21), then a replicator dynamics of any two-strategy game becomes $\frac{dq}{dt} = q(1-q)h(q)$. Compare to Exercise 3.3.

Exercise 9.8. Show that for $m > 3$, there is a (non-generic) two-strategy, m-player matrix game such that $\mathbf{p} = (p, 1-p)$ is an ESS, $h(p) = 0$ and yet $h'(p) = 0$. Compare to Theorem 9.7, part 3 and also to Bukowski and Miękisz (2004, Corollary 1).

Hint. Consider a polynomial $h(q)$ with a root of multiplicity 3 (or more).

Exercise 9.9. Show that for a two-strategy, m-player matrix game with generic payoffs, any mixed ESS must be of level 1. Give an example of a game with non-generic payoffs where a mixed ESS is of a higher level.

Hint. Compare to Exercise 9.8.

Exercise 9.10. Show that a two-strategy, m-player matrix game has always at least one ESS.

Hint. See Theorem 9.7.

Exercise 9.11 (Bukowski and Miękisz, 2004). Consider a two-strategy, three-player matrix game and give conditions for when each possible ESS combination could occur.

Exercise 9.12. Consider the payoffs defined in the matrices (9.26) of Example 9.8. Find the ESSs of the game if:
(a) $\alpha_{11} = \alpha_{22} = \frac{13}{96}, \alpha_{13} = \alpha_{20} = \frac{3}{32}$, (b) $\alpha_{11} = \alpha_{22} = -\frac{7}{24}, \alpha_{13} = \alpha_{20} = -\frac{3}{8}$.

Exercise 9.13 (Haigh and Cannings, 1989, Theorem 2). Show that the expected payoff in a multi-player war of attrition game described in Section 9.2.1 is 0.

Exercise 9.14 (Haigh and Cannings, 1989, Example 1, p.69). Find the candidate for an ESS in a four-player war of attrition game where the rewards V_k for k^{th} place are given by:
(a) $V_1 = 12$, $V_2 = 9$, $V_3 = 6$, $V_4 = 0$, (b) $V_1 = 12$, $V_2 = 6$, $V_3 = 9$, $V_4 = 0$.

Exercise 9.15 (Broom et al., 2000c). Consider the knockout contests games of Section 9.3.1. By considering the rewards attained for victory and defeat in a given contest, show that the payoffs for the kth round are given by (9.48). Using this matrix show that a necessary condition for a unique Evolutionarily Stable Option in each round is given by (9.49).

Hint. For a matrix game, a (generic) matrix has a unique ESS if it satisfies the negative-definiteness condition (6.45).

Chapter 10

Extensive form games and other concepts in game theory

Most evolutionary game-theoretic models that we have considered fall into the class of *normal form* games (see Section 2.1) with *perfect information*. In this chapter we consider two key departures from this structure which are still highly relevant to biological modelling. First we consider games in *extensive form*. Although they apply to a wider class, such games can be most usefully thought of as modelling interactions where individuals make sequential decisions, so the second player can choose a strategy based upon the observed choice of the first player. An example is a revised version of the game of brood care and desertion introduced in Section 8.4.2, which we consider below. Secondly we consider games where individuals do not have perfect information; either no player has perfect information, or some players do but others do not. Thus in the asymmetric Hawk-Dove game of Section 8.4.1 where individuals vary in size, perfect information would involve knowing the size of both yourself and your opponent. It is possible that an individual knows its own size, but that it has only an estimate of its opponent's size. Alternatively, in the kleptoparasitism models of Chapter 19 it is possible that food varies in quality (or in how much remains to be consumed) and the handler of the item may know this and the attacker may not (Broom and Rychtář, 2009). We then consider a type of game which can be thought to be both in extensive form and with imperfect information, repeated games.

10.1 Games in extensive form

A game in *extensive form* is a game governed by a sequence of moves, where each move is decided upon by one of the players of the game, or by chance, if appropriate. Thus common games such as chess, bridge and backgammon can be thought of as games in extensive form. Chess is a two-player game in extensive form with perfect information, and moves alternate between the players. We shall define an extensive form game using four central concepts (see e g van Damme, 1991).

DOI: 10.1201/9781003024682-10

10.1.1   Key components

10.1.1.1   The game tree

Throughout this book, a simple (un)directed *graph* will mean an ordered pair $G = (V, E)$ where V is a set of *vertices* and E is a set of (un)ordered pairs of distinct vertices (each such pair is called an *edge*). A *path* between two vertices v and v' is a finite sequence $\{v_i\}_{i=1,\ldots,n}$ of vertices such that $v_1 = v$, $v_n = v'$, and that there is an edge between v_i and v_{i+1} for all $i = 1, \ldots, n-1$. A graph is called *connected* if there is a path between any two vertices of the graph.

A tree is a simple connected graph which contains no cycles, so there is a unique path between two points. A tree is called a *rooted tree* if one vertex has been designated the root. In a rooted tree, the edges have a natural orientation, towards or away from the root. The *game tree* is a directed rooted tree which starts at a unique starting vertex, the root, from which there is a unique path to any other vertex. We shall assume that the tree, and so all the paths within it, are finite. The game proceeds along a path of the tree, where each vertex $x \in X$ represents a situation (e.g. a position in a game of chess). If $x \in X$ is a vertex, the set of immediately subsequent vertices, the *successors* of x, will be denoted by $S(x)$. A vertex $x \in X$ with no successors ($S(x) = \emptyset$) is called a *terminal vertex*.

10.1.1.2   The player partition

The player partition divides the set of nonterminal vertices of the game tree into disjoint sets $P_1, P_2, \ldots, P_m$, where m is the number of players. For example in a game of chess $m = 2$ and along every path of the tree if $x \in P_1$ all immediately subsequent vertices $S(x)$ are in P_2. Backgammon is also a two-player game with perfect information, but at some vertices in the game tree choices are made at random, not by the players. Thus we can add an extra player P_0 to the player partition, responsible for random moves. In the case of backgammon this is the roll of two dice, which then decides which options for moves the player to play has.

10.1.1.3   Choices

At each $x \in X$ the next step along the path, the choice $c(x)$, is decided by player i if $x \in P_i$. For a given $x \in P_i$, the possible choices that can be made are the elements of $S(x)$, i.e. $c(x) \in S(x)$, representing in a game of chess the moves the player to move can make in that position. We shall initially assume that all players (or at least the player to make the choice) know which vertex the game occupies with certainty at any time. This is clearly true in a game of chess, for example.

10.1.1.4 Strategy

A (pure) strategy of player i, ϕ_i, is the allocation of a choice for every $x \in P_i$. A mixed strategy as we know is a probability distribution over a set of pure strategies. We could think about a general mixed strategy which is a probability distribution over all possible pure strategies, or alternatively think about a probability distribution over the choices available at a particular vertex. For a generic game with perfect information we do not need to consider mixed strategies, since as we will see in Section 10.1.2 backwards induction leads to a unique pure solution to the game.

10.1.1.5 The payoff function

At the end of the path there is a terminal vertex where the game is over. At any terminal vertex, rewards $r_1, \ldots, r_m$ are allocated to each player. In traditional scoring in a game of chess, for example, the two rewards always add to 1, and there are only three possible payoff sets $(1,0)$ (white wins), $(0,1)$ (black wins) and $(1/2, 1/2)$ (a draw).

Example 10.1 (Sequential game of brood care and desertion, compare to Section 8.4.2). Consider a male-female pair that is just about to have offspring. A male chooses whether to care or desert (in most circumstances, a male can make the decision even before the offspring are born). Then, already knowing whether the male will care or not, the female makes her decision to care or desert.

The game is illustrated in Figure 10.1. The analysis of extensive form games, in general, is discussed in Section 10.1.2. In this specific example, we can see that the optimal strategy for the male is to desert (because if he would care, the female would desert) and the optimal strategy for the female is to care.

In general, it is possible to think of an adapted sequential version of most asymmetric games, since once the roles of the individuals can be distinguished, it can be assumed that the individual in one role has to choose his strategy before the other. Thus in the Owner-Intruder game of Section 8.3 we could imagine an intruder choosing his strategy first (and this may even include whether to attempt a challenge or not). The kleptoparasitism game of Section 19.5 can also be thought of in terms of extensive form games, as contests involve an individual which initially challenges (or not) and then an individual which responds to a challenge once it is made, and game trees similar to the one in Figure 10.1 are used in the analysis of these models.

10.1.2 Backwards induction and sequential equilibria

How do we find solutions to extensive form games? For example can we find a Nash equilibrium? Note that we will, in general, not discuss ESSs in this chapter. Finding ESSs is more complex in extensive form games, and

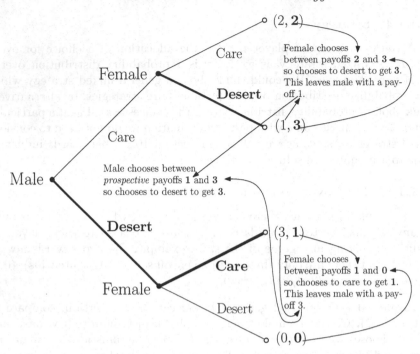

FIGURE 10.1: The sequential game of brood care and desertion from Example 10.1. The game tree, with the root the vertex furthest to the left, and the analysis via backwards induction. Optimal choices are denoted by bold lines.

this is best left to a work dedicated to the study of extensive form games; see Cressman (2003). We thus mention ESSs here only in some very specific circumstances.

It is easy to see that any vertex x in the tree K is the root of a subtree of K, denoted K_x. Starting from x we thus have a new game, a *subgame* of our original game. Note that when we do not have perfect information (Section 10.2) every subtree does not correspond to a subgame, as we shall see.

If a player knows her payoff from all of the choices in $S(x)$, she can choose the largest of these, and make that choice (which, in turn, gives the payoff at x). Thus at any point in the tree we can find the optimal choice at x conditional on knowledge of the optimal choices at all vertices which follow it. We can thus follow an inductive method starting at the terminal vertices and work backwards, *backwards induction*. Note that we have come across a similar procedure in Section 4.2.5 and Chapter 9, and we shall also see it again in Chapter 11.

Backwards induction will find *equilibrium paths*, and in particular, the so-called *subgame perfect equilibrium path*, the unique path along which the game

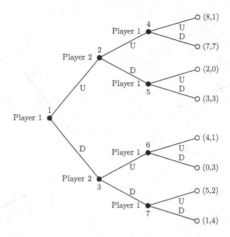

FIGURE 10.2: A game tree for the extensive form game from Example 10.2. Nonterminal nodes are labelled $1, \ldots, 7$.

must travel if all players always act rationally. In an extensive form game with perfect information, if the payoffs are generic (see Section 2.1.3.3), e.g. if all payoffs for all players are distinct, then there is a unique equilibrium path; see Exercise 10.1.

Example 10.2. Consider the two-player game described in Figure 10.2. The game is played over three rounds. Player 1 starts by choosing up (U) or down (D). Then player 2 chooses U or D and finally player 1 chooses again after which the payoffs are distributed. Find the equilibrium strategy.

We analyse this game using backwards induction as shown in Figure 10.3(a). Player 1 knows all the payoffs for successors of vertices $4, 5, 6, 7$. This allows her to make the optimal choice at those vertices and, in turn, yield the payoffs at the vertices. Subsequently, player 2 knows all of the payoffs for the successors of vertices 2 and 3 which allows him to choose optimally there and this, in turn, yield payoffs at vertices 2 and 3. Hence this allows player 1 to choose optimally at vertex 1. Thus the optimal moves are player 1 plays D at vertex 1, player 2 plays D at vertex 3 and player 1 plays U at vertex 7 and the rewards to players 1 and 2 are 5 and 2, respectively.

We note that as defined a strategy should specify the play at every vertex, and not just those that are reached by optimal play. Thus even if others make a mistake and we arrive at a vertex that should not be reached, a player should choose optimally at this point. A *sequential equilibrium* is one that plays the equilibrium strategy at every vertex, i.e. in every subgame.

If player 2 makes the mistake of playing U instead of D in Example 10.2 at vertex 3 it is important that player 1 makes the choice U with reward 4 rather

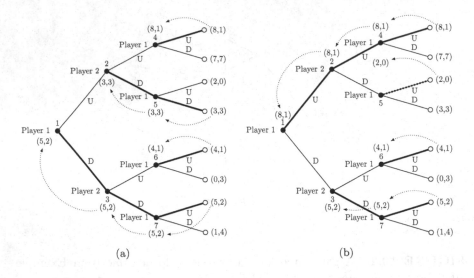

FIGURE 10.3: (a) The game tree and analysis for the extensive form game from Example 10.2. Optimal choices are denoted by bold lines. (b) Analysis when where we assume player 1 plays U at vertex 5 (for whatever reason).

than D with reward 0. The full strategy combination, yielding the optimal moves at each vertex, is thus described by $\{1_D 2_D 3_D 4_U 5_D 6_U 7_U\}$.

It turns out that there may be many equilibria of a game which do not have the sequential property, i.e. do not induce an equilibrium in every subgame. Such equilibria require, in some sense, illogical play (or the threat of illogical play) at some vertices, and so are deemed not sensible (Selten, 1965; Selten, 1975). We will not go into details here (e.g. see van Damme, 1991) but will consider one example.

Returning to Example 10.2, suppose that player 1 would play U and not D at vertex 5. This would then make U the best choice for player 2 at vertex 2 and thus U best for player 1 at vertex 1. This would yield the play to be UUU with player 1 receiving 8 and player 2 receiving 1 (the underlying strategy if logical play occurred at other vertices would be $\{1_U 2_U 3_D 4_U 5_U 6_U 7_U\}$). This is a Nash equilibrium because no unilateral change is optimal for the players concerned; see Exercise 10.2. Thus the threat of playing U at vertex 5 has improved the reward to player 1. The situation is described in Figure 10.3(b). However, this is not a credible threat, as if vertex 5 was reached, player 1 should logically play D.

The threat of playing U at vertex 5 is not credible for rational players. But what about for animal populations? What if the choice 5_U was inherited? Assuming all strategies i.e. all combinations of U or D at each of vertices $1, \ldots, 7$ are possible, and that occasionally individuals in the game end up playing a different strategy than intended (the trembling hand idea from Section 3.2.4) then the game may end up at vertex 5 even if optimal play at every vertex would not lead it there (see Figure 10.3) because player 2 may occasionally play D at vertex 2 by mistake. Hence a population playing $\{1_U 2_U 3_D 4_U 5_U 6_U 7_U\}$ would be invaded by $\{1_U 2_U 3_D 4_U 5_D 6_U 7_U\}$ since they do equally well everywhere except for the rare occasions when vertex 5 is reached, when $\{1_U 2_U 3_D 4_U 5_D 6_U 7_U\}$ do better. Thus $\{1_U 2_U 3_D 4_U 5_D 6_U 7_U\}$ would gradually replace $\{1_U 2_U 3_D 4_U 5_U 6_U 7_U\}$.

This, in turn, would make $\{1_U 2_U 3_D 4_U 5_D 6_U 7_U\}$ unstable (2_D is now optimal not 2_U), and so the sequential strategy combination $\{1_D 2_D 3_D 4_U 5_D 6_U 7_U\}$ would eventually prevail.

Note that Selten (1975) considers games where every available choice is made with some (small) non-zero probability (a version of the trembling hand), and under these circumstances only sequential equilibria are possible (called *perfect* equilibria).

10.1.3 Games in extensive form and games in normal form

We have so far discussed games in extensive form as a distinct type of game, as opposed to one in normal form. It is, in fact, possible to write any game in extensive form as a game in normal form, by listing the strategies available to each player and computing the payoffs directly assuming any combination of strategies. For example, the game in Example 10.1 has the strategies Care or Desert available to player 1 and the strategies (Care, Care), (Care, Desert), (Desert, Care) or (Desert, Desert) available to player 2, where the first (second) element in the pair is the choice made if player 1 Cares (Deserts). The payoffs for this game can be summarised as follows in a payoff bimatrix:

$$
\begin{array}{c}
\\
\text{Care} \\
\text{Desert}
\end{array}
\begin{array}{cccc}
\text{(Care, Care)} & \text{(Care, Desert)} & \text{(Desert, Care)} & \text{(Desert, Desert)} \\
\left(\begin{array}{cccc}
(2,2) & (2,2) & (1,3) & (1,3) \\
(3,1) & (0,0) & (3,1) & (0,0)
\end{array} \right).
\end{array}
\tag{10.1}
$$

Thus we can always think of an extensive form game as a game in normal form. However, as we see, converting extensive form games to normal form usually introduces a high degree of degeneracy. In our example, the payoffs for Care against (Desert, Care) and Care against (Desert, Desert) are equal, since the two-player, two-strategies are identical against the player 1 strategy actually played. There are four distinct pairs in the extensive form game, against eight in the matrix. This degeneracy increases quickly with the complication of the game; see Exercise 10.6. Many of our results depend on the payoffs being generic, including the general methodology for finding the ESSs of matrix and bimatrix games from Chapters 3 and 6 (as does the existence of the unique equilibrium path for extensive form games; see Exercise 10.1). Thus introducing non-generic payoffs often proves fatal to our analysis, and we should more properly think of games in extensive form as a distinct type of game for the purposes of evolutionary game theory.

10.2   Perfect, imperfect and incomplete information

In this section we first return to the idea of games in normal form, and in particular matrix games. There are a number of ways that a player might not have full information about a game. She might not know the opponent (or opponents in the case of multi-player matrix games) that she might play or perhaps even how many opponents there are. She might see a new opponent that she has never faced before, but be uncertain about its strength, which will then affect the expected payoffs from a given contest. Such an uncertainty about payoffs is termed *incomplete information*. This uncertainty about pay-offs is common in biology, for example with regard to assessing the size and strength of opponents as we saw in Section 8.4.1, and this is what we focus on first, before considering the alternative concept of imperfect information in extensive form games, where it is the precise location on the game tree that is sometimes not known.

10.2.1   Disturbed games

A *disturbed game* is a particular type of game with incomplete information which we describe below.

A disturbed game can be defined for an arbitrary normal form game (Harsanyi, 1973; van Damme, 1991), but we shall introduce it simply for a two-player bimatrix game with the payoff bimatrix (A, B). Let X_k be a random vector with distribution $f_k(X_k)$ for $k = 1, 2$ and x_1, x_2 be a pair of independent random observations of X_1, X_2. Player 1 (player 2) only knows the actual value of its own vector x_1 (x_2), and only knows the distribution of the vector of the other player. If player 1 plays strategy i and player 2 plays strategy j, then player 1 receives payoff $a_{ij} + x_{1,ij}$ and player 2 receives payoff $b_{ji} + x_{2,ji}$). Thus in a disturbed game an individual knows its own payoffs but not those of other players.

Example 10.3. Consider a Hawk-Dove contest over a food item where the value of the food item depends on the individual's level of hunger that is unknown to an opponent. Set up a model and analyse the game.

We will model the above scenario as a disturbed bimatrix game. Let the cost of a Hawk-Hawk contest always be C and let V be the intrinsic value of an item. Let X be a random variable representing the hunger level. Assume that X is uniformly distributed in the range $[-V, V]$ and that the value of a food item for an individual with hunger level x is $V + x$. A strategy is a function $p : [-V, V] \to [0, 1]$ that dictates a probability of playing Hawk when at hunger level $y \in [-V, V]$.

The focal individual with hunger level y has the following payoff matrix:

$$
\begin{array}{cc}
 & \begin{array}{cc} \text{Hawk} & \text{Dove} \end{array} \\
\begin{array}{c} \text{Hawk} \\ \\ \text{Dove} \end{array} & \begin{pmatrix} \dfrac{V + y - C}{2} & V + y \\ \\ 0 & \dfrac{V + y}{2} \end{pmatrix}
\end{array}.
\tag{10.2}
$$

Now, assuming that everybody in the population adopts a strategy p, we want to determine the best strategy q a focal individual can adopt in this population. If the focal individual has a hunger level y and plays against an individual with hunger level x, he will get

$$
E[q(y); p(x)] = q(y)p(x)\frac{V + y - C}{2} + q(y)(1 - p(x))(V + y) \cdots \tag{10.3}
$$
$$
+ 0 + (1 - q(y))(1 - p(x))\frac{V + y}{2}
$$
$$
= (1 + q(y) - p(x))\frac{V + y}{2} - q(y)p(x)\frac{C}{2}. \tag{10.4}
$$

Each individual opponent will play its $p(x)$ based upon its observed value of x, so the expected reward for playing q is just

$$
E[q(y); p] = (1 + q(y) - E_X(p(X)))\frac{V + y}{2} - q(y)E_X(p(X))\frac{C}{2} \tag{10.5}
$$
$$
= (1 - E_X(p(X)))\frac{V + y}{2} + q(y)\left(\frac{V + y}{2} - E_X(p(X))\frac{C}{2}\right). \tag{10.6}
$$

Hence, there is a unique critical value $y_c = y_c(p)$ satisfying

$$
\frac{V + y_c}{2} - E_X(p(X))\frac{C}{2} = 0 \tag{10.7}
$$

such that the best response to p is given by

$$
q(y) = \begin{cases} 1 & \text{if } y \geq y_c, \\ 0 & \text{otherwise.} \end{cases} \tag{10.8}
$$

Since an ESS must be a best response to itself, it must be of the form as in (10.8), i.e. there must be an x_c such that

$$
p(x) = \begin{cases} 1 & \text{if } x \geq x_c, \\ 0 & \text{otherwise.} \end{cases} \tag{10.9}
$$

Hence, using the distribution of X, we obtain

$$
E_X(p(X)) = \frac{V - r_s}{2V}, \tag{10.10}
$$

and because for the ESS we need $y_c = x_c$, we get from (10.7) that x_c must satisfy

$$\frac{V + x_c}{2} = \frac{V - x_c}{2V}\frac{C}{2},\qquad(10.11)$$

which means that

$$x_c = V\frac{C - 2V}{C + 2V}.\qquad(10.12)$$

For similar analysis and further developments on this model, see Binmore and Samuelson (2001).

We note that there is still a symmetry about the above games, since all players know their own payoffs but not those of others. Games can also be asymmetric, where one individual possesses more information than the other, e.g. in the kleptoparasitism model of Broom and Rychtář (2009) where when a challenge occurs, the handler knows the precise value of the food item competed for, but the challenger does not.

10.2.2   Games in extensive form with imperfect information— The information partition

In extensive form games the idea of disturbed games, with unknown payoffs, is still relevant, see Exercise 10.5, but we shall focus on an alternative situation without perfect information which is particular to extensive form games, where the players are uncertain about which vertex of the game tree that they occupy. Here we need to extend, and modify a little, the ideas that we introduced in Section 10.1.

The *information partition* divides each set P_i further into subsets, information sets, which we label U_{ij}. The partition of P_i is denoted by $U_i = \{U_{i1}, \ldots, U_{iJ_i}\}$ so that $P_i = \bigcup_{j=1}^{J_i} U_{ij}$, and we have the full information partition $U_1, U_2, \ldots, U_m$. An information set indicates a set of vertices which the player to make the decision cannot distinguish between. The player will know which information set he is in, but not which vertex within that set. For perfect information, each information set contains only a single vertex, so that there is no such uncertainty. Otherwise we have *imperfect information* (as distinct from incomplete information, as described above). The partitions can arise quite naturally. For instance, extending Example 10.2, when an individual knows it is his move but does not know which move was chosen by the opponent, we get the game as in Figure 10.4.

The information partition must satisfy the following requirements. Firstly each vertex in a given information set must have the same number of successors. This is forced by the fact that an individual should be able to make the same choices in any vertex of a given information set (otherwise, he would be able to distinguish between the vertices, which he cannot). Also, with suitable ordering of the subsequent steps, each player must make the same choice at all elements of any information set. Consequently, no path can contain two

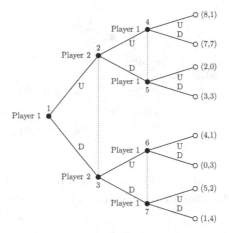

FIGURE 10.4: The game tree for the extensive form game from Example 10.2 where we assume that the players do not know their opponent's moves. Thus player 2 does not know whether he is in vertex 2 or 3. Similarly, player 1 cannot distinguish vertex 4 from 5 and vertex 6 from 7. However, since she remembers her first move, she can distinguish vertices 4 or 5 from 6 or 7. Indistinguishable vertices are connected by a dotted line.

vertices within the same information set (otherwise, there would be an infinite path).

A *subgame* is defined as in Section 10.1, but with the extra condition that the root of the subtree is in an information set with only a single element, i.e. it is a part of the tree where the player to choose knows its exact location.

Example 10.4. Following the sequential game of brood care and desertion from Example 10.1 let us suppose that player 2 cannot observe whether player 1 chooses to desert before making his own decision (this is another way of formulating the simultaneous decision case). Thus the information sets are $U_1 = [\{1\}]$ and $U_2 = [\{2, 3\}]$. Player 2 must make the same choices at vertices 2 and 3, and so only has the two strategies, Care and Desert, as player 1. The situation is shown in Figure 10.5.

With simultaneous play we have the payoff (bi)matrix

$$
\begin{array}{c}
 \\
\text{Care} \\
\text{Desert}
\end{array}
\begin{array}{cc}
\text{Care} & \text{Desert} \\
\begin{pmatrix} (2, 2) & (1, 3) \\ (3, 1) & (0, 0) \end{pmatrix}
\end{array} . \tag{10.13}
$$

Assuming that individuals always know which role they are in, we have two pure ESSs (Care, Desert) and (Desert, Care); see Exercise 10.8.

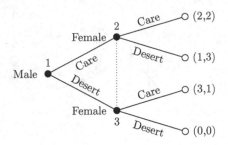

FIGURE 10.5: The game tree for the game of brood care and desertion with incomplete information from Example 10.4.

In the above example for simultaneous play we have two pure ESSs, instead of a unique path in the sequential game. Thus the result that there is a unique equilibrium path when we have perfect information does not carry over to when information is not perfect. If the moves are simultaneous, the second player cannot use the choice that the first has made, and consequently the first player cannot rely on the second player playing optimally conditional upon its move. Note that we have not observed a mixed ESS here, but that mixed solutions will be possible in related games if we had a version of the game involving nonlinear payoffs, similar to Section 8.4.2.2.

In general, the analysis of sequential games with imperfect information is more complex than that for perfect information. The same backwards induction procedure applies, except that each subgame must be analysed as a single unit. Within the subgame the analysis will not be backwards induction, but the subgame may be reformulated in normal form as in (10.13)). The backwards induction procedure for the whole game then works through a sequence of vertices which are the initial vertices of the subgames.

For a biological example of an extensive form game with imperfect information, we consider a brood parasite laying eggs in the nest of a host in Chapter 20. This example follows a sequence of decisions where a parasite may (or may not) have laid an egg in the nest of a host bird, when the host has to make a choice about whether to take either, both or neither of a sequence of two defensive steps against the parasite (Planque et al., 2002) (see also Harrison and Broom, 2009).

10.3   Repeated games

Here we show that repeated games could be analysed in the framework of extensive form games. Consider a Prisoner's Dilemma game where a pair of

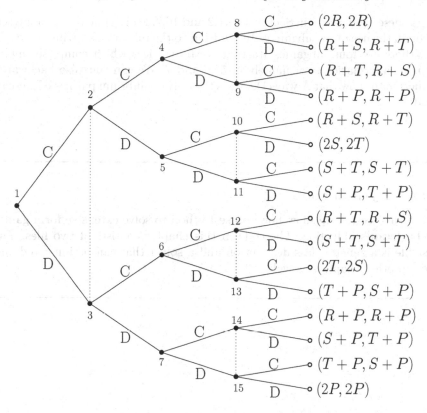

FIGURE 10.6: The Iterated Prisoner's Dilemma game as an form extensive game.

players play each other in two successive rounds. Listing the order of moves as player 1 round 1, player 2 round 1, player 1 round 2 and player 2 round 2 we can write the game as a sequential game with all paths of length 4 and 16 possible end vertices. With the conventional payoffs $T > R > P > S$ from Section 4.2 we obtain the game tree in Figure 10.6.

Note that since in reality in each round both players will play their moves simultaneously, the information sets are $U_1 = [\{1\}, \{4\}, \{5\}, \{6\}, \{7\}]$ and $U_2 = [\{2,3\}, \{8,9\}, \{10,11\}, \{12,13\}, \{14,15\}]$. Information sets with a single element, and thus the roots of subgames, occur whenever it is player 1 to move, but never when it is player 2.

Thus we can think of such a repeated game as a single extensive form game with alternating moves by player 1 and player 2, where the information sets for player 2 are such that they cannot discern the move of player 1 immediately preceding their move and the final payoff is just the sum of the payoffs from the sequence of interactions. In fact, we shall generally express such a game in a different way, as we see in Chapter 15 (see also Section 4.2.5).

As described earlier in Sections 10.1.2 and 10.2.2 this game can be analysed through the use of its subgames. In addition to the full game starting at vertex (1), there are four subgames starting at the vertices which comprise single-element information sets, namely 4, 5, 6 and 7. We can thus use backwards induction to show that Always Defect is the only equilibrium path; see Exercise 10.10.

10.4   Python code

In this section we show how to use Python to solve extensive form games by backwards induction. The code in this chapter consists of two files. The first file is a general class description and a solver that can be imported and used in other chapters as well.

```python
1   """ Builder and solver for extensive form games
2
3   This script provides a class 'node' that can be used
4   to define and store the game in the extensive form.
5
6   It also provides the function solveExtensiveGame
7   to solve the game
8
9   The game is coded as a rooted tree.
10  The user has to specify the nodes, starting from the root.
11  For each root, give the label, player's name and player's choices
12  To give the choice, specify the name and the label of the target
13  Terminal nodes have no players but we have to specify payoffs
14
15  CAUTION - the way the game tree is built causes an implicit
16           preference for actions: in every node, if two (or more)
17           actions yield the same payoff to the focal player,
18           the action that was created earlier is preferred.
19
20  """
21
22  class node:
23      " Defines the storage of the  game as a rooted tree"
24
25      def __init__(self, label=None, player=None, Players = [],
26                   parent = None):
27          self.label   = label      # Label of the node
28          self.player  = player     # Player at this node
29          self.Players = Players    # List of all players in the game
30          self.choices = []         # Available choices at this node,
31                                    # Initially set to empty,
32                                    # Add them as provided
33          # If a player chooses choices[a], the game will go
34          # to the node children[a]
35          # Initially set to empty, add them as provided
```

```python
36              self.children = []
37              self.best     = None     # Initially no choice is best
38              # Store payoffs ALL players are getting if at this node
39              self.payoffs  = [None for p in Players]
40              self.parent   = parent  # Node that leads to this one
41
42      def add_choice(self, choiceName, targetLabel):
43          """Adds a choice at the node
44              Returns a node where this choice leads """
45          # Create a new node; remember this is its parent
46          newNode = node(label=targetLabel, parent=self,
47                              Players = self.Players)
48          # Add the choice to the list of choices at current node
49          self.choices.append(choiceName)
50
51          # Add the new node to the children of the current node
52          self.children.append(newNode)
53          return newNode
54
55
56  ## Solve the game
57  def solveExtensiveGame(node,  printing = False):
58      """ Solves the game with the root given by node 'node'
59      Uses recursive algorithm / backward induction to solve it
60      During the solution, it stores optimal solution in the node
61      At the end, it prints out the optimal branch
62      If printing is set out to True, it prints out details
63          of how the solution was found"""
64      if printing: print('*** solving game at node ' + node.label)
65
66      # If there is a player, i.e. not a terminal node
67      if node.player is not None:
68          if printing:
69              print('player is ' + node.player)
70              print('choice are ')
71              print(node.choices)
72
73          # Initialize list of possible payoffs for all actions
74          posPay = [None for c in range(len(node.children))]
75
76          # Determine all payoffs the player can get
77          for c in range(len(node.children)):
78              if printing: print('at node ' + node.label +
79                                  ' player ' + node.player +
80                                  ' trying choice '+ node.choices[c]
81                                  + ' yielding to node with label '
82                                  + str(node.children[c].label))
83
84              # See what payoffs one can get with choice[c]
85              posPay[c] = solveExtensiveGame(node.children[c],
86                                              printing)
87
88          # Determine where the maximal payoff occurs
89          # In case of multiple max payoffs, the next line returns
90          # the FIRST occurence (i.e. order of actions matters)
91          ind = posPay.index(max(posPay))
92
```

```
93           # Get the corresponding best choice
94           node.best = node.choices[ind]
95
96           # Update payoffs at this node by payoffs from best choice
97           node.payoffs = node.children[ind].payoffs
98           if printing:
99               print('at the node label ' + str(node.label) +
100                   ' the possible payoffs are')
101              print(posPay)
102              print(node.choices)
103              print('the best choide is ' + node.best)
104              print('the payoffs at node ' + node.label + ' are ')
105              print(node.payoffs)
106      else:   # If there is no player, i.e. terminal node
107          if printing: print(' terminal node ')
108
109      # Determine what will the function return
110      if node.parent is not None: # If not at the root of the game
111          # Get the player whose choice lead here
112          lastPlayer = node.parent.player
113
114          # Get the payoff the last player would get here
115          payoff_to_lastPlayer = ...
116                  node.payoffs[node.parent.Players.index(lastPlayer)]
         if printing:
117              print('The player ' + lastPlayer +
118                  ' would get ' + str(payoff_to_lastPlayer) )
119              print('*** game at node ' + node.label + ' solved ')
120
121          return payoff_to_lastPlayer
122
123      else:   # We are at the root of the game
124          if printing: print('*** game at node ' + node.label
125                              + ' solved ')
126          # Print out the solution. Print sequence of actions first.
127          # Start at the current node
128          place = node
129          # While there is a choice to make/place to go
130          while place.best is not None:
131              # Identify where the optimally-played game goes
132              nextPlace = ...
                     place.children[place.choices.index(place.best)]
133              print('at node ' + place.label +
134                  ' player ' + place.player +
135                  ' chooses ' + place.best +
136                  ' which leads to node ' + nextPlace.label)
137              # Move to the next node of the game
138              place = nextPlace
139
140          # Now we are at the terminal node
141          print ('at the terminal node ' + place.label + ':')
142
143          # Printout payoffs to all players there
144          for p in range(len(place.Players)):
145              print('Player ' + place.Players[p] + ' gets '
146                          + str(place.payoffs[p]))
```

The second file contains an example of the extensive form game, then shows how to define it and how to solve it.

```
1   """ Extensive form game
2   It defines and solves the game corresponding to Figure 10.2
3   two players, three rounds, two options in each round
4   """
5
6   ## Import the class and a solver for extensive games
7   # See comments in the file ExtensiveGames
8   from ExtensiveGames import node, solveExtensiveGame
9
10  ## Define the game
11  # Specify the root of the game and who playes at the root
12  # and who are all players in the game
13  n1 = node(label = '1', player = 'Player 1',
14            Players = ['Player 1', 'Player 2'])
15
16  # Add choices at the root
17  n2 = n1.add_choice(choiceName = 'U', targetLabel = '2')
18  n2.player = 'Player 2'
19
20  n3 = n1.add_choice(choiceName = 'D', targetLabel = '3')
21  n3.player = 'Player 2'
22
23  # Add choices at node n2
24  n4 = n2.add_choice(choiceName = 'U', targetLabel = '4')
25  n4.player = 'Player 1'
26  n5 = n2.add_choice(choiceName = 'D', targetLabel = '5')
27  n5.player = 'Player 1'
28
29  # Add choices at node n3
30  n6 = n3.add_choice(choiceName = 'U', targetLabel = '6')
31  n6.player = 'Player 1'
32  n7 = n3.add_choice(choiceName = 'D', targetLabel = '7')
33  n7.player = 'Player 1'
34
35  # Add choices at node n4, n5, n6, n7
36  n8 = n4.add_choice(choiceName = 'U', targetLabel = '8')
37  n9 = n4.add_choice(choiceName = 'D', targetLabel = '9')
38  n10 = n5.add_choice(choiceName = 'U', targetLabel = '10')
39  n11 = n5.add_choice(choiceName = 'D', targetLabel = '11')
40  n12 = n6.add_choice(choiceName = 'U', targetLabel = '12')
41  n13 = n6.add_choice(choiceName = 'D', targetLabel = '13')
42  n14 = n7.add_choice(choiceName = 'U', targetLabel = '14')
43  n15 = n7.add_choice(choiceName = 'D', targetLabel = '15')
44
45  # Specify payoffs at the terminal nodes
46  n8.payoffs  = [8, 1]   # payoff to P1, payoff to P2
47  n9.payoffs  = [7, 7]   # payoff to P1, payoff to P2
48  n10.payoffs = [2, 0]   # payoff to P1, payoff to P2
49  n11.payoffs = [3, 3]   # payoff to P1, payoff to P2
50  n12.payoffs = [4, 1]   # payoff to P1, payoff to P2
51  n13.payoffs = [0, 3]   # payoff to P1, payoff to P2
52  n14.payoffs = [5, 2]   # payoff to P1, payoff to P2
```

```
53  n15.payoffs = [1, 4]  # payoff to P1, payoff to P2
54
55  # Solve the game and print out the way how we solved it
56  solveExtensiveGame(n1, printing = True)
57
58
59  """
60  # Another example game is below
61  # Just a simple game, 2 players, two moves
62  GT = node(label = '0', player = 'Alice', Players=['Alice','Bob'])
63  print(GT.label)
64  n1 = GT.add_choice('left', '1')
65  n2 = GT.add_choice('right', '2')
66  n1.player = 'Bob'
67  n2.player = 'Bob'
68  n3 = n1.add_choice('up', '3')
69  n4 = n1.add_choice('down', '4')
70  n5 = n2.add_choice('up', '5')
71  n6 = n2.add_choice('down', '6')
72  n3.payoffs = [1, 7]  # payoff to Alice, payoff to Bob
73  n4.payoffs = [6, 5]  # payoff to Alice, payoff to Bob
74  n5.payoffs = [4, 3]  # payoff to Alice, payoff to Bob
75  n6.payoffs = [0, 1]  # payoff to Alice, payoff to Bob
76  solveExtensiveGame(GT, printing = True)
77  """
```

10.5   Further reading

See van Damme (1991) and also Laraki et al. (2019) for more information on extensive form games. The topic of extensive form games has been widely discussed, see especially Selten (1975), other useful references being Kreps and Wilson (1982) and van Damme (1984). Cressman (2003) is a whole book which studies extensive form games from the point of view of evolutionary dynamics, see also Chamberland and Cressman (2000) and Gatti et al. (2013). For evolutionary games in extensive form in the context of multi-agent learning see Gatti and Restelli (2016). For disturbed games see Harsanyi (1973) and also van Damme (1991).

Biological interactions involving a sequence of decisions, such as a brood parasite and its host, can be modelled using extensive form games (Planque et al., 2002, Harrison and Broom, 2009). Games with imperfect or incomplete information are relevant to competition for a territory or food item (e.g. see Broom and Rychtář, 2009) where the owner may possess more information than the intruder about the value of the territory.

Disturbed games are relevant to the idea of biological competition for food or territory, where an individual knows its own level of hunger or strength but not that of its opponent, although in such a case an individual's payoff would be directly affected by the strength of its opponent (e.g. see Enquist and Leimar, 1983 for an example with fighting deers and Hofmann and Schild-berger, 2001 for an example with fighting crickets).

10.6 Exercises

Exercise 10.1. Show that in an extensive form game with perfect information and generic payoffs, there is always a unique equilibrium path. Contrast this to the situation with imperfect information, Exercise 10.9.

Exercise 10.2. Consider the extensive form game from Figure 10.2. Show that the strategy $\{1_U 2_U 3_D 4_U 5_U 6_U 7_U\}$ is a Nash equilibrium (see also Figure 10.3 where player 1 insists on playing U at the vertex 5).

Hint. Player 1's strategy is to play U at every vertex. Show that player 2's strategy to play U at vertex 2 and D at vertex 3 is the best response to player 1's strategy. Similarly, show that player 1's strategy is a best response to player 2's strategy.

Exercise 10.3. Consider a subgame of the game described in Figure 10.2 that starts at vertex 2. Note that optimal play yields $(3, 3)$, whereas the players could get payoff $(7, 7)$ if they played irrationally. Write this subgame in normal form and compare it to the Prisoner's Dilemma.

Exercise 10.4. Find the normal form version of the extensive form game from Example 10.2.

Exercise 10.5. Consider a sequential game of brood care and desertion from Example 10.1 as in Figure 10.1 but where payoffs to the deserting individual are disturbed by the addition of a random variable X uniformly distributed on $[-2, 2]$ and known only to the individual. Find the optimal strategy.

Exercise 10.6. Consider a two-player extensive form game played over $2n$ rounds where players make alternate moves. Let each player have k choices at any vertex. How many strategies would each player have if the game was written in normal form (as in Section 10.1.3)?

Exercise 10.7. Find the ESS for the extensive form game with incomplete information shown in Figure 10.4.

Exercise 10.8. Solve the bimatrix game with payoffs given by (10.13).

Exercise 10.9. In the game shown in Figure 10.7(a) find the unique path for the sequential game. Also, consider the game shown in Figure 10.7(b) and construct the payoff matrix for the simultaneous game, and hence show that there are two pure ESSs.

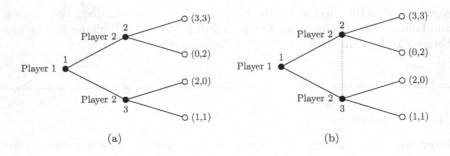

(a)　　　　　　　　　　　　　　　(b)

FIGURE 10.7: (a) An extensive game with complete information and unique ESS, (b) An extensive game with incomplete information and two ESSs.

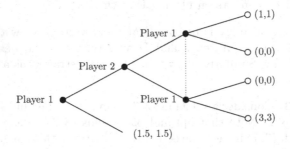

FIGURE 10.8: Game from Exercise 10.11. Determine the Nash equilibria.

Exercise 10.10. Solve the game shown in Figure 10.6. Recall that we assume $T > R > P > S$.

Exercise 10.11. Consider the game in Figure 10.8. Determine the Nash equilibria.

Hint. Consider each possible sequence of plays, and decide when it will be in the interest of either player to change. In particular, can you decide whether it is best for player 1 to change from the choice of playing D in the first move?

Chapter 11

State-based games

In previous chapters we have considered games where there may have been a large number of strategies available to the players, but the number of distinct situations where they had to make a choice were restricted. There may have been many individuals, but the position of all within the population was essentially identical. Within some of the m-player contests of Chapter 9 there could be distinctions based upon results achieved, but this was restricted to competition within groups of size m only and prior to the selection of competing groups all individuals were identically placed.

Thus, in the Hawk-Dove game (see Section 4.1), there is only the choice of Hawk or Dove, in a single type of interaction. In the Owner-Intruder game (see Section 8.3) there is the same range of choices, but they must be made in either the Owner or the Intruder position. The kleptoparasitism game (see Section 19.5) is a lot more complicated because of its non-linearity, but the set of scenarios where a strategy needs to be employed, and the strategies available, namely challenge or not when the opportunity arises and defend or not when challenged, are just as simple.

In Chapter 10 we considered *extensive form games* and saw situations where each interaction is a series of choices for the participants, and so the individual games themselves can be complex. In this chapter and the next we consider games which are still simple in form, but where complexity arises from the fact that there are multiple scenarios which a player can face, and where a complete strategy involves a choice in every conceivable position that an individual can face.

We begin by considering situations where the different scenarios are governed by internal properties of the individual itself, such as its state of hunger, or by properties of nature, such as the time of year, but which are not explicitly governed by the composition of the population, which is still effectively infinite and well mixed. In Chapter 12 we will then move on to consider finite population games both for their own interest and as an introduction to games where the structure of the population is vital, in particular games on graphs.

DOI: 10.1201/9781003024682-11

11.1 State-based games

We have usually previously assumed that all individuals are identical, and rewards and costs are the same for any pair of individuals when they meet. When this has not been the case, such as in Chapter 8 when individuals could occupy distinct roles, and their strategy depends upon the role occupied, all situations consisted of one individual in each role. In fact, an animal's decisions often depend upon certain temporary characteristics of the animal, such as its level of hunger or the proximity to the onset of winter. A hungry animal should be willing to take more risks than a satiated one, for example. Such temporary characteristics are termed the *state* of an individual. It should be noted that under some circumstances a property of an individual, e.g. its size, may be considered as a fixed parameter (e.g. for an adult that has stopped growing) and as a state variable in others (e.g. a juvenile which is still growing); we consider size in Section 11.2. The concept of state is described in detail in Houston and McNamara (1999) and we will only give a brief flavour of this subject.

Optimal decisions may be made in response to nature only, and so be in the realm of decision theory rather than game theory, and we shall start by thinking about that case, considering a simple optimal foraging problem.

11.1.1 Optimal foraging

Optimal foraging theory is an area of mathematical modelling in biology, concerning how best to search for food (for a distinct class of search games, see Section 20.6). We shall not discuss this except in the briefest of terms. There has been debate about the best way for individuals to search an environment for food items. So-called Levy walks (a type of random walk) have been proposed as the best way for individuals to search for food in natural environments where food dispersal can happen in a variety of ways (see e.g. Sims et al., 2008) but is generally not uniformly distributed (in the deterministic or random sense). In general, optimal foraging may depend not only upon how food is spread out, but also upon what an individual is trying to maximise.

The simplest assumption is that an individual may wish to maximise its expected foraging return. In healthy individuals with little mortality risk due to starvation, this may be true. Alternatively, individuals may follow a different strategy, supposing they need to reach a sufficient foraging threshold to survive. As we see, this can be a more conservative or more risky strategy, depending upon the threshold required. If survival is likely, but not certain, then achieving the threshold value requires a conservative strategy. If survival is unlikely, see for example Pitchford and Brindley (2001) and Pitchford

TABLE 11.1: Foraging rewards from Example 11.1. An individual spending k units of time in area B, receives a reward $2E$ from area B and total reward X total (with probability p).

k	E	x	p
0	0	3	1
1	0	2	0.7
1	1	4	0.3
2	0	1	0.49
2	1	3	0.42
2	2	5	0.09
3	0	0	0.343
3	1	2	0.441
3	2	4	0.189
3	3	6	0.027

et al. (2005), where only a small number out of thousands of young survive to adulthood, a much more risky strategy is needed.

Example 11.1 (Optimal foraging and risk). An animal has two available feeding areas. In area A it can feed at a constant rate 1 per unit time, at no risk. In area B it either receives 0 with probability 0.7 or 2 with probability 0.3 per unit time. Our animal has three time units to forage. How should it divide its time, if it needs to reach a resource level x_c to survive?

If our individual spends k time units in area B then it obtains resource $X = 3 - k + 2E$, where E is binomially distributed with parameters k and 0.3. This leads to Table 11.1. It is left as Exercise 11.10 that the optimal choice of k (assuming the least variable option is chosen if expectations are tied), when the required resource is x_c, is as shown in (11.1).

$$\begin{array}{c|cccccc} x_c & 1 & 2 & 3 & 4 & 5 & 6 \\ \hline k & 0 & 0 & 0 & 1 & 2 & 3 \end{array} \tag{11.1}$$

In the above game, suppose that $x_c = 4$. As we see the optimal choice of k is 1. This is assuming a fixed strategy, where k is picked in advance. However, what if we considered a more dynamic strategy, where the choice could be updated after each period? In that case, when playing a mixture of risky and safe options, it is best to pick the riskier options first, as we see in Section 11.1.3. Thus an animal should forage in area B first, and if it succeeds it can switch to area A, safely achieving the critical level 4. If it fails to find food, it can then stay at area B until the end of foraging, and will survive with probability $0.3 + 0.7(0.3)^2 = 0.363$ instead of 0.3 achieved with the static strategy. In the sections that follow, such dynamic strategies are central to the models.

11.1.2   The general theory of state-based games

We shall now move on to look at the theory of state more generally, following the methodology of Houston and McNamara (1999). The state of an animal is described by a vector **x** potentially consisting of a number of elements. An animal which is currently in state **x** at time t has an expected reward $V(\mathbf{x}, t)$. For example, this may be the expected number of offspring produced from t until death, which is *Fisher's reproductive value*, Fisher (1930). The animal is faced with a potential choice of actions at each successive time point. The chosen action u influences the fitness in three ways.

(i) *The direct contribution*: as a result of the action taken, in combination with the state of the individual and the time, the individual might make a direct contribution to its overall fitness, e.g. by giving birth. This is labeled $B(\mathbf{x}, t; u)$.

(ii) *The survival probability*: the action taken, state and time will affect the probability that the individual survives to time $t + 1$. For example giving birth may strongly influence survival for an animal in a poor state, but affect it much less in a healthy individual. This is labeled $S(\mathbf{x}, t; u)$.

(iii) *Change in state*: each action will potentially affect the state of an individual. Giving birth will weaken the state, whereas concentrating on feeding will likely improve the state for the next time point. The new state is denoted $\mathbf{x}'_{\mathbf{u}}$.

Assuming that we know how the state $\mathbf{x}'_{\mathbf{u}}$ will contribute to reproductive success in the future, $V(\mathbf{x}'_{\mathbf{u}}, t + 1)$, we can combine the terms above to give the expected reward to our individual, given it is in state **x** and chooses action u at time t as

$$H(\mathbf{x}, t; u) = B(\mathbf{x}, t; u) + S(\mathbf{x}, t; u)V(\mathbf{x}'_{\mathbf{u}}, t + 1). \qquad (11.2)$$

More generally if we do not know in what state $\mathbf{x}'_{\mathbf{u}}$ the individual ends up after an action u (because, for example, assuming some stochastic effects the individual can end up in many possible states) but know the expectation $E_u[V(\mathbf{x}'_{\mathbf{u}}, t + 1)]$ of making choice u, then we obtain

$$H(\mathbf{x}, t; u) = B(\mathbf{x}, t; u) + S(\mathbf{x}, t; u)E_u[V(\mathbf{x}'_{\mathbf{u}}, t + 1)]. \qquad (11.3)$$

We can then find the optimal choice of action by choosing u^* so that

$$H(\mathbf{x}, t; u^*) = \max_u H(\mathbf{x}, t; u), \qquad (11.4)$$

although as we saw in Section 11.1.1 (see also Section 11.1.3, expectations alone may not always be enough to find the best strategy. Assuming that this optimal choice is made, we obtain

$$V(\mathbf{x}, t) = H(\mathbf{x}, t; u^*). \qquad (11.5)$$

Thus, in general, the reproductive value of an individual is obtained by assuming it behaves optimally.

11.1.3   A simple foraging game

Houston and McNamara (1999) give the following simple example, again a foraging scenario as in Section 11.1.1. Suppose that an animal with current energy reserves (its state) x can either forage or rest in cover at every time step. If it forages (chooses an action u_f) it will increase its energy level from x to $x+1$ but risks death by predation, surviving with probability $1-z$. If it rests (chooses an action u_r) its energy level does not increase, but there is no predation risk. Thus we have

$$H(x,t;u_f) = (1-z)V(x+1,t+1), \tag{11.6}$$
$$H(x,t;u_r) = V(x,t+1). \tag{11.7}$$

Assuming that there are no ties, so one strategy is always clearly better, the optimal strategy is to pick u_f if and only if $H(x,t;u_f) > H(x,t;u_r)$ which is equivalent to

$$z < z_c = \frac{V(x+1,t+1) - V(x,t+1)}{V(x+1,t+1)}, \tag{11.8}$$

so that our individual should forage if and only if the predation risk is less than a critical value z_c.

In the above example, $V(x,t)$ is a function of both time t and reserves x, and hence the optimal foraging strategy also depends upon them. McNamara et al. (1994) consider a model of small birds foraging in daytime in the winter. The aim for each day's foraging is to minimise risk, at the same time achieving sufficient reserves at the end of the day to survive until the following morning.

Supposing that the day ends at t_{max}, the energy gain per unit time e is equal to 1 and reserves need to be at least as large as some critical value x_c, we obtain the boundary condition

$$V(x,t_{\max}) = \begin{cases} 1, & \text{if } x \geq x_c, \\ 0, & \text{if } x < x_c, \end{cases} \tag{11.9}$$

where payoff 1 represents survival and 0 represents death. It can be shown, see Exercise 11.1, that

$$V(x,t) = \begin{cases} 1 & \text{if } x_c - x \leq 0, \\ (1-z)^{x_c-x} & \text{if } 0 < x_c - x \leq t_{\max} - t, \\ 0 & \text{if } x_c - x > t_{\max} - t. \end{cases} \tag{11.10}$$

Thus survival is certain if the critical food level has already been reached (in this case the animal should just rest) and impossible if there is insufficient time to reach x_c even with constant foraging (in this case it does not matter what the animal does). Survival has probability $(1-z)^{x_c-x}$ (representing the risk of death in a series of $x_c - x$ foraging periods) otherwise; in this case since any strategy involving foraging $x_c - x$ times out of the next $t_c - t$ time steps yields the same reward, it does not matter what the animal does at the next

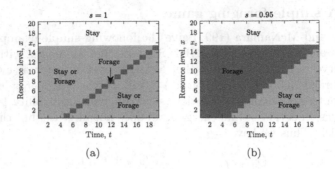

(a)   (b)

FIGURE 11.1: Optimal foraging strategies at time t and resource level x; in all cases $t_{\max} = 20$, $z = 0.01$, the boundary condition is given by (11.10) with $x_c = 16$.

step (unless $(x_c - x) = t_{\max} - t$ in which case the animal should forage). See Figure 11.1a.

In reality foraging is stochastic, so reserves do not necessarily increase if foraging occurs. We can account for this by modifying (11.6) into

$$H(x, t; u_f) = (1 - z)\big(sV(x + 1, t + 1) + (1 - s)V(x, t + 1)\big), \qquad (11.11)$$

where s is the probability of successful foraging. Even if we keep everything else the same, when $s < 1$ the optimal strategy becomes to forage well before it is absolutely necessary, i.e. for t smaller than the critical time $t_{\max} - (x_c - x)$. When $t_{\max}$ and x_c/e are not too big, the mere fact that $s < 1$ will make the optimal strategy to simply be to forage at a high rate early on (to guard against a potential run of bad luck that may come if foraging is at a constant rate during the day) and then cease foraging if reserves have been built up to a high enough level; see Figure 11.1(b).

11.1.4   Evolutionary games based upon state

We now revisit the game of brood care and desertion (Webb et al., 1999) which was introduced in Section 8.4.2. Recall that we consider birds which have to choose to raise their brood or abandon them, so that they could take advantage of a second mating opportunity during a breeding season. In Section 8.4.2 the male and the female chose separately, independently of the other. The game was also developed further in Section 8.4.2.2 to allow the rewards for desertion to depend upon the number of males and females available to mate in the second round, which, in turn, depends upon the strategies selected in the first round by the whole population. We thus considered a sequence of rounds, where the choices of the players were made simultaneously within a round.

Suppose now that decisions are not made simultaneously but sequentially, for example with the male choosing first. Houston and McNamara (1999) consider the following example payoffs to the male and the female to explore the difference between simultaneous and sequential decision-making,

$$
\begin{array}{cc}
& \text{Female cares} \quad \text{Female deserts} \\
\begin{array}{c} \text{Male cares} \\ \text{Male deserts} \end{array}
\left(\begin{array}{cc} (5,5) & (3,6) \\ (4,4) & (2,3) \end{array} \right).
\end{array}
\tag{11.12}
$$

We write the payoffs in the standard bimatrix form with the payoff to the male first. If there are simultaneous decisions, the ESS strategy is for the male to care and the female to desert. See Exercise 11.4.

Now suppose the choices are made sequentially but with the male making the first choice. We discussed this concept of sequential decisions in Chapter 10. If the male cares then the female will desert, and if the male deserts then the female will care. The payoff to the male out of these two options is largest if he deserts, and so the ESS is for the male to desert and the female to care. Thus the result of the game is completely different in the two different cases of simultaneous and sequential play.

We assume in the following that decisions are sequential, with the male choosing first.

If both parents care at time t, then it is assumed that S_{mf} offspring survive to maturity from time $t + \tau_{\text{care}}$ when care ceases. At that point each of them has reproductive value $R_{\text{off}}(t + \tau_{\text{care}})$. As the degree of relatedness of offspring to parent is $1/2$, the contribution to each parent of the brood is thus (see for example Section 15.1 on the relatedness principle)

$$
R_{\text{b}}(C, C; t) = \frac{1}{2} S_{mf} R_{\text{off}}(t + \tau_{\text{care}}).
\tag{11.13}
$$

Similarly, by S_m and S_f we denote the number of surviving offspring if only the male or female cares. If neither cares, we assume that the brood dies and so has contribution 0. So the brood contribution to their parents is thus given by

$$
R_{\text{b}}(C, D; t) = \frac{1}{2} S_m R_{\text{off}}(t + \tau_{\text{care}}),
\tag{11.14}
$$

$$
R_{\text{b}}(D, C; t) = \frac{1}{2} S_f R_{\text{off}}(t + \tau_{\text{care}}),
\tag{11.15}
$$

$$
R_{\text{b}}(D, D; t) = 0.
\tag{11.16}
$$

Each parent also has another contribution to its fitness. We denote the expected reward to a male that is free to mate again at time t by $W_m(t)$. A male that does not desert will only be free again at time $t + \tau_{\text{care}}$, whereas one that deserts will be free at time t. Thus writing the reward to a male that cares when his partner also cares at time t as $H_m(C, C; t)$ (and similarly, if

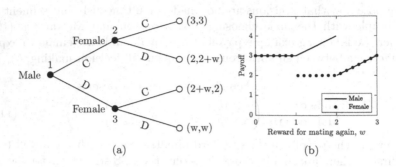

(a)   (b)

FIGURE 11.2: A two-round sequential brood care and desertion game: (a) scheme of the game, (b) payoffs to males and females when optimal strategies are played.

either or both do not care) we obtain male rewards

$$H_m(C,C;t) = R_b(C,C;t) + W_m(t + \tau_{\text{care}}), \qquad (11.17)$$
$$H_m(C,D;t) = R_b(C,D;t) + W_m(t + \tau_{\text{care}}), \qquad (11.18)$$
$$H_m(D,C;t) = R_b(D,C;t) + W_m(t), \qquad (11.19)$$
$$H_m(D,D;t) = R_b(D,D;t) + W_m(t). \qquad (11.20)$$

We can obtain female rewards in a similar way.

Example 11.2 (Two-round sequential brood care and desertion game). Let us consider a sequential brood care and desertion game with two rounds $t = 1, 2$ only, similar to those we considered in Chapter 10. Let the male play first in each round. Suppose that $W_m(1) = W_f(1) = w$, for some constant $w > 0$, and $W_m(2) = W_f(2) = 0$ since remating is not possible after round 2. Similarly, suppose that $\tau_{\text{care}} = 1$ so that caring in round 1 removes any remating possibilities. Finally let the values of the broods from the first round be given by

$$R_b(C,C;1) = 3, R_b(D,C;1) = R_b(C,D;1) = 2, R_b(D,D;1) = 0, \quad (11.21)$$

see Figure 11.2a. Note: the values of $W_m(t)$ in (11.17)–(11.20), and consequently w above, will in general depend upon the number of free males/females in the population at time t (and at later times), in a similar way to the game described in Section 8.4.2.2.

We can see that the solution of the above game depends on the value of the parameter w. If $0 < w < 1$ the female will care irrespective of the male strategy, and hence the male will care. This yields the payoffs $(3,3)$. If $1 < w < 2$ the female will care if the male deserts and desert if the male cares. Hence the male will desert, yielding the payoffs $(2+w, 2)$. If $2 < w$ the female will desert irrespective of the male strategy, and hence the male will desert, yielding the payoffs (w, w).

Rewards at time t depend upon future values of the payoff functions, but not on past ones (we note that it is easy to think of games where this is not the case, for example where rewards depend upon a state variable related to resources). Thus if there is a finite end to the time period (the end of the breeding season in our case) we can start from the final time point t_{max} and use backwards induction based upon assuming the correct strategies are employed in all later time steps. In this way we find a unique pure ESS for the whole game over the time period, analogously to that in Section 10.1.2. Note that if decisions are made simultaneously, then there may have been different optimal strategies at any given time point, as we saw in the simple two round game of Section 8.4.2.2. To find the full set of solutions in this case, it is necessary to work backwards from each possible solution, in turn, and this may generate yet more cases earlier in time as each solution can lead to more than one earlier solution, see Section 10.2. Note that this process is analogous to the multi-player games involving structures of pairwise contests from Chapter 9. Following such a process will find all the ESSs of the game. A solution found in this way is only a candidate ESS, however, as the possibility of it being invaded by a strategy which differs from it in more than one place has not been ruled out, and this must, in general, be investigated.

A general principle of the work of Houston and McNamara (1999) is that rewards and costs are not fixed things, but are properties that emerge from the game. They can be highly complex and non-intuitive, and so this approach is quite different from that of the fixed reward and cost approach that is used in most evolutionary games. Similarly, the order of decisions can have a profound effect. We shall return to this theme when we briefly consider the important concept of life history theory in Section 11.3.

11.2 A question of size

The state of an individual can be something that fluctuates over short timescales, like food reserves, but can also be something that is more long lasting, such as size. Once an individual reaches adulthood, its size is (we shall assume) fixed, but as it grows from birth, its size will follow a path from small to large, and its behaviour may depend strongly upon its size. For instance, there may be predators which are dangerous when an individual is small, but not when it is sufficiently large; zebras of all sizes are at risk from lions, for example, but baboons can only kill foals.

Example 11.3 (Size game, Maynard Smith, 1982, Chapter 11). Let us assume that members of a population have a juvenile phase and an adult phase. During the juvenile phase individuals grow but cannot reproduce; during the adult phase they stop growing but are able to reproduce instead. Every year there is some chance of death, which depends upon age, but for simplicity is assumed to be independent of which phase an individual is in, conditional upon its age. An individual's fecundity is greater the longer it delays breeding. Thus there is a trade-off between the advantages of early or late onset of adulthood, namely a greater quantity versus a greater quality of available breeding years. How successful an individual is, once it becomes an adult, also depends upon the strategies of other members of the population, as late onset adults will typically be tougher opponents. Model the scenario to find out the ESS value of the time that adulthood should be reached.

11.2.1 Setting up the model

An individual strategy is the age at which adulthood begins. By T, we will denote the strategy when the individual enters the adult phase at age T. Following Maynard Smith (1982), we shall assume a discrete system where at each year in the juvenile stage the individual grows, and at each year in the adult phase the individual has the chance to reproduce. Let s_t denote the probability of survival from year t to $t+1$.

The expected probability of survival to adulthood of an individual playing T is

$$A_T = \prod_{t=0}^{T-1} s_t. \tag{11.22}$$

Once it reaches adulthood, it has an expected number of years of breeding given by

$$Y_T = 1 + s_T + s_T s_{T+1} + \ldots = \sum_{j=T-1}^{\infty} \prod_{t=T}^{j} s_t. \tag{11.23}$$

To account for the frequency dependence of the success of the strategies, assume that in a population which is comprised of individuals all playing strategy T, an individual playing strategy T' has $H_T(T')$ offspring per year. To account for the fact that fecundity is greater the longer it delays breeding, the function $H_T(T')$ should be increasing in T'.

Thus the fitness of an individual playing T' in a population playing T is given by

$$\mathcal{E}[T', \delta_T] = A_{T'} Y_{T'} H_T(T'). \tag{11.24}$$

Whilst we use the above formula to define the fitness, we note that this can be a problematic measure; see Exercise 11.12 for a potential weakness of formula (11.24) when the population size is not constant. This issue of varying population size was known a hundred years ago and is approached using the Euler-Lotka equation; see Sharpe and Lotka (1911), and see also Section 11.3.

11.2.2 ESS analysis

We are interested in whether any mutant individual can invade the population. In particular, can a mutant playing $T' = T + 1$ or $T' = T - 1$ invade?

If neither of these two alternatives can invade, this will usually be enough to prevent invasion by any alternative strategy for reasonable parameters, but see Exercise 11.7.

Note that

$$\mathcal{E}[T+1, \delta_T] = A_{T+1} Y_{T+1} H_T(T+1) \tag{11.25}$$

$$= (A_T s_T) \sum_{j=T}^{\infty} \prod_{t=T+1}^{j} s_t H_T(T+1) \tag{11.26}$$

$$= A_T (Y_T - 1) H_T(T+1). \tag{11.27}$$

Thus strategy $T+1$ cannot invade strategy T if

$$H_T(T+1) < \frac{Y_T}{Y_T - 1} H_T(T). \tag{11.28}$$

Similarly,

$$\mathcal{E}[T-1, \delta_T] = A_{T-1} Y_{T-1} H_T(T-1) \tag{11.29}$$

$$= (A_T / s_{T-1}) \sum_{j=T-2}^{\infty} \prod_{t=T-1}^{j} s_t H_T(T-1) \tag{11.30}$$

$$= A_T (Y_T + 1/s_{T-1}) H_T(T-1). \tag{11.31}$$

This implies that $T-1$ cannot invade T if

$$H_T(T-1) < \frac{s_{T-1} Y_T}{s_{T-1} Y_T + 1} H_T(T). \tag{11.32}$$

Thus the breeding advantage of waiting an extra year cannot be too large, and the breeding disadvantage of starting a year early must be large enough.

11.2.3 A numerical example

Suppose that $s_t = s$ for all t. Then $Y_t = 1/(1-s)$ for all t, and the condition for T to be an ESS is given by

$$H_T(T) > \max\left(s H_T(T+1), \frac{1}{s} H_T(T-1)\right). \tag{11.33}$$

To show this is Exercise 11.6.

Now, assume that the rewards are given by

$$H_t(t) = t - 1 + \alpha\beta, \tag{11.34}$$
$$H_t(t-1) = t - 1, \tag{11.35}$$
$$H_t(t+1) = t - 1 + \alpha. \tag{11.36}$$

These correspond to a situation where every individual has $t-1$ offspring per year, except for a proportion β of individuals, the largest ones, which get an

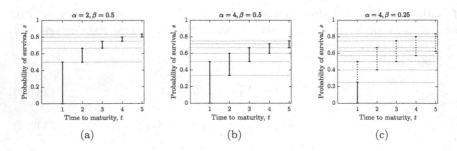

FIGURE 11.3: ESS in the size game from Section 11.2.3. Depending on parameter values, there can be (a) only pure and unique ESSs, (b) only pure, but possibly multiple ESSs, and (c) mixed ESSs (dotted lines).

extra α. It is assumed that all individuals that mature later are larger than those that mature earlier, and all who mature at the same time are equally likely to be amongst the largest. Thus T is a pure ESS if

$$\frac{T-1}{T-1+\alpha\beta} < s < \frac{T-1+\alpha\beta}{T-1+\alpha}. \tag{11.37}$$

There can be a mixed ESS involving T and $T+1$ (and potentially other strategies) if each invades the other, i.e.

$$\frac{T-1+\alpha\beta}{T-1+\alpha} < s < \frac{T}{T+\alpha\beta}. \tag{11.38}$$

We will now consider some parameter values of α, β.
(a) For the case with $\alpha = 2, \beta = 1/2$, T is an ESS if

$$\frac{T-1}{T} < s < \frac{T}{T+1}. \tag{11.39}$$

(b) For the case with $\alpha = 4, \beta = 1/2$, there is sometimes a unique ESS, sometimes there is more than one.
(c) For the case with $\alpha = 4, \beta = 1/4$, the only possible pure ESS is 1; but for any T, there can be a mixed ESS involving that strategy. See Figure 11.3.

11.3   Life history theory

The size game of Section 11.2 can be thought of as an example of a game incorporating a population with a *life history*. In most of the models in our book, games are played between already grown adults, with no mention of their growth phase as juveniles. Life history theory studies the evolution of populations where the growth phase from infant to adult is treated as a central component in the behaviour of the individual. In reality this variability of the size of individuals within populations, and the varying stages of the lives of individuals, is a key feature of animal populations.

We consider a population of individuals with various size classes C_i, $i = 0, 1, 2, \ldots$, the number in size class C_i at time t being denoted by $N_i(t)$. The index i can be thought of as the age of the individual (sometimes a class M, the maximum age allowable in the population, is included, where on reaching this age the individual dies before reproducing). An individual in age class C_i has a number of offspring f_i and, if it survives (with probability s_i), it progresses to age class C_{i+1} at the end of the year. From a mathematical modelling point of view the change in the population composition from time t to time $t + 1$ can be represented by the matrix equation

$$\mathbf{N(t + 1)} = L\mathbf{N(t)}, \tag{11.40}$$

where $\mathbf{N(t)} = (N_0(t), \ldots, N_{M-1}(t))^T$, and L is the *Leslie matrix*,

$$L = \begin{pmatrix} f_0 & f_1 & \cdots & f_{M-2} & f_{M-1} \\ s_0 & 0 & \cdots & 0 & 0 \\ 0 & s_1 & \cdots & 0 & 0 \\ \vdots & \vdots & \ddots & \vdots & \vdots \\ 0 & 0 & \cdots & s_{M-2} & 0 \end{pmatrix}. \tag{11.41}$$

Using the Leslie matrix we can find the long-term growth rate of the population (found from the dominant eigenvalue) and the age class distribution of the population (the eigenvector associated with the dominant eigenvalue).

The Leslie matrix that we have considered above features constant values of f_i and s_i, corresponding to a purely demographic model. In applications of life history these parameters are not constant, but functions which depend upon terms such as resource allocation into growth repair or reproduction, which can be thought of as strategies. It is beyond the scope of this book to discuss this properly, and we refer the reader to, for example, Roff (1992) and Caswell (2000).

Typically, however, life history theory does not involve game theory. Although the terms s_i and f_i are potentially functions of strategies, optimal choice does not depend upon the strategies of other individuals, but there is no reason why this has to be the case. The size game of Section 11.2 is a good simple example of how a game can be incorporated into the theory. We can see that the terms s_i correspond exactly, and for an individual playing strategy m in a population playing n we obtain

$$f_i = \begin{cases} 0 & i < m, \\ H_m(n) & i \geq m. \end{cases} \tag{11.42}$$

Game theory has recently been incorporated into life history theory in Argasinski and Broom (2021) where a new modelling framework combining replicator dynamics and an age-structured population model has been introduced.

In particular, this allows models of populations which are made up of individuals with competing strategies who change over their life cycle (they illustrate this by a sex ratio example, following earlier work in Argasinski (2012)).

11.4 Python code

In this section we show how to use Python to produce Figure 11.1 and the solution of a stochastic foraging game from Section 11.1.3.

```python
1   """ Stochastic foraging game
2   Birds look for food in a given time interval 0 ... maxT
3   At the end of the interval they need to have enough energy
4   reserves to survive.
5   They can be killed during foraging and may not always find food.
6
7   The script outputs optimal behaviour as a plot:
8   Stay/Rest (Red) or
9   Forage (Green).
10  Black is for "will not survive for sure"
11  """
12
13  ## Define parameters
14  z = 0.01      # Prob of being killed during foraging
15  c = 0.9       # Prob of increasing energy reserves during foraging
16                # c above 0.9 could cause rounding errors for small t
17                # especially when maxT is large
18  maxT = 30     # Maximal time for foraging
19  maxE = 20     # Maximal energy reserves
20  threshold = 15  # Energy must be above threshold to survive
21
22  ## Import basic packages
23  import matplotlib.pyplot as plt
24  from matplotlib import colors
25
26  ## Auxiliary variables
27  # Give numerical values to ``actions''
28  Forage = 2
29  Rest = 1
30  CannotSurvive = 0 # Technically not an action, but happens
31
32  # Initialize an action given the resources and the time
33  # action[t][e] will store action at time t and energy level r
34  action = [[None for t in range(maxT)] for e in range(maxE+1)]
35
36  # Set boundary condition for having enough resources
37  for t in range(maxT):
38      action[maxE][t] = Rest # No need to forage when at max level
39
40  # Initialize survival probability
41  # Include even energy level maxE+1 to simplify code later
```

```
42   #    (birds will not forage if at the max level, but will "try")
43   survival = [[0 for t in range(maxT+1)] for e in range(maxE+2)]
44
45   # Set boundary conditions for survival at the end of the day
46   for e in range(maxE+2):
47       if e≥threshold:
48           # If enough energy reserves to survive
49           # Survives for sure
50           survival[e][maxT] = 1
51       else:
52           # If  not enough reserves to survive
53           # Does not survive for sure
54           survival[e][maxT] = 0
55
56   ## Analysis by backwards induction
57   # Start at the last time birds make decisions and move back
58   for t in range(maxT-1, -1, -1):
59       for e in range(maxE):    # For every energy level
60           # Determine the survival probability if resting
61           p_rest = survival[e][t+1]
62
63           # Determine survival prob if  foraging
64           p_forage = (survival[e+1][t+1]*c + p_rest*(1-c))*(1-z)
65
66           # Determine what action to take
67           if p_forage > p_rest:
68               # Foraging gives higher survival probability
69               # Record the probability at this level and time
70               survival[e][t] = p_forage
71
72               # Record the best action at this level and time
73               action[e][t] = Forage
74
75           elif p_forage == p_rest:
76               # Foraging is same as resting
77               # Record the survival probability
78               survival[e][t] = p_forage    # same as for resting
79
80               # To assign an action, we will distinguish two cases
81               if p_forage == 0:
82                   # Sure not to survive, record it
83                   action[e][t] = CannotSurvive
84               else:
85                   # This case should not happen in generic games
86                   # But is happening due to rounding
87                   action[e][t] = Forage
88
89           else:
90               # Stay is better than forage
91               # Record survival probability at this level and time
92               survival[e][t] = p_rest
93               # Record the best action at this level and time
94               action[e][t] = Rest
95
96   ## Plot the actions
97   # Create discrete color scheme
98   # It is for:               No survival    Rest    Forage
```

```
99  cMap = colors.ListedColormap(['black',    'red',    'green'])
100
101 fig, ax = plt.subplots()  # Get handles of figure and the axis
102 # Show the array `action' as an image with the given color coding
103 actionPlot = ax.imshow(action, origin='lower', aspect='auto',
104                        cmap = cMap)
105 # Create labels
106 plt.xlabel('Time')
107 plt.ylabel('Energy reserve level')
108 plt.show()
```

11.5   Further reading

For a comprehensive discussion of many aspects of state-based evolution see Houston and McNamara (1999); also see McNamara et al. (1994) and Webb et al. (1999). For the size game see Maynard Smith (1982) and also Mirmirani and Oster (1978) and Parker (1979). For a new review paper which considers the more realistic evolutionary models in general, see McNamara (in press) (see also McNamara and Leimar, 2020).

Argasinski and Kozlowski (2008) (see also Argasinski, 2006; Argasinski and Broom, 2018a,b) developed models which incorporate population dynamics and evolutionary games in a more realistic way, and this work has led to Argasinski and Broom (2021) which incorporates the two usually separate fields of life history theory and game theory into a complementary model.

11.6   Exercises

Exercise 11.1. Consider the foraging game described in Section 11.1.3 with boundary condition (11.9). Show (11.10) is the only solution with $V(x,t) = \max_u H(x,t,u)$ where H is given by (11.6)–(11.7).

Exercise 11.2. Find the optimal foraging strategy in the stochastic foraging game from Section 11.1.3, i.e. solve $V(x,t) = \max_u H(x,t,u)$ with boundary condition (11.9) where $H(x,t;u_r)$ is given by (11.7) and $H(x,t;u_f)$ is given by (11.11).

Hint. See Figure 11.1(b).

Exercise 11.3. Consider the foraging game described in Section 11.1.3 and assume that the boundary condition is given by $V(x,t_{\max}) = x$. Find the optimal foraging strategy in this case.

Exercise 11.4. Find the ESS of the bimatrix brood care and desertion game with payoffs given by (11.12).

Exercise 11.5. Consider the game as in Example 11.2, but suppose that the reward for desertion depends upon the choices made by the population in the previous round, and if a fraction x males and y females deserted, the reward to a male (female) is wy (wx). Construct the equivalent scheme to that in Figure 11.2(a) and analyse the game.

Exercise 11.6. Consider the size game from Section 11.2. For the case where $s_i = s$ for all i, show that $Y_t = 1/(1-s)$, and that invasion by $T' = T - 1$ and $T' = T + 1$ is prevented if and only if (11.33) holds.

Exercise 11.7. Show that for the size game from Example 11.3 there are parameters (values of s_t) such that T cannot be invaded by $T - 1$ nor $T + 1$, but T is still not an ESS.

Hint. Consider strategy T within a population where the survival rate becomes very high for time $T + 1$ onwards.

Exercise 11.8. For the size game from Exercise 11.6, find the optimal strategy (for all possible s) for the case where $H_T(T') = \beta T'$, for some positive constant β.

Hint. Note that this is a case where the payoffs are independent of the strategies within the population.

Exercise 11.9. Verify the solutions for the game in Section 11.2.3 for the three pairs of parameters (a), (b) and (c).

Exercise 11.10. Verify the solutions to the Example 11.1 from Section 11.1.1 given in (11.1).

Exercise 11.11. Show that the age class distribution of the population described in Section 11.3 is given by the matrix equation (11.40). Verify that the size game of Example 11.3 can be considered as an example of a game-theoretical version of this with parameters given by (11.42).

Exercise 11.12. Discuss the fitness function of (11.24) in terms of the growth rate of the population. Under what circumstances is it a "true" measure of fitness, and when will it over/underestimate the fitness of individuals which delay the onset of adulthood?

Chapter 12

Games in finite populations and on graphs

In this chapter we consider finite population games and see that we need some new concepts distinct from those for infinite population games. Finite games also provide a useful introduction to games where the structure of the population is vital, in particular games on graphs.

12.1 Finite populations and stochastic games

Up until now we have considered a population of (effectively) infinite size. Here we consider a population of finite size N. The games in this context have been studied extensively, see for example Nowak (2006a, Chapters 6-9).

12.1.1 The Moran process

We shall start by assuming individuals have a fixed fitness, depending upon type, but independent of interactions with other players. In terms of the payoffs of a matrix game, this is equivalent to

$$a_{ij} = r_i \quad \forall i, j \tag{12.1}$$

for some positive constants r_i. In previous sections, where populations consisted of an infinite number of individuals playing a resident strategy, and a relatively small (but still infinite) number comprising a fraction ε of the population playing a mutant strategy, this situation was not interesting, as the fitter strategy is bound to win such a contest, and the solution to any such "game" is that there is a unique pure ESS S_i, when $r_i > r_j$ for all $j \neq i$. If two or more strategies have the same fitness $\max_j(r_j)$, then evolution does not favour any of these strategies over the others in this non-generic case, and there is no ESS.

For finite populations the situation is different. In this case dynamic considerations are particularly important. The standard dynamics applied to this population is as follows. At each time step an individual is chosen for reproduction at random with a probability proportional to its fitness and its offspring replaces a randomly chosen individual (which could be its parent). This is called the *Moran process* (Moran, 1958, 1962).

We shall start by considering the neutral fitness case, where $r_i = 1$ for all S_i, as in the original Moran process. Suppose we have N individuals, made up of m_i individuals of type i, $i = 1, \ldots, n$. The population is thus described by a (row) vector $\mathbf{m} = (m_i)$ with $\sum_i m_i = N$. At each time point a random individual is chosen to give birth, and another to die, selected independently of each other. Let $\mathbf{e}_i$ be a (row) vector with 0s everywhere except on the ith place where there is a 1. The possible transitions in the population, together with the probabilities of those transitions, are as follows:

$$P(\mathbf{m} \to \mathbf{m}^*) = \begin{cases} \dfrac{m_i}{N} \dfrac{m_j}{N} & \mathbf{m}^* = \mathbf{m} + \mathbf{e}_i - \mathbf{e}_j, i \neq j, \\ \displaystyle\sum_{i=1}^{n} \left(\dfrac{m_i}{N}\right)^2 & \mathbf{m}^* = \mathbf{m}, \\ 0 & \text{otherwise.} \end{cases} \tag{12.2}$$

Thus at any time t, $\mathbf{m}(t)$ only depends upon $\mathbf{m}(t-1)$ and no earlier time points are relevant; thus the Moran process is a *Markov process* (Karlin and Taylor, 1975).

Now suppose that not all of the values of r_i are equal. We note that there are many ways we can incorporate this in the process. As described above, we shall make the birth rate depend upon fitness and the death rate not, although alternatively the death rate could depend upon fitness and the birth rate not, or we could use a combination of the two. When we consider the fixation probabilities (see below in Section 12.1.2) for the Moran process, this is actually not too important; see Exercise 12.4. On the other hand, the fixation probabilities get affected for evolution on graphs; see Exercise 12.8. Moreover, the absorption and fixation times (see Section 12.1.5) are affected even for the Moran process.

In our model the probability of giving birth is proportional to the fitness of the individual. One way of thinking of the fitness of an individual is as the number of offspring that it will have that will survive to adulthood (see Chapter 5). We can thus perhaps think that at any given time step, and for every type S_i there are $m_i r_i$ offspring of that type that may be born. An offspring of type S_i will thus be born with probability $m_i r_i / (\sum_l m_l r_l)$. The transition probabilities now become

$$P(\mathbf{m} \to \mathbf{m}^*) = \begin{cases} \dfrac{m_i r_i}{\sum_{l=1}^{n} m_l r_l} \dfrac{m_j}{N} & \mathbf{m}^* = \mathbf{m} + \mathbf{e}_i - \mathbf{e}_j, i \neq j, \\ \displaystyle\sum_{i=1}^{n} \dfrac{m_i r_i}{\sum_{l=1}^{n} m_l r_l} \dfrac{m_i}{N} & \mathbf{m}^* = \mathbf{m}, \\ 0 & \text{otherwise.} \end{cases} \tag{12.3}$$

It is easy to see that multiplying all values of r_i by a constant leaves all of the probabilities unchanged, so that without loss of generality we can set one of our fitnesses to be equal to 1. Irrespective of the (finite) population size, and the number of strategies available in the population, there is a non-zero probability that any given type will reach fixation under the Moran process, since $r_i > 0$ for all i, and it is certain that one type will eventually do so.

12.1.2 The fixation probability

The long-term outcome of the process described above, in the absence of the introduction of new mutations, is a population consisting of just a single type. The important question is, which type is likely to dominate, i.e. how likely is each such fixation to occur? Thus the probability of fixation, the *fixation probability*, is the single most important property of a finite evolutionary system. This is usually considered as the probability of fixation of a single mutant in a population otherwise entirely made up of a resident type.

It thus makes sense for us to consider the case where we have two types of individuals only, type A with fitness r, and type B with fitness 1. The state of the population is described by a single number, N_A, the number of individuals of type A. We can find an expression for the probability of the population containing i mutants at time $t + 1$, $\pi_i(t + 1)$, in terms of the probabilities of occupying the different population sizes at time t and these transition probabilities using the equation

$$\pi_i(t+1) = \sum_j p_{j,i}\pi_j(t), \tag{12.4}$$

where $p_{i,j} = Prob((i, N - i) \to (j, N - j))$ is the probability that $N_A = j$ at time point $t + 1$ given that $N_A = i$ at time point t. By (12.3), we get

$$p_{i,i+1} = \frac{ir}{ir + N - i}\frac{N - i}{N}, \tag{12.5}$$

$$p_{i,i-1} = \frac{N - i}{ir + N - i}\frac{i}{N}, \tag{12.6}$$

$$p_{i,i} = \frac{ir}{ir + N - i}\frac{i}{N} + \frac{N - i}{ir + N - i}\frac{N - i}{N} \tag{12.7}$$

$$= 1 - \frac{ir}{ir + N - i}\frac{N - i}{N} - \frac{N - i}{ir + N - i}\frac{i}{N}. \tag{12.8}$$

In the terminology of Markov processes, (12.4) are the *Chapman-Kolmogorov forward equations* and we can use standard techniques (see e.g. Karlin and Taylor, 1975) to find various important properties of our random process, including the fixation probability.

Denoting P_i as the fixation probability of A given $N_A = i$, we obtain the following difference equations,

$$P_i = P_{i-1}p_{i,i-1} + P_i p_{i,i} + P_{i+1}p_{i,i+1}, \tag{12.9}$$

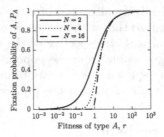

FIGURE 12.1: The Moran fixation probability (12.11) for various N.

with the obvious boundary conditions, the fixation probabilities on the absorbing states, $P_0 = 0, P_N = 1$. We can solve the above equations to obtain

$$P_i = \begin{cases} \dfrac{1 - (1/r)^i}{1 - (1/r)^N} & r \neq 1, \\ i/N & r = 1. \end{cases} \qquad (12.10)$$

This, in turn, gives the fixation probability of a single mutant of type A in a population of type B as

$$P_A = P_1 = \begin{cases} \dfrac{1 - (1/r)}{1 - (1/r)^N} & r \neq 1, \\ 1/N & r = 1. \end{cases} \qquad (12.11)$$

This is the *Moran probability*, see Figure 12.1, and is the benchmark against which fixation probabilities in more complex structured populations are compared.

By symmetry, the fixation probability of a single mutant of type B in a population of type A if $r \neq 1$ is

$$P_B = \frac{r - 1}{r^N - 1}. \qquad (12.12)$$

Example 12.1 (Moran process). Consider a well-mixed population of three individuals. Find the fixation probability of a mutant with fitness r.

At any given time, the population is described by the number of mutants, $i = \{0, 1, 2, 3\}$. If $i = 0$, then mutants will never fixate and thus $P_0 = 0$. If there is $i = 1$ mutant at time t, there will be $i' \in \{0, 1, 2\}$ mutant(s) at time $t' = t + 1$. There will be no mutant ($i' = 0$) if either one of the residents reproduces and the mutant dies, which happens with probability (compare to (12.6))

$$p_{1,0} = \frac{2}{r + 2} \frac{1}{3}. \qquad (12.13)$$

There will be $i' = 2$ mutants if the mutant reproduces and either of the residents dies, which happens with probability (compare to (12.5))

$$p_{1,2} = \frac{r}{r+2}\frac{2}{3}. \tag{12.14}$$

There will be $i' = 1$ mutants if neither of the above changes happens, i.e. with probability

$$p_{1,1} = 1 - p_{1,0} - p_{1,2}. \tag{12.15}$$

Similarly, if there are $i = 2$ mutants at time t, there will be $i' \in \{1,2,3\}$ mutants at time $t' = t+1$ and we get the transition probabilities as

$$p_{2,3} = \frac{2r}{2r+1}\frac{1}{3}, \tag{12.16}$$

$$p_{2,1} = \frac{1}{2r+1}\frac{2}{3}, \tag{12.17}$$

$$p_{2,2} = 1 - p_{2,1} - p_{2,3}. \tag{12.18}$$

Once $i = 3$, the mutants have fixated and thus $P_3 = 1$. Putting this all together, as in (12.9), we have that the probabilities of $i \in \{1,2\}$ mutants to fixate in a population of three individuals have to satisfy

$$P_1 = p_{1,0} \cdot 0 + p_{1,1}P_1 + p_{1,2}P_2, \tag{12.19}$$

$$P_2 = p_{2,1}P_1 + p_{2,2}P_2 + p_{2,3} \cdot 1. \tag{12.20}$$

The above system is easy to solve and one can verify that P_1 is given as in (12.11).

12.1.3   General Birth-Death processes

The recurrence relation (12.9) is, in fact, equivalent to the *Birth-Death process*, or equivalently the random walk, with the birth rate β_i and death rate δ_i when the population is at state i (Karlin and Taylor, 1975). The general equations are given by

$$x_i = \delta_i x_{i-1} + (1 - \beta_i - \delta_i)x_i + \beta_i x_{i+1}, \tag{12.21}$$

where the probability of fixation (i.e. reaching state N) starting from state i is denoted by x_i, and so $x_0 = 0, x_N = 1$.

A solution to (12.21) is

$$x_i = \frac{1 + \sum_{j=1}^{i-1}\prod_{k=1}^{j}\frac{\delta_k}{\beta_k}}{1 + \sum_{j=1}^{N-1}\prod_{k=1}^{j}\frac{\delta_k}{\beta_k}}. \tag{12.22}$$

We note that Karlin and Taylor (1975) discussed this process without upper limit N, so that the summation in the denominator of (12.22) went to infinity.

We recover the Moran probability by appropriate substitution in the solution for x_1. In general, (12.22) can be directly applied to some cases of evolutionary processes including games on graphs.

12.1.4   The Moran process and discrete replicator dynamics

Here we show that there is a direct link between the discrete replicator dynamics (see Section 3.1.1) and the Moran process.

Let the population consist of two types of individuals, A and B. We thus have $\mathbf{m} = (m_A, m_B)$ and we consider also the replicator dynamics for $\mathbf{p} = (p_A, p_B) = \frac{1}{N}\mathbf{m}$ which is given by the equation

$$p_A(t+1) = p_A(t)\frac{f_A(\mathbf{p}(t))}{\bar{f}(\mathbf{p}(t))}, \tag{12.23}$$

where

$$f_A(\mathbf{p}(t)) = r, \tag{12.24}$$

$$f_B(\mathbf{p}(t)) = 1, \tag{12.25}$$

$$\bar{f}(\mathbf{p}(t)) = \frac{ir + N - i}{N}, \tag{12.26}$$

noting that in the discrete replicator dynamics $t+1$ refers to generation $t+1$.

Substituting (12.24)–(12.26) into (12.5) and (12.6) we get

$$p_{i,i+1} = \frac{f_A(\mathbf{p}(t))}{\bar{f}(\mathbf{p}(t))}p_A(t)(1 - p_A(t)), \tag{12.27}$$

$$p_{i,i-1} = \frac{f_B(\mathbf{p}(t))}{\bar{f}(\mathbf{p}(t))}p_A(t)(1 - p_A(t)). \tag{12.28}$$

The expected change in the A population at the next time step is

$$E[m_A(t+1) - m_A(t)] = p_{i,i+1} - p_{i,i-1} \tag{12.29}$$

$$= \left(\frac{f_A(\mathbf{p}(t))}{\bar{f}(\mathbf{p}(t))} - 1\right)p_A(t). \tag{12.30}$$

Thus assuming that $\mathbf{p}$ is approximately constant over the next N time steps, including neglecting the possibility of fixation or extinction in those time steps, we get

$$m_A(t+N) - m_A(t) \approx N\left(\frac{f_A(\mathbf{p}(t))}{\bar{f}(\mathbf{p}(t))} - 1\right)p_A(t), \tag{12.31}$$

which, after dividing both sides by N, is approximately (12.23) once we realise that increasing the time by 1 in (12.23) represents replacing the population by a new generation of individuals, and N time steps in the Moran process is a good measure of one generation.

12.1.5 Fixation and absorption times

Another important question in such a population is, how long does it take for a mutant to fixate? There is always some chance that the mutant will be eliminated even if $r > 1$ and it is fitter than the resident, so we must distinguish between two times. We define T_i, the *absorption time*, as the expected time until the population contains only one type of individual, either A or B, given there are currently $N_A = i$ A individuals in the population. The *fixation time* F_i is the expected time to mutant fixation, conditional on fixation occurring.

12.1.5.1 Exact formulae

As seen above, at any time step, the population moves from state $N_A = i$ (for $1 \leq i \leq N - 1$) to state $N_A = i+1$ with probability $p_{i,i+1}$, to state $N_A = i - 1$ with probability $p_{i,i-1}$, and remains at state $N_A = i$ with probability $1 - (p_{i,i+1} + p_{i,i-1})$. This yields the following system of equations

$$T_i = 1 + p_{i,i+1}T_{i+1} + p_{i,i-1}T_{i-1} + (1 - p_{i,i+1} - p_{i,i-1})T_i, \qquad (12.32)$$

for $1 \leq i \leq N - 1$ with boundary conditions $T_0 = T_N = 0$. We leave it for the reader as Exercise 12.3 to solve the system.

The situation with fixation times is somewhat more difficult as it is an absorption conditional on fixation. The equations are

$$P_iF_i = P_i + p_{i,i+1}P_{i+1}F_{i+1} + p_{i,i-1}P_{i-1}F_{i-1} + (1 - p_{i,i+1} - p_{i,i-1})P_iF_i, \quad (12.33)$$

for $1 \leq i \leq N - 1$ with boundary condition $F_N = 0$ (Landauer and Buttiker, 1987; Antal and Scheuring, 2006). Following Antal and Scheuring (2006) this can be derived as follows. Define $A_i(t)$ as the probability of fixation occurring at time t, given that the population is in state i at time 0. Then clearly $P_i = \sum_{t=0}^{\infty} A_i(t)$ and

$$F_i = \frac{\sum_{t=0}^{\infty} tA_i(t)}{\sum_{t=0}^{\infty} A_i(t)} = \frac{\sum_{t=0}^{\infty} tA_i(t)}{P_i}. \qquad (12.34)$$

Similarly to (12.9) $A_i(t)$ satisfies

$$A_i(t) = A_{i-1}(t - 1)p_{i,i-1} + A_i(t - 1)p_{i,i} + A_{i+1}(t - 1)p_{i,i+1}. \qquad (12.35)$$

Multiplying both sides of (12.35) by t, summing from 0 to ∞ and using the fact that

$$\sum_{t=0}^{\infty} tA_i(t - 1) = \sum_{t=0}^{\infty}(t + 1)A_i(t) = P_i(F_i + 1), \qquad (12.36)$$

we obtain (12.33).

12.1.5.2   The diffusion approximation

For a very large population, an alternative approach can be taken. In this case, the Moran process can be approximated by a continuous stochastic process. We can think of the composition of the population following a sequence of small changes (a proportion $1/N$ of the population) in an equivalently short time interval (since the standard unit of time is measured in generational changes, and so a change happens in $1/N$ of a unit of time, similarly to what we saw in Section 12.1.4). As N tends to infinity this tends to a continuous stochastic process changing in real time. This process is known as the *diffusion approximation* and was extensively developed by Kimura and Crow (1963); Kimura (1964). The mathematical consequences of the process is that the Chapman-Kolmogorov forward equations (12.4) become a partial differential equation in their limit. The fraction of mutants $\rho = i/N$ then satisfies

$$c(\rho, t + \delta t) = R(\rho - \delta\rho)c(\rho - \delta\rho, t) + L(\rho + \delta\rho)c(\rho + \delta\rho, t) \cdots$$
$$+ \left(1 - R(\rho) - L(\rho)\right)c(\rho, t), \qquad (12.37)$$

where $c(\rho, t)$ is the probability that the mutant density is ρ at time t, δt is an increment in time, $\delta\rho(= 1/N)$ is an increment/decrement in ρ and $L(\rho)$ and $R(\rho)$ are the probabilities of a decrease/increase in the size of the mutant population (Sood et al., 2008). Expanding (12.37) to second order in $\delta\rho$ gives the following partial differential equation

$$\frac{\partial c(\rho, t)}{\partial t} = -\frac{\partial}{\partial \rho}\left(v(\rho)c(\rho, t)\right) + \frac{\partial^2}{\partial \rho^2}\left(D(\rho)c(\rho, t)\right), \qquad (12.38)$$

where $v(\rho)$ and $D(\rho)$ are, respectively, called the drift and diffusion terms, and are given by

$$v(\rho) = \frac{\delta\rho}{\delta t}\left(R(\rho) - L(\rho)\right), \qquad (12.39)$$

$$D(\rho) = \frac{1}{2}\frac{(\delta\rho)^2}{\delta t}\left(R(\rho) + L(\rho)\right). \qquad (12.40)$$

In a well-mixed population where

$$L(\rho) = R(\rho) = \rho(1 - \rho), \qquad (12.41)$$

we obtain

$$\frac{\partial c(\rho, t)}{\partial t} = \frac{1}{N}\frac{\partial^2}{\partial \rho^2}\left(\rho(1 - \rho)c(\rho, t)\right). \qquad (12.42)$$

This partial differential equation turns out to be the *Fokker-Planck equation* from physics (Fokker, 1914; Planck, 1917; Kolmogoroff, 1931). Kimura (1964) found complete solutions for the probability density of the population composition at general time intervals for neutral evolution (compare to Waxman, 2011).

Another application of this approach for large populations, again based upon the diffusion approximation, but which has been extended to consider interactions on graphs as well as well-mixed populations, is in Houchmandzadeh and Vallade (2010, 2011) (see also Shakarian et al., 2012).

12.2 Games in finite populations

Taylor et al. (2004) extended the Moran process approach above to playing games in finite populations. For two types of individuals, mutants M and residents R, we consider the familiar 2×2 payoff matrix

$$\begin{array}{c} \\ M \\ R \end{array} \begin{array}{c} M \quad R \\ \left(\begin{array}{cc} a & b \\ c & d \end{array} \right). \end{array} \tag{12.43}$$

The average payoffs to a mutant individual in a population where there are $m_M = i$ mutants in total is thus

$$E_{M,i} = \frac{a(i-1) + b(N-i)}{N-1}, \tag{12.44}$$

and similarly, the average payoff to a resident individual in such a population is

$$E_{R,i} = \frac{ci + d(N-i-1)}{N-1}. \tag{12.45}$$

Note that we assume above that an individual cannot play a game with itself and hence we have the factors $(i-1)$ and $(N-i-1)$.

If a strategy is an ESS in an infinite population, it is fitter than any mutant playing an alternative strategy in a population comprising a mixture of a sufficiently small group playing the mutant strategy and the remainder playing the resident strategy. For a finite population the smallest possible group is a single individual. Thus a natural extension for the definition of a pure ESS is that a mutant should be less fit than a resident in a population of one mutant and $N-1$ residents (*selection opposes M invading R*) i.e. $E_{M,1} < E_{R,1}$ which is

$$b(N-1) < c + d(N-2). \tag{12.46}$$

However, as argued by Taylor et al. (2004), (12.46) is insufficient for stability, since it is possible that if the population increases a little by chance then the mutant fitness can suddenly change to be greater than that of the residents e.g. if a is large, and so they introduce a second condition, that *selection opposes the replacement of R by M*, i.e. that

$$P_M < \frac{1}{N}, \tag{12.47}$$

so that the fixation probability P_M of the mutant is less than it would be if it was identical to the resident type.

Therefore, Nowak (2006a) proposed the following definition.

Definition 12.2. *For a finite population size N and a 2×2 matrix game (12.43), a pure strategy R is called an* evolutionarily stable strategy, ESS_N, *if (12.46) and (12.47) hold.*

Similarly, as in (12.22) P_M is given by

$$P_M = \frac{1}{1 + \sum_{j=1}^{N-1} \prod_{k=1}^{j} \frac{p_{k,k-1}}{p_{k,k+1}}}, \qquad (12.48)$$

where $p_{k,k+1}$ and $p_{k,k-1}$ are the transition probabilities given by (12.5) and (12.6). Since the formulae (12.5) and (12.6) use fitness, we need to find a way to translate the payoffs of the game $E_{M,i}$ and $E_{R,i}$ into the fitness of the respective types $f_{M,i}$ and $f_{R,i}$.

If we assume that the fitness is equal to the payoff, i.e. $f_{M,i} = E_{M,i}$ and $f_{R,i} = E_{R,i}$, the stability condition (12.47) is very complex. Thus Taylor et al. (2004) considered the idea of intensity of selection. Assume that the fitness is given by

$$f_{M,i} = 1 - w + w E_{M,i}, \qquad (12.49)$$
$$f_{R,i} = 1 - w + w E_{R,i}, \qquad (12.50)$$

where $0 < w \leq 1$ is the intensity of selection. A small w represents *weak selection* and means that the game has a small effect on the process of evolution (the contribution of the game to fitness is small compared to that of other sources e.g. lone foraging), whereas $w = 1$ is the case where the fitness is equal to the payoff. This idea is essentially the same as adding a constant $(1-w)/w$ to all of the payoffs of a game, i.e choosing background fitness $\beta = (1-w)/w$ (see Section 3.1.1.1), with weak selection meaning the constant is arbitrarily large.

The transition probabilities (12.5) and (12.6) now become

$$p_{i,i+1} = \frac{i(1 - w + w E_{M,i})}{i(1 - w + w E_{M,i}) + (N - i)(1 - w + w E_{R,i})} \frac{N - i}{N}, \qquad (12.51)$$

$$p_{i,i-1} = \frac{(N - i)(1 - w + w E_{R,i})}{i(1 - w + w E_{M,i}) + (N - i)(1 - w + w E_{R,i})} \frac{i}{N}. \qquad (12.52)$$

Using approximations

$$(1 + x)(1 + y) \approx 1 + x + y \text{ for } x, y \approx 0, \qquad (12.53)$$
$$(1 + x)/(1 + y) \approx 1 + x - y \text{ for } x, y \approx 0, \qquad (12.54)$$

we first get that for small $w \approx 0$

$$\frac{p_{i,i-1}}{p_{i,i+1}} \approx 1 + w(E_{R,i} - E_{M,i}), \qquad (12.55)$$

and thus, after substituting into (12.48) we get that, for small w, the stability condition (12.47) is equivalent to

$$a(N-2) + b(2N-1) < c(N+1) + d(2N-4) \qquad (12.56)$$

(see Nowak et al., 2004).

It should be noted that for $w \approx 0$, the fixation probabilities of all individuals are arbitrarily close to $1/N$, so that any advantages found are small. One justification of using weak selection is that it makes the mathematics easier, whilst at the same time representing biologically realistic cases. The mathematical results do not generalise to cases where selection is not weak, but they can give some insight into such cases.

For a large population the conditions for ESS_N (12.46) and (12.56) reduce to

$$b < d, \text{ and} \qquad (12.57)$$
$$a + 2b < c + 2d. \qquad (12.58)$$

The first of these two conditions is the standard condition for an ESS in an infinite population. The addition of the second condition leads to what Taylor et al. (2004) called the *rule of 1/3*. The rule says that if $a > c$ and $b < d$ (so in an infinite population there are two pure ESSs) selection favours M replacing R, so the stability condition (12.56) does not hold for weak selection, if the unstable internal equilibrium value is less than a third i.e.

$$\frac{d-b}{a-c+d-b} < \frac{1}{3}. \qquad (12.59)$$

We have noted that in a finite population eventual fixation of a single type is certain. Even if selection pressure is toward an equilibrium i.e. when $a < c$ and $d < b$, such as in the Hawk-Dove game with $V < C$, chance will eventually mean that the population moves sufficiently far from equilibrium to eliminate one of the types. However, the population may spend a very long time near an apparently stable mixed population, and thus the time to reach such fixation may be long. Thus simulations may demonstrate apparent stability, in contrast to the theoretical results. This is perhaps an instance of the simulations giving a more accurate picture, and we may consider what is called a quasi-stationary distribution, which is the stationary distribution of the Markov chain conditional on non-absorption, for example see Nåsell (1991, 1996).

12.3 Evolution on graphs

In Section 12.1 we considered a finite, yet well-mixed, population where every individual could (and did) interact with every other. In this section, we

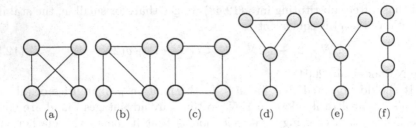

FIGURE 12.2: Connected undirected graphs with 4 vertices. The following cases are of special interest: a complete graph (a), a circle (c), a star (e) and a line (f).

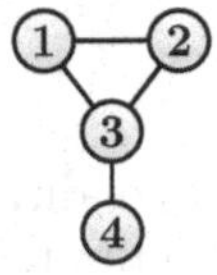

$$W = \begin{pmatrix} 0 & 1/2 & 1/2 & 0 \\ 1/2 & 0 & 1/2 & 0 \\ 1/3 & 1/3 & 0 & 1/3 \\ 0 & 0 & 1 & 0 \end{pmatrix}$$

(a)   (b)

FIGURE 12.3: (a) A graph with four vertices and equally weighted edges. (b) Its corresponding weighting matrix W.

assume that a population consists of N individuals and that each individual occupies a vertex in a given graph $G = (V, E)$. We moreover assume that every vertex is occupied, and by one individual only. Two individuals can interact only if they are connected by an edge of the graph.

G is thus a finite and undirected graph, which we assume is connected and simple, i.e. no vertex is connected to itself and there are no parallel edges. We show some simple examples of typical graphs in Figure 12.2.

The graph structure is represented by a matrix $W = (w_{ij})$, where w_{ij} is the probability of replacing a vertex j by a copy of a vertex i, given that vertex i was selected for reproduction. $w_{ij} = 0$ if there is no edge between vertices i and j. For connected vertices, it is often the case that $w_{ij} = 1/e_i$ where e_i is the degree of vertex i; see Figure 12.3. The well-mixed population that we have considered up until now is a special case of this. It is represented by the *complete graph*, the graph where every pair of vertices are connected, with all weights equal. We note that this treatment with weights w_{ij} is sufficiently general to allow us to consider directed graphs or graphs where edges carry a different weight (see Lieberman et al., 2005).

We suppose that the population evolves according to an evolutionary dynamics and the evolutionary process can be represented as a discrete time Markov chain. Supposing that $C \subseteq V$ is the set of vertices occupied by mutants, then at the next time step the set occupied by mutants will become either

(1) $C \cup \{j\}$, $j \notin C$, provided (a) a vertex $i \in C$ was chosen for reproduction and (b) it placed its offspring into vertex j; or

(2) $C \setminus \{i\}$, $i \in C$, provided (a) a vertex $j \notin C$ was selected for reproduction and (b) it placed its offspring into i; or

(3) C, provided an individual from C $(V \setminus C)$ replaces another individual from C $(V \setminus C)$.

The states $\emptyset$ and V are the absorbing points of the dynamics. It is usually assumed that at the beginning of the evolutionary process, all vertices are occupied by residents and then one vertex is chosen uniformly at random and replaced by a mutant; this is called *random initialisation*, and we shall assume that here. An alternative method, *temperature initialisation* (e.g., see Adlam et al., 2015) is also used; in this case on the assumption that mutations arise only in new offspring, the new mutant is placed upon the graph proportionally to the frequency at which the individuals on the site are replaced.

We have outlined the possible transitions in the Markov chain, but not the probabilities. Whilst the possible transitions are generally the same for any of the evolutionary dynamics commonly used, the transition probabilities are not and depend upon a choice of evolutionary dynamics. We shall initially assume the *Birth Death Birth* (BDB) dynamics, also called the *Invasion Process* (IP), where an individual is selected to give birth proportional to its fitness, and then copies itself into one of its neighbours (usually at random, with equal probability). This dynamics is often written Bd, where the capital B signifies that this is the step at which selection occurs. An example of one step of the BDB dynamics is shown in Figure 12.4.

Example 12.3 (Transition probabilities for a graph). For the graph shown in Figure 12.4, find the probability that the mutant will be eliminated before it can copy itself into any of its neighbours, assuming that $r = 3$.

At any given time step, one of the following three options happens (and a transition happens once either the first or the second option occurs).

1. The mutant is selected for reproduction (and then copies itself into one of its neighbours). This occurs with probability

$$\frac{r}{3+r} = \frac{1}{2}. \tag{12.60}$$

2. A neighbour of the mutant is selected for reproduction and replaces the mutant. This occurs with probability

$$\frac{1}{3+r} \left(1 + \frac{1}{2} + \frac{1}{2} \right) = \frac{1}{3}. \tag{12.61}$$

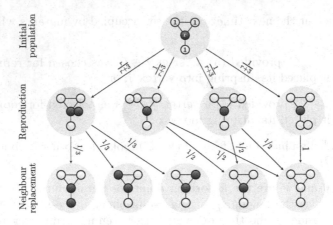

FIGURE 12.4: One step of the BDB dynamics (Invasion Process).

3. A neighbour of the mutant is selected for reproduction and replaces another neighbour, in which case the process continues in the next time step. This occurs with probability

$$\frac{1}{3+r}\left(0 + \frac{1}{2} + \frac{1}{2}\right) = \frac{1}{6}.\tag{12.62}$$

Thus the probability we require is

$$\frac{\frac{1}{3}}{\frac{1}{3} + \frac{1}{2}} = \frac{2}{5}.\tag{12.63}$$

12.3.1 The fixed fitness case

Let us first assume that individuals do not really interact and that the fitness of a mutant is r while the fitness of a resident is 1 (regardless of its position on the graph and on the composition of its neighbourhood).

Let $f_i \in \{1, r\}$ denote the fitness of an individual at vertex i. Under the rules of the BDB dynamics, the individual at vertex i is selected for reproduction with probability

$$s_i = \frac{f_i}{\sum_{j \in V} f_j}.\tag{12.64}$$

Following Lieberman et al. (2005), the BDB dynamics yields

$$P_C = \frac{\displaystyle\sum_{i \in C}\sum_{j \notin C}\left(r w_{ij} P_{C \cup \{j\}} + w_{ji} P_{C \setminus \{i\}}\right)}{\displaystyle\sum_{i \in C}\sum_{j \notin C}\left(r w_{ij} + w_{ji}\right)}\tag{12.65}$$

with $P_\emptyset = 0$ and $P_V = 1$, where P_C denotes the probability of mutant fixation given C is the set currently occupied by mutants.

Note that the system (12.65) of linear equations is very large. The size is typically of the order of 2^N equations in many cases. Consequently, the general analytical solution of (12.65) is only known in a few cases. We note that the system is also sparse (since from any state C, one can go to at most N other states). The graphs where analytical progress has been made have essentially been of three types (see Broom and Rychtář, 2008):

(i) regular graphs, where the size of the system can be reduced by symmetries in the equations as shown in Section 12.3.1.1,

(ii) graphs where the greatest degree of the vertices is two (lines and circles —although a circle is a regular graph), which means that mutants must always be in the form of a line segment, significantly reducing the number of states that need to be considered,

(iii) graphs with a high degree of symmetry, such as bipartite graphs (Voorhees, 2013); with the star (see Figures 12.2(e) and 12.5) as a special example, where many of the states are isomorphic, so again the number of effective states is relatively small.

To see how symmetries can be used to reduce the size of the system, see Figure 12.5.

12.3.1.1 Regular graphs

A graph is *regular* if the number of edges e_i is constant for all vertices. It follows that $T_i^- = \sum_j w_{ji}$, the *in-temperature* (or just the temperature) of vertex i, is constant, which is known as the *isothermal property* . We note that there is also the *out-temperature* $T_i^+ = \sum_j w_{ij}$, which is also constant for regular graphs, but which in general is distinct from the in-temperature.

Thus if we assume that P_C depends only on the size of set $C \subseteq V$, we can replace P_C by the probability of fixation from $|C|$ mutants, $x_{|C|}$, and we get that for regular graphs the system (12.65) reduces to

$$x_{|C|} = \frac{r}{r+1}x_{|C|+1} + \frac{1}{1+r}x_{|C|-1}, \qquad (12.66)$$

with boundary conditions $x_0 = 0, x_N = 1$. The equation (12.66) is a special case of (12.21) and the solution is thus given by (12.22) as

$$x_1 = \begin{cases} \dfrac{1 - (1/r)}{1 - (1/r)^N} & r \neq 1, \\ 1/N & r = 1, \end{cases} \qquad (12.67)$$

which is the Moran fixation probability (12.11). Substituting this solution into (12.65) shows that this does indeed solve the system of equations. Since we have found a solution to (12.65), and this solution must be unique, it also follows that our assumption that we can replace P_C by $x_{|C|}$ must be correct.

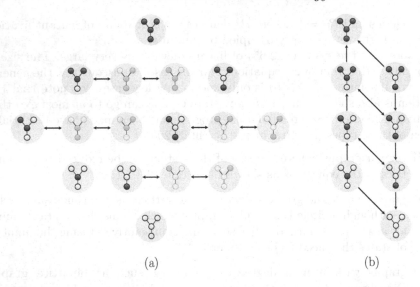

(a)   (b)

FIGURE 12.5: Symmetries allow us to reduce the count of distinct states, and consequently reduce the size of the system (12.65) such as in the dynamics on a star graph. (a) All distinct states, (b) states relevant to the system (12.65).

12.3.1.2 Selection suppressors and amplifiers

Lieberman et al. (2005) considered a variety of graphs, generating some novel and surprising results. The *burst* is a directed graph where there is a single central node, with edges directed to a number of leaves. It is thus like a star with edges directed away from the centre. The burst (see Figure 12.6) completely suppresses selection, so that the fixation probability of a randomly placed mutant is $1/N$, irrespective of the fitness of the individuals (the type of the individual at the centre always fixates). We note that a burst is not a connected graph, as there is no way to reach the central vertex from other vertices, and no connected graph can completely suppress selection in this way.

At the other extreme Lieberman et al. (2005) (see also Díaz et al., 2013) found a number of graph families where the probability of selection of a random mutant tends to 1 as $N \to \infty$ for any advantageous mutant ($r > 1$). In particular, a *superstar* of level k with N individuals has approximate fixation probability

$$\frac{1 - (1/r)^k}{1 - (1/r)^{kN}},\tag{12.68}$$

i.e. equivalent to an individual of fitness r^k in a well-mixed population. Another example is the *funnel*. See Figure 12.6.

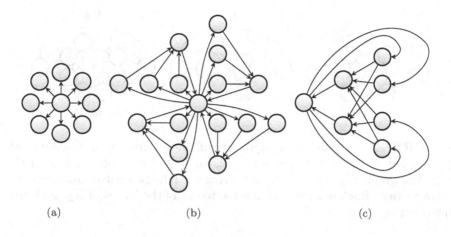

(a)   (b)   (c)

FIGURE 12.6: (a) The burst (and any one rooted graph in general) is a selection suppressor; (b) the superstar and (c) the funnel are selection amplifiers.

In general, for equally weighted graphs, regular graphs tend to have the lowest fixation probability for a randomly placed advantageous mutant, while highly irregular graphs such as the star have the highest. There is a high degree of correlation between the average fixation probability of a single mutant and the variance of vertex degrees on the graph (Broom et al., 2011). For an intriguing connection with Randic indices, see Estrada (2010).

Graphs that increase (decrease) the fixation probability of the advantageous mutants compared to the well-mixed population are called *amplifiers (suppressors) of natural selection* (Lieberman et al., 2005). As noted in Tkadlec et al. (2020), under the BDB dynamics, many amplifying families of graphs have been constructed, such as the Star graph (Chalub, 2016; Hadjichrysanthou et al., 2011), the Complete Bipartite graph (Monk et al., 2014) and the Comet graph (Pavlogiannis et al., 2017), as well as families that guarantee fixation, such as the superstar and the funnel mentioned above, in the limit of large population size (Giakkoupis, 2016; Galanis et al., 2017; Pavlogiannis et al., 2018). Extensive computer simulations on small populations have also shown that many graphs have amplifying properties (Hindersin and Traulsen, 2014, 2015; Tkadlec et al., 2019; Möller et al., 2019).

Cuesta et al. (2017) presented suppressors of selection. They found (by searching through all graphs of order less than 10) a unique suppressor of selection of order 6, namely the graph they called ℓ_6 and constructed the graphs ℓ_8 and ℓ_{10} (as well as the whole ℓ-family) from this initial example, see Figure 12.7.

Adlam et al. (2015) showed that whether the graph is an amplifier or not depends on the seeding of the mutants. They considered the question of

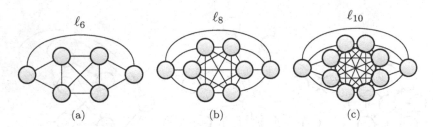

FIGURE 12.7: Examples of supressors of the selection from Cuesta et al. (2017). For $N = 2n + 2$, where $n \geq 2$, an ℓ_N-graph is obtained from the complete graph K_{2n} by dividing its vertex set into two halves and adding 2 extra vertices. Each of them is connected to one of the halves of K_{2n} and with the other extra vertex.

whether there was a structure that could amplify selection for any mutant initialisation. By considering *adversarial placement*, the (artificial) mechanism where the opposing type gets to choose where a new mutant is seeded (with the aim of making the fixation probability as small as possible) they showed that at most limited amplification could be achieved. In particular, they showed that the effective level of fitness (i.e. the fitness following the Moran formula that would lead to the observed fixation probability) could not go up by more than 1, independently of the actual fitness r. They also demonstrated that the most well-known amplifier for uniform initialisation is not an amplifier for temperature initialisation. They went on to find a set of structures that are amplifiers for both uniform and temperature initialisation (as well as combinations of the two).

Results are different for other dynamics, and in Section 12.3.2 we discussed a range of dynamics in detail. One commonly used dynamics is the DBB dynamics, where death is followed by birth, and selection happens at the birth event. Hindersin and Traulsen (2015) showed that most undirected random graphs are suppressors of selection for DBB dynamics. Tkadlec et al. (2020) showed that any amplification under DBB updating is necessarily bounded and transient. The boundedness result states that even if a population structure does amplify selection, the resulting fixation probability is close to that of the well-mixed population. The transience result states that for any population structure there exists a threshold r^* such that the population structure ceases to amplify selection if the mutant fitness advantage r is larger than r^*.

Allen et al. (2020) uncovered the first known examples of transient amplifiers of selection (graphs that amplify selection for a particular range of fitness values) for the DBB process. They also found new families of "reducers of fixation", which decrease the fixation probability of all mutations, whether beneficial or deleterious.

12.3.2 Dynamics and fitness

We consider six of the commonly used natural extensions of the Moran process that we studied in Section 12.1.1. They vary in the order of birth and death and whether selection happens at the birth or at the death stage. We note that sometimes other dynamics are used, for example "imitation dynamics" (Ohtsuki and Nowak, 2006) and (the undirected version of) "link dynamics" (Antal et al., 2006).

1. *BDB, Bd or IP dynamics* (Lieberman et al., 2005)—an individual is chosen for reproduction with probability proportional to its fitness and its offspring replaces a randomly chosen neighbour.

2. *BDD or bD process* (Masuda, 2009)—an individual is chosen for reproduction at random and its offspring replaces a neighbouring individual which is chosen with probability inversely proportional to its fitness.

3. *DBD, Db or Voter model* (Antal et al., 2006)—an individual first dies with a probability inversely proportional to its fitness and is then replaced by the offspring of a randomly chosen neighbour.

4. *DBB or dB process* (Ohtsuki et al., 2006)—an individual first dies at random and is then replaced by an offspring of a neighbour that is chosen with probability proportional to its fitness.

5. *LB process* (Masuda and Ohtsuki, 2009)—each edge is considered separately in each direction, and weighted proportionally to its undirected weight and the fitness of the origin vertex. A weighted edge is then selected at random, with the origin individual replacing the destination one.

6. *LD process* (Masuda and Ohtsuki, 2009)—each edge is considered separately in each direction, and weighted proportionally to its undirected weight and the inverse fitness of the destination vertex. A weighted edge is then selected at random, with the origin individual replacing the destination one.

We summarise the dynamics above in Table 12.1, where f_i is the fitness of individual i and $\mathfrak{r}_{ij}$ is the probability that i replaces j. We note that the LB and LD dynamics yield the same fixation probability for the fixed fitness case we are currently considering but this is not the case when games are considered. Several dynamics may also be combined into a single mixed dynamics, see for example Zukewich et al. (2013); Tkadlec et al. (2020). Different updating mechanisms can yield different evolutionary outcomes, see for example Wu et al. (2015).

We saw in Section 12.3.1.1 that graphs with the isothermal property had the Moran probability as the fixation probability for all r for the BDB process. In general, the graphs which have the Moran fixation probability differ

TABLE 12.1: Dynamics defined using the replacement weight as in Pattni et al. (2015). In each case, B (D) is appended to the name of the dynamics if selection happens in the birth (death) event.

Dynamics	$P(i$ replaces $j)$	Dynamics	$P(i$ replaces $j)$
BDB	$\mathfrak{r}_{ij} = \dfrac{f_i}{\sum_n f_n} \dfrac{w_{ij}}{\sum_n w_{in}}$	BDD	$\mathfrak{r}_{ij} = \dfrac{1}{N} \dfrac{w_{ij} f_j^{-1}}{\sum_n w_{in} f_n^{-1}}$
DBD	$\mathfrak{r}_{ij} = \dfrac{f_j^{-1}}{\sum_n f_n^{-1}} \dfrac{w_{ij}}{\sum_n w_{nj}}$	DBB	$\mathfrak{r}_{ij} = \dfrac{1}{N} \dfrac{w_{ij} f_i}{\sum_n w_{nj} f_n}$
LB	$\mathfrak{r}_{ij} = \dfrac{w_{ij} f_i}{\sum_{n,k} w_{nk} f_n}$	LD	$\mathfrak{r}_{ij} = \dfrac{w_{ij} f_j^{-1}}{\sum_{n,k} w_{nk} f_k^{-1}}$

TABLE 12.2: Different subclasses of graphs as subsets of W, the set of all strongly connected graphs on N vertices.

Symbol	Name	Condition
W_H	Uniform weight	$w_{ij} = 1/N$ for all i, j
W_C	Circulations	$T_n^+ = T_n^-$ for all n
W_I	Isothermal	$T_i^+ = T_j^-$ for all i, j
W_R	Right stochastic	$T_n^+ = 1$ for all n
W_L	Left stochastic	$T_n^- = 1$ for all n
C_N	Equally weighted cycles	$w_{l_i l_{i+1}} = w_{l_{i+1} l_i} = 1/2$ for $i = 1, \dots, N$ for some labelling $l_1, l_2, \dots, l_N$ of vertices (assuming $l_{N+1} = l_1$)

depending upon the dynamics used. In Pattni et al. (2015) a complete classification of such graphs for the above dynamics was provided.

Table 12.2 shows different classes of all strongly connected graphs on N vertices (i.e. graphs for which there is a path with non-zero probability from any vertex i to any vertex j) considered in Pattni et al. (2015).

Figure 12.8 shows the classes of graphs that are equivalent to the Moran process (under a given dynamics). For example M_{BDB}, the set of graphs where BDB dynamics is equivalent to a Moran process, is the set of all graphs that map onto an element of the circulations under $f_R = (w_{ij}) \to (w_{ij}/\sum_n w_{in})$ which maps W to W_R. Similarly, M_{LB} is precisely the set of circulations.

One thing that we observe from Figure 12.8 is that it is far easier for matrices to satisfy the conditions for the Moran fixation probability for the BDB and DBD dynamics than it is for BDD and DBB; in particular, for these latter processes positive self weights $w_{ii} > 0$ are required.

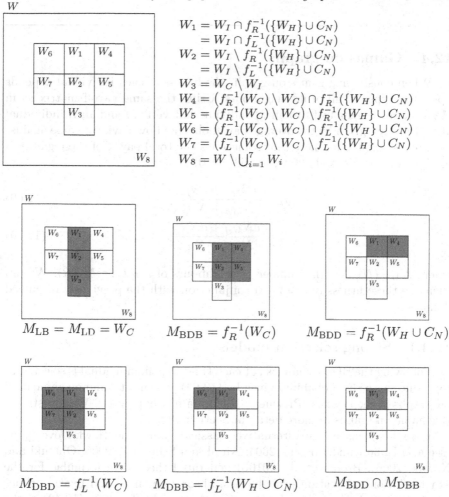

FIGURE 12.8: The diagram on top shows eight partitions W_i of W (labelled using their index $i = 1, 2, \ldots, 8$). Here $f_R = (w_{ij}) \to (w_{ij}/\sum_n w_{in})$ which maps W to W_R and $f_L = (w_{ij}) \to (w_{ij}/\sum_n w_{nj})$ which maps W to W_L. A combination of these partitions make up the sets M_*, the sets of graphs where the dynamics $*$ is equivalent to the Moran process.

Whilst Pattni et al. (2015) was the first complete presentation of these results, some of them were shown in previous work, in particular by Lieberman et al. (2005) (including the classic *Isothermal theorem*, see also Section 12.3.1.1, and *Circulation Theorem*) and also by Masuda (2009).

12.4   Games on graphs

When considering games on graphs the fitness of each individual depends upon the types of all of its neighbours. Using the same payoff matrix as in (12.43) above, the payoffs to an M individual at vertex i and an R individual at vertex j are given by the average payoff obtained by playing a game against each of its neighbours (note that sometimes the total payoff of these games is used, e.g. Szolnoki et al., 2008; Voelkl, 2010)

$$f_i = \frac{aN_{M,i} + bN_{R,i}}{N_{M,i} + N_{R,i}}, \qquad (12.69)$$

$$f_j = \frac{cN_{M,j} + dN_{R,j}}{N_{M,j} + N_{R,j}}, \qquad (12.70)$$

where $N_{M,i}$ ($N_{R,i}$) is the number of neighbours of i of type M (R). We can then use these fitness functions in conjunction with the previously discussed dynamics.

12.4.1   Strong selection models

Following the BDB dynamics , at each time step an individual is selected to reproduce with the probability given by (12.64). The most common game used on graphs is the classical Prisoner's Dilemma of Chapter 4.2. We investigate this game on graphs in more detail in Section 15.6.

Here we focus on an alternative classical game, the Hawk-Dove game (see also Hauert and Doebeli, 2004; Antal and Scheuring, 2006; Ohtsuki and Nowak, 2006). Broom et al. (2010b) generated theoretical formulae for the exact solutions of fixation probabilities, absorption times and fixation times for the star and the circle.   In one example the payoff matrix (12.43) of the Hawk-Dove game becomes

$$\begin{array}{cc} & \begin{array}{cc} \text{Hawk} & \qquad \text{Dove} \end{array} \\ \begin{array}{c} \text{Hawk} \\ \text{Dove} \end{array} & \begin{pmatrix} a = (15 - C)/2 & b = 10 \\ c = 5 & d = 15/2 \end{pmatrix}, \end{array} \qquad (12.71)$$

which is equivalent to a reward $V = 5$, an arbitrary cost C, plus a background fitness of 5 (recall the relationship between background fitness and intensity of selection from 12.2). In Figure 12.9(a) we see how varying the cost C in the Hawk-Dove game affects the fixation probability of a single mutant Hawk in a population of Doves in the three graphs, the star, the circle and the complete graph.

As we can see, changing the value of C has a gradual effect on the fixation probability on the circle, a sudden and dramatic effect on the fixation probability on the complete graph and almost no effect on the fixation probability

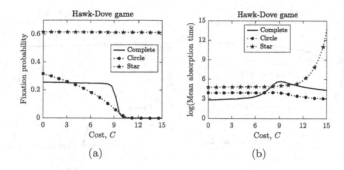

FIGURE 12.9: (a) Fixation probability and (b) the mean time to absorption when a mutant Hawk invades into a resident population of Doves on different graphs of size $N = 100$. The payoff matrix is (12.71).

on the star. In particular, comparing the complete graph and the circle, it is clear that the fixation probability can be significantly different for different regular graphs (as opposed to the fixed fitness case, where the fixation probability on all regular graphs is the Moran probability).

As seen in Figure 12.9(b), the time to absorption is hardly affected by the value of C on the circle or star (except as C approaches 15 and so the payoff of a mutant against another mutant approaches 0, leading to continual swapping of the central individual and large absorption times). The larger times for intermediate C on the complete graph corresponds to when the equivalent infinite population has an ESS corresponding of roughly equal numbers of Hawks and Doves.

In general, the dynamics are similar for regular graphs. For instance, the fixation probability for the IP and the VM are identical (Antal et al., 2006), see Exercise 12.4, and while the fixation probabilities for the BDD and DBB processes are different for small population sizes, these differences disappear for sufficiently large populations. However, the choice of a dynamics can be very important for irregular graphs as seen below (with the exception of the case of $r = 1$, see Exercise 12.5).

Consider evolution on the star for the Hawk-Dove game as in Hadjichrysanthou et al. (2011). Figure 12.10 shows the fixation probability of a single Hawk entered into a population of Doves as a function of varying cost C, where the parameters are as in payoff matrix (12.71) above. In all of the processes except the BDB, Hawks have lower fixation probabilities on the star than on the complete graph for low costs, whereas the reverse is true for the BDB. The differences between the fixation probabilities for the different processes are very large for medium to large values of the population size N. Thus the dynamics used can have a profound effect on the fixation probability, and indeed it can have similar effects on other properties such as the fixation time.

It should be noted that a similar figure can be obtained for the fixed fitness case as well. This raises an interesting question, namely what do we mean by

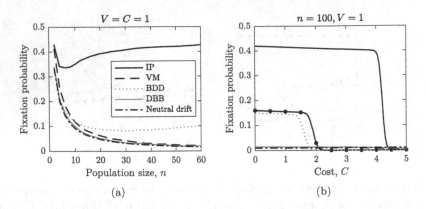

(a)   (b)

FIGURE 12.10: The average fixation probability of a single mutant Hawk on a star graph under various dynamics in the Hawk-Dove game (12.71) in the case where (a) $V = C = 1$ and n varies, (b) $n = 100, V = 1$ and C varies. Background fitness is 2 and $w = 1$. The line with circles represents the respective case on the complete graph.

fitness in this context? In Section 5.4 we discuss fitness, and whereas there are complications about the definition, it is possible to think of fitness as the expected number of offspring (raised to adulthood) of an individual in an effectively infinite well-mixed population. Whilst in our more complex populations we cannot neatly find the fitness in this way because of complications linked to the structure and stochastic effects of the finite population, the simplistic definition of fitness used in the current chapter can be far away from fitness in the above meaning. In particular, one might suppose that a mutant with 0 fitness should have 0 probability to fixate, and that an individual with infinite fitness would be certain to fixate within a population of unit fitness. We note that the BDB and DBD processes are consistent with this interpretation, but that the BDD and DBB processes are not, at least when we do not have self-loops (i.e. we have $w_{ii} = 0$) as is usually assumed in models. We saw in Section 12.3.2 that for a graph to be equivalent to the Moran process for either of BDB and DBD process, self-loops were necessary. Thus it is perhaps more accurate to say that "true" fitness is an increasing function of r, rather than r itself.

12.4.1.1   Theoretical results for strong selection

It is known that the problem of finding the fixation probability for strong selection is an NP hard problem for general graphs (Ibsen-Jensen et al., 2015; Allen et al., 2017), and so progress can only be made on specific classes of graphs.

The case of complete graphs was studied in Taylor et al. (2004), see Section 12.2. Altrock et al. (2017) studied games on circles. They focus on an exponential payoff-to-fitness mapping (Traulsen et al., 2008), i.e. if the payoff of an individual from the game is E, then the fitness is given by $f = e^{\beta E}$ where the parameter β measures the intensity of selection, similar to w as used in Section 12.2. As discussed in Traulsen et al. (2008) the fitness increases with the payoff, and β can take any positive value, as opposed to w in the formula $f = 1 - w + wE$ from Section 12.2 (although note that sometimes the alternative formula $f = 1 + wE$ is used). The case with $\beta = 0$ represents neutral drift, and when β is small, the exponential function is well approximated by a linear function, and so we recover the usual weak selection results as shown in Section 12.2 (see Exercise 12.14). For large β, we have strong selection, where negative payoffs yield a fitness close to zero and positive payoffs yield large fitnesses; one advantage is that we do not need to worry about payoffs becoming negative as we do with either of the alternative formulae involving w.

Altrock et al. (2017) found that the transitions from i to $i+1$ mutants occur with probability (see Exercise 12.13)

$$p_{i,i+1} = \begin{cases} \frac{e^{\beta 2b}}{e^{\beta 2b} + 2e^{\beta(c+d)} + (N-3)e^{\beta 2d}}, & i = 1, \\ \frac{e^{\beta(a+b)}}{2e^{\beta(a+b)} + (i-2)e^{\beta 2a} + 2e^{\beta(c+d)} + (N-i-2)e^{\beta 2d}}, & 1 < i < N-1, \\ \frac{e^{\beta(a+b)}}{2e^{\beta(a+b)} + (N-3)e^{\beta 2a} + e^{\beta 2c}}, & i = N-1, \\ 0, & i = 0, N, \end{cases} \quad (12.72)$$

and the transitions from i to $i-1$ mutants happen with probability (again see Exercise 12.13)

$$p_{i,i-1} = \begin{cases} \frac{e^{\beta(c+d)}}{e^{\beta 2b} + 2e^{\beta(c+d)} + (N-3)e^{\beta 2d}}, & i = 1, \\ \frac{e^{\beta(c+d)}}{2e^{\beta(a+b)} + (i-2)e^{\beta 2a} + 2e^{\beta(c+d)} + (N-i-2)e^{\beta 2d}}, & 1 < i < N-1, \\ \frac{e^{\beta 2c}}{2e^{\beta(a+b)} + (N-3)e^{\beta 2a} + e^{\beta 2c}}, & i = N-1, \\ 0, & i = 0, N. \end{cases} \quad (12.73)$$

Thus we have a one-dimensional problem where there is a unique transition probability for movements leading to an increase in the population size, and similarly, a unique probability for movements that lead to a decrease in the population size. We can apply the same methods as in the well-mixed case, and follow the usual recurrence relation formulation to obtain a formula for the fixation probability of i mutants x_i using

$$x_i = p_{i,i+1}x_{i+1} + (1 - p_{i,i+1} - p_{i,i-1})x_i + p_{i,i-1}x_{i-1}, \quad (12.74)$$

leading to the formula (12.48) for fixation probability; see figure 12.11. Similarly, the methodology from Section 12.1.5 can be used to obtain fixation and absorption times.

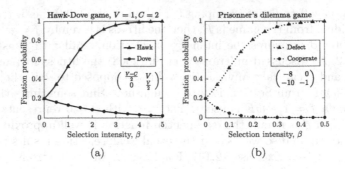

FIGURE 12.11: Fixation probability of a single mutant on a cycle with $N = 5$ as the selection intensity β varies.

12.4.2 Weak selection models

To make theoretical progress for general graphs, often weak selection models (see Section 12.1.4) are used. In the following we shall consider a methodology developed in Fu et al. (2009).

Consider a population of Resident (R) and Mutant (M) individuals playing the standard two-player, two-strategy game with payoff matrix (12.43). To follow the evolutionary process, we need to consider the distribution of individuals and their connections within the population. In particular, the population can only change if an individual replaces/ is replaced by a neighbour, so we need to condition on the specific neighbours an individual has. In what follows we consider the payoff to a player of arbitrary strategy Y, where the four potential payoffs are $a_{YM} = a, a_{YR} = b$ if $Y = M$ and $a_{YM} = c, a_{YR} = d$ if $Y = R$. The expected total payoff of a Y player with a given M individual as its neighbour is

$$E_{Y,M,i} = \left(1 + \frac{k_Y(i) - 1}{k_Y(i)} n_Y^M(i)\right) a_{YM} + \left(\frac{k_Y(i) - 1}{k_Y(i)} n_Y^R(i)\right) a_{YR}, \quad (12.75)$$

where $k_Y(i)$ is the average degree of Y individuals and $n_Y^X(i)$ is the expected number of XY links of a Y individual, when there are i type M individuals in the population. In particular, writing $N_{XY}(i)$ as the total number of XY links in the population, we have that

$$k_M(i) = \frac{2N_{MM}(i) + N_{MR}(i)}{i} = n_M^M(i) + n_M^R(i), \quad (12.76)$$

$$k_R(i) = \frac{2N_{RR}(i) + N_{MR}(i)}{N - i} = n_R^R(i) + n_R^M(i). \quad (12.77)$$

These can be approximated for any particular structure through simulation of the neutral case. Similarly, the payoff to such an individual with a given R

neighbour is

$$E_{Y,R,i} = \left(\frac{k_Y(i) - 1}{k_Y(i)} n_Y^M(i) \right) a_{YM} + \left(1 + \frac{k_Y(i) - 1}{k_Y(i)} n_Y^R(i) \right) a_{YR}. \quad (12.78)$$

The corresponding fitnesses are then $r_{Y,M,i} = 1 - w + w E_{Y,M,i}$ and $r_{Y,R,i} = 1 - w + w E_{Y,R,i}$, respectively. The evolutionary process then occurs along the interface between the two types, and we need to follow the numbers of individuals and links involved in this.

Denote the number of M and R individuals in the neighbourhood of a site where an R (M) is replaced as $k_R^M(i)$ $(k_M^M(i))$ and $k_R^R(i)$ $(k_M^R(i))$, respectively, and the number of $M(R)$ individuals along the interface by $n_M(i)(n_R(i))$. We leave it for the reader as Exercise 12.15 to show that, for $1 \leq i < N - 1$,

$$k_R^M(i) = N_{MM}(i+1) - N_{MM}(i), \quad (12.79)$$

$$k_R^R(i) = N_{RR}(i) - N_{RR}(i+1). \quad (12.80)$$

For $i = N - 1$ we have only a single resident, thus giving

$$k_R^M(N-1) = N_{MR}(N-1), \quad (12.81)$$

$$k_R^R(N-1) = 0. \quad (12.82)$$

Similarly (when an M is replaced), for $1 \leq i < N - 1$,

$$k_M^M(i+1) = k_R^M(i), \quad (12.83)$$

$$k_M^R(i+1) = k_{RR}(i), \quad (12.84)$$

and

$$k_M^M(1) = 0, \quad (12.85)$$

$$k_M^R(1) = N_{MR}(1). \quad (12.86)$$

The numbers of M and R types along the interface are then $n_M(i) = N_{MR}(i)/k_M^R(i)$ and $n_R(i) = N_{MR}(i)/k_R^M(i)$, respectively.

The probability of an increase in the number of mutants at the next step is

$$p_{i,i+1} = \frac{n_R(i)}{n_M(i) + n_R(i)} \frac{k_R^M(i) r_{M,R,i}}{k_R^M(i) r_{M,R,i} + k_R^R(i) r_{R,R,i}}. \quad (12.87)$$

Similarly, the probability of a decrease is

$$p_{i,i-1} = \frac{n_M(i)}{n_M(i) + n_R(i)} \frac{k_M^R(i) r_{R,M,i}}{k_M^M(i) r_{M,M,i} + k_M^R(i) r_{R,M,i}}. \quad (12.88)$$

We note that the above equations do not formally require weak selection, except that the number of links of the different types $N_{MM}(i), N_{MR}(i)$ and $N_{RR}(i)$, and so all of the terms derived from them, follow from the neutral model simulations, which are only valid under weak selection.

As above we can apply (12.48) to find the fixation probability and formulae from Section 12.1.5 to obtain fixation and absorption times.

12.4.2.1 The structure coefficient

There is a general condition under weak selection for M to be favoured by selection over R, namely that the fixation probability of a single M in an R population ρ_M is greater than the equivalent fixation probability of a single R in an M population ρ_R (contrast this with the distinct conditions for an ESS_N discussed in Section 12.2). This occurs if

$$\sigma a + b > c + \sigma d, \tag{12.89}$$

where σ is the *structure coefficient* of the process, as demonstrated in Tarnita et al. (2009b). The value of σ depends both upon the graph and the update rule, but not the payoffs a, b, c and d. For example, for regular graphs with degree k and $N \gg k$, we have $\sigma = (k+1)/(k-1)$.

As we shall see in Section 13.4.1, there is only a single structure coefficient here because games are pairwise, but in more complex multi-player scenarios there are more coefficients. Further in Section 15.6 we shall see the conditions for the evolution of cooperation on a regular graph which relates directly to the structure coefficient argument above.

12.5 Python code

In this section we show how to use Python to calculate numerical solutions to equations (12.9), (12.32) and (12.33), for a game in a finite population from Section 12.2. We also recommend a recently developed Python package DyPy (Nande et al., 2020).

```python
1  """ Games in finite population
2  This script provides numerical solutions to the 2 by 2 game.
3  It returns mean fixation probabilities, mean absorption times
4  and mean fixation times.
5  It also runs stochastic simulation to simulate the above.
6  """
7
8  ## Import basic packages
9  import numpy as np
10 from numpy.random import seed, rand
11 seed(1)
12
13 ## Define parameters
14 N = 5    # Number of individuals
15 w = 1    # Intensity of selection, between 0 and 1
16 a = 3    # Payoff to mutant when playing mutant,     must be ≥0
17 b = 1    # Payoff to mutant when playing resident,   must be ≥0
18 c = 6    # Payoff to resident when playing mutant,   must be ≥0
19 d = 0    # Payoff to resident when playing resident, must be ≥0
20
21 ## Auxiliary functions
```

```
22  def transitionProb(N, Payoffs, w=1):
23      """ Returns transition probabilities between different states
24
25          Inputs:
26          population size N
27          Payoff matrix a, b, c, d
28          intensity of selection w
29
30          Returns: pInc, pDec, pStay, arrays of size N-1,
31              pInc[m-1] = probability mutants increase from m to m+1
32              pDec[m-1] = probability mutants decrease from m to m-1
33              pStay[m-1] = probability mutants number stays the same
34      """
35
36      # Unpack Payoffs
37      a,b,c,d = Payoffs
38
39      # Mutant count
40      m = np.arange(1,N)  # [1, 2, 3, ..., N-1]
41      # It starts at 1, we have to do `-1' for indexing
42
43      # Set EM[m-1], average payoff to mutants when there are
44      #       m mutants and N-m residents
45      EM = (a*(m-1)+ b*(N-m))/(N-1)
46
47      # Set ER[m-1], average payoff to residents when there are
48      # m mutants and N-m residents
49      ER = (c*m + d*(N - m - 1))/(N-1)
50
51      # Incorporate the intensity of selection
52      rM = 1-w+w*EM
53      rR = 1-w+w*ER
54
55      # pInc[m-1] is prob of mutants to increase from m to m+1
56      pInc = (m*rM / (m*rM + (N-m)*rR)) * (N-m)/N
57
58      # pDec[m-1] is probability to decrease from m to m-1
59      pDec = ((N-m)*rR / (m*rM +(N-m)*rR)) * m/N
60
61      # pStay[m-1] is probability to stay on m mutants
62      pStay= 1-pInc-pDec
63
64      return pInc, pDec, pStay
65
66  def exactSolution(N, Payoffs, w=1):
67      """ Provides solutions for the fixation probabilities,
68      absorption times and fixation times
69      For fix probabilities P, it defines and solves appropriate
70      AP = B, it defines and solves similar system for the times
71      """
72
73      # Get transition probabilities
74      pInc, pDec, pStay = transitionProb(N, Payoffs, w)
75
76      # Main diagonal of A will be pStay-1 with added 1 at the end
77      mainDiag = np.append(pStay-1,1)
78
```

```
79      # Subdiagonal is pDec without first element and 0 at the end
80      # Remove first element and add 0 at the end
81      subDiag = np.append(np.delete(pDec,0),0)
82
83      # The superdiagonal will be pInc exacly
84      # Create the matrix with the three specified diagonals
85      A = np.diag(mainDiag) + np.diag(pInc,1) + np.diag(subDiag,-1)
86
87      # The right hand side will have 0's but at the last entry 1
88      B = np.zeros(N)
89      B[N-1] = 1
90
91      # Fixation probability solves AP = B
92      P = np.linalg.solve(A, B)
93
94      # Absorption times solve AT = B
95      absorptionTime = np.linalg.solve(A, B-1)
96
97      # Fixation times solve A*(PF)= (B-1)P
98      PF = np.linalg.solve(A, (B-1)*P)
99      fixationTime = PF/P
100
101     return P, absorptionTime, fixationTime
102
103
104 def stochasticSimulation(N, Payoffs, w, mInit = 1, reps = 100000,
105                     maxUpdates = 10000):
106     """ Performs stochastic simulation
107     By default starts with 1 mutant and repeats 100 000 times
108     To make sure we do not get stuck each repetition, it  will
109         run at most maxUpdates updates per repetition
110     """
111     # Simulated fixation probabilities will be
112     # 1 if mutant won, 0 if lost, -1 if undecided, 0 if lost
113     simP = -np.ones(reps)  # Undecided initially
114
115     # Simulated fixation times will be 0 initially
116     simT = np.zeros(reps)
117
118     # Get transition probabilities
119     pInc, pDec, pStay = transitionProb(N, Payoffs, w)
120
121     # Repeat the simulation reps times
122     for rep in range(reps):
123         # Set initial mutant count
124         m = mInit
125
126         # We are going to do the first update
127         update = 0
128
129         # While there is at least 1 mutant and at least 1 resident
130         # update the mutant counts (more than maxUpdates times)
131         while m>0 and m<N and update<maxUpdates :
132             # Increase the update count
133             update += 1
134
135             # Generate a random number to see what happens next
```

```
136            r = rand(1)
137
138            # Determine what happens next:
139            if r<pDec[m-1]: # W
140                # Mutant count decreases with prob pDec[m-1]
141                m -= 1
142            elif r<pDec[m-1]+pInc[m-1]:
143                # Mutant count increases with prob pInc[m-1]
144                m += 1
145            else:
146                pass          # Mutant count stays the same
147
148        # Now there is either no mutant, no resident
149        # or we already updated too many times
150
151        # Record how long it took for the simulation to stop
152        simT[rep] = update
153
154        # Determine the outcomes of the simulation
155        if m == N:
156            # Only mutants in the population. Mutants won
157            simP[rep] = 1
158        elif m == 0:
159            # Only residents in the population. Mutants lost
160            simP[rep] = 0
161        else:
162            pass    # Undecided but simP is already -1
163
164    # Consider only decided contests, i.e. where simP is 0 or 1
165    # Remove times for undecided contests first
166    simT = simT[simP>-1]
167
168    # Now remove probabilities from undecided contests
169    simP = simP[simP>-1]
170
171    # Record how long it took when mutants won
172    simF = simT[simP==1]
173
174    return simP, simT, simF
175
176 ## Run and print out the exact solutions
177 P, T, F = exactSolution(N, [a,b,c,d], w)
178 # In the printouts, remove the Nth entry as it is always
179 # P[N]=1, T[N]=0, F[N]=0
180 print('Mean fixation probabilities are ', P[0:N-1])
181 print('Mean absorption times are ', T[0:N-1])
182 print('Mean fixation times are ', F[0:N-1])
183
184 ## Run the simulation and printout the means of solutions
185 simP, simT, simF = stochasticSimulation(N, [a,b,c,d], w,
186                                  reps = 100000)
187 print('100,000 simulations for 1 initial mutant')
188 print('Simulated mean fix probability is ', np.mean(simP))
189 print('Simulated mean absorption time is ', np.mean(simT))
190 print('Simulated mean fixation time is ', np.mean(simF))
```

12.6   Further reading

A classical book on general stochastic processes is Karlin and Taylor (1975), although there are many others. For more on the diffusion approximation read Kimura (1964), and Waxman (2011). Zheng et al. (2011) consider the diffusion approximation in the context of evolutionary games and the Moran process.

For more on games in a well-mixed finite population, see Taylor et al. (2004) who considered various properties of evolutionary processes to illustrate the difference between finite and infinite population games; see also Lessard (2011) who considered multi-player games. For dynamics on graphs, see Lieberman et al. (2005); Nowak (2006a), see also Ohtsuki et al. (2007) for a generalisation; Allen et al. (2012) for dynamics with mutations; Cuesta et al. (2022) for dynamics where more than one vertex can be updated at the same time. Szabo and Fath (2007) is a very good review article. For a much more recent survey of work on evolutionary graph theory, see Díaz and Mitsche (2021). For work specifically on directed graphs see Masuda and Ohtsuki (2009); Masuda (2009); and also Shakarian and Roos (2011), Voorhees and Murray (2013). Ohtsuki and Nowak (2006) study different updating rules for games on circles, and Ohtsuki and Nowak (2008) consider ESSs on graphs. For methods to approximate stochastic processes on graphs, see Overton et al. (2019). For more on fixation probabilities, absorption and fixation times, see Antal and Scheuring (2006); Sood et al. (2008). The speed of the process for an Hawk-Dove game in groups of primates is investigated in Voelkl (2010).

A new and interesting way to calculate absorption times using martingales is shown in Monk and van Schaik (2021). Related papers about fixation and martingales are Monk et al. (2014); Monk (2018). Fixation probabilities in network structured meta-populations are studied in Yagoobi and Traulsen (2021).

Fast numerical calculations of fixation probability and times are given in Hindersin et al. (2016). Möller et al. (2019) developed a genetic algorithm which identifies graphs that have high or low fixation probabilities and short or long fixation times, studying them to find common themes. The fixation probability, fixation time and their relationship is also investigated in Tkadlec et al. (2019, 2021).

A recent paper that develops a methodology for considering eco-evolutionary dynamics on network structures is given in Pattni et al. (2021).

12.7 Exercises

Exercise 12.1. Solve the system (12.9) to obtain (12.10).

Exercise 12.2 (Karlin and Taylor, 1975). Solve the system (12.21) to obtain (12.22).

Exercise 12.3 (Traulsen and Hauert, 2009). Solve the system (12.32).

Exercise 12.4 (Antal et al., 2006). On a complete graph, consider the IP dynamics, the BDD process, the voter model and the DBB process. Show that the IP dynamics and the voter model have the same fixation probability for general mutant fitness r. Show that all four processes have the same fixation probability in the limit as the population size tends to infinity.

Hint. Formulate each process as a general birth-death process (12.21) and solve it using (12.22).

Exercise 12.5 (Antal et al., 2006). On a general graph, consider the IP dynamics, the BDD process, the voter model, and the DBB process (described in Section 12.3.2) and show that the different dynamics have no effect on the fixation probability when $r = 1$.

Hint. Show that transition probabilities are the same for all the processes.

Exercise 12.6. Repeat Example 12.3 for a mutant starting at any of the alternative vertices on the graph.

Exercise 12.7. For the graph shown in Figure 12.4, find the fixation probability of the mutant assuming that $r = 3$.

Exercise 12.8 (Hadjichrysanthou et al., 2011). Show that the dynamics described in Exercise 12.4 yield different outcomes for irregular graphs such as a star.

Exercise 12.9 (Broom and Rychtář, 2008). A graph is called *isothermal* if $\sum_j w_{ji}$ is constant as a function of i. Show that for equal weights a graph is isothermal if and only if it is regular. Also, show that in general a graph is isothermal if and only if the matrix $W = (w_{ij})$ is double stochastic (Lieberman et al., 2005), i.e. $\sum_j w_{ji} = 1$.

Exercise 12.10 (Broom and Rychtář, 2008). A star is a non-directed graph with N vertices labelled $0, 1, \ldots, N-1$ where the only edges are between vertices 0 and $i \in \{1, \ldots, N-1\}$. Show that the dynamics (12.65) reduces to a system of in the order of $2N$ equations and solve that system.

Exercise 12.11 (Broom and Rychtář, 2008). A line is a non-directed graph with N vertices labelled $1, 2, \ldots, N$ where the only edges are between vertices i and $i + 1$. Show that the dynamics (12.65) reduces to a system of in the order of $N^2/4$ equations.

Exercise 12.12 (Pattni et al., 2015). For which of the six evolutionary dynamics described in Table 12.1 are the following graph weights equivalent to the Moran process?
(i) $w_{i,j} = 1/4$ when $i - j \mod N \in \{-2, -1, 1, 2\}$ and $w_{i,j} = 0$ otherwise, for $N \geq 5$.
(ii) $w_{i,j} = 1/i$ for all i, j.

Exercise 12.13 (Altrock et al., 2017). For the strong selection model of Section 12.4.1.1 show the transition probability formulae for the circle graph given in (12.72) and (12.73).

Exercise 12.14 (Altrock et al., 2017). Find the transition probabilities from Exercise 12.13 for the weak selection case of small β and compare this to the results from Section 12.2.

Exercise 12.15 (Fu et al., 2009). Show that the formulae for the neighbourhood numbers k_R^M, k_M^M, k_R^R and k_M^R needed for (12.87) and (12.87) in Section 12.4.2 are as given directly before (12.87).

Exercise 12.16. Consider the Hawk-Dove game from (12.71) played on a regular graph with degree k as in Section 12.4.2.1. For what values of C and k is Hawk favoured by selection over Dove under weak selection?

Chapter 13

Evolution in structured populations

13.1 Spatial games and cellular automata

Spatial games and cellular automata usually take place on a regular grid. In one dimension this is just a line graph if there are boundaries, or a circle if the space "wraps around". Similarly, in two dimensions it can be a grid where every individual has either four edges (vertical and horizontal neighbours) or eight edges (vertical, horizontal and diagonal neighbours) which are in the form of a square with boundaries, or a torus if it "wraps around". The circle and torus are regular graphs and so have the isothermal property, for instance, but the line and square are not.

An important distinction between the games that we have been considering and cellular automata, is that usually in cellular automata all individuals update simultaneously. We note that this is not always the case, and that there are also asynchronous versions as we discuss below. For synchronous updating, it is supposed that each vertex updates its type based upon the configuration of its neighbours, including its initial type. An update rule is a choice of new type for each of the possible configurations. Thus if there are n types of individuals, then the number of possible update rules for a regular graph of degree k is $n^{n^{k+1}}$; see Exercise 13.1.

Thus in the simplest possible case of interest, a one-dimensional linear or circular space with two types of individual (which Wolfram, 2002 refers to as Black and White), the number of rules is $2^{2^3} = 256$, numbered from 0 to 255 (Wolfram, 2002, Chapter 3). This can lead to many interesting patterns, as well as very simple ones. For instance, starting with a single Black individual on an infinite line (thus equivalent to our a mutant as discussed throughout Chapter 12), under some rules (e.g. rules 0 and 128 under Wolfram's numbering system) all the cells become White, while in others (e.g. rule 255) all become Black. There are also rules such as 7 and 127 in which all cells alternate between Black and White on successive steps. See Figure 13.1 for rule 18.

Many rules exhibit behaviour where a pattern consisting of a single cell or a small group of cells persists. Sometimes this pattern remains stationary (e.g. rules 4 and 123), but in other cases (e.g. rules 2 and 103) it moves to the left or right. The basic structure of the cellular automata discussed here implies

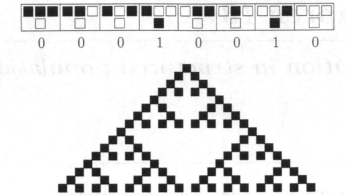

FIGURE 13.1: Wolfram's rule number $18 = (00010010)_2$. The possible values of the three neighbouring cells are shown in the top row of each of the 8 panels, and the resulting value the central cell takes in the next generation is shown below in the centre. The resulting pattern after several generations is shown below.

that the maximum speed of any such motion must be one cell per step, and in many rules, this maximum speed is achieved. However, under other rules (e.g. 3 and 103) the average speed is instead only half a cell per step.

Patterns often remain of a fixed size, but sometimes the patterns grow forever. These may be purely repetitive ones (e.g. rules 50 and 109) but they can be more complex, displaying what Wolfram (2002) calls *nested patterns*. Under rule 225, the width of the pattern does not grow at a fixed rate, but rather is proportional to the square root of the number of steps on average.

Southwell (2009) and Southwell and Cannings (2013) consider best response games (see Section 3.1.3) played on cellular automata, where individuals play games against all of their neighbours as in Section 12.4 and update their type according to which would give them the highest payoff. In one dimension this yields a subset of the rules given by Wolfram (2002), but they also consider two-dimensional processes which can lead to very complex behaviour, as you might expect since it is a two-dimensional version of the process described above.

A good example of this type of process can be found in Nowak and May (1992b), where they consider the spatial struggle of Cooperate and Defect in a Prisoner's Dilemma on a two-dimensional lattice. Individuals play a one-shot Prisoner's Dilemma against all of their neighbours, and then change their strategy by copying their neighbour with the largest payoff (imitation dynamics, see Cressman, 2003). We note that for this process, the update depends upon the types of an individual's neighbours' neighbours, so this cellular automaton lies outside of Wolfram's framework as described above. Just as in Southwell (2009), this game can produce many interesting spatial patterns, although the dynamics used is crucial for this (for example, best response dynamics, see Section 3.1.3, would simply yield all defection after a single step).

Given there is spatial correlation between the type of individual, cooperators are more likely to be next to other cooperators than defectors, and this can be enough to beat the dilemma. There are circumstances where cooperators completely dominate the space, others when they are eliminated, and there can be coexistence, but with the cooperators clustering together, with the location and shape of cooperator regions changing with time.

It should be noted that the synchronous updating of the above examples is different in character to the asynchronous updating described in evolutionary processes on graphs (although we note that synchronous updating has also been used on graph models, Santos et al., 2006a). Similarly, the evolutionary graph theory models from the previous chapter can of course have a lattice as the underlying graph, and there are various works that consider this, for example Szabó et al. (2005); Vukov et al. (2006), Doebeli and Hauert (2005).

One disadvantage is that asynchronous updating is a lot less computationally efficient, and these processes often involve large simulations. The results though can be very different; for instance the regular patterns of Wolfram would not occur with randomly chosen asynchronous updating.

An interesting method that applies a more rigorous mathematical approach to evolutionary games on lattices (where the underlying updating rule is asynchronous) was given in Durrett (2014). Here the approach is to show that in a suitably limiting scenario, the evolutionary process can be solved using a partial differential equation. The model was extended to 3x3 evolutionary games (Durrett (2014) had only considered 2x2 games) in Nanda and Durrett (2017). This was applied, in particular, to models for cancer evolution, and we shall return to this work in Section 22.2, where we consider evolutionary cancer models more generally.

13.2 Theoretical developments for modelling general structures

In Allen et al. (2017) the authors approached evolutionary dynamics in a novel way. Consider the usual evolutionary graph models from Chapter 12. How the population evolves depends upon the specific mix of types across the graph, and they approach this using coalescent theory (Kingman, 1982).

This methodology considers a modern genealogy and tries to reconstruct ancestral lineages through random walks. The authors use this approach to model evolution on a graph by considering the time to when the individuals at two vertices are descended from a common ancestor. In particular, mutant

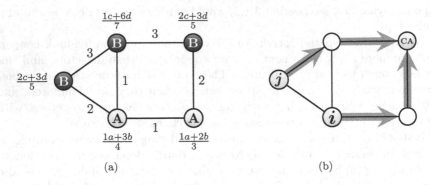

(a)   (b)

FIGURE 13.2: (a) Population structure is represented by a graph with edge weights which are shown next to the edges. The individual play a standard matrix game, their weighted payoffs are shown next to the vertices. (b) The coalescence time τ_{ij} is the expected meeting time of random walks from i and j, representing time to a common ancestor (CA).

fixation in the population occurs precisely at the first time when all individuals are descended from the initially introduced mutant.

Recalling that w_{ij} is the edge weight from vertex i to vertex j, consider a random walk with probability of movement from vertex i to vertex j as $p_{ij} = w_{ij}/\sum_k w_{ik}$. The *coalescent time* τ_{ij} is the expected number of steps for two independent random walks starting from vertices i and j to meet (Cox, 1989; Liggett, 2012), see Figure 13.2(b). The coalescent times solve the following system of equations

$$\tau_{ij} = \begin{cases} 1 + \frac{1}{2}\sum_k(p_{ik}\tau_{jk} + p_{jk}\tau_{ik}), & i \neq j, \\ 0, & i = j. \end{cases} \qquad (13.1)$$

To show this is Exercise 13.2.

If T is the absorption time (i.e. the time to fixation or extinction of the mutant), then the expected time during which vertices i and j have the same type is $T - \tau_{ij}$. Thus using the pairwise interactions and knowledge of the payoff values we can find whether cooperation is favoured or not.

Denote t_n as the mean coalescent time for the individuals at the ends of an n-step random walk, with the initial vertex chosen proportional to the sum of their weights (we note that $w_{ij} = w_{ji}$, so we do not need to specify in-weights or out-weights). In particular

$$t_n = \sum_{i,j} \pi_i p_{ij}^{(n)} \tau_{ij}, \qquad (13.2)$$

where $\pi_i = \sum_k w_{ik}/\sum_{j,k} w_{jk}$ and $p_{ij}^{(n)}$ is the transition probability of going from i to j in n steps of the random walk, using the usual terminology. In Allen et al. (2017) it is shown that the key terms to consider are t_1, t_2 and t_3.

Allen et al. (2017) considered the donation game where individuals are either cooperators or defectors. In every interaction cooperators pay a cost c and any individual of either type receives benefit b when paired with a cooperator, leading to a Prisoner's dilemma with payoffs $R = b - c, T = b, S = -c$ and $P = 0$, see Figure 13.2(a) (this game is described more fully in Section 15.5). We note here that the "T" mentioned in this paragraph is the temptation payoff from the classical Prisoner's dilemma, not to be confused with the absorption time, which it represents elsewhere in this section.

In particular, they prove that for weak selection, cooperation is favoured over defection if and only if

$$- c(T - t_0) + b(T - t_1) > -c(T - t_2) + b(T - t_3). \qquad (13.3)$$

This inequality has a nice intuitive explanation. The first term, $-c(T - t_0)$, is the cost for being a cooperator, paid for the whole time T since $t_0 = 0$. The second term, $b(T - t_1)$, is the average benefit that the cooperator gets from its neighbours (the time that a random neighbour is a cooperator is simply $T - t_1$). The terms on the right-hand side of the inequality similarly correspond to the average payoff to a random individual two steps away Thus the inequality represents the payoff to a cooperator being higher than that of a random individual two steps away. This is fundamental since such individuals are the direct competitors of our cooperator for reproduction into neighbours, i.e. one step away. Note that it is not necessary to calculate T, because it cancels out in (13.3).

It follows that cooperation is favoured if and only if $t_3 < t_1$ (this does not always have to be the case; for example for bipartite graphs including stars, $t_1 = t_3$) together with

$$\frac{b}{c} > \frac{t_2}{t_3 - t_1}. \qquad (13.4)$$

For a regular graph with uniform edge weights, we have $t_1 = N-1, t_2 = N-2$, and $t_3 = N-3+N/k$. Thus (13.4) becomes $b/c > (N-2)/(N/k-2)$, recovering the $b/c > k$ rule in the large N limit.

The above results extend to more general two strategy games. Recall that the condition for strategy A to be favoured over strategy B under weak selection is $\sigma a_{11} + a_{12} > a_{21} + \sigma a_{22}$ for structure coefficient σ (see Section 12.4.2.1; here we write the payoffs a_{ij} in their general form to avoid duplicate meanings for b and c). We have that (see Exercise 13.3)

$$\sigma = \frac{-t_1 + t_2 + t_3}{t_1 + t_2 - t_3}. \qquad (13.5)$$

We note that whilst the results above relate to a graph structure where individuals interact with their neighbours in the same way as described in Chapter 12, the coalescent methodology is more general than this, since it works on the graph for the matrix of replacement weights w_{ij}. This is what makes the methodology potentially very powerful.

The replacement weights and the payoffs for the game could come from any general process, for example that of Section 13.3 below (although there it will only work under very specific assumptions, such as for the independent model for the class of multi-player games which are simply averages of pairwise interactions).

13.3   Evolution in structured populations with multi-player interactions

Broom and Rychtář (2012) introduced a new framework for modelling the evolution of a population in a structured environment. This framework was based upon three core components: (1) the population (environment) structure; (2) the interactions between the individuals (evolutionary games); and (3) the evolutionary dynamics of the population. The framework can accommodate a variety of structures and evolutionary games; moreover, the evolutionary games are multi-player games, and they may include interactions in groups of varying size, including lone individuals. The initial paper Broom and Rychtář (2012) defined the modelling framework and discussed several example structures, but it did not include the evolutionary dynamics which were introduced in Broom et al. (2015b).

13.3.1   Basic setup

A population of N individuals $I_1, \ldots, I_N$ can move to M distinct places $P_1, \ldots, P_M$. The distribution of the individuals over the places at time t is defined by a binary matrix $X(t) = (X_{n,m}(t))$ of size $N \times M$ given by

$$X_{n,m}(t) = \begin{cases} 1, & \text{if the individual } I_n \text{ is at the place } P_m \text{ at time } t, \\ 0, & \text{otherwise,} \end{cases}$$

and the probability distribution of the population at time t was written as

$$P(X(t) = x)(x_{<t}) = P(X(t) = x \mid X(1) = x_1, \ldots, X(t-1) = x_{t-1}),$$

where x is any particular value that the matrix X may take. Similarly,

$$p_{n,m,t}(x_{<t}) = P(X_{n,m}(t) = 1)(x_{<t})$$

denotes the probability of the individual I_n being at the place P_m at time t given the history $x_{<t}$.

The population distribution thus follows a random process, which could depend upon its entire history. This allows for a vast range of possible models,

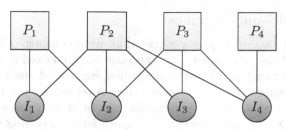

FIGURE 13.3: Illustration of an independent process. The node I_i is connected to the node P_j only if the probability $p_{i,j}$ of the individual I_i visiting the place P_j is not 0.

and therefore Broom and Rychtář (2012) defined several specific scenarios, each of which included some simplifications.

If the population distribution is independent of the history of the process:

$$P(X(t) = x)(x_{<t}) = P(X(t) = x),$$

the model is called *history-independent*. A history-independent process is called *homogeneous* if $P(X(t) = x) = P(X = x)$.

If the probability of an individual visiting a place does not depend upon the simultaneous movements of other individuals (although it may depend upon the history):

$$P(X_{n_1,m_1}(t) = 1 \ \& \ X_{n_2,m_2}(t) = 1)(x_{<t}) = p_{n_1,m_1,t}(x_{<t})p_{n_2,m_2,t}(x_{<t}),$$

then the model is called *row-independent*.

If the model is history-independent, homogeneous, and row-independent, then we have the following independent movement probabilities:

$$p_{n,m,t}(x_{<t}) = p_{n,m} \quad \text{for all } n, m, t, x_{<t},$$

and in this case the model is called simply *independent*. A model with history dependence was considered in Pattni et al. (2018) and Figure 13.3 provides a sample illustration of an independent process.

13.3.2 Fitness

An important component of any evolutionary model is the fitness of the individuals within the population. The reward for an individual I_n at time t was denoted $R(n, x, t, x_{<t})$. Assuming only the current distribution affects the reward, the *mean reward* can be considered as the fitness

$$f_n = \sum_x P(X(t) = x)R(n, x),$$

where $R(n, x)$ is the reward to I_n given the population distribution x (this often only depends upon the group the individual is in).

13.3.3 Multi-player games

We present two multi-player games that were considered in Broom and Rychtář (2012) and later papers (they also considered a fixed fitness game that we omit). Each game has two strategies, which we label A and B, respectively.

The multi-player Hawk–Dove game. Here A represents the Hawk strategy and B the Dove strategy. All individuals that meet at a place play a multi-player Hawk–Dove game, where they compete for a single reward of value V. If all individuals are Doves, they share the reward. If there are one or more Hawks, the Doves flee and the Hawks fight, with the winner getting the reward and the rest paying a cost C. All individuals also receive a background reward of value R, representing the reward gained from other activities. Thus in a group of a Hawks and b Doves, the average payoffs for the Hawks and the Doves, respectively, are

$$E^A_{a,b} = R + \frac{V - (a-1)C}{a}, \tag{13.6}$$

$$E^B_{a,b} = \begin{cases} R, & \text{if } a > 0, \\ R + \frac{V}{b}, & \text{if } a = 0. \end{cases} \tag{13.7}$$

The public goods game. Here A represents Cooperate and B represents Defect. Each cooperator pays a cost C to provide a benefit V, which is shared by all other individuals in the group (but not the providing cooperator); note that the cost is paid even by lone cooperators. All players also receive a background payoff R. Thus for a group of a Cooperators and b Defectors, the average payoffs to Cooperators and Defectors, respectively, are

$$E^A_{a,b} = \begin{cases} R - C, & \text{if } a = 1, \\ R - C + \frac{a-1}{a+b-1}V, & \text{if } a \geq 2, \end{cases} \tag{13.8}$$

$$E^B_{a,b} = R + \frac{a}{a+b-1}V. \tag{13.9}$$

We should also note that the public goods game presented above is just one version of a social dilemma, and there are many other types of games that can involve multiple (and indeed a variable number of) players, which pick cooperative versus non-cooperative strategies. These have been treated in a variety of papers (e.g. see Archetti and Scheuring, 2012), and considering the variability of the number of players, in particular, was an important focus of Peña and Nöldeke (2016) and Broom et al. (2019b), as we discuss in Section 13.4.2.

13.3.4 Evolutionary dynamics

The evolutionary dynamics of structured populations are based on a Moran process to keep the population size constant, and six main updating mechanisms are used in evolutionary graph theory as shown in Section 12.3.2. (Masuda, 2009; Pattni et al., 2015, 2017).

Following Pattni et al. (2015), these dynamics can be applied in this framework once the correct evolutionary weights are selected. Here, $w_{n_1 n_2}$ are the *replacement weights*, which can be thought of as the expected proportion of time the individual I_{n_1} spends interacting with the individual I_{n_2}. The replacement weights are computed as

$$
w_{n_1 n_2} = \begin{cases} \sum_{m=1}^{N} \sum_{G:n_1,n_2 \in G} \dfrac{\chi(m,G)}{|G|-1}, & \text{if } n_1 \neq n_2, \\ \sum_{m=1}^{N} \sum_{G:G=\{n_1\}} \chi(m,G), & \text{if } n_1 = n_2, \end{cases}
\tag{13.10}
$$

where $\chi(m,G)$ is the probability of a group G meeting at the place P_m. In particular, an individual selected for reproduction may replace itself with its own offspring; the population composition does not change in this case.

13.3.5 The Territorial Raider model

Broom and Rychtář (2012) introduced one special case of an independent model, the *territorial raider* model. In this model, individuals $I_1, \ldots, I_N$ each occupy their own place $P_1, \ldots, P_N$, and the places are arranged as vertices of a connected graph. Individuals may either stay at their own place, or move to one of the neighbouring places (although returning to their own place for the start of the next move). Each individual follows its own distribution. The movement patterns within this model may still be complex, and Broom et al. (2015b) introduced a single *home fidelity* parameter h, which governed movement in the entire population. The home fidelity parameter could take any nonnegative value, and any individual was h times as likely to stay at its own place than to move to any particular neighbouring place.

One of the key results of Pattni et al. (2015) was that the temperature $T_n = \sum_{i \neq n} w_{in}$, $1 \leq n \leq N$, was a good predictor for the average fixation probability for small networks (size 4). We note that since $w_{in} = w_{ni}$ as defined above, T_n is equal to both T_n^- and T_n^+ as defined in Section 12.3.1.1 except that the self-weights w_{nn} have been removed (often in those models the self-weights were 0, though they are not here). Larger populations and a wider variety of network structures were considered in Schimit et al. (2019). While the general thesis of the earlier investigations was confirmed, this more detailed study showed that the relationship between the mean temperature and fixation probability is complex, and in particular, it depends on the game being played.

High (low) temperatures generally enhanced (suppressed) the strength of selection. This observation agrees with results of evolutionary graph theory, see for example Möller et al. (2019); Broom et al. (2010c, 2011). In particular, finding these temperatures provides a computationally efficient method of estimating the mutant fixation probabilities (Shakarian and Roos, 2011), and it may potentially lead to some general analytical insight.

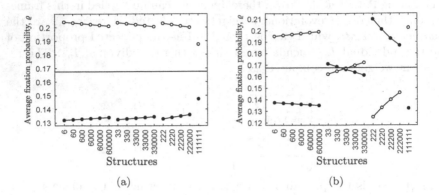

(a)　　　　　　　　　　　　　　　　　(b)

FIGURE 13.4: Comparing average fixation probability for different complete structures where figure (a) uses DBD dynamics and figure (b) uses DBB dynamics. Each number indicates a subpopulation of a certain size. For example 60 is a complete structure with 2 subpopulations of size 6 and 0, respectively; 2220 has three subpopulations of size 2 and one of size 0. We see that in figure (a) for the DBD dynamics, cooperators perform poorly in all cases. In figure (b), cooperators do better for small groups (greater than one). Increasing the number of empty places is beneficial for defectors.

Additionally, Pattni et al. (2017) considered an extension of the territorial raider model where each place was home to a subpopulation of individuals rather than a lone individual. The DBB and DBD dynamics were used for the evolution of this population playing the public goods game. This led to a natural extension of the notion of temperature; the *subpopulation temperature* of subpopulation Q_m is, $T_{Q_m} = \sum_{n_1 \notin Q_m, n_2 \in Q_m} w_{n_1 n_2}, 1 \leq n_1, n_2 \leq N$ which is a better predictor than the standard temperature for this type of population. The conclusion was that for the DBB dynamics low subpopulation temperatures were conducive to the evolution of cooperation in the public goods game, while cooperation could never evolve for the DBD dynamics. Moreover, small subpopulations (but not subpopulations of single individuals) were particularly helpful for cooperation to evolve. This is illustrated in Figure 13.4.

In Pattni et al. (2018) and Erovenko et al. (2019) a population was considered where individuals could move over an underlying graph, accumulating payoffs over a number of steps before returning to their home vertex for the replacement phase. We note that for these mobile populations the crucial effect of the dynamics is diminished. For the complete graph Pattni et al. (2018) considered the BDB and DBB dynamics, and found that cooperation could evolve in either case with no appreciable difference between the two. In Erovenko et al. (2019) it was shown that for the BDB dynamics, the complete graph (as considered before) promotes stability, with populations of cooperators or defectors being hard to invade. The star graph, on the other hand, promotes instability, and often neither type of population can resist invaders. The circle graph generally led to either cooperation or defection being favoured, whatever the composition of the population. It was hypothesised that network topology, including key properties like the clustering coefficient and degree centralisation, were key drivers of this behaviour, but this has yet to be properly explored.

13.4   More multi-player games

Multi-player games have been considered in conjunction with evolutionary dynamics and structured populations in different contexts, and a good review is Perc et al. (2013), which in particular focuses on public goods games. As they explain, the models often reduce to two important questions: how are groups of individuals formed to determine the payoffs which, in turn, lead to the fitnesses of individuals, and how do these fitnesses contribute to how the population reproduces. Thus a number of ways have been developed to incorporate structure into the population.

One natural way to include a spatial structure is through regular lattices, as discussed in Section 13.1. In addition to the pairwise games previously discussed, individuals can also play multi-player games against all of their neighbours; here an individual will also be involved in similar games centred upon each of its neighbours, and so it will be involved in some games involving all of its neighbours' neighbours as well. The involvement of such non-direct neighbours can have important consequences, as shown in Szolnoki et al. (2009). We also note that increasing group size does not automatically lead to mean field behaviour (Szolnoki and Perc, 2011; Szabó and Szolnoki, 2009).

As discussed in Section 13.3, there are models where individuals are mobile, rather than fixed at a particular position. Such movement can be random (Cardillo et al., 2012), or influenced by interactions with others (Jiang et al., 2010; Xia et al., 2012), for example they might move with a higher probability for unfavourable group compositions (leading to lower payoff) from interactions, as in Pattni et al. (2018); Erovenko et al. (2019).

13.4.1   Structure coefficients and multi-player games

For finite structured populations, we saw in Section 12.4.2 (see also Tarnita et al., 2009a), that when considering two-player games an important property, for example in assessing whether cooperation can evolve in the case of the Prisoner's dilemma, is the structure coefficient. How does this extend to multi-player games? This question was addressed by Wu et al. (2013); McAvoy and

Hauert (2015) and Peña et al. (2016a) (see also Richter, 2019) who considered the specific coefficients to be used; a more general type of structural coefficient is needed here, and typically $m - 1$ are needed for an m player game. This has only been developed for specific types of graph so far.

In particular, Wu et al. (2013) showed that for evolution involving m-player games under weak selection, the condition for strategy A to be selected over strategy B, is given by

$$\sum_{j=0}^{m-1} \sigma_j(a_j - b_{m-1-j}) > 0 \qquad (13.11)$$

where a_j (b_j) is the payoff to an A (B) individual playing against $m - 1$ opponents, j of which are of type A, for some structure coefficients $\sigma_0, \sigma_1, \ldots, \sigma_{m-1}$. We note that in Wu et al. (2013) σ_j was written k_j and then a further scaling happened to give their σ_j values, reducing the number of coefficients by one (the assumption of the existence of a non-zero coefficient allowed for division by this coefficient). We use our version for consistency with the below.

McAvoy and Hauert (2015) extended this theory to multi-player games with more than two strategies, and they give the following formula for the number of structure coefficients required (noting that in general, as above, their figure included one more structure coefficient than most previous work):

$$\phi_n = \sum_{j \in J} \left(m(j) + \sum_{k=1}^{m(j)-1} (m(j) - k)q(k, n - 2) \right), \qquad (13.12)$$

where n is the number of strategies, and $q(a, b)$ is the number of partitions of a with at most b parts. The precise form of ((13.12)) depends upon how games are played on the particular structure. In the simplest case where every player plays a game against all subsets of size $m - 1$ of their neighbours, e.g. all neighbours if there is a regular graph of degree $m - 1$, this simplifies to the form

$$\phi_n(m) = m + \sum_{k=1}^{m-1} (m - k)q(k, n - 2). \qquad (13.13)$$

Here McAvoy and Hauert (2015) used this formula to confirm existing formulae generated by a number of other authors for special cases. Thus they obtain two structure coefficients for the two-player, two-strategy game of Tarnita et al. (2009b) (as previously mentioned, this is one more than from that paper), three structure coefficients for two players and more than two-strategies (Tarnita et al., 2011) and m structure coefficients for the m-player two, strategy game (Wu et al., 2013). Equation (13.13) also gives $m(m+1)/2$ coefficients for m players and 3 strategies; see Figure 13.5.

For some simple structures these structure coefficients have nice forms. Peña et al. (2016b) discuss some from previous works. For instance the structure coefficients of the Moran process with m-player interactions (Gokhale

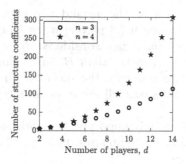

FIGURE 13.5: The number of distinct structure coefficients vs. the number of players, d, in an irreducible d-player interaction in a population of size $N = d$.

and Traulsen, 2010) are given by

$$\sigma_{m-1} = \frac{N-m}{N},$$ (13.14)

with $\sigma_j = 1$ for $0 \le j \le m - 2$, and those for the DBB process on the cycle (van Veelen and Nowak, 2012) are

$$\sigma_j = \begin{cases} 1, & j = 0, \\ \frac{2N}{N+1}, & j = 1, \dots, m-3, \\ \frac{2N-1}{N+1}, & j = m-2, \\ \frac{3(N-m)}{N+1}, & j = m-1. \end{cases}$$ (13.15)

In general, Peña et al. (2016b) consider the set of all games with particular properties, for instance that are "cooperation games" (a broader class of the games that we discuss in Section 15.4) and that favour the cooperative strategy for a particular structure. They define Q_1 (Q_2) as the set of games where cooperation is favoured under population structure S_1 (S_2) and consider two concepts; *containment order* which gives an ordering for sets Q_1 and Q_2 only if one is a subset of the other, and *volume order* where the ordering is given by the volume of the region for which Q_1 and Q_2, respectively hold. These are explored in general terms and also for particular examples.

13.4.2 Games with variable group sizes

The effect of variable group sizes was considered in Peña and Nöldeke (2016) (see also Peña and Nöldeke, 2018; Broom et al., 2019b) who proved general results for two strategy multi-player games based upon the *gain sequence*.

Considering a game with m players and two strategies A and B, and defining the payoff to an A-player (B-player) when k others play A and $m - k - 1$ others play B as a_k (b_k), they considered $d_k = a_k - b_k$, which they termed the *gain* of playing A rather than B. In Section 9.1 the importance of this function (and more generally the incentive function, the difference between the expected payoffs over all group compositions) for ESS analysis was shown.

For variable group sizes it is important to work out the correct way of interpreting the distribution of group sizes. Peña and Nöldeke (2016) defined the experienced group size as

$$\hat{p}_m = \frac{p_m m}{E_p[M]} \qquad (13.16)$$

where p_m is the proportion of groups of size m; this represents the proportion of individuals who are in groups of size m and is the distribution that needs to be considered when analysing payoffs to individuals. Note that Broom et al. (2015b, 2019b) called the experienced group size the *group size from the individual's perspective* and the group size as the *group size from the observer's perspective*.

A central concept to the work of Peña and Nöldeke (2016) is *variability order*. A probability distribution q is more variable than distribution p, written $q \geq_v p$, if for all convex functions $\phi : \mathbb{R} \to \mathbb{R}$ we have that, considering X with distribution p and Y with distribution q, $E_q[\phi(Y)] \geq E_p[\phi(X)]$.

For $q \geq_v p$, it is not sufficient that $E_q[Y] = E_p[X]$ and $Var_q[Y] \geq Var_p[X]$. What is required is that the sequences $p_0, p_1, \ldots$ and $q_0, q_1, \ldots$ cross exactly twice with $q_s > p_s$ for all s sufficiently small and large, with the reverse for intermediate values.

Following Chapter 9 we see that the expected gain for a population where a proportion of individuals playing A is x is

$$d(x, m) = \sum_{k=0}^{m-1} \binom{m-1}{k} x^k (1-x)^{m-1-k} d_k \qquad (13.17)$$

for m-player games, and so the incentive function is the expectation of this over all games,

$$h(b, \hat{p}) = E_{\hat{p}}[d(x, \hat{M})] = \sum_m \hat{p}_m d(x, m). \qquad (13.18)$$

Peña and Nöldeke (2016) then proved a series of results based upon the above concepts. If the gain sequence is convex (concave) then the experienced-variability effect is positive (negative); thus for example, for convex gain functions the more variable distributions (in the way defined above) have gain functions which are larger everywhere than the equivalent gain function for less variable distributions.

They considered two examples, the Volunteer's Dilemma (as defined in Section 15.4.2) and a public goods game with discounting (related to but distinct from the variable Prisoner's Dilemma from Section 15.4.2). From the point of view of the cooperative strategy, the gain sequences for these two games are both convex. Thus here cooperators do better in comparison to defectors when the experienced distribution is more variable.

We shall see more on the subject of variability in group size in Section 15.4.2. There we shall follow Broom et al. (2019b) who considered a large number of different public goods games from the literature and again considered the effect of group size variability.

13.5 Evolving population structures

There are also models that are co-evolutionary in nature, where as the composition of the population changes, the structure changes too. For example if links can be adjusted based upon payoffs received as in Wu et al. (2009b) or other related factors (Wu et al., 2009c; Zhang et al., 2011b) this can strongly promote cooperation. Below we firstly consider a model where there is birth and death within the population.

13.5.1 Games with reproducing vertices

We shall start by discussing such games where the vertices on a graph reproduce and die based on their success levels in local interactions (see Southwell and Cannings, 2009, 2010). Our focus is on the deterministic case where the condition of each vertex is updated simultaneously.

A system state is an undirected graph, where each vertex is associated with a pure strategy. The system is updated over discrete time steps. A given update consists of two stages, the *reproduction stage*, and then the *killing stage*. In the reproduction stage, each vertex simultaneously produces an offspring vertex, with the same strategy as the parent. There are eight possible models for how the connections of the offspring depends upon the parent, which we describe below. For example, in the $m = 1$ model the offspring just inherit their parent's neighbourhood (i.e., the offspring are born connected to the same individuals as the parent is connected to). After the reproduction stage, the killing stage consists of destroying all vertices that have a total payoff (the sum of the payoffs from a game with each neighbour) that is below the fitness threshold.

Each of the eight reproduction models is specified by three binary numbers r_0, r_1, r_2 in $\{0, 1\}$. The parents always retain their pre-existing connections, but the offspring's connections are specified as follows:

- If $r_0 = 1$ the offspring are connected to their parents' neighbours. Otherwise they are not.

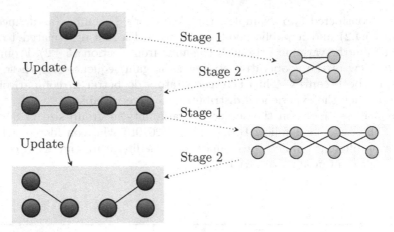

FIGURE 13.6: Updating a graph G under the degree capped reproduction model consists of performing two stages for $m = 1$ model and $Q = 2$. Stage 1, Reproduce: Simultaneously give every vertex of G an offspring vertex which is born with the same neighbourhood as its parent. Stage 2, Cull: Remove every vertex of degree greater than 2. The figure shows the updates when starting with the complete graph on two vertices. After two updates, we end up with two copies of the original graph and two isolated vertices.

- If $r_1 = 1$ the offspring are connected to their parents. Otherwise they are not.

- If $r_2 = 1$ the offspring of connected parents are connected to one another. Otherwise they are not.

The model specified by the binary string $r_2r_1r_0$ is called model number $m = r_02^0 + r_12^1 + r_22^2$. In models with $m = 0, 2, 4, 5, 6$ and 7 it is possible to gain significant understanding of the dynamics of general games (Southwell and Cannings, 2009, 2010).

One case that was studied in detail (Southwell and Cannings, 2009, 2010) is where there is just one strategy, which scores a payoff of -1 against itself, and the fitness threshold is $-Q$. In this case an individual survives the killing stage if and only if it has a degree less than or equal to Q, and so these systems correspond to "degree capped reproduction" where individuals reproduce, but die from overcrowding once they get more than Q neighbours. When reproduction takes place according to the $m = 1$ or $m = 3$ models these systems can generate complex self replicating structures.

Figure 13.6 shows a simple example of a self replicating structure under the $m = 1$ model with degree cap $Q = 2$. When Q is larger much more interesting dynamics can occur, because the graph can grow larger before the highly connected vertices die. In many cases simple initial graph structures, like two vertices connected by a single edge, evolve to generate large collections of different connected components, which grow, change, and self replicate. For $Q = 8$ the degree capped reproduction model (using reproduction method $m = 1$) eventually produces 723 different self replicating structures. More structures are produced for higher degree cap values, and there are many unanswered questions about these systems. In particular, it is unknown whether systems with a finite degree cap Q can generate connected components of arbitrary size or generate arbitrarily large numbers of non-isomorphic connected components. Studying these questions may give insight into how complex self replicating structures arise from simple biologically inspired rules.

Other types of model where the network structure changes as a result of strategic interaction were considered in Richter (2016, 2017). Also, the remarkably complex dynamics observed in Southwell and Cannings (2009) inspired the further search for simple network evolution rules inducing complexity (Southwell et al., 2012).

13.5.2 Link formation models

A different type of model of network evolution was considered in Broom and Cannings (2013), within which individuals form or break links at random, according to their degree. This forms an interesting model of sociability, or behavioural responses to epidemics. The basic idea is that each vertex has a pre-allocated minimum and maximum allowed degree. The system evolves via a Markov chain so that, at any given time, a vertex is chosen at random. If the vertex has more links than its maximum allowed degree then one of its links is deleted randomly. If the vertex has less links than its minimum allowed degree then a link is formed to a randomly selected unconnected individual. Otherwise no action is taken. Analysis revealed that these systems evolve to a closed class of structures, the *minimal set*, which Broom and Cannings (2013) shows how to find the stationary distribution of, focusing on the special case where the maximum and minimum degrees are the same, i.e. each individual has a unique target.

Later work (Broom and Cannings, 2015, 2017; Cannings and Broom, 2020; Aizouk and Broom, in press) explored the nature of the minimal set, classifying the vertices into four categories depending upon whether they always achieved their target, were never over target, never under target or could be under or over target. Interestingly, some vertices could be always on target even for a continuously changing process. They also investigated game theoretic generalisations of this model, where formation and breaking of links also depend upon strategies employed by the vertices. In Broom and Cannings (2017), in particular, we can see that such games are inherently very complex, even for small populations.

The models that we considered here are of a specific type, but they relate to a number of models by other authors. In particular, there is a close connection to the economic models of Jackson and colleagues (Jackson, 2003, 2010, 2011;

Jackson and Wolinsky, 1996) where individuals form connections to others and their payoffs are determined by the network. Links are formed in different ways in these models; in particular, Bala and Goyal (2000); Dutta and Jackson (2000) have links which are formed unilaterally, as in Broom and Cannings (2013).

13.6 Python code

In this section we show how to use Python to simulate games with reproducing vertices (Southwell and Cannings, 2010) discussed in Section 13.5.1.

```python
1  """ Games with reproducing vertices
2  This script is based on Some Models of Reproducing Graphs:
3  III Game Based  Reproduction.
4  For a given maximum degree Q and a starting graph G
5  it runs a given number of iteration of the "m=1" process.
6  It plots the initial and final graph.
7  """
8
9  ## Import basic packages
10 import networkx as nx
11 import matplotlib.pyplot as plt
12 import copy
13
14 ## Define the basic parameters
15
16 # Set up the initial graph
17 G = nx.Graph()
18 G.add_nodes_from(['a', 'b'])
19 G.add_edge('a','b')
20
21 # Set the parameter Q (max degree)
22 Q = 5
23
24 # Set the number of iterations to run
25 numIter = 4
26
27 # Plot the initial graph G
28 plt.figure(1)
29 nx.draw(G, with_labels=True)
30
31 # Iterate a given number of times
32 for iteration in range(numIter):
33
34     # Copy G for stage 1 process
35     G1 = copy.deepcopy(G)
36
37     # For all vertices in the graph
38     for parent in G:
```

```
39
40          # Name the offspring
41          offspring = parent+str(iteration)
42
43          # Add it to the graph
44          G1.add_node(offspring)
45
46          # Connect it to its parent's neighbors
47          [G1.add_edge(offspring,nbr) for nbr in ...
                G.neighbors(parent)]
48
49      # Create a copy of the graph for stage 2 process
50      G2 = copy.deepcopy(G1)
51
52      # Cull the vertices with degree more than Q
53      for node in G1:
54          if G1.degree(node)>Q+1:
55              G2.remove_node(node)
56
57      # Copy G2 back to G and restart the cycle
58      G = copy.deepcopy(G2)
59
60  # Plot the final graph G
61  plt.figure(2)
62  nx.draw(G, with_labels=True)
63
64  # Display the plots
65  plt.show()
```

13.7 Further reading

For cellular automata see Wolfram (2002, 1986). A fascinating cellular automaton is given in the *Game of Life* see Gardner (1970) and also Sigmund (1993). For more on cellular automata and their use in modelling, see Ermentrout and Edelstein-Keshet (1993); Chopard and Droz (1998); Deutsch and Dormann (2005); Schiff (2008).

For multi-player games on graphs, see Santos et al. (2008) (see also Li et al., 2014, 2016) where each individual's payoff came from multi-player games involving it and all of its neighbours. For two strategy best response multi-player games on graphs see Cannings (2009); Haslegrave and Cannings (2017).

For a good review of models with structured populations and evolutionary dynamics up to the time of its publication, see Perc et al. (2013). This review focuses on multi-player games, in particular Public Goods Games (see also Szolnoki et al., 2009; Szolnoki and Perc, 2011; Szabó and Szolnoki, 2009; Isaac and Walker, 1988). For a more recent review, centred upon the framework from Section 13.3, see Broom et al. (2021).

An alternative flexible model of structure populations to Broom and Rychtář (2012) is given in Tarnita et al. (2009a); here individuals are distributed over sets, and interactions occur between individuals within the same set.

Alternative structured population models include island models (Constable and McKane, 2014), with evolution in near isolated communities with low migration, or community-structured populations (Wang et al., 2011) which involve interactions occur at multiple levels.

For an area of research related to the dynamic models from Section 13.5.2, concerning biological markets and partner choice, see Noë (2001); Noë and Hammerstein (1994); Hammerstein and Noë (2016). For co-evolutionary models, again where the structure changes, see Pacheco et al. (2006b,a); Wu et al. (2009b,c); Zhang et al. (2011a,b).

13.8 Exercises

Exercise 13.1. Consider a cellular automata with m types of individuals on a regular graph of degree k. Show that there are $m^{m^{k+1}}$ different updating rules that prescribe how a type of an individual will be updated based on her current type and the type of her neighbours.

Exercise 13.2 (Allen et al., 2017). Show that the random walk described in Section 13.2 satisfies (13.1).

Exercise 13.3. Use (13.4) and the condition on the structure coefficient for cooperation to be favoured that follows it to show inequality (13.5).

Exercise 13.4 (Broom and Rychtář, 2012). Consider the territorial raider model from Section 13.3.5 on a well-mixed graph with n places and $p_{n,m} = 1/n$ for all n, m. Find the distribution of the group size for any given individual, i.e. the group size from the *individual's perspective*. Find the expectation of the group from both the individual and the observer's perspective and compare them.

Exercise 13.5 (Peña and Nöldeke, 2016). Consider the methodology from Section 13.4.2. Show that if $q \geq_v p$, then $E_q[Y] = E_p[X]$.

Hint. $\phi(x) = x$ and $\phi(x) = -x$ are convex.

Exercise 13.6. Consider the models from Section 13.5.1 with $m = 1, m = 2$ and $m = 4$ for $Q = 2$ and analyse the long-term behaviour in each case, starting from a single individual.

Exercise 13.7. Repeat Exercise 13.6 when the initial population is a line graph with three individuals.

Exercise 13.8. The code from Section 13.6 considers the model from Section 13.5.1. Modify this code to work for processes where the value of m is greater than 1.

Chapter 14

Adaptive dynamics

14.1 Introduction and philosophy

The replicator dynamics considers evolution using a fixed set of strategies and only allows the frequencies of existing strategies to change. It is not possible for new strategies to emerge. There is thus selection between the competing strategies in the population following the replicator equation, but there is no mutation. So we see that one of the key elements of evolution is missing, and such a population could not necessarily reach an ESS if the initial mixture of strategies would not allow it, nor hope to adapt to a changing environment.

The theory of adaptive dynamics was developed to allow the population to evolve by allowing mutations. It is assumed that mutations are small in size (in the sense that their phenotypic effect is small), so that any new strategy is very close to existing strategies and the population strategies can only change gradually through time. We note that in the single trait case that we mainly concentrate on, this small mutation effect is often not necessary (depending upon the specific fitness functions), but that for multiple traits this assumption is more fundamental to some of the methodology used.

Suppose that we have some continuous trait x, e.g. height. This could either be a pure strategy from a continuous strategy set, for example as in the war of attrition, or alternatively it could represent a mixed strategy, e.g. the probability of playing Hawk in a Hawk-Dove game. If it is a mixed strategy, then it is one that can be played by an individual (as opposed to a population average), and in general the available traits in our population form a continuous set. We assume that the whole population plays x, except a small mutant group which have changed to playing a different strategy $x + h$. If this mutant group can invade x, the population may then move to the new strategy $x + h$.

The population then changes under evolution following a succession of such changes. As mentioned in Section 3.1.4, it is assumed that the introduction of new mutations is sufficiently slow, that the competition between the resident and mutant strategies has been completed before the next mutant is introduced, i.e. that mutation and selection happen on different timescales with mutation occurring on a slower timescale to selection.

We also note that the concept of adaptive dynamics can easily be extended to consider a number of different traits, represented by a vector $\mathbf{x}$.

DOI: 10.1201/9781003024682-14

The practical application of this extension is more difficult mathematically, and also there are some issues relating to the biological assumptions in this multi-dimensional case, which we discuss at the end of this chapter. For most of what follows, we restrict ourselves to the single trait case only. For readers who want a detailed and precise account of the underlying assumptions, see Metz (2012).

14.2 Fitness functions and the fitness landscape

Imagine an individual with a particular trait y in a population consisting entirely of individuals (except for our focal individual) with trait x. The fitnesses of traits are as defined elsewhere in the book, so that the fitness of our focal individual is labelled

$$\mathcal{E}[y; \delta_x], \tag{14.1}$$

following notation established in Chapter 2. We shall, in particular, consider the fitness advantage of an individual with trait y in a population of individuals with trait x represented by $s(y, x)$ so that

$$s(y, x) = \mathcal{E}[y; \delta_x] - \mathcal{E}[x; \delta_x] \tag{14.2}$$

and so $s(x, x) = 0$ (we note that Waxman and Gavrilets (2005) use $s(y, x)$ for $\mathcal{E}[y, \delta_x]$ and assume $s(x, x) = 1$, i.e. they treat $s(y, x)$ as a fitness, not a fitness advantage). In general, in adaptive dynamics $s(y, x)$ can be thought of as the exponential growth rate of a mutant group of infinitesimal proportion within the resident population (Metz et al., 1992; Metz, 2008); note that this interpretation is consistent with the replicator dynamics for the population. One advantage of this is that it can be applied to general ecological scenarios, and the fitness from (14.1) does not need to be explicitly defined at all. Assuming that $y = x + h$ for some small h, y will replace x in the population if and only if

$$s(y, x) = s(x + h, x) > 0. \tag{14.3}$$

Assuming that $s(y, x)$ is differentiable with respect to y, this is equivalent to

$$\left. \frac{\partial}{\partial y} s(y, x) \right|_{y=x} > 0 \tag{14.4}$$

if $h > 0$ and

$$\left. \frac{\partial}{\partial y} s(y, x) \right|_{y=x} < 0 \tag{14.5}$$

if $h < 0$. Assuming that the derivative of $s(y, x)$ with respect to y is non-zero, it means that either only mutants with $y > x$ or only mutants with $y < x$ can

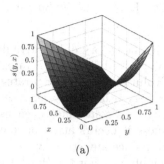
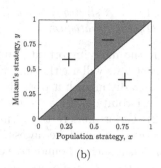

(a) (b)

FIGURE 14.1: (a) Invasion fitness landscape and (b) pairwise invasibility plot (the sign of $s(y, x)$) for $s(y, x) = (y - x)(1 - 2x)$ from Example 14.1. A mutant using y can invade a population with trait x if $s(y, x) > 0$, i.e. if a point (x, y) is in a white region.

invade, so that in whichever order mutants appear, the direction of change of the population is certain, with an increase in x if (14.4) occurs, and a decrease if (14.5) occurs.

We can define the fitness advantage $s(y, x)$ for all pairs x and y. From this we obtain the *invasion fitness landscape*, which is a three-dimensional plot of x, y and $s(y, x)$; see Figure 14.1.

Example 14.1 (Hawk-Dove game). For example suppose that x is the probability of playing Hawk in the Hawk-Dove game with payoff matrix (4.1), with parameters $V = 2$ and $C = 4$. We get

$$\mathcal{E}[y; \delta_x] = -xy + 2(1 - x)y + (1 - x)(1 - y) \tag{14.6}$$
$$= 1 - x + y(1 - 2x) \tag{14.7}$$

and thus

$$s(y, x) = (y - x)(1 - 2x). \tag{14.8}$$

Since

$$\frac{\partial}{\partial y} s(y, x) = 1 - 2x \begin{cases} > 0, & \text{if } x < 0.5, \\ < 0, & \text{if } x > 0.5, \end{cases} \tag{14.9}$$

it is easy to see that if $x \neq 0.5$ then the population will evolve towards 0.5.

We can thus follow the evolution of the population on the invasion fitness landscape. The important part of the landscape is that part of it within a small distance of the line $y = x$. As described above, if the derivative of $s(y, x)$ with respect to y is non-zero the population moves through the trait-space according to the invasion fitness landscape, until it reaches a point where this derivative is zero, i.e.

$$\left. \frac{\partial s(y, x)}{\partial y} \right|_{\substack{y=x^* \\ x=x^*}} = 0 \tag{14.10}$$

Definition 14.2. *A strategy x^* is called an* Evolutionarily Singular Strategy *(ess) when (14.10) is satisfied.*

These are the most important points in the theory of adaptive dynamics, and consequently we will use the term Evolutionarily Singular Strategy a lot in this chapter. To distinguish it from the Evolutionarily Stable Strategy (ESS), we denote an Evolutionarily Singular Strategy using the lower case term ess (Metz, 2012). Unfortunately the term ESS has sometimes been used. It is clear that every ESS must be an ess (unless it occurs at a boundary of the allowable traits, if such a boundary exists), since any strategy that is not an ess can be invaded by some of its near neighbours. For the Hawk-Dove game from Example 14.1, we have a unique ess at the ESS $x^* = 1/2$.

Unless noted otherwise, x^* will represent an ess in this chapter.

14.2.1   Taylor expansion of $s(y, x)$

The Taylor series expansion of $s(y, x)$ at (x^*, x^*) is

$$s(y, x) = s(x^*, x^*) + \cdots \tag{14.11}$$

$$\left. \frac{\partial s(y, x)}{\partial y} \right|_{\substack{y=x^* \\ x=x^*}} \cdot (y - x^*) + \left. \frac{\partial s(y, x)}{\partial x} \right|_{\substack{y=x^* \\ x=x^*}} \cdot (x - x^*) + \cdots$$

$$\frac{1}{2} \left[\left. \frac{\partial^2 s(y, x)}{\partial y^2} \right|_{\substack{y=x^* \\ x=x^*}} \cdot (y - x^*)^2 + \left. \frac{\partial^2 s(y, x)}{\partial x^2} \right|_{\substack{y=x^* \\ x=x^*}} \cdot (x - x^*)^2 + \right.$$

$$\left. + 2 \left. \frac{\partial^2 s(y, x)}{\partial x \partial y} \right|_{\substack{y=x^* \\ x=x^*}} \cdot (y - x^*)(x - x^*) \right] + \text{h.o.t.}$$

By (14.2), $s(x, x) = 0$ for all x and thus

$$0 = \left. \frac{\partial s(y, x)}{\partial y} \right|_{\substack{y=x^* \\ x=x^*}} + \left. \frac{\partial s(y, x)}{\partial x} \right|_{\substack{y=x^* \\ x=x^*}}, \tag{14.12}$$

$$0 = \left. \frac{\partial^2 s(y, x)}{\partial y^2} \right|_{\substack{y=x^* \\ x=x^*}} + 2 \left. \frac{\partial^2 s(y, x)}{\partial x \partial y} \right|_{\substack{y=x^* \\ x=x^*}} + \left. \frac{\partial^2 s(y, x)}{\partial x^2} \right|_{\substack{y=x^* \\ x=x^*}}. \tag{14.13}$$

Since x^* is an ess, we have that the first of the derivatives from (14.12) is equal to zero, and thus the second of the derivatives is also equal to zero. Consequently, it follows from (14.13) that

$$s(y, x) = \frac{1}{2} \left[\left. \frac{\partial^2 s(y, x)}{\partial y^2} \right|_{\substack{y=x^* \\ x=x^*}} \cdot (y - x^*)^2 + \left. \frac{\partial^2 s(y, x)}{\partial x^2} \right|_{\substack{y=x^* \\ x=x^*}} \cdot (x - x^*)^2 \cdots \right.$$

$$\left. - \left(\left. \frac{\partial^2 s(y, x)}{\partial y^2} \right|_{\substack{y=x^* \\ x=x^*}} + \left. \frac{\partial^2 s(y, x)}{\partial x^2} \right|_{\substack{y=x^* \\ x=x^*}} \right) \cdot (y - x^*)(x - x^*) \right]. \tag{14.14}$$

14.2.2 Adaptive dynamics for matrix games

In general, in matrix games with two strategies

$$s(y, x) = \mathbf{y}A\mathbf{x}^T - \mathbf{x}A\mathbf{x}^T, \tag{14.15}$$

which is linear in y and so the second derivative of $s(y, x)$ with respect to y is zero, and if there is an internal equilibrium the situation is exactly as in Example 14.1. It is left as Exercise 14.1 that the conditions for x^* to be an ESS are equivalent to

$$s(y, x^*) \leq s(x^*, x^*) = 0 \text{ for all } y \tag{14.16}$$

and

$$s(x^*, y) > s(y, y) = 0 \tag{14.17}$$

whenever equality holds in (14.16). If $0 < x^* < 1$ then (14.16) always holds so the ESS condition is reduced to (14.17). For matrix games $s(y, x)$ is quadratic, so if (14.17) holds for all y close to x^*, it will hold globally. Thus a non-boundary ess is an ESS for a two-strategy matrix game if and only if (14.17) holds. We will see later that satisfying (14.17) is termed as x^* "can invade" its close neighbours.

Also, the second derivative of $s(y, x)$ with respect to y at $y = x^*, x = x^*$ (and in fact everywhere) is equal to zero, and thus we see that from the point of view of adaptive dynamics, matrix games are in some way non-generic because of the flat landscape in the direction of the y-axis. Moreover,

$$\left. \frac{\partial^2 s(y, x)}{\partial x^2} \right|_{\substack{y=x^* \\ x=x^*}} = 2(a_{11} + a_{12} + a_{21} - a_{22}), \tag{14.18}$$

and thus by Haigh's condition (6.45) the second derivative from ((14.18)) is greater than zero for a mixed ESS. We will see later that it means that a mixed ESS in a two-strategy matrix game is a *convergent stable* ess. Exercise 13.2 shows that a pure ESS is generally not an ess.

Matrix games with more than two strategies are considered in Exercise 14.12.

As long as our payoffs are linear on the left (see Chapter 7) we obtain $\mathcal{E}[y; \delta_x] = yf(x)$ and thus

$$s(y, x) = (y - x)f(x) \tag{14.19}$$

for some $f(x)$ (in the case of matrix games $f(x)$ is just a linear function). Thus $\frac{\partial}{\partial y} s(y, x) = f(x)$ and so x^* is an ess precisely when $f(x^*) = 0$. Moreover,

$$\frac{\partial^2}{\partial y^2} s(y, x) = 0. \tag{14.20}$$

Thus the non-genericity that we noted in the case of matrix games, in fact, applies to a much wider class of games, namely all games which are linear on the left (see Exercise 14.4).

14.3 Pairwise invasibility and Evolutionarily Singular Strategies

The behaviour at an Evolutionarily Singular Strategy x^* varies depending upon the nature of the fitness landscape immediately around it. There are four key properties that an ess may possess in various combinations, which we describe below.

14.3.1 Four key properties of Evolutionarily Singular Strategies

14.3.1.1 Non-invasible strategies

The ess is called *non-invasible* if no near neighbour can invade. This happens if within a population of x^* individuals there is a local fitness maximum at $y = x^*$, so that x^* is non-invasible if

$$\left. \frac{\partial^2 s(y,x)}{\partial y^2} \right|_{\substack{y=x^* \\ x=x^*}} < 0 \qquad (14.21)$$

and only if $\left. \frac{\partial^2 s(y,x)}{\partial y^2} \right|_{\substack{y=x^* \\ x=x^*}} \leq 0$. We note that for generic games, x^* is non-invasible if and only if (14.21) holds. We are exceptionally interested in non-generic games here because of the matrix games (and other games which are linear on the left) which we consider elsewhere in the book.

The idea of non-invasibility is related, although not identical, to the idea of the ESS. If the population is non-invasible, then it is resistant to invasion by all nearby mutants (which are all that are allowed in the framework of adaptive dynamics), although it may be able to be invaded by a mutant that is "far away"; see Exercise 14.3. We note that whatever local properties that an ess may possess, it is not enough to guarantee that the strategy is an ESS, since there is the possibility of invasion by a distant mutant.

It follows that under adaptive dynamics, the non-invasible x^* will persist throughout time once reached. In fact, it is of especial interest when an ess is invasible; see Section 14.3.2.3.

14.3.1.2 When an ess can invade nearby strategies

We say that an ess *can invade* if it is able to invade all near neighbours, i.e. that $s(x^*, x) > 0$ for x close to x^*. Recalling that $s(x^*, x^*) = 0$, x^* can thus invade its near neighbours (Kisdi and Meszéna, 1993) for the generic case if and only if

$$\left. \frac{\partial^2 s(y,x)}{\partial x^2} \right|_{\substack{y=x^* \\ x=x^*}} > 0. \qquad (14.22)$$

14.3.1.3 Convergence stability

An ess is called *convergence stable* if evolution sends the population towards x^* so that alternative strategies are only invaded by those closer to x^*. This occurs if the derivative of the fitness gradient is negative, Eshel (1983) (see also Meszéna et al., 2001), i.e.

$$\frac{d}{dx}\left(\left.\left(\frac{\partial s(y,x)}{\partial y}\right|_{y=x}\right)\right|_{x=x^*} = \left.\frac{\partial^2 s(y,x)}{\partial y^2}\right|_{\substack{y=x^* \\ x=x^*}} + \left.\frac{\partial^2 s(y,x)}{\partial x \partial y}\right|_{\substack{y=x^* \\ x=x^*}} < 0. \quad (14.23)$$

Hence, by (14.13), x^* is convergence stable for the generic case if and only if

$$\left.\frac{\partial^2 s(y,x)}{\partial x^2}\right|_{\substack{y=x^* \\ x=x^*}} > \left.\frac{\partial^2 s(y,x)}{\partial y^2}\right|_{\substack{y=x^* \\ x=x^*}}. \quad (14.24)$$

14.3.1.4 Protected polymorphism

An ess is called a *protected polymorphism* if two strategies close to, and on either side of, x^* can each invade the other. Thus y_1 and y_2 form a protected polymorphism if $s(y_1, y_2) > 0$ and $s(y_2, y_1) > 0$ which is equivalent to $s(y,x)$ having a local minimum at $y = x = x^*$ along the line $(y - x^*) = -(x - x^*)$. It follows from (14.14) that x^* is a protected polymorphism for the generic case if and only if

$$\left.\frac{\partial^2 s(y,x)}{\partial x^2}\right|_{\substack{y=x^* \\ x=x^*}} + \left.\frac{\partial^2 s(y,x)}{\partial y^2}\right|_{\substack{y=x^* \\ x=x^*}} > 0. \quad (14.25)$$

We will discuss these properties in more detail later, in the context of concrete examples.

14.3.2 Classification of Evolutionarily Singular Strategies

It is not possible to have all combinations of the above four properties from Section 14.3.1. For example, we cannot have a protected polymorphism that is non-invasible and cannot invade. Out of the 16 potential combinations, only 8 can occur in practice.

In fact, using (14.14) we see that to classify all Evolutionarily Singular Strategies, we just need to consider the function

$$s(y,x) = \frac{1}{2}\left(ax^2 + by^2 - (a+b)xy\right), \quad (14.26)$$

which has an ess at $x^* = 0$. Note that we introduce the factor $1/2$ to make our example match the definitions below in (14.29). The function (14.26) has roots given by

$$y = x, \quad (14.27)$$

$$y = \frac{a}{b}x, \quad (14.28)$$

TABLE 14.1: Eight possible combination for Evolutionary Singular Strategies based on the values of a, b, c from (14.29). NI = non-invasible, CS = convergence stable, CI = can invade, PP = protected polymorphism. In the last column, we see a unified condition on $\alpha = \text{sign}(a)$, $\beta = \text{sign}(b)$ and $\gamma = \text{sign}(c)$.

Case	Condition	NI	CS	CI	PP	(α, β, γ)
1	$a < 0 < b, a + b > 0$	×	×	×	✓	$(-1, +1, -1)$
2	$0 < a < b$	×	×	✓	✓	$(+1, +1, -1)$
3	$0 < b < a$	×	✓	✓	✓	$(+1, +1, +1)$
4	$b < 0 < a, a + b > 0$	✓	✓	✓	✓	$(+1, -1, +1)$
5	$b < 0 < a, a + b < 0$	✓	✓	✓	×	$(+1, -1, -1)$
6	$b < a < 0$	✓	✓	×	×	$(-1, -1, -1)$
7	$a < b < 0$	✓	×	×	×	$(-1, -1, +1)$
8	$a < 0 < b, a + b < 0$	×	×	×	×	$(-1, +1, +1)$

and the eight possible combinations follow based on the sign of $\frac{a}{b}$ and its relationship to 1 or -1.

Alternatively, following Waxman and Gavrilets (2005), which combination occurs depends on the signs of

$$a = \left.\frac{\partial^2 s(y, x)}{\partial x^2}\right|_{\substack{y=x^* \\ x=x^*}}, b = \left.\frac{\partial^2 s(y, x)}{\partial y^2}\right|_{\substack{y=x^* \\ x=x^*}}, \tag{14.29}$$

$$c = |a| - |b|.$$

Those combinations are summarised in Table 14.1 and shown in Figure 14.2.

14.3.2.1   Case 5

In case 5 we see that x^* is non-invasible, convergence stable, can invade but is not a protected polymorphism. This is perhaps the simplest case. A function which is an example of this case is given by (14.30)

$$s(y, x) = x^2 - 2y^2 + xy. \tag{14.30}$$

If the population starts with x sufficiently close to x^* (and, if $s(y, x)$ is a quadratic function given by (14.26), this is any value of x) convergence stability ensures that the population gets closer and closer to x^*. For any value of x close to x^*, if a mutant appears such that x^* lies between x and the mutant, if the mutant can invade it will completely replace x, since x^* is not a protected polymorphism. Thus the population eventually settles at x^*.

14.3.2.2   Case 7

Now consider the following function, which is an example of case 7,

$$s(y, x) = -2x^2 - y^2 + 3xy. \tag{14.31}$$

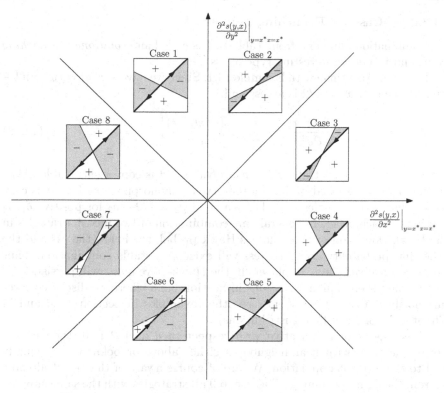

FIGURE 14.2: All possible pairwise invasibility plots $(\text{sign}(s(y,x)))$ for the function given by $s(y,x) = (ax^2 + by^2 - (a+b)xy)/2$. Cases are numbered as in Table 14.1. The bold lines are $y = x$, the other lines are $y = \frac{a}{b}x$.

Here if the population starts at x^*, it will stay there indefinitely, since x^* is non-invasible. However, if it starts at any other point it cannot reach x^*. Since x^* is not convergence stable, if the population is sufficiently close to x^*, in the generic case it will move away from x^*, not towards it. If the function $s(y,x)$ is quadratic as given by (14.26), then "sufficiently close" again means any value of x, and if the population starts at any positive value of x it will move towards $+\infty$, whereas if it starts at negative x it will move towards $-\infty$. Thus we have a stable strategy which cannot be reached. Note that we have already considered a similar situation in Chapter 9, where there was an internal ESS which could not be reached by the sequential introduction of new strategies, because each pure strategy was also an ESS. This occurred because the games in question were multi-player, and payoffs were thus nonlinear. Nonlinearity is needed here as well, as it is not possible to have a stable strategy which cannot be reached for matrix games under adaptive dynamics (see Exercise 14.10).

14.3.2.3   Case 3—Branching points

Combination number 3 from Table 14.1 is called an *evolutionary branching point* and is a very interesting type of ess.

Let us first continue with Example 14.1. Since we have $s(x,y) = (y-x)(1-2x)$, we see that $x^* = 1/2$ is an ess and

$$\frac{\partial^2 s(y,x)}{\partial y^2}\bigg|_{\substack{y=x^* \\ x=x^*}} = 0, \quad \frac{\partial^2 s(y,x)}{\partial x^2}\bigg|_{\substack{y=x^* \\ x=x^*}} = 4. \tag{14.32}$$

Since $4 > 0$ then our strategy $1/2$ can invade and is convergent stable. Also, since (14.25) is satisfied we have a protected polymorphism at $1/2$. It is easy to see that two strategies $y_1 = 1/2 + \delta_1$ and $y_2 = 1/2 - \delta_2$ for positive δ_1, δ_2 can invade each other. In general, any combination of two mixed strategies in the Hawk-Dove game where one has Hawk probability below the ess and the other has probability above the ess will exist in a stable combination. Thus mixtures of mixed strategy players in the Hawk-Dove game will persist.

We have seen similar situations in Section 3.2.3 where we discussed Evolutionarily Stable Sets (ESSet) (see other examples in Section 4.1.6 and in Chapter 19 for Ideal Free Distributions).

Since the second derivative with respect to y at $1/2$ is 0, then it is a borderline case, with near neighbours either above or below doing equally well to the resident population. We are of course aware of this result already, as from the Bishop-Cannings Theorem 6.9 all strategies with the same support as an ESS do equally well when the population consists (essentially) only of ESS players.

We now look at an example of the form (14.26) to illustrate an evolutionary branching point.

Example 14.3. Consider a function

$$s(y, x) = 2x^2 + y^2 - 3xy. \tag{14.33}$$

It is left as Exercise 14.5 that $x^* = 0$ is the (unique) Evolutionarily Singular Strategy that is a branching point. Starting at any other strategy it is easy to see that the population will move towards 0 by a series of small mutational steps, and it will get as close as we like to 0 (we know this, since 0 is convergence stable). At some point, when we are sufficiently close to 0, there will be a mutant which is at the other side of 0. The population will then settle at an equilibrium mixture between them (0 is a protected polymorphism). Let us label these two values x_1 and x_2. What happens when new mutants are introduced? We cannot answer this question, without considering the fitness of a mutant in a population comprised of the equilibrium mixture of x_1 and x_2; we shall denote the advantage of mutants over the equilibrium mixture to be $s(y; x_1, x_2)$. It is shown in Geritz et al. (1998) that an approximation of this function is given by

$$s(y; x_1, x_2) = \frac{1}{2} \left. \frac{\partial^2 s(y, x)}{\partial y^2} \right|_{\substack{y=x^* \\ x=x^*}} (y - x_1)(y - x_2). \qquad (14.34)$$

For $s(y, x)$ given by (14.33) we obtain $s(y; x_1, x_2) = (y - x_1)(y - x_2)$.

Thus for case 3 where the second derivative with respect to y from ((14.21)) is positive, it is clear than any mutant y from within the interval (x_1, x_2) cannot invade, but any mutant outside the interval can invade. When such a y invades it is likely that it will replace the value of x_1 or x_2 which is closest to it and so a sequence of such invasions divides the population into two types that move away from 0 and each other over time. The population thus splits into two sub-populations with distinct traits.

14.4 Adaptive dynamics with multiple traits

We have so far only looked at a single trait, but the definitions of the key properties convergence stability, non-invasibility and the ability to invade can be extended to multiple dimensions; see Exercises 14.11 and 14.12. However, unless mutations are independent, consideration of convergence stability can be complex (Leimar (2009)). In addition the idea of a protected polymorphism is more complex in general, since the polymorphism could potentially be between pairs which differ in any non-empty subset of the available traits, although whether any given pair of traits forms a protected polymorphism can be analysed along a straight line between them.

Following the central emphasis of our book, we have concentrated on the static properties of adaptive dynamics, and the classifications of the different types of Evolutionarily Singular Strategy. Whilst these are certainly of great interest, adaptive dynamics is however, as the name indicates, a dynamic process. Thus we should give some thought to what happens when the population is some distance from an ess. For a single trait behaviour is generally relatively simple, but for multiple traits the situation can be much more complicated. Above we have considered a mutant $\mathbf{y}$ invading $\mathbf{x}$, where $\mathbf{y}$ can potentially differ from $\mathbf{x}$ in all positions. If this is generally true, and there are correlations between distinct traits, then it is important to consider which combinations are likely to occur. Is every value of $\mathbf{y}$ close to $\mathbf{x}$ equally likely, or are only certain values of $\mathbf{y}$ possible/likely? If there are many potential invading mutants, which are likely to invade in practice? We briefly discuss this in the following section.

Let us assume that all traits evolve completely independently, which is perhaps the simplest case. We can think of these each as a single trait from a distinct species, so that no individual possesses more than one trait, as in

Dieckmann and Law (1996). Their fitnesses depend upon the combination of all traits in the population, but mutations only affect a single species (and so mutations are necessarily independent), and we can define a fitness function for each y_i, $\mathcal{E}(y_i, \delta_\mathbf{x})$. This then leads to the fitness advantage of y_i over the current population value of trait i x_i as

$$s(y_i, \mathbf{x}) = \mathcal{E}[y_i; \delta_\mathbf{x}] - \mathcal{E}[x_i; \delta_\mathbf{x}]. \tag{14.35}$$

The evolution of the population is governed by a separate differential equation for the value x_i of trait i,

$$\frac{dx_i}{dt} = k_i(\mathbf{x}) . \frac{\partial}{\partial y_i} s(y_i, \mathbf{x}) \bigg|_{y_i = x_i}, \tag{14.36}$$

where $k_i(\mathbf{x})$ are coefficients determining the speed of evolutionary change for each trait. These will be governed by the frequency of mutations for a particular trait (for different species this will also be influenced by other factors including the length of the generations and the birth rate for that species, the population size and the size of mutations for that species). Equation (14.36) is known as the *canonical equation of adaptive dynamics*; see Exercise 14.13.

Thus even though we have a polymorphic population, we can sometimes think of it in a *quasi-monomorphic* way. This is perhaps only really plausible biologically in the multiple species case that we discuss above (and in fact, there can be additional complications even then). In general, the canonical equation (14.36) has a more complex form which involves the covariances of the occurrences of the different possible mutants $\mathbf{y}$ close to the current population state $\mathbf{x}$ in the form of the mutational covariance matrix (see e.g. Metz, 2012).

For the case where each trait comes from a distinct species, it is assumed that all values of $\mathbf{y}$ are not possible; only those which are identical to $\mathbf{x}$ in all but one position. It is also assumed that, in common with the strictly monomorphic case (and other game-theoretical models), the conflict between the population trait and the latest mutant is resolved before the introduction of any new mutant. Thus at any one time competition is in one trait only, related to a single differential equation. We note that our differential equations operate on a slower timescale to the resolution of this competition, and so it appears that all traits change at the same time.

Dieckmann and Law (1996) discuss the following example, of a predator-prey system. The evolution of the population size of prey n_1 and predators n_2 are given by the Lotka-Volterra system (Lotka, 1925)

$$\frac{dn_1}{dt} = n_1(r_1 - \alpha n_1 - \beta n_2), \tag{14.37}$$

$$\frac{dn_2}{dt} = n_2(-r_2 + \gamma n_1). \tag{14.38}$$

In the standard formulation the intrinsic per capita growth rate of the prey r_1 and death rate of the predator r_2, as well as the three interaction terms α, β and γ, are constants. Dieckmann and Law (1996) assume that the three interaction terms are not constant, but are governed by a single trait for each species x_1 and x_2, where

$$\frac{\beta(x_1, x_2)}{u} = \exp(-\delta_1^2 + 2c_2\delta_1\delta_2 - \delta_2^2), \tag{14.39}$$

$$\frac{\gamma(x_1, x_2)}{u} = c_1\beta(x_1, x_2), \tag{14.40}$$

$$\alpha(x_1) = c_7 - c_8 s_1 + c_9 x_1^2, \tag{14.41}$$

$$\delta_1 = \frac{x_1 - c_3}{c_4}, \tag{14.42}$$

$$\delta_2 = \frac{x_2 - c_5}{c_6}, \tag{14.43}$$

where $c_1 - c_9, u$ are constants. See Marrow et al. (1992) for a rationale.

Within the population the per capita birth and death rates for rare mutants are thus

$$b_1(y_1, \mathbf{x}, n) = r_1, \tag{14.44}$$
$$d_1(y_1, \mathbf{x}, n) = \alpha(y_1)n_1 + \beta(y_1, x_2)n_2, \tag{14.45}$$
$$b_2(y_2, \mathbf{x}, n) = \gamma(x_1, y_2)n_1, \tag{14.46}$$
$$d_2(y_2, \mathbf{x}, n) = r_2. \tag{14.47}$$

The mutation process occurs so that the rate of mutation of trait i is μ_i and when a mutation occurs, the new value follows a normal distribution with mean x_i and variance σ_i^2. Thus there are many possible paths that the population can follow, and we may be interested in finding the mean path. A description of how to analyse a representative path (though not necessarily the mean path, see Collet et al. (2013)) and also the distribution of traits more generally, is given in Dieckmann and Law (1996).

An interesting question is how does the presence of many traits affect evolutionary branching? In multiple dimensions, a necessary (but not sufficient) condition for branching is that the singular point is convergence stable and invadable from one direction at least (Geritz et al., 2016). If sexual dimorphism is allowed by the model, then this significantly reduces the potential for evolutionary branching (Bolnick and Doebeli, 2003; Van Dooren et al., 2004). The work of Doebeli and Ispolatov (2010) indicates that when there are multiple co-evolving traits, there may be many directions where the conditions for branching are satisfied, and that these will often involve a number of traits that change at the same time. The interactions between the traits are thus important, not just the number of distinct traits. In general, the relationship between the number of traits and the level of branching can be complex, as demonstrated in Débarre et al. (2014).

14.5 The assumptions of adaptive dynamics

As with all mathematical models, there are a number of assumptions associated with adaptive dynamics. We will only look at those that are distinct from assumptions in evolutionary games in general, which have been discussed elsewhere. The first of these assumptions is that all mutations have a small effect, and so only invasion from strategies that are very close to the resident population need be considered. Some of the main results rely on this, and the situation would be different if either invasion from strategies with larger differences were allowed, or there was not a continuum of traits and so very small jumps were not allowed in some circumstances. One consequence would then be that the adaptive landscape would need to be considered more fully; for example for the single trait case this would involve more than the part of the space close to the line $y = x$. A second related consequence would be that consideration of the derivatives of $s(y, x)$ at $y = x = x^*$ would no longer be sufficient. This would, of course, significantly complicate analysis in some cases.

Secondly, as in conventional ESS theory, the mutation rate is small so that any battles between two strategies are completely resolved before a new mutant enters the population. Similarly, random drift is ignored. Whilst there is thus not a qualitative difference in assumptions here, there is certainly a quantitative one. An adaptive dynamics population changes by mutations that are small in size, in an effectively infinite population. The small sizes of mutations means that competing strategies are very closely matched, and so the time for one to replace another is very long. Mutations are sufficiently rare that a second does not occur whilst such contests are occurring, and a sufficient number of small mutations are needed to change the trait by an appreciable amount. The timescales involved to change anything under these assumptions are thus very large, of course. Real evolution, however, will proceed at a quicker pace, not entirely according to these assumptions, although simulations (see e.g. Dieckmann and Law, 1996; Metz et al., 1996) suggest that the central predictions of adaptive dynamics will still be valid under some reasonable relaxation of the assumptions.

For the case of multiple traits it can be assumed mutations only affect a single trait at a time as in Dieckmann and Law (1996) and then evolution proceeds according to (14.36). If mutations can affect more than one trait at the same time, then it may be assumed that mutants test different traits close to **x**, and the direction with the greatest increase in fitness is the one chosen. However, if mutations occur rarely, then under the usual assumptions an advantageous mutant will take over the population completely before a second can occur. If there are multiple potentially advantageous mutations, then whichever occurs first can replace the population strategy. Thus it may be logical that a random advantageous mutant is picked to replace the population. This of course is much more complicated as then the population would evolve in multiple potential directions and the mathematical analysis would need to be more sophisticated (an example of this methodology is given in Dieckmann and Law, 1996).

14.6   Python code

In this section we show how to use Python to produce pairwise invasibility plots.

```python
1   """ Pairwise invasibility plots
2   User defines the fitness and range of strategies 1D strategy
3   The script generates the pairwise invasibility plot
4   """
5
6   ## Define fitness functions
7   def fitness(y,x):
8       "Fitness of individual using y in the population using x "
9       return 2*(x-0.5)**2+(y-0.5)**2-3*(x-0.5)*(y-0.5)
10
11  # Strategy is a single number. Define its max and min value
12  xMin = 0      # minimum x value
13  xMax = 1      # maximal x value
14  numX = 1000   # number of points between xMin and xMax
15
16  ## Import basic packages
17  import matplotlib.pyplot as plt
18  from matplotlib import colors
19  import numpy as np
20
21  ## Auxiliary variables
22  # Vector representing strategies in the population
23  X = np.linspace(xMin, xMax, numX)
24
25  # Strategy for focal individual. Same as X, just vertical
26  Y = X[:, np.newaxis]
27
28  ## Actual calculations
29  # Get the sign of fitness advantage
30  sign = np.sign(fitness(Y,X) - fitness(X,X))
31
32  ## Plot the sign
33  # Define color map           negative   0      positive
34  cMap = colors.ListedColormap(['gray', 'black', 'white'])
35  plt.imshow(sign, origin='lower', aspect='auto',
36             extent = [xMin, xMax, xMin, xMax],
37             cmap = cMap)
38  plt.xlabel('Population strategy')
39  plt.ylabel('Individual strategy')
40  plt.show()
```

14.7　Further reading

The first explicit mention of adaptive dynamics was in Hofbauer and Sigmund (1990), although the idea of convergent stability comes from several years earlier in Eshel (1983) (see also Apaloo, 1997, 2006 for a related idea).

A lot of the key theoretical ideas were developed in Geritz et al. (1998) and much of the development of the theory was due to the same group of authors, see e.g. Metz et al. (1996), Kisdi and Meszéna (1993), Metz et al. (1992), Metz (2011) and Collet et al. (2013). Metz (2011) also includes information on the history of adaptive dynamics. See Metz (2012) for a detailed discussion about the background and assumptions of adaptive dynamics. Meszéna et al. (2001) and Dieckmann and Law (1996) give important results relating to multiple traits (see also Marrow et al. (1992) for related work on "red queen" dynamics). More recently, Geritz et al. (2016) gave sufficient conditions for evolutionary branching for multiple traits. (see also Ito and Sasaki, 2016).

Applications of adaptive dynamics can be found in the books by Dercole and Rinaldi (2008) and Dieckmann et al. (2012). A non-technical review and discussion of the theory of adaptive dynamics can be found in Waxman and Gavrilets (2005). See Kisdi (2020) for a short recent review. A similar method which incorporates population dynamics is the G-function method of Brown and Vincent (1987), which is discussed in Sections 20.1 and 22.1. See also Cohen et al. (1999); Vincent and Brown (2005); Brown and Vincent (1992). A closely related methodology is due to Abrams et al. (1993).

14.8　Exercises

Exercise 14.1 (Meszéna et al., 2001). Show that for a matrix game, x^* is an ESS if and only if (14.16) and (14.17) hold.

Hint. Use Theorem 6.2.

Exercise 14.2. Show that for a matrix game, a pure ESS is not an Evolutionarily Singular Strategy.

Exercise 14.3. Give an example of a fitness function $\mathcal{E}(y, \delta_x)$ where there is an x^* that is a non-invasible Evolutionarily Singular Strategy that is not an ESS, in other words, x^* cannot be invaded by nearby strategies but can be invaded by a strategy coming from far away.

Exercise 14.4. Show that the sex ratio game from Section 4.4 yields

$$s(y, x) = \frac{y}{x} + \frac{1 - y}{1 - x} - 2, \tag{14.48}$$

and that the unique ess is given by $x^* = 1/2$. Classify this ess, and comment.

Exercise 14.5 (Evolutionary Branching Point). If $s(y, x) = 2x^2 + y^2 - 3xy$ show that $x^* = 0$ is an ess that can invade, is convergent stable and a protected polymorphism but invasible, i.e. an evolutionary branching point.

Exercise 14.6. For the remaining cases 1,2,4,6,8 from Table 14.1 that we have not discussed in the text, find an example function which yields an ess of that type.

Exercise 14.7. Suppose that $s(y, x) = x^2 - 3y^2 + 2xy + 8x - 8y$. Show that $x^* = -2$ is an ess, and classify x^* according to the cases in Table 14.1.

Exercise 14.8 (Nowak, 1990a). Let the payoff of strategy y against strategy x ($x, y \in [0, 1]$) be given by

$$\mathcal{E}[y; x] = ay^2 + byx + cx^2 + dy + ex + f, \qquad (14.49)$$

for some constants a, b, c, d, e, f. Assume that $a < 0$. Show that, provided that $0 \leq -d/(2a + b) \leq 1$, the strategy $\hat{x} = -d/(2a + b)$ is an ESS.

Hint. Show that $\mathcal{E}[y; \hat{x}] < \mathcal{E}[\hat{x}; \hat{x}]$ for all $y \neq \hat{x}$.

Exercise 14.9 (Nowak, 1990a). In the notation of Exercise 14.8, show that if $2a + b > 0$, then $\hat{x}$ is an inaccessible ESS in the sense that if $x < \hat{x}$ ($x > \hat{x}$), then δ_x can only be invaded by mutants who play $y < x$ ($y > x$) (i.e. the population goes even further away from $\hat{x}$).

Hint. The maximum of the function $\mathcal{E}[y; x]$ (for fixed x) is $y = -(bx + d)/2a$ (i.e. the best mutant would use this strategy y) and we have that if $2a + b > 0$, then $-(bx + d)/2a < x$ if and only if $x < -d/(2a + b) = \hat{x}$.

Exercise 14.10. Use Haigh's condition (6.45) and the definition of convergence stability (13.24) to show that in the single variable case any mixed strategy ESS must be a convergent stable ess.

Exercise 14.11 (Multiple traits, Meszéna et al., 2001). Assume there are now multiple traits $i = 1, 2, \ldots, n$ and for for $\mathbf{x} = (x_i)$ and $\mathbf{y} = (y_i)$ defined as above

$$s(\mathbf{y}, \mathbf{x}) = \mathcal{E}[\mathbf{y}; \delta_{\mathbf{x}}] - \mathcal{E}[\mathbf{x}; \delta_{\mathbf{x}}]. \qquad (14.50)$$

A strategy x^* is an ess if

$$\left[\frac{\partial}{\partial y_i} s(\mathbf{y}, \mathbf{x}) \bigg|_{y=x=x^*} \right]_i = \mathbf{0}. \qquad (14.51)$$

Show that the conditions for non-invasibility, ability to invade and convergence stability are as follows (note the equivalent conditions for the single trait case given in Section 14.3):

1. An ess is non-invasible if the matrix

$$\left[\frac{\partial^2 s(y,x)}{\partial y_i \partial y_j} \bigg|_{y=x=x^*} \right] \tag{14.52}$$

is negative definite, see Meszéna et al. (2001).

2. An ess can invade if the matrix

$$\left[\frac{\partial^2 s(y,x)}{\partial x_i \partial x_j} \bigg|_{y=x=x^*} \right] \tag{14.53}$$

is positive definite, see Meszéna et al. (2001).

Exercise 14.12 (Matrix game with multiple traits). Consider a situation of n traits and an $n \times n$ matrix game with a payoff matrix A. Show that the matrix for the non-invasibility condition from (14.52) is degenerate, as all of the elements of the matrix are zero. Also, show that the conditions for an ess to invade and to be convergence stable are identical, and the same as Haigh's condition (6.45) for the same strategy to be an ESS.

Hint. For $\mathbf{x} = (x_i)$ we can write $x_n = 1 - x_1 - \ldots - x_{n-1}$ and thus get

$$s(\mathbf{y}, \mathbf{x}) = \mathbf{y}A\mathbf{x}^T - \mathbf{x}A\mathbf{x}^T \tag{14.54}$$

$$= \sum_{i=1,j=1}^{n-1} (y_i - x_i)x_j(a_{ij} - a_{in} - a_{nj} + a_{nn}). \tag{14.55}$$

The (i,j) entry in the matrix in (14.53) is thus $-(a_{ij} + a_{ji} - a_{in} - a_{jn} - a_{ni} - a_{nj} + 2a_{nn})$, which should be compared to Haigh's condition (6.45).

Exercise 14.13. Use the canonical equation of adaptive dynamics (14.36) to show the long-term evolution of the population in the two trait case where $s(y_1, \mathbf{x}) = a(2y_1 x_2 - y_1^2)$, $s(y_2, \mathbf{x}) = a(2y_2 x_1 - y_2^2)$ when at the start of the process (a) $a = 1, x_1 = 1, x_2 = -1$, (b) $a = -1, x_1 = -1, x_2 = 2$, (c) $a = -1, x_1 = 1, x_2 = 1$.

Chapter 15

The evolution of cooperation

In Section 4.2 we considered the Prisoner's Dilemma game, which had at its heart the idea of cooperation. A player will make a choice which (in the short term at least) gives the player a smaller reward than it would otherwise receive, but which helps the other player by increasing its payoff. We saw in the standard game how such a strategy is not a good one, but how in the iterated game cooperation could emerge. We will consider this scenario in more detail later in this chapter.

Alternatively, we can think of a *cooperator* as an individual who pays a cost C, so that another individual could receive a benefit B. A *defector* has no cost and does not deal out benefits. When modelled as a simple game (see e.g. (15.2) below), we soon realize that defectors again do better. However, cooperation and altruistic behaviour between animals, including of course humans, is widespread and occurs in many taxa (see e.g. Dugatkin, 1997a). Individuals perform actions which seem against their own (short-term) interests, and which at first sight seem to contradict the laws of natural selection and the logic of game theory that we have expounded in this book. An example of such a behaviour may be food sharing by vampire bats *Desmodus rotundus* (Wilkinson, 1984). Bats go out every night to find a blood source such as a cow or similar animal. Not every bat finds an animal to feed on during the night, and those that return having fed then feed those who have not.

As we have described above, there is a natural implication that $B > 0$ and $C > 0$, but there are of course four distinct kinds of behaviour defined depending upon the signs of the values B and C (Hamilton, 1964; West et al., 2007b).

- $B > 0, C < 0$ is *mutual benefit*, where the act of helping also helps the focal individual.

- $B < 0, C < 0$ is *selfishness*, where the focal individual performs an action which helps itself at the expense of the other.

- $B > 0, C > 0$ represents *altruism*, where the focal individual pays a cost to help the other.

- $B < 0, C > 0$ is *spite*, where an individual pays a cost to harm another individual.

The first two scenarios are the usual situations that you expect in game-theoretical models, where individuals act to increase their own benefit. The

DOI: 10.1201/9781003024682-15

third $B > 0, C > 0$ is the usual assumption associated with kin selection and the evolution of altruistic behaviour.

There are many mechanisms for the evolution of cooperation, see Nowak (2006b); Taylor and Nowak (2009); Nowak (2012); West et al. (2007b). Here we will discuss the following in more details.

1. *Kin selection* that operates when the donor and the recipient of an altruistic act are genetic relatives.

2. *Greenbeard genes* are a special form of kin selection.

3. *Direct reciprocity* requires repeated encounters between the same two individuals.

4. *Indirect reciprocity* is based on reputation; a helpful individual gets a better reputation and an individual with a good reputation is more likely to receive help.

5. *Punishment* is a way to enforce cooperation.

6. *Network reciprocity* is a reciprocity based on some type of spatial or other clustering.

7. *Multi-level selection*, alternatively known as group selection, is the idea that competition is not only between individuals but also between groups.

15.1   Kin selection and inclusive fitness

Until now we have assumed that individuals should choose the strategy which maximises their individual payoff, irrespective of its effect on other members of the population. However, if other individuals are related to our focal individual, if it can help them then it receives an indirect benefit as these related individuals share a proportion of genetic material with it. *Hamilton's rule* (Hamilton, 1964, 1970) says that making a choice which is (potentially) costly to oneself but beneficial to a relative is favoured by natural selection if

$$rB > C, \tag{15.1}$$

where B and C are as above and r is the coefficient of relatedness between the individual and its relative, i.e. the probability that they have the same allele at a given locus for genetic reasons, for example $r = 1/2$ for father and son. The *inclusive fitness* of the individual is increased by $rB - C$. We can see a simple proof of Hamilton's rule below framed by asymmetric games.

Assume there are two possible versions A_1 and A_2 of an allele in the population competing at a single locus. Assume that an allele A_1 is such that whenever a special pair of individuals (such as a father and a son, which we shall call the giver G and the recipient R, respectively) meet, it can induce an action of G that results in a fitness cost $C > 0$ to G but a fitness benefit $B > 0$ for the receiver R. We would like to know whether this allele can spread (or be stable) in the population.

The game described above can be written in the form of an asymmetric game as described in Chapter 8. A player "chooses" a strategy—an allele. The player can be in one of the two roles (i.e. being carried by giver or recipient). A player with an allele (or strategy) A_2 is able to accept a potential benefit if in a recipient role. A player with an allele A_1 will also benefit as a recipient, but in addition it provides a benefit if it is a giver. Assuming A_2 does not give any benefit, we get that the payoffs from pairwise interactions are given by the following bimatrix:

$$
\begin{array}{cc}
 & \begin{array}{cc} A_1 \text{ in R} & A_2 \text{ in R} \end{array} \\
\begin{array}{c} A_1 \text{ in G} \\ A_2 \text{ in G} \end{array} & \left(\begin{array}{cc} (-C, B) & (-C, B) \\ (0, 0) & (0, 0) \end{array} \right).
\end{array}
\tag{15.2}
$$

Assuming the players have probability $\rho = 1/2$ of being in either role, we get

$$
\mathcal{E}[A_i; \Pi] = \frac{1}{2} \mathcal{E}_G[A_i; \Pi] + \frac{1}{2} \mathcal{E}_R[A_i; \Pi],
\tag{15.3}
$$

where $\mathcal{E}_G[A_i; \Pi]$ and $\mathcal{E}_R[A_i; \Pi]$ are payoffs to A_i when in a role of a giver and recipient. By (15.2) we get

$$
\mathcal{E}_G[A_1; \Pi] = -C,
\tag{15.4}
$$

$$
\mathcal{E}_G[A_2; \Pi] = 0 \text{ and}
\tag{15.5}
$$

$$
\mathcal{E}_R[A_i; \Pi] = r_i(p)B,
\tag{15.6}
$$

where $r_i(p)$ is the probability that a recipient with an allele A_i will have a giver with an allele A_1 in a population where the frequency of A_1 is p. Giver and recipient can both possess A_1 either because the A_1 in the recipient is a direct copy of the one in the giver or because A_1 has a certain abundance in the population and they are independent alleles. Let r be the coefficient of relatedness as defined above. We obtain

$$
r_1(p) = r + (1 - r)p.
\tag{15.7}
$$

Similarly, we get

$$
r_2(p) = 1 - \big(r + (1 - r)(1 - p)\big)
\tag{15.8}
$$

$$
= r_1(p) - r.
\tag{15.9}
$$

Thus

$$
\mathcal{E}[A_1; \Pi] - \mathcal{E}[A_2; \Pi] = \frac{1}{2}\Big(-C + r_1(p)B - \big(r_1(p) - r\big)B\Big)
\tag{15.10}
$$

$$
= \frac{1}{2}(rB - C).
\tag{15.11}
$$

A_1 is an ESS if and only if $\mathcal{E}[A_1; \delta_{A_1}] > \mathcal{E}[A_2; \delta_{A_1}]$, i.e. if and only if

$$rB - C > 0. \tag{15.12}$$

Moreover, it follows from (15.10) that A_1 can also invade an A_2 population if and only if (15.12) holds. Thus here there is a unique pure ESS, either A_1 or A_2.

Note that we have a different situation than we had in asymmetric games from Chapter 8. The players can be in different roles, but cannot change strategies when changing the roles. Also, a player in a recipient role faces different interactions depending on its strategy (this is captured by (15.7)-(15.9)).

15.2 Greenbeard genes

Hamilton (1964) (see also Dawkins, 1976, 1999) describes a form of kin selection, the hypothetical example of a *greenbeard gene* that makes its bearers grow a green beard and also help any other bearer of a green beard (but nobody else). Since then, other examples of greenbeard effects have been analysed in Gardner and West (2010) (see also Queller, 1984; Gardner and West, 2004). First, the greenbeard could be either helping other greenbeards, or harming other non-greenbeards. Second, the greenbeard could act to help or harm without any discrimination, but the help could benefit only other greenbeards or the harm could be caused only to other non-greenbeards. We thus have four different kinds of greenbeard.

- *Facultative-helping greenbeards*: when a greenbeard meets another greenbeard it will help at a cost c to itself, and benefit b to the other greenbeard.

- *Obligate-helping greenbeards*: when a greenbeard meets another individual it will give help at a cost c to itself, but only other greenbeards can make use of the benefit of value b.

- *Facultative-harming greenbeards*: when a greenbeard meets a non-green beard it harms the non-greenbeard at a cost a to itself and d to the non-greenbeard.

- *Obligate-harming greenbeards*: when a greenbeard meets another individual it will perform a harmful act at a cost a, but only non-greenbeards are harmed, paying cost d.

We note that a, b, c, d from above naturally link to B, C considered at the beginning of the chapter; see for example Exercise 15.1.

TABLE 15.1: Summary of greenbeard effects. p is the proportion of a green-beard allele A_1 in the population. Where appropriate, a is the cost of harming, b is the benefit of helping, c is the cost of helping and d is the damage of harming.

	Greenbeard type			
	Facultative-helping	Obligate-helping	Facultative-harming	Obligate-harming
Payoff matrix	$\begin{pmatrix} b-c & 0 \\ 0 & 0 \end{pmatrix}$	$\begin{pmatrix} b-c & -c \\ 0 & 0 \end{pmatrix}$	$\begin{pmatrix} 0 & -a \\ -d & 0 \end{pmatrix}$	$\begin{pmatrix} -a & -a \\ -d & 0 \end{pmatrix}$
f_{A_1}	$p(b-c)$	$pb - c$	$-(1-p)a$	$-a$
f_{A_2}	0	0	$-pd$	$-pd$
A_1 favoured if	$\frac{b}{c} > 1$	$\frac{b}{c} > \frac{1}{p}$	$\frac{d}{a} > \frac{1-p}{p}$	$\frac{d}{a} > \frac{1}{p}$

We can again model this scenario as an asymmetric game as was done above in Section 15.1. However, since a particular individual can act as a giver and a receiver at the same time, as is the case for the toxin-producing bacterium *Photohabdus luminescens* (Forst and Nealson, 1996), we will model this scenario by a symmetric (matrix) game as follows.

Assume that in the case(s) of helping greenbeards there is a single locus and only two versions of the allele A_1, A_2 such that A_1 is a greenbeard and A_2 is not with p being the proportion of A_1. The population is well mixed and when individuals meet, they both can act as a giver as well as a receiver. For example, if A_1 and A_2 meet, then the payoff to A_2 will be 0 because either A_1 will not provide any help if it is facultative, or A_1 will provide help as an obligate, but A_2 will not be able to use it. At the same time, the payoff to A_1 will be either 0 (if facultative) or $-c$ if obligate. If A_1 meets another A_1, then it pays the cost c for helping but also receives a benefit of b from the other, so the total payoff to the focal A_1 is $b - c$.

This situation and its analysis is described in Table 15.1. We can see from Table 15.1 that the condition for the spread of facultative-helping greenbeards, the type usually considered (see e.g. Dawkins, 1976, 1999), is independent of frequency, and so they can be beneficial even in small numbers. The other three types depend on frequency and are always favoured when making up a sufficiently large proportion of the population (so are stable), but are not favoured when in a small proportion (so can never invade). Thus for any of the last three types to become established, some extra feature of the population such as spatial clustering is needed.

One problem is, how do we observe greenbeard genes in action? Once the greenbeards fixate, they are hard to observe, since their distinctive feature of helping/harming some individuals but not others will be obscured by the absence of two types in the population. The helping actions will continue, but there will be no observation of helping only to greenbeards. If facultative-harming greenbeards fixate, there will be no observed harming behaviour as there will be nobody to harm. If obligate-harming greenbeards fixate, their harming behaviour will continue but without a harming effect. However, despite the above problems, all four kind of greenbeards have been observed in nature (see Gardner and West, 2010, Fig. 3, and references therein).

It should be noted that greenbeards can be vulnerable to cheating. "False-beard" genes which are distinct from the greenbeard gene, but mimic its appearance (in the case of facultative greenbeards) or its ability to make use of the benefit/resist the harm (in the case of obligate greenbeards), can invade, since they will receive the benefit/not incur the harm without bearing the cost. Thus for instance for facultative-helping greenbeards, the falsebeard will trick greenbeards into paying the cost and so will receive the benefit, but will never pay the cost itself to help others. It is easy to see that if it is a perfect mimic so that it is indistinguishable from greenbeards to other greenbeards, then it will invade a population of greenbeards irrespective of the values of b, c and p (see e.g. Pepper and Smuts, 2002). We have seen a similar situation before; for example in Exercise 2.2, the toxin-producing bacteria acts like an obligate harming greenbeard that can successfully wipe out any sensitive bacterial strain; but it is also prone to invasion by a resistant bacteria that is immune to the toxin but does not produce it. In turn, this can lead to a Rock-Scissors-Paper game from Example 2.1. A good paper that considers the conditions under which greenbeard genes can be stable is Jansen and van Baalen (2006).

Also, we note that Gardner and West (2010) achieved the conclusions from Table 15.1 in a different way by considering an alternative definition of the relatedness coefficient r. For the case of two alleles A_1 and A_2 with population frequencies p and $1 - p$, they defined the relatedness of A_1 with another A_1 as 1, as they are identical. If r is the relatedness of A_2 to A_1, we obtain mean relatedness of the population to A_1 as $p \cdot 1 + (1-p) \cdot r$. The mean relatedness to A_1 in the population is set as 0. This makes the definition logically equivalent to before as, for example, the relatedness between father and son is just the average of the relatedness of a gene to an identical copy and to a random member of the population, which is $1/2$. Thus we obtain the relatedness of A_2 to A_1 as

$$r = -\frac{p}{1 - p}. \tag{15.13}$$

The conclusions from Table 15.1 were then drawn by evaluating B and C for different kinds of greenbeards and by an application of Hamilton's rule $rB > C$, see also Exercise 15.1. We note that the relatedness of A_2 to A_1 may thus be different to the relatedness from A_1 to A_2. This at first seems illogical but is, in fact, reasonable when we consider the relatedness to an individual being a comparative measure. If $p \approx 0$ then the relatedness of A_2 to A_1 is approximately 0 (almost all individuals are A_2, so A_2 is a typical member of the population). However, if $p \approx 1$ then almost all individuals are identical to

A_1, but A_2 is very different, and hence the relatedness of A_2 to A_1 is large and negative.

15.3 Direct reciprocity: developments of the Prisoner's Dilemma

We have already looked into the evolution of cooperation in Chapter 4 when we considered the Iterated Prisoner's Dilemma (IPD). In this chapter we discuss this, and other important modelling ideas, in more detail.

15.3.1 An error-free environment

We saw in Section 4.2.5 that repeated interactions between individuals could foster cooperation. Recall that a Prisoner's Dilemma game is a matrix game with the payoff matrix

$$
\begin{array}{cc}
 & \text{Cooperate} \quad \text{Defect} \\
\begin{array}{c} \text{Cooperate} \\ \text{Defect} \end{array} &
\left(\begin{array}{cc} R & S \\ T & P \end{array} \right),
\end{array}
\tag{15.14}
$$

where the payoffs satisfy

$$T > R > P > S \text{ and} \tag{15.15}$$
$$2R > S + T. \tag{15.16}$$

We note that a commonly used simplification of the Prisoner's Dilemma involves cooperators paying a cost c and individuals playing with cooperators receiving benefit $b > c$, giving $R = b - c, S = -c, T = b, P = 0$, see (15.36). The simple but effective strategy Tit for Tat, which combined a propensity to cooperate with a willingness to punish but also to forgive was an ESS when the chances of a repeat interaction w were sufficiently high and when the strategy space was not too large.

Mesterton-Gibbons (2000) (see also Nowak, 2006a) considers contests between a number of various strategies such as All Defect (ALLD), Tit for Tat (TFT), Tit for two Tats (TF2T), Suspicious Tit for Tat (STFT), Generous Tit for Tat and Grim (GRIM). The description of the strategies is summarised in Table 15.2 and the results in the pairwise contests are shown below in (15.17) (see Example 15.1 for some more detailed analysis).

TABLE 15.2: Several strategies for the Iterated Prisoner's Dilemma.

Strategy	Description
ALLD	Always play D
GRIM	Play C but permanently switch to D once the opponent defects
TFT	Start with C and then repeat the opponent's last move
STFT	Start with D and then repeat the opponent's last move
TF2T	Play C unless an opponent played D in both the last two moves
WSLS	Start with C and then play C if and only if the last payoff from the last round was R or P
GTFT	Start with C and then play C either with probability 1 if the opponent cooperated on the last move, or with probability $\min\{1 - \frac{T-R}{R-S}, \frac{R-P}{T-P}\}$ if the opponent defected

$$
\begin{array}{c}
 & \text{ALLD} & \text{GRIM} & \text{TFT} & \text{STFT} & \text{TF2T} \\
\text{ALLD} & \frac{P}{1-w} & T+\frac{Pw}{1-w} & T+\frac{Pw}{1-w} & \frac{P}{1-w} & \frac{(1-w^2)T+w^2P}{1-w} \\
\text{GRIM} & S+\frac{Pw}{1-w} & \frac{R}{1-w} & \frac{R}{1-w} & S+wT+\frac{w^2P}{1-w} & \frac{R}{1-w} \\
\text{TFT} & S+\frac{Pw}{1-w} & \frac{R}{1-w} & \frac{R}{1-w} & \frac{S+wT}{1-w^2} & \frac{R}{1-w} \\
\text{STFT} & \frac{P}{1-w} & T+wS+\frac{w^2P}{1-w} & \frac{T+wS}{1-w^2} & \frac{P}{1-w} & T+\frac{Rw}{1-w} \\
\text{TF2T} & \frac{(1-w^2)S+w^2P}{1-w} & \frac{R}{1-w} & \frac{R}{1-w} & S+\frac{Rw}{1-w} & \frac{R}{1-w}
\end{array}
\tag{15.17}
$$

Example 15.1 (*IPD contests*). Find the sequence of plays and the payoffs in contests involving the strategies STFT and GRIM.

We obtain the sequences of plays for the three possible contests as below:

$$
\begin{array}{c|l}
GRIM & CCCCC\ldots \\
GRIM & CCCCC\ldots \\
\hline
STFT & DDDDD\ldots \\
STFT & DDDDD\ldots \\
\hline
GRIM & CDDDD\ldots \\
STFT & DCDDD\ldots
\end{array}
\tag{15.18}
$$

Thus using the summation of an infinite geometric series, it is easy to see that we obtain the payoffs shown for these games in (15.17).

It follows from (15.17) that ALLD is not an ESS if we treat the IPD as a matrix game because it is vulnerable to invasion by STFT by drift (see Section 5.2 and Chapter 12). Moreover, once the population consists of a mixture of

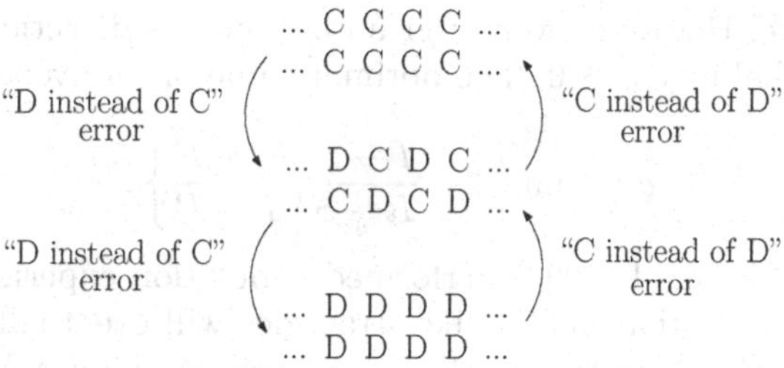

FIGURE 15.1: Game between two TFT players in an error-prone environment (with mistakes possible but rare).

STFT and ALLD (note that everybody always plays D in such a population), TF2T can invade.

We note that TF2T does better than TFT against STFT (see Exercise 15.3). In fact, Axelrod entered TF2T for the second of his tournaments (Axelrod, 1984) because he realized that TF2T would have won the first tournament. Unfortunately the introduction of a sufficient number of aggressive strategies in the second tournament meant that TF2T did not win the second.

15.3.2 An error-prone environment

TFT proved to be quite successful in promoting cooperation as seen above. However, so far we have only studied the ideal situation where individuals interpret their opponent's move correctly and also correctly perform their chosen action (see e.g. Ohtsuki, 2004). If there is even a tiny chance of error that will cause the intent to cooperate to be (or be perceived as to be) executed as a defection, the pairwise contests between TFTs will be full of mutual and originally unprovoked retaliation. Indeed, after the first error, the game between TFT switches from mutual cooperation to alternating cooperation and defection. Assuming that errors can also be made in the other direction, when intention to defect leads to cooperation, a second error can revert the game back to mutual cooperation, but it can also lead to mutual defection; see Figure 15.1.

The TF2T strategy is more forgiving and it is, in fact, unlikely that rare errors would damage mutual cooperation between two such players. However, as seen above, TF2T is too generous and can be exploited by defectors. Molander (1985), Nowak (1990b) and Nowak and Sigmund (1990) demonstrated that strategies that mix the TFT approach with a simple Cooperate (unconditionally) receive payoffs close to the ideal R even in an error-prone environment. A strategy GTFT—generous Tit for Tat—was shown to correct mistakes by simply being more forgiving than TFT. Unlike TFT or other previously studied strategies that were deterministic, GTFT is a stochastic strategy. GTFT starts by cooperating and also always cooperates if the opponent cooperated

on the last move. However, even after an opponent's defection, GTFT cooperates with probability $q > 0$. The optimal value of forgiveness q was shown
to be

$$q = \min\left\{1 - \frac{T-R}{R-S}, \frac{R-P}{T-P}\right\}. \tag{15.19}$$

Nowak and Sigmund (1993) performed simulation experiments that were
aimed at confirming that GTFT-like strategies will eventually dominate the
field. However, they discovered that a surprising winner WSLS emerged.
WSLS stands for Win Stay, Lose Shift. It starts with C and then plays C
if and only if the payoff from the last round was R or P. WSLS, also known
as Pavlov (Kraines and Kraines, 1989) performs very well in the IPD game
(Kraines and Kraines, 2000), although as Nowak (2006a, Section 5.5) suggests,
WSLS can sometimes be replaced by ALLD.

15.3.3   ESSs in the IPD game

We have seen many different strategies in the IPD game. Usually, once a
game is defined, we try to find the ESSs. So what are the ESSs of the IPD
game? This is a question which is difficult to answer and which may not have
many practical consequences. The IPD is a nice model game that shows that
the concept of an ESS depends on what allowable strategies there are in the
population. For example, if only ALLD and ALLC would be available, then
ALLD would be an ESS.

In many instances, there may be no ESS (Boyd and Lorberbaum, 1987;
Farrell and Ware, 1989; Lorberbaum, 1994).

On the other hand, Lorberbaum et al. (2002) give the following three ESSs
in the case where strategies only have a one-move memory and are forced to
be stochastic and can cooperate or defect with probabilities in $[e, 1-e]$ with
sufficiently small $e > 0$.

1. ALLD-e : always defects with probability $1 - e$.

2. WSLS-e : on the first move or after receiving payoff R or P cooperate
   with probability $1 - e$. Otherwise, defect with probability $1 - e$.

3. Grudge-e : cooperate with probability $1 - e$ on the first move or after
   receiving a payoff R. Otherwise, defect with probability $1 - e$. This is a
   stochastic version of GRIM.

Interestingly, although as we saw above, mistakes shook the ideal picture
of TFT and the IPD game, Boyd (1989) shows that mistakes allow a strategy
called Contrite Tit for Tat (CTFT) (Sugden, 1986, p. 110) to be an ESS.
Rather than its own and its opponent's moves, CTFT tracks its own and
its opponent's standing as follows. An individual starts in good standing. A
player is in good standing if she has cooperated in the previous round or if
she has defected while provoked (i.e. while she was in good standing and the

other player was not). In every other case defection leads to bad standing. The strategy CTFT begins with a cooperative move and cooperates except if provoked; see Exercise 15.14.

15.3.4 A simple rule for the evolution of cooperation by direct reciprocity

Consider a simplified strategy space of the IPD, considering only ALLD and TFT (which in this case is equivalent to GRIM), ; see for example Taylor and Nowak (2009) or Mesterton-Gibbons (2000, p. 185). If two cooperators (TFT) meet, they cooperate all the time. If two defectors (ALLD) meet, they defect all the time. If a cooperator meets a defector, the cooperator cooperates in the first round and defects afterwards, while the defector defects in every round. The payoff matrix is thus given by the following submatrix of (15.17):

$$
\begin{array}{cc}
 & \begin{array}{cc} TFT & ALLD \end{array} \\
\begin{array}{c} TFT \\ ALLD \end{array} &
\begin{pmatrix} \frac{R}{1-w} & S + \frac{wp}{1-w} \\ T + \frac{wP}{1-w} & \frac{P}{1-w} \end{pmatrix}.
\end{array}
\tag{15.20}
$$

Note that defection, ALLD, is always an ESS (because $P > S$). Cooperation (TFT) is an ESS if

$$
w > \frac{T - R}{T - P}.
\tag{15.21}
$$

When the game is as in (15.2) or more precisely as in (15.36), then (15.21) is analogous to Hamilton's rule (15.1).

15.3.5 Extortion and the Iterated Prisoner's Dilemma

We shall now return to the standard Iterated Prisoner's Dilemma, and consider a specific simple class of strategies, those with memory 1. Here players choose their strategy only based upon the outcome of the previous round. In any given round an individual can obtain any of the four payoffs R, T, S or P, and then play a mixed strategy conditional upon this outcome. Letting p_X be the probability that an individual cooperates given its payoff from the last round was X, a strategy is defined by the vector (p_R, p_T, p_S, p_P), together with the choice of starting strategy. We shall only consider the case where $w = 1$ and so the game continues for an infinite number of interactions. In such circumstances (Sigmund, 2010) the choice of initial strategy is irrelevant for the long-term payoff (this can have an effect for very specific pure strategy choices if the Markov transition matrix over the states is not irreducible, as we discuss below, but practically in the analysis below this can be ignored).

Such strategies received much interest following the work of Press and Dyson (2012). They defined the class of zero determinant (ZD) strategies, being those which have the following probabilities

$$p_R = 1 + \alpha R + \beta R + \gamma, \tag{15.22}$$

$$p_T = \alpha T + \beta S + \gamma \tag{15.23}$$

$$p_S = 1 + \alpha S + \beta T + \gamma, \tag{15.24}$$

$$p_P = \alpha P + \beta P + \gamma, \tag{15.25}$$

for some values α, β and γ. It was shown that if player 1 uses the above strategy, then irrespective of player 2's strategy, the payoffs E_1, E_2 to players 1 and 2 satisfy

$$\alpha E_1 + \beta E_2 + \gamma = 0. \tag{15.26}$$

With suitable choice of α, β and γ, namely $\gamma = -(\alpha+\beta)P$, and $-\beta/\alpha > 1$, player 1 can ensure that payoffs satisfy

$$E_1 - P = \chi(E_2 - P) \tag{15.27}$$

where $\chi = -\beta/\alpha$. Thus player 1 has a surplus over the mutual punishment reward P which is a simple multiple (with factor greater than 1) of player 2's surplus. We will denote this strategy by S_χ. To show this equality is Exercise 15.7.

Against an opponent playing S_χ the evolutionary response for player 2 is to maximise its own payoff, which accedes the largest advantage for player 1. Press and Dyson (2012) called such strategies *extortion strategies*. Thus a rational player can gain an advantage against an evolutionary opponent, and here the IPD can be considered as a type of ultimatum game.

We note that this interesting mix of rational and evolutionary players has potentially important applications. For example, in Chapter 22 we consider the modelling of cancer as an evolutionary process. Here there is a game between the evolving cancer and the physician, who is a rational player. The consequences of this scenario, in particular, is discussed in Section 22.4.

The above terminology was that used in Hilbe et al. (2013), slightly different from Press and Dyson (2012), and it is their work that we mainly follow in this section, as unlike that original work, they considered a fully evolutionary scenario more in the spirit of classical evolutionary games and the approach of this book. Can such extortion strategies evolve?

Hilbe et al. (2013) considered the donation game where in all interactions cooperators pay a cost c to provide a benefit of value b for the other player, giving $R = b - c, T = b, S = -c$ and $P = 0$ (we first met this game in Section 13.2; see Section 15.5 for a more detailed description). They then analyse how the strategy S_χ performs against some classic IPD strategies TFT, WSLS, ALLC and ALLD. They obtained the following payoff matrix for this game, in a similar vein to how we considered the games in Section 15.3.1.

$$
\begin{array}{c}
\quad\quad\quad \text{TFT} \quad\quad \text{WSLS} \quad\quad\quad S_\chi \quad\quad\quad \text{ALLC} \quad\quad \text{ALLD} \\
\begin{array}{c}
\text{TFT} \\
\text{WSLS} \\
S_\chi \\
\text{ALLC} \\
\text{ALLD}
\end{array}
\left(
\begin{array}{ccccc}
\frac{b-c}{2} & \frac{b-c}{2} & 0 & b-c & 0 \\[2mm]
\frac{b-c}{2} & b-c & \frac{b^2-c^2}{b(1+2\chi)+c(2+\chi)} & \frac{2b-c}{2} & \frac{-c}{2} \\[2mm]
0 & \frac{(b^2-c^2)\chi}{b(1+2\chi)+c(2+\chi)} & 0 & \frac{(b^2-c^2)\chi}{b\chi+c} & 0 \\[2mm]
b-c & \frac{b-2c}{2} & \frac{b^2-c^2}{b\chi+c} & b-c & -c \\[2mm]
0 & \frac{b}{2} & 0 & b & 0
\end{array}
\right).
\end{array}
$$

$$(15.28)$$

We note here that payoffs are somewhat different to how they appeared in earlier sections. For example the payoff of TFT versus TFT is $(b-c)/2$, whereas for the TFT strategy as we had first described it in Section 4.2.5 players would start by cooperating and so cooperate every round and thus receive a reward of $b - c$. In particular, the starting strategy would appear to be crucial. However, it is perhaps unrealistic to assume that a player can play perfectly for ever, and if there is even the smallest probability of error (see Sections 15.3.2 and 15.3.3) so that the intended strategy is played with probability $1 - \epsilon$ not 1, we obtain the payoff matrix as above (note that if errors in different directions had different probabilities, e.g. I am more likely to make a mistake trying to cooperate than trying to defect, the payoffs would be different).

As we see in the above payoff matrix, S_χ cannot be a pure ESS whenever $b > c$, which is generally assumed in this game. The payoff matrix (15.28) is (very) non-generic, since there are only three free parameters that generate 25 entries, and for example the replicator dynamics have continuous families of fixed points. Thus such analysis is problematic. Instead Hilbe et al. (2013) followed the evolution of various mixed population of the above five strategies, see Figure 15.2.

In general, extortion strategies can act as catalysts to facilitate the evolution of strategies that could otherwise not enter the population, and so have a significant impact on how a population evolved. In particular, if the population starts with a large fraction of ALLD and smaller fractions of WSLS and S_χ, WSLS can evolve because of the presence of S_χ, when it would not have been able to if there was only ALLD and WSLS.

Allowing all memory one strategies by introducing mutations (Imhof and Nowak, 2010) they found that extortion strategies could not be the outcome of evolution themselves for large populations, but for smaller populations they could evolve. This was also true for a second interesting class of strategies *equalizer strategies*, which set the other player's score at a fixed value irrespective of the choice that they make. Finally we note that they also considered a

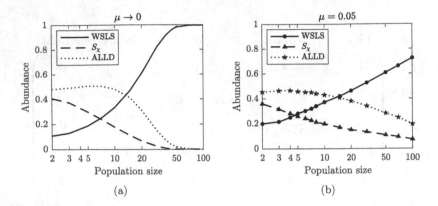

FIGURE 15.2: Evolutionary competition in the IPD. For various population sizes, the graphs show the frequency of each strategy in the mutation-selection equilibrium. Adding ALLC only leads to minor changes in the stationary distribution, which now slightly favours S_χ. For the copying process, Hilbe et al. (2013) assume that individuals A and B are chosen randomly. A switches to B's strategy with a probability given by $(1 + \exp(s(P_A - P_B)))^{-1}$, where P_A and P_B are the corresponding payoff values and $s \geq 0$ is the selection strength. The parameters are $b = 3, c = 1, s = 1$, and $\chi = 2$. (a) the limit of rare mutations ($\mu \to 0$) and (b) $\mu = 0.05$.

two species model, and in this extortioners had more scope to evolve than for the single population case.

15.4   Public Goods games

We shall firstly consider an example of a *public goods game* where individuals in a group can invest to provide collective resources for the group, or not invest and keep the resources for themselves.

Example 15.2 (*Public Goods Game*, Fehr and Gachter, 2002). Four players are given 20 monetary units (which we shall henceforth refer to as dollars) each. They decide whether and how much to invest in a common pool; any money that is invested increases in value by 60% and is then divided among the four players, so that the return for each invested dollar is 40 cents to each of the four players.

For example, assuming that k players invest all \$20, and $4 - k$ players invest nothing, then an investing player will receive \$8k, and a non-investing player will receive \$8k plus having the original \$20. Thus an investing player only receives back $8 for the invested $20, and not investing is the optimal strategy of the Public Goods Game, although everybody would do much better if everybody invested. Hence we get the familiar situation of the single-shot Prisoner's Dilemma where mutual defection is an ESS. In fact, real humans playing this game for the first time are slow to realise this, and a number of rounds (with newly formed groups) are required, during which the investment level continuously declines.

15.4.1 Punishment

The most common idea of how cooperation is enforced through repeated single-shot games is through the addition of *punishment*.

A development of the above experiment introduces a second round that includes punishment as follows.

Example 15.3 (*Public Goods Game with punishment*, Fehr and Gachter, 2002). After everyone makes their investment choice and gets the return back as in Example 15.2, players observe the choices made by the others, and can punish offenders (i.e. those that did not invest fully). Players pay a fee to impose a fine upon offenders from $0 to $10. Every dollar paid in fees for the fine means a fine of $3 for the receiver of the fine.

In real experiments individuals played a series of games (with newly formed groups) and investment levels rose, so that in contrast to very low investment without punishment, there was very high investment with punishment. However, since groups are reformed so that no individual benefits from its own punishment decisions, the analysis is similar to before, and we see that optimal play involves no punishment and consequently no investment. Thus the behaviour of human players is not always rational, and such games are of significant interest to psychologists for this reason.

Indeed, suppose that we have a group of m players with V to invest. Any investment is multiplied by c, $m > c > 1$, and shared. Individuals can pay a unit cost to punish any defectors, which makes them pay a penalty P. We consider the possible payoffs to a single individual, which we shall term our focal individual, given the strategies of the other players. Suppose that out of $m - 1$ other players, k invest and punish all defectors, l defect (and do not punish), and the remaining $m - k - l - 1$ invest but do not punish. Then the rewards to our focal individual will be

$$E = \begin{cases} \frac{(m-l)cV}{m} & \text{if it invests but does not punish,} \\ \frac{(m-l)cV}{m} - l & \text{if it invests and punishes,} \\ \frac{(m-l-1)cV}{m} + V - kP & \text{if it defects.} \end{cases} \tag{15.29}$$

The optimal play for our focal individual is

$$\begin{array}{ll} \text{Defect} & \text{if } V\left(1 - \frac{c}{m}\right) > kP, \\ \text{Invest but do not punish} & \text{otherwise.} \end{array} \tag{15.30}$$

It is never best to invest and punish, since investing without punishing is always better. Thus any stable population must involve games where $k = 0$, and so the only stable strategy involves defection by all individuals.

We note that the above situation is similar to the invasion of greenbeard genes by false greenbeards discussed in Section 15.2.

An interesting possibility is "second order" punishment, where punishment is not only applied to those who do not contribute, but also those who contribute but do not punish. Let us continue the above Example 15.3, but with repeated rounds of potential punishment until no players punish. Consider three strategies: *cooperators* that invest and punish everyone who does not play exactly as they do (but do not repeatedly punish the same aberrant individuals in subsequent punishment subrounds), *defectors* that do not invest and do not punish, *investors* that invest but do not punish. Now, fix a focal individual and assume that out of the remaining $m - 1$ individuals k are cooperators, l are defectors and the remaining $m - k - l - 1$ investors. The rewards to our focal individual in this case will be

$$
E = \begin{cases}
\frac{(m-l)cV}{m} - kP & \text{if an investor,} \\
\frac{(m-l)cV}{m} - (m - k - 1) & \text{if a cooperator,} \\
\frac{(m-l-1)cV}{m} + V - kP & \text{if a defector.}
\end{cases}
\tag{15.31}
$$

The optimal play for our focal individual is

$$
\begin{array}{ll}
\text{Defect} & \text{if } V\left(1 - \frac{c}{m}\right) > kP - (m - k - 1), \\
\text{Cooperate} & \text{otherwise.}
\end{array}
\tag{15.32}
$$

Thus defect is always stable and invest and punish is stable if $V(1 - c/m) < (m - 1)P$.

There are still some problems with how such punishment can emerge, and how it can persist in the population. Sigmund (2007) discusses how punishment can evolve, based upon either group selection or spatial arguments. Often if human players believe that their play is observed rather than anonymous, this will lead to higher levels of cooperation, as defecting may harm their reputation (see Section 15.5). An alternative recent idea is that of pooled punishment, where individuals contribute to a levy in order to facilitate the punishment of defectors, the equivalent of society setting up a police force (Sigmund et al., 2010). An advantage of this system is that second order free-riders who cooperate but do not contribute to punishment are immediately exposed even in the absence of defection, and so it is not possible for such a strategy to gradually creep into the population. Thus there is stability, although at some cost, since the punishment cost is paid even in the complete absence of defectors; the police are paid to deter crime, but must be paid even if this deterrence is completely successful.

15.4.2 General social dilemmas

There are actually many multi-player games where the kind of situation described in the previous section are considered. These games are termed the social dilemmas.

Consider a situation where there are two available strategies, Cooperate and Defect, and the payoffs to an individual who plays Cooperate or Defect in a group of $1 + c + d$ players where c of the others play Cooperate and d of he others play Defect are of the form

$$E_C(c, d) = P_C(c)u_C(c, d)V - k_C(c)K, \tag{15.33}$$
$$E_D(c, d) = P_D(c)u_D(c, d)V. \tag{15.34}$$

Here $P_*(c)$ is the production function which gives how much of a public good of basic value V is produced when our focal individual is present with c other co-operators (the number of extra defectors is irrelevant), $k_C(c)$ is a cost function which gives the proportion of the cost value K paid by the focal cooperator (a defector would pay no cost) and $u_*(c, d)$ is the proportion of the public good that the focal individual receives.

In Broom et al. (2019b) the following games, which is a compilation of appropriate public goods games used by different authors, each of this form, were considered. The public goods game defined at the start of this section in example 15.2, is equivalent to the Prisoner's Dilemma from Table 15.4.2.

As discussed in Chapter 13, Peña and Nöldeke (2016) proved results for games of the above type where specific conditions were assumed. Broom et al. (2019b) considered the games above, which mostly did not satisfy those conditions, and analysed them for variable group sizes following a negative binomial distribution, looking at the effect of changing the group size variance whilst holding the group size mean fixed.

The results differed for the different types of game considered, which could be divided into three broad types. For non-threshold public goods games, such as the (multi-player) Prisoner's Dilemma, it was found that the more variable the group size, the greater the incentive function (the payoff to a cooperator minus that to a defector), and so the more cooperation was favoured. For threshold public goods games, like the fixed stag hunt, it was found that the situation was more complicated. This depended upon the relationship between the threshold value and the mean group size, for example, as for a mean group size above the threshold value, low variability ensures that the threshold can be met, but if the mean group size is below the threshold value the opposite is true.

The third class of game is a commons dilemma, an example of which is the Hawk-Dove game, which is a different form of social dilemma to the public goods game, and does not feature in Table 15.4.2. The distinction here is that for a public good dilemma the resource is produced by the cooperating individuals, and so the addition of extra cooperators can often be beneficial,

TABLE 15.3: Summary of public goods dilemmas used in this section. For each game the values of the key payoff terms from equations (15.33) and (15.34) are given, as functions of c and d. In each case $P_D(c) = P_C(c-1)$.

	$p_C(c)$	$u_C(c,d)$	$u_D(c,d)$	$k_C(c)$
Prisoner's Dilemma	$c+1$	$\frac{1}{c+d+1}$	$\frac{1}{c+d+1}$	1
Prisoner's Dilemma with variable production function	$\sum_{k=0}^{c}\omega^k$	$\frac{1}{c+d+1}$	$\frac{1}{c+d+1}$	1
Charitable Prisoner's Dilemma	$c+1$	$\frac{c}{c+1}\frac{1_{c>0}}{c+d}$	$\frac{1_{c>0}}{c+d}$	1
Volunteer's Dilemma	1	1	1	1
Snowdrift	1	1	1	$\frac{1}{c+1}$
Stag Hunt	$(c+1)1_{c+1\geq L}$	$\frac{1}{c+d+1}$	$\frac{1}{c+d+1}$	1
Fixed Stag Hunt	$1_{c+1\geq L}$	$\frac{1}{c+d+1}$	$\frac{1}{c+d+1}$	1
Threshold Volunteer's Dilemma	$1_{c+1\geq L}$	1	1	1
Threshold Snowdrift	$1_{c+1\geq L}$	1	1	$\frac{1_{c+1<L}}{L}+\frac{1_{c+1\geq L}}{c+1}$

but for a commons dilemma there is a limited resource to be shared, and whilst being with cooperators is preferable to being with defectors, additional individuals of either type are not beneficial. Here it was found that for given reward value V, the incentive function for more variable cases starts lower and ends up higher than for less variable cases. The probability of cooperating in the ESS is higher when the group size variability is higher when V is low, but it is lower for high variability for high V; see Figure 15.3.

Finally, we note that at the start of this chapter we discussed various different factors (kin selection, direct reciprocity, indirect reciprocity, population assortment and multi-level selection) that affect the evolution of cooperation. None of these occur in the work we have discussed in this chapter, so what is it that helps cooperation evolve?

It should perhaps be noted that some of the games considered here, such as the "Prisoner's Dilemma" allow cooperators to benefit from their own contribution. This means that cooperation can even be favoured for fixed group size. The Charitable Prisoner's Dilemma, on the other hand, is more in the spirit of the two-player PD game. Here a cooperator cannot benefit from its own contribution, and it was found that Defect is the unique ESS.

Thus whilst variability in group size can play a role in the evolution of cooperation, as we have seen, it is not a new mechanism as such. It cannot lead to the evolution of cooperation on its own in the absence of the previously mentioned mechanisms, but could be a facilitator to enhance their effect in

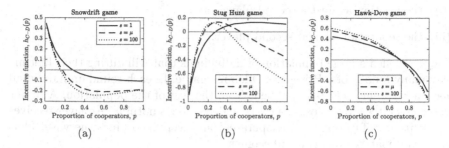

FIGURE 15.3: Incentive function $h_{C-D}(p) = E_C - E_D$ as a function of the proportion of cooperators in the population; in each case $K = 1, \mu = 5$. (a) Snowdrift game with $V = 1.5$, (b) Stag hunt game $L = 2, V = 8$, (c) Hawk Dove game $V = 1$.

particular in games which are, or resemble, non-threshold public goods games, which are the most commonly used.

15.5 Indirect reciprocity and reputation dynamics

We have seen above that in the IPD game the strategy TFT promotes cooperation by punishing defectors in subsequent contests, and the strategy CTFT does the same whilst making use of the concept of reputation. Reputation can be thought of as information about an individual's past behaviour, which can be used to inform a player how to respond in an interaction with that individual (e.g. if reputation is simply good or bad, you may choose to co-operate if and only if an individual's reputation is good). Nowak and Sigmund (1998b) considered a model of indirect reciprocity involving reputation. They showed that, in an analogue to Hamilton's rule (15.1), in their model indirect reciprocity can only promote cooperation if the probability, q, of knowing someone's reputation exceeds the cost-to-benefit ratio

$$q > \frac{c}{b}. \tag{15.35}$$

For a fuller description, see Exercise 15.13.

Nowak and Sigmund (1998a) further elaborated on the idea of individuals having a good or bad reputation, which is widely known to other members of the population, and which is affected by their actions. This idea was more fully developed by Ohtsuki and Iwasa (2004) who detailed all of the possible ways of assigning a reputation to a player given the observation of

(i) their current reputation,

(ii) their action against a second player, and

(iii) the reputation of that second player.

They modelled a large population of individuals that will during their lifetime play many rounds of a Prisoner's Dilemma game. In each round an individual plays a randomly selected opponent from the rest of the population. A player who cooperates pays a cost c, whilst a defector pays nothing. A player receives a benefit $b > c$ if its opponent cooperates, but receives nothing otherwise. Thus the payoff matrix (15.14) now becomes

$$\begin{array}{cc} & \begin{array}{cc} C & D \end{array} \\ \begin{array}{c} C \\ D \end{array} & \begin{pmatrix} b-c & -c \\ b & 0 \end{pmatrix}. \end{array} \tag{15.36}$$

In a given round when two players meet, each is aware of the reputations of both, either good (G) or bad (B). There are thus four possible pairs of reputations (listing the focal individual first): GG, GB, BG and BB.

Each individual has its own strategy $\mathbf{p} = p(i,j)_{i,j \in \{B,G\}}$ where e.g. $p(B,G) \in \{C,D\}$ specify how to play against an individual in good standing if currently the player is in bad standing. There are a total of $2^4 = 16$ strategies.

Reputation can change over time following the observations of previous contests. It is assumed that the population has a *social norm* which indicates how reputations change. This norm can assign a new reputation to an individual, either G or B, based upon the reputation pair and the action chosen. In its simplest form when only the last action is considered, there are 8 combinations of reputation pair and strategy (GGC, GGD, GBC, GBD, BGC, BGD, BBC, BBD), and so $2^8 = 256$ ways of updating individuals. Each updating method is indicated by $\mathbf{d} = \{d(i,j,k); i,j \in \{B,G\}, k \in \{C,D\}\}$. For example, $d(B,G,D) \in \{B,G\}$ will dictate the reputation of an individual who in bad standing defected against an individual in good standing.

Ohtsuki and Iwasa (2004) allowed for two types of errors in the population. These were "execution errors" where an individual tried to cooperate but was forced to defect by mistake, and "assignment errors" where observers allocated the wrong reputation (in either direction) based upon the actions undertaken.

As the reputations of individuals change, so may the strategies they find employed against them. The payoff to a player can thus be affected through a change of opponent's strategy caused by its change in reputation.

Out of $256 \times 16 = 4096$ possible pairs $(\mathbf{d}, \mathbf{p})$, Ohtsuki and Iwasa (2004) identified 25 nontrivial pairs such that under the social norm $\mathbf{d}$ a rare mutant with strategy $\mathbf{p}' \neq \mathbf{p}$ cannot invade the population (ALLD is an ESS under any norm so it was not considered). Among those 25 pairs, the frequency of cooperation ranged from 0.101 to slightly over 0.94 when $b = 2, c = 1$ and the probability of each type of error was 0.02. Eight pairs had the frequency 0.94 or larger and they termed them the *leading eight*; the remaining had frequency of cooperation below 0.84.

TABLE 15.4: The leading eight. Each of the *s can be either G or B, giving $2^3 = 8$ possibilities and ** is sometimes C and sometimes D, depending upon the particular entries replacing the *s.

d	GG	GB	BG	BB
C	G	*	G	*
D	B	G	B	*
p	C	D	C	**

The leading eight are summarised in Table 15.4 and Ohtsuki and Iwasa (2006) investigated the character of these leading eight social norms further.

There are a number of features common to all of the leading eight. These are:

- $p(GG) = C, d(GGC) = G$. These two properties ensure that (in the absence of errors) full cooperation is maintained when it exists.

- $d(GGD) = B, d(BGD) = B$. These ensure that an individual which defects against one with a good reputation, acquires or keeps a bad reputation.

- $p(GB) = D, d(GBD) = G$. There is no point in labelling such defectors bad if there are no consequences. Players with a bad reputation should therefore be defected against by "good" players. An individual which carries out the punishment above according to the rules, should maintain her good reputation.

- $p(BG) = C, d(BGC) = G$. A player with a bad reputation should still reward those with a good reputation, and in return his reputation is restored. This last point is particularly important where there are errors in the population (as described above) so that "good" players who play by the rules may acquire a bad reputation; they must be able to regain their good reputation.

The above specifies five of the eight values of the function d and the three remaining can take either value. Thus different societies can have different specific rules, as long as they have certain features in common.

For these rules to work they must be stable. Thus in a population where (almost) all abide by them no invader can do better. This essentially reduces to the advantage of having a good reputation outweighing the short-term advantage of any decision that would mean acquiring a bad reputation. The immediate benefit of defecting against a good player c must be outweighed by the price required to regain a good reputation from a bad reputation (the loss of b).

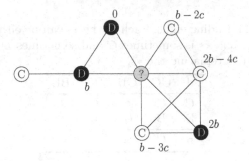

FIGURE 15.4: Payoffs for the DBB Process (see Section 12.3.2) in the PD game for the individuals competing for an empty vertex in the graph that was formerly occupied by a defector.

15.6 The evolution of cooperation on graphs

In Section 12.3 we considered evolution on graphs. Much of the work done in this area has used the Prisoner's Dilemma as a basis, using the notation of costs and benefits as in (15.36). Thus models involve graphs with individuals on vertices that are either cooperators or defectors (see e.g. Ohtsuki et al., 2006). Each cooperator will cooperate with all of its neighbours, and thus pay a cost of c times the degree of its vertex. Every individual, whether cooperator or defector, will receive a benefit for each of its cooperating neighbours; see Figure 15.4. The payoff to an individual connected to a total of k neighbours, l of which are cooperators, will thus be

$$E = \begin{cases} lb & \text{if it is a defector and} \\ lb - kc & \text{if it is a cooperator.} \end{cases} \tag{15.37}$$

Ohtsuki et al. (2006) considered the Prisoner's Dilemma game using the DBB process dynamics described in Section 12.3.2. In any scenario it appears at first sight best to defect. However, because the population is structured, any individual plays the same individuals over and over again, and just as in the IPD, cooperation can be stable if defection leads to a sufficiently high level of clustering of defectors and of cooperators. On regular graphs Ohtsuki et al. (2006) found that for a large population with weak selection a fixation probability (see Section 12.1.2) of a single cooperator is greater than $1/N$ and of a mutant defector is less than $1/N$ if

$$\frac{b}{c} > k, \tag{15.38}$$

where k is the degree of the graph. Thus the more limited the connections on the graph, the easier it is for cooperation to spread, and in particular, the well-mixed population of the complete graph is the hardest regular graph of all for the spread of cooperation.

The rule given by (15.38) can be intuitively understood as follows. Suppose that an individual has just died and a cooperator and a defector are competing for the empty site. The fitnesses f_C and f_D of the cooperator and defector will be

$$f_C = b(k-1)q_{C|C} - ck, \tag{15.39}$$

$$f_D = b(k-1)q_{C|D}, \tag{15.40}$$

where the conditional probability of finding a cooperator next to a cooperator is $q_{C|C}$ and that of finding a cooperator next to a defector is $q_{C|D}$. Cooperators are more likely to be next to cooperators than defectors are, and Ohtsuki et al. (2006) show that

$$(k-1)(q_{C|C} - q_{C|D}) = 1 \tag{15.41}$$

for the weak selection case. This means that $f_C - f_D = b - ck$ which yields (15.38).

Ohtsuki et al. (2006) also observed that their simple rule does not fit that well for scale-free networks (Barabási and Albert, 1999), presumably due to their large variance of degree distribution. Nevertheless, scale-free networks are powerful promoters of cooperation (Santos and Pacheco, 2005; Santos et al., 2006b; Santos and Pacheco, 2006).

The evolution of cooperation on graphs has received a significant amount of attention since the work of Ohtsuki et al. (2006). For instance Santos et al. (2006a) consider the evolution of cooperation on heterogeneous graphs, and Voelkl and Kasper (2009) consider the emergence of cooperation among primates using a graph-based model. A good review of work (published prior to the review) on the evolution of cooperation on graphs, as well as evolutionary graph theory and games on graphs in general, is given by Shakarian et al. (2012).

15.7   Multi-level selection

We introduced the idea of group selection in Chapter 1 and discussed different levels of selection in Chapter 5. Traulsen and Nowak (2006) introduced a model of multi-level selection, where a population consists of k groups, each of which consists of at most m individuals; see also Taylor and Nowak (2009), who we shall now follow.

Interactions happen within groups, each individual playing the Prisoner's Dilemma against a random opponent to ascertain their fitnesses, which is given by $1 - w + wE$, where E is the payoff from the game, and w is the intensity of selection as in Section 12.2. At any time point an individual from the whole population is selected with probability proportional to its fitness to reproduce, its offspring joining the same group. Thus groups with fitter individuals will grow more quickly than others. When any group reaches maximum size, it splits to form two groups with probability p. Whenever a new group is formed in this way, a group, selected at random, dies. If the group does not split, then a random individual within the group is selected to die, so that the group size will remain at $m - 1$. There is thus selection within groups but also between groups, as groups with fitter individuals increase in size and can split more often.

For p sufficiently small, evolution happens within groups faster than between groups so that when a single mutant is introduced into the population, its probability of fixation is simply the probability of it fixing within the group, multiplied by the probability of a single group completely comprised of mutants fixing in a population of groups completely comprised of resident individuals.

Taylor and Nowak (2009) show that, under weak selection, the above model effectively simplifies to consideration of a matrix game with the following payoff matrix:

$$
\begin{array}{cc}
 & \begin{array}{cc} C & \qquad D \end{array} \\
\begin{array}{c} C \\ D \end{array} & \left(\begin{array}{cc} (m+k)R & mS + kR \\ mT + kP & (m+k)P \end{array} \right).
\end{array} \tag{15.42}
$$

Thus Cooperate is an ESS if

$$
\frac{k}{m+k} > \frac{T-R}{T-P}. \tag{15.43}
$$

A wider theory of multilevel selection has been developed, and in general the application of this methodology is increasingly common, including in the modelling of human behaviour. Beyond the simple example above, and the earlier discussion in Chapter 5, consideration of this is beyond the scope of this book. A recent book which discusses a lot of these ideas in detail is Hertler et al. (2020).

15.8   Python code

In this section we show how to use Python to implement strategy CTFT for the IPD game. It could then be used in conjunction with the program in Chapter 4.

```
1  """ This script provides class definition for CTFT strategy
2  Import this definition into the code for IPD tournament in ch4
3  Around line 70 of file ch4 put: from ch15 import CTFT
```

```
4    and update a list of participating strategies accordingly
5    """
6
7    # Import class Strategy and function IPDgame
8    from IPD import Strategy, IPDgame, Defect, Cooperate
9
10   # Auxiliary variable
11   Good = 1    # good standing
12   Bad = 2     # bad standing
13
14   class CTFT(Strategy):
15       name = "CTFT"
16
17       MyStanding = Good    # Initially good standing
18       OppStanding = Good   # Initially good standing
19
20       def Intended(self, opponent):
21         if len(self.History)==0:
22               # Cooperate on the first move
23               return Cooperate
24         else:
25           # If not first move, get my last move
26           myLast = self.History[-1]
27
28           # Get opponent's last move
29           oppLast = opponent.History[-1]
30
31           # Update my standings
32           # Self is in good standing only if in the last round
33           # (1) cooperated and opponent was in good standing; or
34           # (2) cooperated and it was in bad standing; or
35           # (3) defected and was in good standing, while opponent
36           #         was in bad standing
37           # Otherwise, self is in bad standing
38           if ((myLast == Cooperate)and(self.OppStanding == Good))\
39               or((myLast == Cooperate)and(self.MyStanding == Bad))\
40                 or((myLast == Defect) and (self.OppStanding == Bad)):
41               myNewStanding = Good
42           else:
43               myNewStanding = Bad
44
45           # Update opponent's standing
46           if ((oppLast == Cooperate)and(self.MyStanding == ...
                  Good)) \
47               or((oppLast == Cooperate)and(self.OppStanding ==Bad))\
48                 or((oppLast == Defect) and (self.MyStanding == Bad)):
49               oppNewStanding = Good
50           else:
51               oppNewStanding = Bad
52
53           # Record my new standing
54           self.MyStanding = myNewStanding
55
56           # Record opponent's new standing
57           self.OppStanding = oppNewStanding
58
59           # Return intended move
```

```
60      # Defect only if self in good and opp in bad standing
61      if (self.MyStanding == Good) and (self.OppStanding==Bad):
62          return Defect
63      else:
64          return Cooperate
```

15.9　Further reading

The special issue "Evolution of cooperation" was published by the *Journal of Theoretical Biology*, Vol. 299, 2012; see Nowak (2012). Perc and Szolnoki (2010) is a review on how the evolution of cooperation is affected by coevolutionary rules. For reviews on the evolution of cooperation see Sachs et al. (2004), West et al. (2007a), Marshall (2011). Nowak et al. (2010) is a (rather controversial) paper on kin selection. The original work on reciprocal altruism is Trivers (1971). Fletcher and Zwick (2006) show that inclusive fitness and reciprocal altruism have the same underlying mechanism (see also Gardner and West (2010)). Taylor et al. (2007) consider the use of direct fitness versus inclusive fitness when studying kin selection.

Doebeli and Hauert (2005) is a review on models of cooperation based on the Prisoner's Dilemma and the Snowdrift game. Fu et al. (2010) study the difference between spatial PD and the Snowdrift game. For studies on the evolution of cooperation in spatially heterogeneous populations see for example Hutson and Vickers (1995); Ferriere and Michod (1996). For a lot more on the IPD see Sigmund (2010), Nowak et al. (1995) and Leimar (1997). For work on important IPD strategies see Kraines and Kraines (1989), Neill (2001), Wakano and Yamamura (2001) (WSLS/Pavlov), Boerlijst et al. (1997) (CTFT). Nowak and Sigmund (2005) is a review on the evolution of indirect reciprocity (see also Leimar and Hammerstein (2001), Panchanathan and Boyd (2003), Milinski et al. (2001)). Schmid et al. (2021) shows unexpected connections between direct and indirect reciprocity while highlighting important differences regarding their evolvability.

Grafen (2007) and Lehmann et al. (2007) study inclusive fitness on networks. Shakarian et al. (2012) is a good review of evolution on graphs. Tarnita et al. (2009a) studies evolution of cooperation in set structured populations. Brännström et al. (2010) investigates the consequences of fluctuating group size for the evolution of cooperation.

For a review of cooperation amongst animals see Dugatkin (1997a); see Voelkl and Kasper (2009) who investigate cooperation in networks of primates.

Sigmund (2007) gives a good review of experiments and explanations underlying the idea of punishment. Hauert et al. (2007) propose a method to have punishment using volunteering. Rand and Nowak (2011) cast some doubt on this, yet it may be resolved by reputation (Hilbe and Traulsen, 2012) or by more selective punishment (García and Traulsen, 2012). Punishment is also considered in Wu et al. (2009a); Zhang et al. (2014) and more recently by Liu et al. (2017, 2018) among others.

Other mechanisms can help cooperation to evolve. Heterogeneity and variation of behaviour by itself was shown to promote cooperation in the IPD game (McNamara et al., 2004). The role of diversity has been investigated in Santos et al. (2012); Mesterton-Gibbons and Sherratt (2011). If individuals are allowed to leave the group (i.e. play Cooperate, Defect or Leave) cooperation can be fostered (Joyce et al., 2006; see also Eshel, 1982) and Zhang et al. (2016), similarly if individuals can cut short a repeated game (Aktipis (2004) considered the strategy *Walk away* which cooperates, but stops the interaction after the first defection; see also Izquierdo et al. (2010)). The option of leaving the game if mutual defection occurs can also reinforce the use of punishment (Fowler, 2005). Sasaki and Uchida (2013) considered the evolution of cooperation by social exclusion. The evolution of cooperation in mobile populations was studied in Erovenko (2019); Erovenko and Rychtář (2016).

Riolo et al. (2001) show that signalling can promote cooperation, Santos and Pacheco (2005); Santos et al. (2006b) and Fletcher and Doebeli (2009) show that preferential interactions between cooperators also promotes cooperation. A multi-scale eco-evolutionary model of cooperation is introduced in Abs et al. (2020).

15.10   Exercises

Exercise 15.1 (Gardner and West, 2010). Show that by defining the relatedness of A_1 to A_1 as $r = 1$ and the relatedness of A_2 to A_1 as $r = -\frac{p}{1-p}$, where p is the proportion of A_1 in the population, we can recover the conclusions of Table 15.1 from Hamilton's rule (15.1).

Hint. For example, consider an obligate-harming greenbeard. In every greenbeard-somebody interaction, the greenbeard pays the cost $C = a$. In a fraction p of all greenbeard-somebody interactions, the other individual is also a greenbeard (and thus receives no harm). In the remaining fraction $(1 - p)$, the other individual is a non-greenbeard, thus receiving harm d. The average benefit to the recipient is $B = -(1 - p)d$, while the cost is $C = a$.

Exercise 15.2. Describe the contests between the pairs of strategies ALLD, GRIM, TFT, STFT and TF2T of the IPD (apart from those given in example 15.1) and show that the payoff matrix is indeed given by (15.17).

Exercise 15.3. Show that for sufficiently large w, TF2T performs better than TFT against STFT (see Section 15.3.1).

Exercise 15.4. Consider the strategy WSLS and add the payoffs of this strategy (and against this strategy) into the matrix (15.17).

Exercise 15.5. Consider the strategies ALLD-e, WSLS-e and Grudge-e introduced in Section 15.3.3. Describe the contests between these strategies in the IPD.

Exercise 15.6 (Molander, 1985). Study the game of two TFT players in the IPD game where the probability of error is ε. Assume that ε is small enough so that effectively at most one player makes a mistake at any given time, and also assume w is large enough so that there are many mistakes in any game. Evaluate the payoffs to the TFT players.

Hint. See Figure 15.1 to establish that payoffs in the game are the same as if the two individuals were choosing the strategy at random.

Exercise 15.7 (Press and Dyson, 2012). Show that given (15.26), we obtain (15.27) assuming that α, β and γ satisfy $\gamma = -(\alpha + \beta)P$ and $-\beta/\alpha > 1$.

Exercise 15.8. For the Prisoner's dilemma game from Section 15.3.5, assume that the payoffs are $R = 3, T = 5, S = 0, P = 1$. Find all of the strategies (p_R, p_T, p_S, p_P) that ensure that payoffs satisfy (15.27) where $\chi = 2$.

Exercise 15.9. Give a multi-player payoff matrix of the Public Goods Game from Example 15.2 and show that not investing is an ESS.

Hint. See Section 9.1.1.

Exercise 15.10. Analyse the Public Goods Game with Punishment from Example 15.3 and show that not investing and not punishing is the only ESS.

Exercise 15.11. Give detailed analysis of the Public Goods Game with second order punishment as stated in Section 15.4.1.

Exercise 15.12. Consider a DBB process and a Prisoner's Dilemma game with a payoff matrix (15.36) on a circle graph with n vertices that are inhabited by $n - 1$ cooperators and a single defector.

1. Find the probability that a defector will be replaced by a cooperator.

2. Find the probability that a cooperator will be replaced by a defector.

Exercise 15.13 (Nowak and Sigmund, 1998b). Consider a single-round Prisoner's Dilemma game with payoff matrix (15.14) and a population of two types of individuals ALLD and conditional cooperators (CC) who cooperate unless they know the reputation of the other player is as a defector. Let q denote the probability of knowing the reputation of another individual. Show that we obtain the payoff matrix

$$
\begin{array}{cc}
 & \begin{array}{cc} CC & \quad\quad ALLD \end{array} \\
\begin{array}{c} CC \\ ALLD \end{array} &
\begin{pmatrix} R & (1-q)S + qP \\ (1-q)T + qP & P \end{pmatrix}
\end{array} \tag{15.44}
$$

and find its ESSs.

Exercise 15.14. Explain the difference between strategies TFT and CTFT.

Hint. They are not the same although both defect after they are provoked. You may want to distinguish between an error-free and an error-prone environment in your analysis.

Chapter 16

Group living

16.1 The costs and benefits of group living

Many animals spend substantial amounts of time in groups. By this we mean that they spend a significant period of time living in close proximity with animals of the same species. For a more detailed discussion of what and what does not constitute an animal group, the reader is referred to *Living in Groups*, Krause and Ruxton (2002).

There are a number of reasons why animals may choose to spend time in groups. Krause and Ruxton (2002) list seven distinct benefits.

1. *Anti-predator vigilance* is one of the most important benefits. If many individuals can watch out for predators, it allows group-mates to spend more time foraging without exposing themselves to great risk. We look at this phenomenon in detail in Section 16.3.

2. *The dilution effect* is the consequence of the fact that even if no individual from the group sees a predator before the attack, then assuming that it can kill only one member of the group, the larger the group, the smaller the chance that any particular individual will be killed. This is a key component of the models in Section 16.3.

3. *Predator confusion.* When attacking large groups which move together, predators often find it hard to single out an individual to attack. This phenomenon occurs in large fish shoals or large flocks of birds that are facing a predator, see e.g. Neill and Cullen (1974).

4. *Foraging benefits.* Groups of predators such as lions may be able to tackle larger prey than single individuals, or may be able to hunt so that fleeing prey are chased towards other group members. Alternatively individuals may forage for food in different directions, and when food is found this can be communicated to the whole group.

5. *Finding a mate.* Some animals may spend much of their lives separately, but come together to mate, for example in leks (Höglund and Alatalo, 1995, Chapter 5). We consider this case when modelling the formation of dominance hierarchies in Section 16.2.

DOI: 10.1201/9781003024682-16

6. *Conserving heat,* for example by huddling.

7. *Conserving energy.* Animals can save energy by moving in groups and reducing air or water resistance, e.g. a skein of geese in flight or a shoal of fish.

Weighing against these benefits, there are a number of costs to living in groups.

1. *Increased attack rate.* Although the dilution effect reduces the chance of an individual being killed in any given attack, large groups will attract predators, so that the rate of attack is likely to increase.

2. *Kleptoparasitism* and interference by conspecifics more generally (see Chapter 19).

3. *Reduction in the local food supply* caused by the foraging of the group itself.

4. *Increased rate of aggression.* We see later in Section 16.2 the importance of dominance hierarchies in some situations, and they typically involve some aggression both in their formation and their maintenance. Very damaging aggressive behaviour in the form of infanticide can also be a problem in some groups, see e.g. Borries and Koenig (2000); Pusey and Packer (1994).

Thus there are significant costs as well as benefits to group formation, and the relative size of each is important to whether group formation is worthwhile, and if it is, how large the group should be. In this chapter we will not look at all possible effects of animal grouping, or even all amenable to mathematical analysis. For instance the investigation of the flocking behaviour leading to predator confusion that we have described above has received much recent attention using mathematical and computational models (see e.g. Sumpter, 2006; Ballerini et al., 2008; Sumpter, 2010; Hildenbrandt et al., 2010). Rather, we shall look at a small number of important examples amenable to modelling using game theory, and talk about some (but by no means all) of the important models only.

16.2   Dominance hierarchies: formation and maintenance

16.2.1   Stability and maintenance of dominance hierarchies

When groups are formed, social interactions occur which often lead to highly structured dominance relationships. Linear dominance hierarchies, where A dominates all, B dominates all but A, C dominates all but A and B etc. are common; see Figure 16.1. There are advantages to the group to be structured in this way, one being that food can be allocated within the group without costly fights.

This benefit is not evenly shared, however, and sometimes it may be beneficial to an individual occupying a low position to disrupt the hierarchy if she can, or even leave. Such hierarchies conceal a range of possible contests which could occur between their members. For any such contest, the individuals must balance the potential benefits such as an increase in rank against costs such as injuries. Stability will occur when lower ranked individuals are rarely willing to challenge those of higher rank. Linear hierarchies are often stable; experiments with fowl (Klopfer, 1973) show that if an individual is removed from a hierarchy, it often automatically returns to its previous position without conflict if it is subsequently reintroduced.

Starting with Vehrencamp (1983), important theoretical work has considered *reproductive skew*, the level of inequality in resource allocation within groups. The higher the reproductive skew, the greater the advantage of occupying the top position(s) in the hierarchy. In classical reproductive skew models, choices are made according to Hamilton's rule (see also Section 15.1; Hamilton, 1964). The choice i is favoured over choice j if

$$B_{1,i} - B_{1,j} + r(B_{2,i} - B_{2,j}) > 0, \tag{16.1}$$

where $B_{1,i}$ is the reward to the focal individual if it makes choice i, $B_{2,i}$ is the reward to the other individual and r is the relatedness of the two individuals.

Example 16.1 (Dominant-subordinate hierarchy contest, Reeve and Ratnieks, 1993). Consider the colony of a social insect that has two queens. The dominant queen takes most resources and gives only a little to the subordinate one. Should the subordinate queen stay, leave or fight to replace the dominant one (the loser of such a contest will be killed)? Also, how much should the dominant queen give to the subordinate?

We will now create and analyse a model from Example 16.1, following Reeve and Ratnieks (1993). After normalising, we may assume that the expected reproductive output of a single queen within the colony is 1. Let x be the reproductive output of a solitary queen which has to set up a new colony, k be the combined reproductive output of the pair and p be the fraction of the output of the pair belonging to the subordinate. Let f be the probability that the subordinate would win a fight. The loser in a fight is assumed to be killed, so the winner of the fight will be a single queen within the colony. The two queens play a sequential game where the dominant queen offers a fraction p to the subordinate, which then has the choice to accept the offer (stay), to fight or to leave. This is a type of extensive form game where the dominant has a continuum of possible actions prior to the choice of the subordinate.

TABLE 16.1: The evolutionarily stable values of p for Example 16.1.

ESS value	Conditions of model parameters
p_s	$r(k-1) < x < k-1$ and $f < \frac{x}{1-r}$
p_p	either $\left\{ \ x < r(k-1) \text{ and } f \geq \frac{r(k-1)}{1-r} \ \right\}$
	or $\left\{ \ r(k-1) < x \text{ and } f \geq \frac{x}{1-r} \ \right\}$
0	$x < r(k-1)$ and $f < \frac{r(k-1)}{1-r}$
Pair split	$x > k-1$ and $f < \frac{x}{1-r}$

The subordinate has expected inclusive fitness of

$$E = \begin{cases} k(r + p(1-r)) & \text{if it stays,} \\ x + r & \text{it it leaves,} \\ f + (1-f)r & \text{if it fights.} \end{cases} \qquad (16.2)$$

We shall assume that if the payoffs for staying and leaving, or staying and fighting, are identical, then the subordinate will choose to stay (for the case where the reverse is true, see Exercise 16.3). Comparing the expected rewards and using (16.1), we get that staying is favourable to leaving if $p \geq p_s$, and staying (and cooperating) is favoured over fighting if $p \geq p_p$ where the *staying incentive*, p_s, and the *peace incentive*, p_p are given by

$$p_s = \frac{x - r(k-1)}{k(1-r)}, \qquad (16.3)$$

$$p_p = \frac{f(1-r) - r(k-1)}{k(1-r)}. \qquad (16.4)$$

Naturally, the dominant queen tries to give a fraction p that would maximize her own inclusive fitness. The evolutionarily stable values of p allowed to the subordinate turns out to be the minimum required to prevent the subordinate from either leaving or fighting and are given in Table 16.1. Note that sometimes an individual will stay in the group and help its relative even if it has no direct reward at all, providing r is sufficiently large. Also note that for some parameters, there is no rational strategy p which can keep the group together, and so the subordinate leaves. This occurs when the stay incentive is so large, that the dominant would prefer the subordinate to leave, i.e. $p_s > p_c$, where p_c is the critical value of p where the dominant would have equal payoff whether the subordinate left or stayed (it is never the case that the peace incentive is so large that the dominant would prefer the subordinate to fight),

$$p_c = \frac{k - 1 - xr}{k(1-r)}. \qquad (16.5)$$

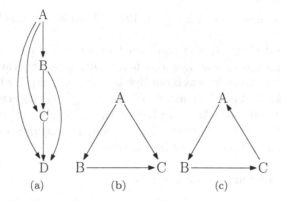

FIGURE 16.1: Dominance hierarchies; (a) linear dominance hierarchy, (b) transitive triad, (c) intransitive triad. An arrow from A to B means that A dominates B.

Situations such as in Example 16.1 are observed in different species of ant such as the Argentine ant, *Linepithema humile*, with colonies involving multiple queens, and the predicted relationships between levels of skew and relatedness, that come out of the models (the higher the relatedness, the higher the skew) is observed in reality (Fournier and Keller, 2001); however some limitations have been observed (see e.g. Nonacs et al., 2006; Liebert and Starks, 2006).

16.2.2   Dominance hierarchy formation

In populations with stable groups, a new individual will often join an existing group following consistent rules, such as a daughter taking the position below its mother. If a group of individuals all meet for the first time, however, there must be a way for them to form a hierarchy from scratch. Important features of any model of this situation include the nature of an individual conflict, the pattern of pairing of individuals to fight (assuming that the hierarchy takes shape out of a series of such contests) and the effect of a given outcome of a fight between two particular individuals on both their dominance and on which conflicts subsequently occur. It is often useful to consider sub-groups of a larger hierarchy, in particular sub-groups of size three, called triads. For a hierarchy to be linear all triads must be transitive (A beats B,C; B beats C) and the number of non-transitive triads is a measure of the distance a hierarchy is from linearity; see Figure 16.1.

Chase and co-workers have carried out important experimental work on hierarchy formation. Chase and Rohwer (1987) considered flocks of Harris' sparrows, and found more triads (95%) were transitive than would be expected by completely random pairwise dominance relations (75%); see Exercises 16.4 and 16.5. In fact, observations of other species, for example chickens (Chase,

1980) and rhesus monkeys (Missakian, 1972), have shown much higher rates of linearity.

Lindquist and Chase (2009) conducted experiments with dominance hierarchies with groups of four Leghorn hens. Dominant individuals generally emerged early on, and there was a relative lack of "pair-flips" where A attacks B, and B attacks A. The phenomenon of "bursting" was observed, where one individual repeatedly attacked another, with the likely aim of establishing dominance over it, and it appears that dominance emerges from a concentrated and purposeful sequence of interactions.

16.2.2.1 Winner and loser models

An important type of model of dominance hierarchy formation is the *winner-loser model*, such as the original models of Landau (1951a,b) and those of Dugatkin (1997b) and Bonabeau et al. (1999). The *winner effect* occurs if winning a contest increases an individual's chances of winning a subsequent contest. The *loser effect* occurs if losing a contest increases an individual's chances of losing a subsequent contest.

The model of Dugatkin (1997b), (see also Dugatkin and Dugatkin, 2007) is a simulation-based model which allows clear predictions regarding the influence of winner and loser effects on dominance, as well as making the model amenable to experimental comparison. The model has only a small number of parameters m, Φ, W, L and RHP_{other} where m is the number of individuals in a group, $\Phi \geq 0$ is the *aggression threshold*, $W \geq 0$ is the winner effect and $L \in [0, 1]$ is the loser effect. In the model individuals decide whether or not to fight based on their own as well as their opponent's *resource holding potential, RHP* (see also Chapter 8). The real value of the RHP depends on an animal's history (the results of previous contests) and by $RHP_{x,T}$ we will denote the RHP of an individual x at time t. We assume that any individual knows accurately its own RHP, but can only estimate the RHP of its opponents; it has no information about a given opponent and simply uses the same estimate for all opponents, RHP_{other}. It is assumed that this estimate is also the same for all individuals x making the estimate, and that this estimate does not depend on time nor on history. This assumption may be reasonable given that any assessment is likely to be based strictly on an individual's size and we are particularly interested in hierarchy formation among individuals that are approximately equal.

Individual x chooses to fight with individual y at time t if

$$\frac{RHP_{x,t}}{RHP_{other}} \geq \Phi, \tag{16.6}$$

and chooses to retreat otherwise. When x meets y, then due to the possible inaccuracy of their opponent's RHP assessment, the following outcomes are all possible:

1. both individuals choose to fight,

2. precisely one chooses to fight, or

3. both choose to retreat.

If both choose to fight, x wins with probability

$$\frac{RHP_{x,t}}{RHP_{x,t} + RHP_{y,t}}. \tag{16.7}$$

At every time step, RHP is updated as follows.

$$RHP_{x,t+1} = \begin{cases} (1+W)RHP_{x,t} & \text{if } x \text{ chose to fight and} \\ & \text{either } y \text{ retreated or } x \text{ won,} \\ (1-L)RHP_{x,t} & \text{if } y \text{ chose to fight and} \\ & \text{either } x \text{ retreated or } y \text{ won,} \\ RHP_{x,t} & \text{if both } x \text{ and } y \text{ retreated.} \end{cases} \tag{16.8}$$

Populations of various sizes were simulated with varying values of W and L, including those where one of the effects was completely absent. The results were different for the models containing winner and loser effects, just winner effects or just loser effects. Clear hierarchies resulted from winner effects alone, whereas for loser effects alone the hierarchy below the top-ranked α individual was less distinct.

The temporal dynamics of dominance hierarchy formation for different combinations of winner and loser effects for the above model was analysed in Kura et al. (2015), and it was shown that a linear hierarchy could be established with only a small amount of knowledge of an opponent's RHP. This model was further developed to include a game theoretical element in Kura et al. (2016). Here, as opposed to Dugatkin (1997b); Dugatkin and Dugatkin (2007); Kura et al. (2015), individuals could have different levels of aggression, and this was a strategic parameter, where each individual could choose its own strategy, independent of their opponent's. They found that a unique pure strategy evolved, as opposed to a mixture of strategies, and that here the hierarchy forms quickly, after which no mutually aggressive contests occur.

Plausible population structures within a group depend upon the size of the group. Mesterton-Gibbons and Dugatkin (1995) considered a model of small groups playing an all play all contest, with individuals being of different strength (having different RHP). When individuals could assess the RHP of their opponents, weaker individuals could give up without a fight and so avoid costs. Thus the outcome of some contests was certain. It was found in simulations that a linear hierarchy can be formed with high probability for a group of size of 7 or 8, providing that there was pre-contest assessment of RHP. Without such assessment, there will be more fights, and linear hierarchies only formed with high probability for very small groups (3 or 4).

Drummond and Canales (1998) found strong winner and loser effects among chicks of the blue-footed booby *Sula nebouxii*, when previous winners and previous losers were paired with individuals who had not previously participated in contests. Interestingly, the winner effects decreased with the passage of time, but the loser effects remained intact throughout the whole ten days of the experiment. Loser effects often dominate winner effects in nature, especially if, as indicated above, they endure for a longer time.

16.2.3 Swiss tournaments

Example 16.2 (Black grouse mating, Höglund and Alatalo, 1995, Ch. 5). On average up to 25 male black grouse *Tetrao tetrix* come together and display in mating arenas called leks. Females visit the lek, choose the male for mating, mate, and leave the lek (to eventually incubate the eggs and attend to the young completely by themselves). Males on the other hand keep displaying and waiting for other females to come. However, not all places in the lek are of equal value; males in a more central position tend to mate with more females and males thus initially fight in order to get to the best possible position. As individuals compete, winners move to the more advantageous centre ground, so that winners will tend to encounter other winners. What is the male's best strategy?

Broom et al. (2000a,c) modelled the above dominance hierarchy formation as a knockout contest. Individuals are paired at random, initially with $m = 2^M$ identical individuals. The important top two positions, occupied by the overall winner and losing finalist, are thus established quickly with relatively little conflict, with the total number of contests being the size of the group minus one. Broom and Cannings (2002) developed this model further as follows.

Example 16.3 (Black grouse mating as a Swiss tournament, Broom and Cannings, 2002). Let there be $m = 2^M$ players of identical strength that engage in M round pairwise contests. Every contest is a generalised Hawk-Dove game so that it ends in a win for one of the players and loss for the other (no draws). A Hawk beats a Dove and if two Hawks or two Doves meet, a winner is determined at random; if two Hawks meet, the loser has to pay a cost C. Within each round, the contests are only between individuals that have won the same number of contests so far (for example, in the second round, a winner from the first round can engage in a contest only with another winner). Let V_i be the reward for winning i times from the M rounds, with $V_0 \leq V_1 \leq \ldots \leq V_M$. What is the optimal strategy?

The above is similar to a game in extensive form; see Section 10.1. The overall strategy for a player is a choice of strategy for every possible position, i.e. $\mathbf{p} = (p_{i,j})$ where $p_{i,j}$ for $j = 0, \ldots, M - 1$ and $i = 0, \ldots, j$ is an individual's probability to play Hawk in a pairwise contest if the individual has so far won i out of j contests. Thus analysis starts with the last round and works backwards (backwards induction, see Section 10.1.2).

Let W_{ij} be the expected reward to a player with a score of i at the end of round j. We know that

$$W_{i,M} = V_i \quad \text{for all } i = 0, \ldots, M. \tag{16.9}$$

Once $W_{i,j+1}$ is known, we can define the following matrix game

$$A_{ij} = \begin{pmatrix} \frac{1}{2}(W_{i+1,j+1} + W_{i,j+1} - C) & W_{i+1,j+1} \\ W_{i,j+1} & \frac{1}{2}(W_{i+1,j+1} + W_{i,j+1}) \end{pmatrix} \tag{16.10}$$

which captures the situation at the onset of the conflict at round j. The matrix game with payoffs A_{ij} has a unique ESS p_{ij} (see Exercise 16.6) given by

$$p_{ij} = \min\left(1, \frac{W_{i+1,j+1} - W_{i,j+1}}{C}\right). \tag{16.11}$$

This, in turn, gives

$$W_{i,j} = \frac{1}{2}\left(W_{i,j+1} + W_{i+1,j+1} - Cp_{ij}^2\right). \tag{16.12}$$

The above procedure yields a unique $\mathbf{p} = (p_{ij})$. If there is an ESS, it must be of this form but there may be no ESS under some conditions (although the above $\mathbf{p}$ does have some stability properties short of being an ESS).

Example 16.4. Consider the Swiss tournament game described above with $m = 4$ players and so $M = 2$ rounds, $V_0 = 0, V_1 = 1, V_2 = 4$ and $C = 2$. Find the candidate ESS $\mathbf{p}$.

We need to find p_{11}, p_{01} and p_{00}. We start at the second round. By (16.11), we get

$$p_{11} = \min\left(1, \frac{4-1}{2}\right) = 1, \tag{16.13}$$

$$p_{01} = \min\left(1, \frac{1-0}{2}\right) = \frac{1}{2}. \tag{16.14}$$

These yield

$$W_{1,1} = \frac{3}{2}, W_{0,1} = \frac{1}{4}. \tag{16.15}$$

Thus

$$p_{00} = \min\left(1, \frac{3/2 - 1/4}{2}\right) = \frac{5}{8}. \tag{16.16}$$

It turns out, see Exercise 16.8, that contests between dominant individuals (those that have high scores) tend to be more violent contests of the Hawk-Hawk type, those between subordinates are less so, with early-stage contests intermediate. As differences in rewards increase, the overall level of violence increases on a sliding scale. Defeat leads an individual to be more passive, whereas victory does not necessarily make it more aggressive. Thus loser effects are emergent from the model, but winner effects are not, which is consistent with real behaviour as in Earley and Dugatkin (2002).

Note that a highly uneven division of rewards here can be thought of as high reproductive skew. In the model of Reeve and Ratnieks (1993) there is a threshold in the level of reproductive skew where it is either best to be aggressive or not (or to leave or not), whereas in the Broom and Cannings (2002) model there is a sliding scale of aggressiveness depending upon the size of the rewards available and how they are divided amongst the population.

16.3   The enemy without: responses to predators

In this section we consider the effect of group living in response to predator attack, concentrating on two key features that we mentioned above, antipredator vigilance and the dilution effect. We will follow the work of McNamara and Houston (1992) and for simplicity ignore the potentially important phenomenon of confusion (see Krakauer, 1995).

Example 16.5 (Foraging and scanning, McNamara and Houston, 1992). Consider a group of individuals that have to forage at a place where there is a risk of predation. Foraging may be terminated at any time due to, for example, bad weather. An individual has to forage for a sufficient time to accumulate enough energy (for example to survive a night, until food is available again) and ultimately to reproduce. However, to achieve that, it must avoid being killed by a predator and it thus has to divide its time between foraging and watching for a predator. What is the optimal proportion of time, u, the individual should forage?

16.3.1   Setting up the game

The scenario described above in Example 16.5 is perfect for game-theoretical modelling due to the following conflict. Every individual would like to forage as much as possible and let other individuals bear the cost of watching out for predators. However, if every individual did just that, nobody would scan for predators. The situation is thus similar to the Prisoner's Dilemma or the tragedy of the commons game. We note that the original model was more complex than the version that we give here. In particular, we do not allow premature termination to foraging, and the distinction between expected future reproductive success (EFRS) when foraging and after it has been terminated can be important. We simply allow the individuals as much time to forage as they wish.

Let there be a group of m foragers, each of which divides its time between foraging and being vigilant for predators. Each animal has a state, e.g. energy reserves of x at any time, and has an EFRS of $R(x)$ if in state x. The function $R(x)$ can be quite general but it is natural to assume that it is increasing in x.

Each animal acquires food at rate γ per unit time spent foraging, i.e. when foraging, the increase of state variable x per unit of time is γ, i.e. $\frac{dx}{dt} = \gamma$. When foraging only for a fraction of the time u, we get $\frac{dx}{dt} = \gamma u$. Attacks by predators will be modelled as a Poisson process; let α be the rate of these attacks.

16.3.1.1 Modelling scanning for predators

In this model if animals engage solely in foraging, they fail to spot predators and can consequently be killed by them. Thus they must potentially assign some proportion of their time to looking out for predators. Let $g(u)$ be the probability an individual fails to spot the predator when foraging for a proportion of time u. The function $g(u)$ should be increasing and concave (e.g. doubling the amount of time spent searching for a predator should increase the chance of spotting it, but not double it). An example of such a function was derived in Pulliam et al. (1982); see also Exercise 16.10.

Individuals spot or fail to spot the predator independently of each other. If no animal detects the predator then a focal individual is killed with probability A_m. Quite often A_m is taken to be $1/m$, i.e. exactly one animal is killed, and which it is selected with equal probability. If the focal animal fails to spot the predator but another does then it dies with probability B_m and if it spots the predator it dies with probability C_m. Assume that $A_m > B_m \geq C_m \geq 0$ (it may be that $B_m = C_m = 0$, so that if any individual spots the predator, they all escape).

16.3.1.2 Payoffs

As usual, we are interested in finding ESSs for this game. Assume that a focal individual plays vigilance strategy u in a population when all others play v. Given v, what is the optimal choice of u? In fact, to find an ESS, we do not need such a general question, we only need to know when u^* is an optimal choice in a population which also plays u^*. This will depend upon the parameters of the population, potentially including its state x.

Let us compare benefits to costs. We note here that this is done to simplify the analysis, and that such an approach is rather against the spirit of the methodology of McNamara and Houston (1992) (see Chapter 11 for more on this), and only works because we neglect the termination of foraging, as we mention above. When an individual forages, the benefit is that its state x increases, i.e.

$$B(u) = \frac{dR(x)}{dt} = \frac{dR(x)}{dx} \cdot \frac{dx}{dt} = R'(x)\gamma u \qquad (16.17)$$

The cost is that the individual can die and risk all of the EFRS it has accumulated so far, i.e.

$$C(u) = \alpha D_v(u) R(x), \qquad (16.18)$$

where

$$D_v(u) = A_n g^{n-1}(v) g(u) + B_n \big(1 - g^{n-1}(v)\big) g(u) + C_n \big(1 - g(u)\big) \qquad (16.19)$$

is the probability that it dies in an attack.

What do we have to optimise? As animals have unlimited time to forage, but death will send their reward to zero, the term that has to be maximised is benefits divided by costs, i.e. to maximise the energy gained for a given risk of death. Thus the optimal choice of u given v is found by maximising

$$\mathcal{E}[u; \delta_v] = \frac{R'(x)\gamma u}{\alpha D_v(u) R(x)}. \qquad (16.20)$$

This payoff function is nonlinear in both the focal player strategy and the population strategy (see Chapter 7).

16.3.2 Analysis of the game

For u^* to be an ESS we need it to be a best reply to itself, so that

$$\mathcal{E}[u^*, \delta_{u^*}] = \max_u \mathcal{E}[u, \delta_{u^*}]. \qquad (16.21)$$

The function g is increasing and concave. This implies that there is a unique best response against any strategy, and in particular, if (16.21) is satisfied, then u^* must be an ESS. Maximising $\mathcal{E}[u, \delta_{u^*}]$ is the same as minimising its reciprocal. We thus consider the derivative

$$\frac{d}{du}\left(\frac{1}{\mathcal{E}[u, \delta_{u^*}]}\right) = \frac{\alpha R(x)}{R'(x)\gamma u^2}\left(u\frac{dD_{u^*}(u)}{du} - D_{u^*}(u)\right). \qquad (16.22)$$

Let us define

$$h(u) = uD'_u(u) - D_u(u) \qquad (16.23)$$
$$= [(A_m - B_m)g^m(u) + (B_m - C_m)][ug'(u) - g(u)] - C_m. \qquad (16.24)$$

The fact that g is increasing and concave also implies that $h(u)$ is increasing in u. Thus $u^*(x) = 1$ (no vigilance) is the unique ESS if and only if $h(1) \leq 0$ and otherwise the ESS, $u^*(x)$, is the unique solution of $h\big(u^*(x)\big) = 0$; see Exercise 16.11.

Thus in the case when foraging cannot be terminated, we see that optimal foraging is independent of α, γ and x. This is not generally true for the more complicated versions of the model of McNamara and Houston (1992).

The above model and analysis were carried out in a more general way in McNamara and Houston (1992) where the authors also considered a number of extensions, generalisations and special cases.

One problem with such foraging groups is that individuals each choose their own foraging level independently. It would be more efficient to have one or two "sentinels" on a rotating basis who were more vigilant than the others, allowing everyone else to concentrate on feeding, as in meerkat colonies (le Roux et al., 2009). Broom and Ruxton (1998b) incorporated this possibility by considering dynamic groups where recent arrivals to a feeding area could have different foraging levels to those that had been there for a time.

Another assumption of these models is that predators focus on one group, and the strategies adopted in a group have no effect on the likelihood of it being attacked. This is perhaps unrealistic, and Jackson et al. (2006) considered a situation where predators could attack more than one group, and based their choices partly on the level of vigilance adopted by the different groups.

16.4 The enemy within: infanticide and other anti-social behaviour

16.4.1 Infanticide

Infanticide, the killing of infants by conspecific males, is a common phenomenon amongst a number of mammals. It has been most observed in primates, for example langurs (Borries and Koenig, 2000) and howlers (Crockett and Janson, 2000) and is also present in some carnivores, particularly lions (Pusey and Packer, 1994), and rodents (Parmigiani et al., 1994). There are several plausible hypotheses for an act of infanticide, but perhaps the most common explanation is *the sexual selection hypothesis* (van Schaik, 2000) of male-male competition for reproduction: when a male kills an unrelated infant it enables the mother to conceive the next infant sooner, which is beneficial if he is likely to be the father of this next infant.

Infanticide is most common in groups with only one breeding male, at the point when the male is replaced by a different male. When infanticide is attempted mothers and other group members often defend the infant, but due to sexual dimorphism in most primates, male defence is more effective, and infanticide is both more difficult and riskier when a defending male is present. There is considerable variation in the prevalence of infanticide within and between different primate populations and species, important factors including the number of males per group, male replacement rates and age at weaning. Low rates of infanticide are common when males enter at the bottom of the male hierarchy, higher rates occurring when new males are immediately dominant.

Example 16.6 (Broom et al., 2004a). Consider a group with two males. One who is father to an infant and the second who has recently joined. Should the new male attempt an act of infanticide and if it does, should the father defend

TABLE 16.2: Summary of the parameters for the model of two male group infanticide from Example 16.6.

Parameter	Description
E_F	Expected number of future surviving offspring (father, if no infanticide)
E_N	Expected number of future surviving offspring (new male, no infanticide)
p	Probability that the next infant is sired by the father if target is killed (if no injuries)
α	Probability that a newborn infant survives to maturity
$\alpha + \beta(t)$	Probability that the target infant aged t survives to maturity if not killed in an attack
ν	Probability that the infant is killed if only females defend
q_1	Probability that the father is injured and the infant is killed
q_2	Probability that neither male is injured, but the infant is killed
q_3	Probability that the new male is injured and the infant survives
$g(t)$	Average number of extra births due to the death of an infant aged t years

the infant? An attempt is a potentially long campaign, so there is only one such "attempt".

We model this problem as a game in an extensive form (see Section 10.1). To set up a model, we consider various scenarios as illustrated in Figure 16.2 with parameters summarised in Table 16.2.

The three possible ESS outcomes are as follows (see Exercise 16.12).

1. The new male attacks, father does not defend, if

$$g(t)p\alpha(\nu - q_2) - (\alpha + \beta(t))(\nu - q_1 - q_2) + q_1 E_F < 0. \qquad (16.25)$$

2. The new male attacks, father defends, if (16.25) does not hold and

$$g(t)\alpha[q_1 + q_2(1-p)] - q_3 E_N > 0. \qquad (16.26)$$

3. The new male does not attack if neither (16.25) nor (16.26) holds.

The model gives predictions for the type of circumstances when infanticide should be prevalent. For example, the new male should attack the infant if his chances of fathering the next infant are large, the chances of killing the infant are large or the average number of extra births following infanticide is large. The father should defend the infant if the target infant has a high chance of survival to maturity, the probability that the infant will be killed is significantly decreased if the father defends or his residual reproductive value E_F is low, e.g. he is old.

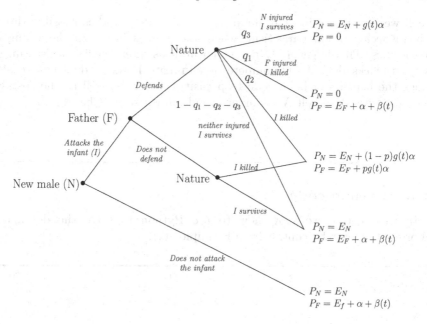

FIGURE 16.2: The game and payoffs of a two-male group infanticide game from Example 16.6. Parameters are summarised in Table 16.2.

As the infant gets older, the benefit of killing it declines, and it may also become of greater value to its father, as the probability of it surviving to adulthood increases. Thus, in general, there are two thresholds where younger infants should be attacked despite defence, older infants should not be attacked irrespective of defence but defence should deter attack in intermediately aged infants.

Infanticide should be more common in populations where the incoming male is dominant, as in Chacma baboons or Hanuman langurs. Thus the observed variation of infanticide in primate multi-male groups may be explained by differences in male migration patterns and the resulting differences in age and dominance rank of competing males. The model might even be able to explain differences across time in a given group or population.

16.4.2 Other behaviour which negatively affects groups

As we saw at the start of this chapter there are other negative effects of grouping and living near conspecifics more generally. Some of these are simply the result of being in the group, such as the attraction of predators or the degradation of the local food supply by group-mates. Other effects are caused by the active behaviours of individuals that deliberately target group-mates, such as in infanticide. An important example is competition between males for mates which is discussed in Chapter 17. Another is kleptoparasitism, which we investigate in more detail in Chapter 19. Another possible threat is intraspecific brood parasitism, where individuals bring up the young of conspecifics. This is generally not as harmful as interspecific parasitism, as parasite chicks do not out-compete or eliminate hosts, and of course also brings the benefits of being able to parasitise others as well as the costs of being parasitised oneself. We look at brood parasitism in Chapter 20.

16.5   Python code

In this section we show how to use Python to solve the dominant-subordinate hierarchy contest from Example 16.1.

```
1   """ Dominant-subordinate hierarchy contest
2   This script has two parts.
3   Part 1 uses the general machinery of the extensive game
4   Part 2 codes specifically this game only
5
6   Because this game has several "best actions".
7   the general machinery discovers only one ESS
8   but the second part discovers all
9   """
10
11  ## Define parameters
12  x = 0.1   # Reproductive output of solitary queen in new colony
13  k = 1.2   # Combined reproductive output of a pair if together
14  r = 0.5   # Relatedness of the two queens
15  f = 0.3   # Probability that subordinate will win the fight
16
17  ## Import basic packages
18  import numpy as np
19  import matplotlib.pyplot as plt
20
21  # Import storage and a solver for extensive games
22  from ExtensiveGames import node, solveExtensiveGame
23
24
25  ## Define and name actions
26  stay  = 0   # Action stay will be in the first row
27  leave = 1   # Action leave in the second row
28  fight = 2   # Action fight in the third row
29  # The order implicitly establishes the preference
30  # In the case two actions yield the same payoff
31  #    action a1 is preferred over action a2>a1
32
33  # The following two arrays allow quick referencing
34  action = np.asarray([stay, leave, fight])
35  Action = ['Stay', 'Leave', 'Fight']
36
37  ## Define fitness functions
```

```python
38  def S_direct(p):
39      """ Returns direct fitness for the subordinate queen
40      Given the dominant allocates p
41      the outcome depends on subordinate's action """
42      return np.asarray(  # return it as an array type
43          [k*p,      # if stay
44           x,        # if leave
45           f])       # if fight
46
47  def D_direct(p):
48      """ Returns direct fitness for the dominant queen
49      Given the dominant allocates p
50      the outcome depends on subordinate's action"""
51      return np.asarray(  # return it as an array type
52          [k*(1-p),    # if stay
53           1,          # if leave
54           (1-f)])     # if fight
55
56  def S_inclusive(p):
57      " returns inclusive fitness for the subordinate "
58      return S_direct(p) + r * D_direct(p)
59
60  def D_inclusive(p):
61      " returns inclusive fitness for the subordinate "
62      return D_direct(p) + r * S_direct(p)
63
64  # Define strategy of dominant queen.
65  # p is a fraction of resources given to subordinate
66  all_p = np.linspace(0,1,1001)
67
68  # Create the root of the game tree
69  D = node(label = 'D', Players = ['Dominant', 'Subordinate'],
70              player = 'Dominant')
71
72  # Initialize list of nodes where Subordinate makes decisions
73  Snode = [[] for p in all_p]
74
75  # Initialize list of terminal nodes
76  # Every allocation of D and every action of S will lead to
77  # a different terminal node
78  TerNode = [[[] for a in action] for p in all_p]
79
80  # Create the actual nodes
81  # for every allocation p of the Dominant queen
82  for j,p in enumerate(all_p): # j is just a counter
83      # After the dominant allocates p, it goes to Snode
84      Snode[j]=(D.add_choice(choiceName = str(p),
85                              targetLabel = 'S'+str(j)))
86
87      # Specify that the Subordinate is the player at that node
88      Snode[j].player = 'Subordinate'
89
90      # For every action of the Subordinate
91      for a in action:
92          # Each action of Subordinate yields to terminal node
93          # Create the actual node
94          TerNode[j][a] = Snode[j].add_choice(Action[a],
```

```
95                                              'T'+str(j)+Action[a])
96            # Add payoffs to the terminal node
97            TerNode[j][a].payoffs = [D_inclusive(p)[a],   # to Dom
98                                     S_inclusive(p)[a]]  # to Sub
99
100  # Solve the game
101  solveExtensiveGame(D)
102  ### the game is solved, the code could stop here
103
104  #######
105  # The part below does backward induction for this specific game
106  # It does not use the general machinery from chapter 10
107  #
108  # In the game, dominant chooses p and subordinate reacts
109  # First get best response of subordinate to any p and then let
110  # dominant choose the best p
111
112  # Initialize for dominant fitness as function of p
113  D_fit = np.zeros(len(all_p))
114
115  # Init best action for Subordinate as -1
116  best_action = -np.ones(len(all_p))
117
118  # For all values of p; track the index j
119  for j, p in enumerate(all_p):
120      # Get action with yielding maximal inclusive fitness
121      # The next line returns the FIRST index,
122      # so order of actions is important
123      best_action[j] = action[np.argmax(S_inclusive(p))]
124
125      # Get fitness of Dominant Queen if choosing p
126      D_fit[j] = D_inclusive(p)[int(best_action[j])]
127
128  # Get allocation(s) that yields max fitness for the Dominant
129  p_ESS = all_p[D_fit == max(D_fit)]
130
131  ## Print the equilibria
132  # Get best actions of subordinate for ESS values of p
133  S_ESS = [action[np.argmax(S_inclusive(p))] for p in p_ESS]
134
135  if len(p_ESS) == 0:
136      # If there is no ESS value of p
137      print('There is no equilibrium')
138  elif len(p_ESS) == 1:
139      # If there is only one ESS value
140      print('the dominant queen allocates ' + str(float(p_ESS)) +
141            ', the subordinate ' + Action[int(S_ESS[0])] + 's')
142  else: # More than one
143      print('there are multiple equilibria, see the plot')
144
145  ## Plot equilibria
146  plt.scatter(all_p, best_action, c='blue', edgecolors='none',
147              s=30, label='Best action')
148  plt.scatter(p_ESS, S_ESS, c='red', edgecolors='none',
149              s=30, label='Equilibrium')
150  plt.yticks(action, Action, rotation=90, va='center')
151  plt.xlabel("Dominant's allocation $p$")
```

```
152  plt.ylabel("Subordinate's best action")
153  plt.legend()
154  plt.show()
```

16.6 Further reading

For a good book on social foraging see Giraldeau and Caraco (2000); Sumpter (2006) shows how complex patterns of behaviour can result from simple behavioural rules. For models of flocking behaviour, see Halloway et al. (2020); Bellomo and Ha (2017) and for predators attacking groups, see Cazaubiel et al. (2017). For general theoretical models of animal aggregation and dispersal, see Broom et al. (2020).

Johnstone (2000) is a very nice review on reproductive skew; for more on reproductive skew see Reeve (2000) and Reeve and Emlen (2000). An alternative model of the division of resources (the "tug of war" model) is described by Reeve et al. (1998). Kokko (2003) revisits some hidden assumptions behind the models and studies whether relaxing them affects the outcomes. van Doorn et al. (2003a,b) study two and multi-player models of the evolution of social dominance. Appleby (1983) gives more on linear dominance hierarchies.

See Chase et al. (1994) and Rutte et al. (2006) for reviews on the winner and loser effect. A game-theoretical model of the winner and loser effect is presented in Mesterton-Gibbons (1999c). A third effect, the *bystander effect*, is discussed by Earley and Dugatkin (2002) with an example of the green swordtail fish *Xiphophorus helleri* (see also Dugatkin and Earley, 2003). The evolution of an aggressive threshold Φ was discussed in Mesterton-Gibbons (1994). To find out more about lekking behaviour, see Wiley (1991).

By studying data from hens and cichlid fish, Lindquist and Chase (2016) (see also Chase et al. (2002); Lindquist and Chase (2009)) found a more complex dynamic relationship than simply linear dominance hierarchies. A population of birds, for example seabirds like the northern gannet, arriving sequentially at a nesting area is given in Broom et al. (1997a, 2000b) and Broom et al. (1996), which leads to a complex multi-player game.

Perhaps the most influential models of anti-predator vigilance, in tandem with dilution, are those of Pulliam et al. (1982) and McNamara and Houston (1992), where the central focus was on the balance between vigilance and foraging. Later models (Proctor and Broom, 2000; Proctor et al., 2003, 2006) in addition considered the effects of the area occupied by the group and the effect of ambiguity regarding missed and false signals (Proctor et al., 2001). For a paper on observations of anti-predator vigilance amongst Burchell's zebra, and blue wildebeest against lions, see Yiu et al. (2021). For more papers on vigilance see Brown (1999); Ale and Brown (2007) and Jackson and Ruxton (2006).

Parmigiani and vom Saal (1994), and van Schaik and Janson (2000) are two books which consider infanticide, see also Cant et al. (2014); Lyon et al. (2011). For an alternative use of game theory regarding group behaviour see Noë and Hammerstein (1994, 1995); Hammerstein and Noë (2016).

16.7 Exercises

Exercise 16.1 (Reeve and Ratnieks, 1993). For the dominant-subordinate hierarchy game from Section 16.2.1, derive the formula (16.2) for the fitness of a subordinate as well as formulae (16.3) and (16.4) for the staying and peace incentives.

Exercise 16.2 (Reeve and Ratnieks, 1993). Derive results from Table 16.1.

Hint. The five rows correspond to the distinct cases $p_c > p_s > \max(p_p, 0)$; $p_p > 0 > p_s$; $p_p > p_s > 0$; $p_s < 0, p_p < 0$; $p_s > p_c, p_s > p_p$ and a plot of the conditions on the f- and x-axes would help.

Exercise 16.3. In the analysis of the dominant-subordinate hierarchy game from Section 16.2.1 we assumed that if the payoffs for staying and leaving, or staying and fighting, are identical, then the subordinate will choose to stay. Analyse the game if ties lead to leaving or fighting.

Exercise 16.4. In a group of size m, show that the expected proportion of transitive triads is $3/4$, assuming that the dominance of any individual over another is completely random.

Exercise 16.5. Assume we have already observed 2 out of 3 dominance relationships in a triad. What is the probability that the triad will be transitive?

Hint. If A beats B,C or A,B beat C, then the probability is different than if the observed dominance relationships are A beats B beats C.

Exercise 16.6 (Broom and Cannings, 2002). Consider a generalised Hawk-Dove game where winners receive W, losers receive L and fighters still have to pay the cost $C > 0$. In such a contest the pay-off matrix becomes

$$\begin{pmatrix} \frac{1}{2}(W + L - C) & W \\ L & \frac{1}{2}(W + L) \end{pmatrix}. \tag{16.27}$$

Find the unique ESS of this matrix game.

Exercise 16.7 (Broom and Cannings, 2002). Consider the Swiss tournament between $m = 2^M$ players described in Example 16.3. Show that the number of individuals that have won i times out of j rounds, $m_{i,j}$ is given by

$$m_{i,j} = \binom{j}{i} 2^{M-j}, \tag{16.28}$$

for $i = 0, \ldots, j$.

Hint. $m_{i,j} = \frac{1}{2} m_{i,j-1} + \frac{1}{2} m_{i-1,j-1}$.

Exercise 16.8 (Broom and Cannings, 2002). In the setting up of the Swiss tournament described in Section 16.2.3, show that if $p_{i(j+1)} = 1$ then $p_{ij} = 1$, or equivalently if $p_{ij} < 1$ then $p_{i(j+1)} < 1$ (although the p_{ij} are not necessarily monotonically decreasing with j).

Exercise 16.9 (Broom and Cannings, 2002). Solve the game from Example 16.3 with $C = 5, M = 4$ and $V_i = 2^i$ for $i = 0, \ldots, 4$.

Exercise 16.10 (Pulliam et al., 1982). As in the Example 16.5, let an individual spend a proportion u of its time feeding and the rest scanning for predators. Assume that it takes a predator time t_a to attack, an individual scans for time t_s and forages between scans for a time taken from a Poisson distribution with appropriate mean to give the foraging proportion u, and that the predator is spotted if and only if the individual is scanning at some point during its attack.

Show that the formula for $g(u)$, the probability the animal fails to spot the predator, will be (approximately)

$$g(u) = \exp\left(-\frac{t_a}{t_s}\left(\frac{1}{u} - 1\right)\right). \tag{16.29}$$

Exercise 16.11. Use (16.23) to identify the ESS of the foraging and scanning game from Example 16.5.

Hint. See Theorem 7.3.

Exercise 16.12 (Broom et al., 2004a). Find the ESSs of the Infanticide game of Example 16.6 in the following cases (each of which use $\alpha = 0.5, \beta(t) = 0.2t, \nu = 0.5, g(t) = 0.5 - 0.25t$).

(a) $q_1 = 0.01, q_2 = 0.39, q_3 = 0.03, p = 0.8, E_F = 1, E_N = 4$ for varying values of t, representing a case where the father is dominant and old, and the new male subordinate and young.

(b) $q_1 = 0.01, q_2 = 0.39, q_3 = 0.03, p = 0.8, E_F = 4, E_N = 1$ for varying values of t, representing a case where the father is dominant and young, and the new male subordinate and old.

(c) $q_1 = 0.03, q_2 = 0.37, q_3 = 0.01, p = 0.2, E_F = 2, E_N = 4$ for varying values of t representing a case where the father is subordinate, and the new male dominant.

Chapter 17

Mating games

17.1 Introduction and overview

Perhaps the most well-known examples of animal conflict involve competition between male animals over access to mates. We can think of aggressive contests such as those in stags (see Chapter 4), elephant seals (McCann, 1981) or hippopotami (Estes, 2012, Chapter 13), as shown on the front cover of this book, or complicated ritual displays such as in the dances of cranes (Panov et al., 2010). The classical model of this type of behaviour is the Hawk-Dove game (see Section 4.1). This is a simple game aimed at emphasising important features of evolutionary games, rather than being an exact model of any particular situation. These features include the existence of strategies of restraint when maximum violence is not used even when individuals compete for very valuable resources and the possibility of stable mixed strategy solutions. However, like the Prisoner's Dilemma, it has been developed in a number of ways to more realistically model behaviour, and we look at one such model below.

Conflict between males can also be more subtle, and two (or more) males can compete without ever meeting. If more than one male mates with a female in a sufficiently short space of time, sperm competition can result. This can involve a simple competition where timing and sperm volume decide the winner. We look at the basis of such a model below in Section 17.3, but there are also more elaborate methods of competition, which we also briefly describe.

In the next two sections females come into their own. Rather than being the passive object of male competition, we investigate situations where both sexes are active players. In Section 17.4 we look at situations where there is conflict between the sexes with regard to the levels of parental investment in bringing up offspring. In Section 18.2 we see a complex interaction where males compete but females select between males, leading to one of the most insightful and important theories relating to evolutionary games. But for now, we return to male versus male conflict.

17.2   Direct conflict

The classical model of direct conflict over resources, in particular mates, is the Hawk-Dove game. As we have seen, this is an idealised game which treats individual contests as independent events, and the rewards and costs also as constant. In reality the value of winning contests, and the cost of losing aggressive contests, can depend upon circumstances and be hard to measure for a contest in isolation. We saw in Example 16.5 in Section 16.3 how McNamara and Houston (1992) introduced models in which payoffs depend upon the state of the animals, and we follow Houston and McNamara (1991) which develops similar reasoning in a model of conflicts over mates using repeated Hawk-Dove contests.

Example 17.1 (Repeated Hawk-Dove game, Houston and McNamara, 1991). Consider a species (for example the jumping spider *Phidippus clarus*, Elias et al., 2008) where males have a series of contests with other males over access to females. During a single contest, a male can either be aggressive (Hawk) or not (Dove). The winner of Hawk-Dove contests is the one playing Hawk; the winner of Dove-Dove or Hawk-Hawk contests is determined at random. The Hawk-Hawk contests are aggressive, and the loser of such a contest can die. If a male wins a contest it mates with the female, and then moves on to search for another female (and the next contest). The loser (if still alive) also moves to search for another female. There can be multiple contests during multiple seasons.

17.2.1   Setting up the model

Assume the opponents are drawn at random from an essentially infinite population. Each male plays a (potentially) mixed strategy $\mathbf{q} = (q, 1 - q)$, where q is the probability of playing Hawk in any round. Because we will eventually perform an ESS analysis, we can assume that everybody (but the focal individual) plays a strategy $\mathbf{p} = (p, 1 - p)$; Houston and McNamara (1991) call p the *Population Hawk Probability (PHP)*.

If a male wins a contest it mates with the female, so the reward for winning is V, a constant expected increase in reproductive success. Let the loser of a Hawk-Hawk contest be either uninjured with probability $1 - z$ or die with probability z. Rather than modelling several seasons in a row, assume that if a male survives until the end of the season, it gets a reward R associated with future rewards from subsequent seasons. We note that, in reality, this future reward would depend upon the strategy of the focal individual and the population as a whole, and thus this assumption is a simplification.

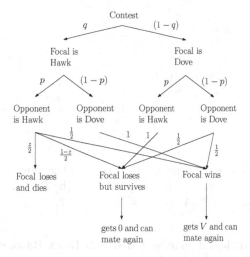

FIGURE 17.1: A single contest in a repeated Hawk-Dove game based on Example 17.1. The focal individual uses **q** in a population where everybody else uses **p**. z is the probability that the loser of a Hawk-Hawk contest dies. During a regular game, an individual accumulates rewards as it goes through a series of rounds of contests until it dies.

17.2.1.1   Analysis of a single contest

A schematic diagram of a single contest is shown in Figure 17.1. It follows that the probabilities of a focal individual winning a contest (w), dying during the contest (δ) and surviving a contest (σ) are given by

$$w = \frac{1}{2}qp + q(1-p) + \frac{(1-q)(1-p)}{2} = \frac{1}{2}(1 + q - p), \tag{17.1}$$

$$\delta = \frac{1}{2}qpz, \tag{17.2}$$

$$\sigma = 1 - \delta = 1 - \frac{1}{2}qpz. \tag{17.3}$$

If a season consists of a single contest only, the payoff to a focal individual using **q** in a population where everybody uses **p** is

$$\mathcal{E}[\mathbf{q}; \delta_{\mathbf{p}}] = wV + (1 - \delta)R \tag{17.4}$$

$$= R + \frac{1}{2}(1-p)V + \frac{1}{2}q(V - zRp). \tag{17.5}$$

Thus the game can be analysed as the standard Hawk-Dove game with reward V, cost Rz and background fitness R; see Exercise 17.1.

17.2.1.2   The case of a limited number of contests per season

Let $K_{\max}$ be the maximal (and fixed) number of contests per season an individual can play. If an animal plays Dove then it will play in them all, but

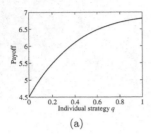 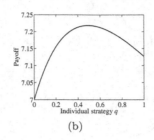 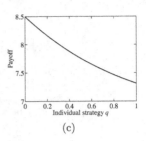

(a)   (b)   (c)

FIGURE 17.2: The payoff function $\mathcal{E}[\mathbf{q}; \delta_{\mathbf{p}}]$ for the game from Example 17.1 where the parameters are as follows. (a) $R = 2$, (b) $R = 4.5$, (c) $R = 6$. Other parameters are $K_{\max} = 25, V = 1, p = 0.8, z = 0.2$.

if $q > 0$ and $p > 0$ then it may be killed in a Hawk-Hawk contest before the end of the season. Let W_k be the additional expected reward from the kth contest for an individual currently in round k. We obtain

$$W_k = \begin{cases} wV + \sigma W_{k+1} & \text{if } k < K_{\max}, \\ wV + \sigma R & \text{if } k = K_{\max}. \end{cases} \qquad (17.6)$$

We note here that w takes the same value in every round because the population is monomorphic. If this was not the case, the Hawk frequency would change from round to round as the more hawkish individuals would have a higher probability of being killed, and so w would change. Thus

$$\mathcal{E}[\mathbf{q}; \delta_{\mathbf{p}}] = W_1 = wE[k]V + \sigma^{K_{\max}} R, \qquad (17.7)$$

where

$$E[k] = \frac{1 - \sigma^{K_{\max}}}{1 - \sigma} \qquad (17.8)$$

is the expected number of contests an individual plays during a season (the number follows a (truncated) geometric distribution).

Thus we can see that the payoff function given by (17.7) is non-linear in both the focal player strategy $\mathbf{q}$ as well as the population strategy $\mathbf{p}$; see Figure 17.2 for possible shapes of the payoff function.

It follows that the best response value of q can be between 0 and 1 even when the population strategy is not an ESS (this is not possible in matrix games); see Figure 17.3 for graphs of best responses.

For $V < zR$ there is a unique ESS $p^* < 1$. For $V > zR$ there is an ESS $p^* = 1$, but sometimes this is not the only ESS. As shown on Figures 17.3(c) and 17.3(d), there are three potential candidates for an ESS, but the middle point is, in fact, not an ESS.

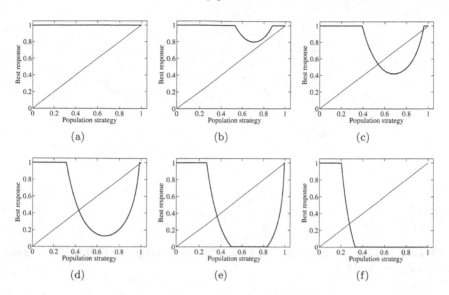

FIGURE 17.3: The best response strategy as a function of the Population Hawk Probability strategy p. (a) $z = 0.2$, (b) $z = 0.25$, (c) $z = 0.3$, (d) $z = 0.35$, (e) $z = 0.4$, (f) $z = 0.5$. Other parameters are $K_{\max} = 25, V = 1, R = 2$. Candidate ESSs occur where the best response curve crosses the straight $p = q$ line.

17.2.2 An unlimited number of contests

There are several assumptions being made in Example 17.1. These include that a male can mate only if he won a contest, and a male can die only if he has lost a contest. These two assumptions are now relaxed in the following example.

Example 17.2 (An alternative repeated Hawk-Dove game, Houston and Mc-Namara, 1991). Consider a situation as in Example 17.1 where males are trying to mate with females. In this case, if a specific male is the only one close to a female, then he mates with her without a contest and then keeps searching for another female. If two males are close to a female, they engage in a Hawk-Dove game as before.

Let $\theta \in (0,1)$ be the probability that a particular female is contested by two males and $s \in (0,1)$ be the probability of surviving between encounters with females. There is no reward from future seasons in this version of the game; other parameters are the same as in Example 17.1, namely the reward for mating with a female is V, the focal strategy is $\mathbf{q}$, the population strategy is $\mathbf{p}$ and the probability that the loser of a Hawk-Hawk fight will die is z.

The schematic diagram of the life of a male is shown in Figure 17.4. Tho payoff to a q-playing individual in a population with Population Hawk

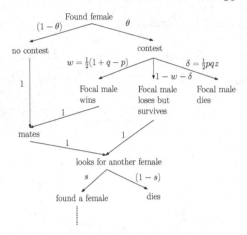

FIGURE 17.4: A diagram of a male's life in the repeated Hawk-Dove game from Example 17.2.

Probability p is given by (see Exercise 17.3)

$$\mathcal{E}[\mathbf{q}; \delta_\mathbf{p}] = \frac{1 - \frac{1}{2}\theta(1 + p - q)}{1 - (1 - \frac{1}{2}\theta qpz)s}V. \tag{17.9}$$

In contrast to the payoff function (17.7) from Section 17.2.1.2, the payoff function $\mathcal{E}[\mathbf{q}; \delta_\mathbf{p}]$ from (17.9) is, for any fixed p, always a monotone function of q.

As illustrated in Figure 17.5, there are three possible scenarios.

1. A unique mixed strategy ESS (for high $zs/(1 - s)$),

2. two ESSs, pure Hawk and a mixed strategy (for intermediate $zs/(1 - s)$, but this only appears for sufficiently large θ), and

3. a pure Hawk ESS only (for low $zs/(1 - s)$).

In general, the pure Hawk ESS is more likely for high θ (most females must be contested), low z (the risk of death in a Hawk-Hawk fight is low), and low s (the probability of survival between contests is low).

17.2.3   Determining rewards and costs

As we discuss in Section 5.6 and also as discussed for example in Houston and McNamara (1991), if there are multiple solutions then two populations subject to the same ecological parameters may exhibit different levels of aggressive behaviour. Members of our two populations would have different lifetime reproductive successes, and so the effective cost of losing a Hawk-Hawk fight within the two populations would be different. Somebody observing the

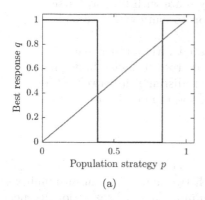
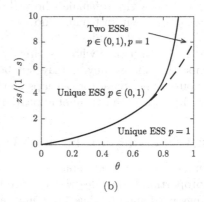

(a)   (b)

FIGURE 17.5: (a) Plot of best responses for the function (17.9) where $\theta = 0.9, z = 0.25, s = 0.965$ and $V = 1$. Candidate ESSs occur where the best response function crosses the straight $p = q$ line. (b) Regions of the ESSs for a game from Example 17.2 as they depend on the parameter values θ and $sz/(1-s)$. The lower curve is $sz/(1-s) = 2\theta/(1 - \theta/2)^2$. The upper curve starts at $\theta = 2/3$ and is $sz/(1-s) = 1/(1-\theta)$.

populations and measuring the cost of defeat can focus on an individual contest and show that the behaviour satisfies the ESS conditions, and the solution would be unique given the cost derived from future contests. Thus we can see that it can be important to take into account the whole sequence of contests.

17.3   Indirect conflict and sperm competition

Males can compete with each other indirectly through efforts to get the female to select them as mates in preference to their rivals, as we shall see later in Section 18.2, and this can involve actively sabotaging the efforts of other male's displays, for instance the destruction of bowers by rival male bowerbirds (Borgia, 1985). An important indirect way that males compete with each other is through the allocation of sperm during matings. Whilst the males themselves do not enter into direct fights, as in the situations described in Section 17.2, if a number of males have mated with the same female, the timing of the matings and the volume of sperm used will affect which of the males will father any offspring.

Example 17.3 (Sperm allocation game, Ball and Parker, 2007). Consider a population of individuals where every female mates with either one or two males. A male can thus encounter a female in one of the three states:

State 0   a virgin female who will only mate with the focal male,

State 1   a virgin female who will go on to mate with a second male, and

State 2   a female who has already mated with another male.

Assume that males may or may not be able to distinguish between a virgin and non-virgin female (or perhaps can even distinguish between state 0 and state 1). What is the optimal allocation of sperm in each case?

17.3.1   Setting up the model

In the modelling and analysis, we follow Ball and Parker (2007). Let q be the proportion of females which mate with two males. The mean number of matings per female is thus $1 + q$ and, assuming an equal sex ratio, the mean number of matings per individual is $1+q$. Thus a proportion $(1-q)/(1+q)$ of encounters will be with females in state 0, and similarly this proportion will be $q/(1 + q)$ for females in state 1 and $q/(1 + q)$ for females in state 2. The strategy of a male is the allocation of an amount of sperm s_i to any female in state i that he encounters, for $i = 0, 1, 2$.

17.3.1.1   Modelling sperm production

Both securing the mating and the actual mating can be costly and males have only a finite amount of energy. Let C be the cost of obtaining a mating with a female, D be the cost per unit of sperm produced and E be the total energy available for reproduction. It is assumed that D and E, like q, are independent model parameters which are fixed. C, however, will depend upon the population strategy, as we see later (although it will be independent of the focal male's strategy conditional on the population strategy).

Given that there is no gain to saving any such energy, if the number of matings that a given male has is k, and the mean volume of sperm used per mating is $\bar{s}$, we have

$$k(C + D\bar{s}) = E. \tag{17.10}$$

A male's strategy is given by a triple (s_0, s_1, s_2), and so

$$\bar{s} = \frac{(1 - q)s_0 + qs_1 + qs_2}{1 + q}. \tag{17.11}$$

Thus the choice of strategy determines $\bar{s}$ and hence k.

17.3.1.2   Model parameters

As usual, we search for ESSs. We are interested in the payoff $W(s_0, s_1, s_2)$ of a single mutant using strategy $\mathbf{s} = (s_0, s_1, s_2)$ in a population where everybody else uses a strategy $\mathbf{s}^* = (s_0^*, s_1^*, s_2^*)$.

The expected number of matings k^* for an average individual in an equilibrium must satisfy $k^* = 1 + q$. By (17.10) we get

$$C(\bar{s}^*) = \frac{E}{1+q} - D\bar{s}^*, \tag{17.12}$$

and thus the constraint (17.10) for a focal individual becomes the following relationship between $\bar{s}$ and k:

$$k\left(\frac{E}{1+q} + D(\bar{s} - \bar{s}^*)\right) = E. \tag{17.13}$$

We should note here that the model is clearly a simplification. To follow the strategy described a male would have to have a fractional number of matings with females of each type. This is clearly not possible, and so the strategy would rather reflect a mean number of matings. But then, for a fixed total amount of sperm, some males would not have enough for the matings they actually have, and some would have too much. In reality, the amount of sperm a male would release would depend on present energy levels and the future expected reward from additional mating.

17.3.1.3   Modelling fertilisation and payoffs

Assume that sperm from the first mating with a particular female is worth more than that from a second mating. This can be incorporated by a discount factor $r < 1$. If the first male contributes s_1 and the second male contributes s_2, the total load is $S = s_1 + rs_2$. Clearly the case when the sperm from the first mating is worth less can be addressed in a similar fashion using $r > 1$.

Following Mesterton-Gibbons (1999a), the probability of a successful fertilisation is $S/(S + \varepsilon)$, for some $\varepsilon \geq 0$. Assuming that the probability of being the male to fertilise the female is proportional to each male's contribution, the probability of fertilisation by male 1 is

$$\frac{s_1}{s_1 + rs_2 + \varepsilon}, \tag{17.14}$$

and the probability of fertilisation by male 2 is

$$\frac{rs_2}{s_1 + rs_2 + \varepsilon}. \tag{17.15}$$

Based on the above, the payoff is given by

$$\begin{aligned}
\mathcal{E}[\mathbf{s}; \delta_{\mathbf{s}*}] &= W(s_0, s_1, s_2) \\
&= \frac{k}{1+q}\left(\frac{s_0(1-q)}{s_0 + \varepsilon} + \frac{qs_1}{s_1 + rs_2^* + \varepsilon} + \frac{qrs_2}{s_1^* + rs_2 + \varepsilon}\right).
\end{aligned} \tag{17.16}$$

17.3.2   The ESS if males have no knowledge

If a male has no knowledge of the female's state, then the individual (population) strategy is indicated by a single choice s (s^*), where $\mathbf{s} = (s, s, s)$

$(\mathbf{s}^* = (s^*, s^*, s^*))$, and the optimal strategy in any population is given by maximising

$$
\mathcal{E}[\mathbf{s}; \delta_{\mathbf{s}*}] = W(s, s, s)
$$
$$
= \frac{k}{1+q}\left(\frac{s(1-q)}{s+\varepsilon} + \frac{qs}{s+rs^*+\varepsilon} + \frac{qrs}{s^*+rs+\varepsilon}\right), \tag{17.17}
$$

given the constraint (17.13). Using the Lagrange multiplier method we get

$$
\lambda D = \frac{1}{1+q}\left(\frac{(1-q)\varepsilon}{(s^*+\varepsilon)^2} + q\frac{2rs^*+(1+r)\varepsilon}{(s^*(1+r)+\varepsilon)^2}\right), \tag{17.18}
$$

$$
W(s^*, s^*, s^*) = (1-q)\frac{s^*}{s^*+\varepsilon} + q\frac{s^*+rs^*}{s^*+s^*r+\varepsilon} = \lambda E, \tag{17.19}
$$

where λ is the Lagrange multiplier, since $k^* = 1+q$. Whilst it was not possible to find an exact formula, in general, for s^*, in the special case where $\varepsilon = 0$ the unique ESS value is given by

$$
s^* = \frac{E}{D}\frac{2rq}{(1+r)^2(1+q)}. \tag{17.20}
$$

This is Exercise 17.4.

17.3.3 The ESS if males have partial knowledge

In the case where males can tell virgin and non-virgin females apart, the male strategy is given by the amount of sperm it allocates to a virgin female, $s_{0,1}$ and to a non-virgin female, s_2. If a mutant can deviate in either $s_{0,1}$ or s_2, we have to maximise

$$
W(s_{0,1}, s_{0,1}, s_2^*) = \frac{k}{1+q}\left(\frac{s_{0,1}(1-q)}{s_{0,1}+\varepsilon} + \frac{qs_{0,1}}{s_{0,1}+rs_2^*+\varepsilon} + \frac{qrs_2^*}{s_{0,1}^*+rs_2^*+\varepsilon}\right), \tag{17.21}
$$

$$
W(s_{0,1}^*, s_{0,1}^*, s_2) = \frac{k}{1+q}\left(\frac{s_{0,1}^*(1-q)}{s_{0,1}^*+\varepsilon} + \frac{qs_{0,1}^*}{s_{0,1}^*+rs_2^*+\varepsilon} + \frac{qrs_2}{s_{0,1}^*+rs_2+\varepsilon}\right). \tag{17.22}
$$

The Lagrange multiplier method yields

$$
(1-q)\frac{\varepsilon}{(s_{0,1}^*+\varepsilon)^2} + q\frac{rs_2^*+\varepsilon}{(s_{0,1}^*+rs_2^*+\varepsilon)^2} = \lambda D, \tag{17.23}
$$

$$
\frac{r(s_{0,1}^*+\varepsilon)}{(s_{0,1}^*+rs_2^*+\varepsilon)^2} = \lambda D, \tag{17.24}
$$

$$
W(s_{0,1}^*, s_{0,1}^*, s_2^*)) = \lambda E. \tag{17.25}
$$

Equations (17.23)–(17.25) cannot be solved directly, but one can obtain numerical results such as in Figure 17.6.

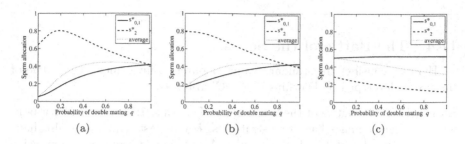

FIGURE 17.6: Evolutionarily stable strategy sperm allocations, $s_{0,1}^*, s_2^*$, when males can discriminate between virgin and mated females, in relation to average probability, q, of double mating by females in the population. Parameter values are $R = 3, D = 1$, and $r = 0.2$ (i.e. strong second male disadvantage) and sperm limitation: (a) $\varepsilon = 0.001$; (b) $\varepsilon = 0.01$; (c) $\varepsilon = 0.1$.

17.3.4 Summary

When there is no information about the state of females, sperm volumes will increase with q as competition from other males increases, unless sperm limitation is high (low R) and the second male's sperm is significantly disfavoured (low r). When males can discriminate between virgins and mated females they may allocate more sperm to virgins in cases where the second male's sperm will be disfavoured and there is sperm limitation. Thus if a species is found where the allocation of sperm to virgin females is significantly greater than to non-virgins, it may be the case that there is both sperm limitation and a significant disfavourment of the sperm of subsequent males in any multiple matings.

The existence of sperm competition has led to a number of developments. In species where matings by females with multiple males commonly occurs, there is often evolution towards larger testes to allow for the production of more sperm (Harcourt et al., 1981). In some species males engage in post-copulatory guarding (Simmons, 2001) where a male stays with the female to prevent other males subsequently mating with her. As well as these obvious adaptations, more elaborate methods of trying to ensure the victory of a male's own sperm have developed. In some species (Hcather and Robertson, 2000) males attempt to remove the sperm of others who have previously mated with the female. In other species males create a post-copulatory plug which can temporarily prevent subsequent matings (Wedell et al., 2002) for long enough to give advantage to his own sperm. There are also a number of chemical methods which can enhance the first male's chances, for instance involving pheromones to reduce the female's attractiveness to other males, and even cases where physical harm is done to the females to prevent them remating (Nessler et al., 2007).

17.4   The Battle of the Sexes

Here we consider the standard game of parental investment, the battle of the sexes, developed by Dawkins (1976, Chapter 9).

Example 17.4 (Battle of the Sexes game, Dawkins, 1976). Consider a population where females have two strategies, coy or fast, and males also have two strategies, faithful and philanderer. A coy female requires an extensive period of courtship, whereas a fast female will mate with a male as soon as he is encountered. Faithful males are prepared to engage in long courtships, and after mating will stay and care for the young. A philanderer refuses to engage in courtship and also leaves immediately after mating. What is the ESS in the population?

Note that a coy female does not mate with a philanderer, i.e. coyness ensures securing a faithful mate.

17.4.1   Analysis as a bimatrix game

To set up a bimatrix game based on Example 17.4, assume that B is the fitness benefit of having an offspring, C_R is the cost (that can be split between both parents) of raising it and C_C is the cost of engaging in a courtship. We have $B, C_R, C_C > 0$. The payoff bimatrix will thus be

$$\begin{array}{ccc}
\text{Male\Female} & \text{Coy} & \text{Fast} \\
\text{Faithful} & \left((B - \frac{C_R}{2} - C_C, B - \frac{C_R}{2} - C_C) \right. & (B - \frac{C_R}{2}, B - \frac{C_R}{2}) \\
\text{Philanderer} & (0,0) & \left. (B, B - C_R) \right)
\end{array}$$

$$(17.26)$$

Let $\mathbf{p}_M = (p, 1-p)$ be the strategy of a male, with p denoting the frequency of being faithful and $\mathbf{p}_F = (q, 1-q)$ be the strategy of a female, with q denoting the frequency of being coy. It follows from Selten's Theorem 8.4 that only pure strategies can be an ESS and we thus just have to check every possible pure combination.

1) Faithful males with coy females is not an ESS because fast females do better.

2) Faithful males with fast females is not an ESS because philanderers do better.

3) If $B > C_R$, then the pair philanderer male and fast female is an ESS, otherwise coy females invade.

4) Philanderer males with coy females are an ESS provided that $B < \min(C_R, C_R/2 + C_C)$, otherwise either faithful males (if $B > C_R/2 + C_C$) or fast females (if $B > C_R$) invade.

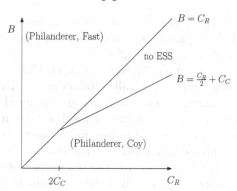

FIGURE 17.7: The ESS in a battle of sexes game from Example 17.4 with payoff bimatrix (17.26) as a function of the model parameters.

Thus there is a unique ESS involving no mating (clearly not a realistic scenario for a real thriving population) if B is sufficiently low, or immediate mating with subsequent male desertion if B is sufficiently large, but for intermediate values of B, when $C_R/2 + C_C < B < C_R$, there is no ESS; see Figure 17.7.

17.4.2 The coyness game

Below is a more realistic model for the battle of the sexes game as presented in McNamara et al. (2009).

Example 17.5 (Coyness game, McNamara et al., 2009). Once a female encounters or is encountered by a male, she inspects him to find out if he is helpful or non-helpful. At any time during the inspection, she may reject him and return to search, mate with him, or keep inspecting him. A helpful male cares for the young together with a female, resulting in "fitter" offspring than if only the female cares. A non-helpful male deserts immediately after mating to search for a new mate but the mother, however, still cares for the offspring. The female strategy is given by two probabilities to either "falsely" reject helpful males or "falsely" accept non-helpful males (the chosen probabilities determine the length of time taken making inspections, so can only be made small at a cost). What is the optimal female strategy and what is the stable fraction of helpful males in the population?

17.4.2.1 The model

Let a unit of time be the time a helpful male cares for young (together with a female). The fitness benefit from an offspring to each parent is B_H (if the male is helpful) or B_N (if the male is non-helpful), where $0 < B_N < B_H$.

When searching for a mate, a male encounters females at rate $\lambda \varrho_F$, where λ is a constant representing the interaction rate and ϱ_F is the proportion of females that are searching for a male. Similarly, a female encounters males at rate $\lambda S \varrho_M$, where S is the sex ratio (of males to females in the population as a whole). λ and S are parameters of the model, the values of ϱ_F and ϱ_M depend on the parameters and strategies.

For the inspection of a male, McNamara et al. (2009) assume that there is an unending sequence of (conditionally) independent identically distributed observations, approximating this as a continuous stream of observations using a diffusion approximation (we saw a similar methodology in Section 12.1.5.2). With this assumption, any optimal inspection strategy is necessarily in the form of a sequential probability ratio test (DeGroot, 2004), where as information accumulates, the female accepts the male when her assessment passes a certain upper threshold, rejects him if her assessment falls below a lower threshold and carries on inspecting if her assessment lies between the two. Thus at any time, she may reject him and return to the search, mate with him or keep inspecting him. Assume that a helpful male is rejected with probability α and a non-helpful male is accepted for mating with probability β. The probabilities α and β correspond to type I and type II errors in statistics and are considered to define the female's strategy. It is derived in McNamara et al. (2009, Appendix 1) that τ_H, τ_N, the expected times of inspections of helpful and non-helpful males, are given by

$$\tau_H = \frac{2}{\nu}\left[(1-\alpha)\ln\left(\frac{1-\alpha}{\beta}\right) + \alpha\ln\left(\frac{\alpha}{1-\beta}\right)\right], \qquad (17.27)$$

$$\tau_N = \frac{2}{\nu}\left[(1-\beta)\ln\left(\frac{1-\beta}{\alpha}\right) + \beta\ln\left(\frac{\beta}{1-\alpha}\right)\right], \qquad (17.28)$$

where ν is a measure of the useful information obtained from the inspection per unit time. Information is gained more quickly for higher values of ν. It follows that, for any reasonable effort to choose helpful males, the smaller α or β are, the larger τ_H and τ_N are.

17.4.2.2   Fitness

An individual's long-term reproduction success is taken as a measure of its fitness. Let p be the proportion of helpful males in the population.

A helpful male spends time $(\lambda \varrho_F)^{-1}$ searching for a female, then τ_H to be inspected and after being accepted to mate (with probability $(1-\alpha)$) he spends a unit of time caring; see Figure 17.8. Thus the proportion of time a helpful male spends searching is given by

$$\varrho_H = \frac{(\lambda\varrho_F)^{-1}}{(\lambda\varrho_F)^{-1} + \tau_H + (1-\alpha)}. \qquad (17.29)$$

Consequently his fitness is given by

$$\gamma_H = \frac{(1-\alpha)B_H}{(\lambda\varrho_F)^{-1} + \tau_H + (1-\alpha)}, \qquad (17.30)$$

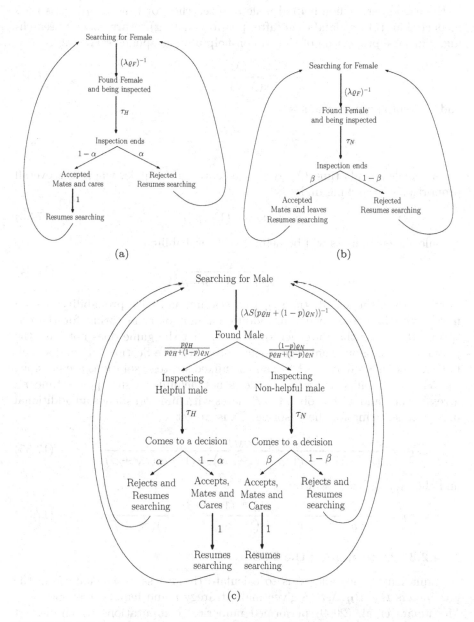

FIGURE 17.8: A model of the life of individuals in the coyness game of Example 17.5. (a) Helpful male, (b) non-helpful male, (c) female. A description of the parameters is in the text.

which is the expected benefit per encounter divided by the expected time of the encounter. A non-helpful male also searches for time $(\lambda \varrho_F)^{-1}$, is then inspected for time τ_N and then (after possibly mating) returns back to searching. Thus the proportion of time a non-helpful male spends searching is

$$\varrho_N = \frac{(\lambda \varrho_F)^{-1}}{(\lambda \varrho_F)^{-1} + \tau_N}, \tag{17.31}$$

and his reproductive fitness is

$$\gamma_N = \frac{\beta B_N}{(\lambda \varrho_F)^{-1} + \tau_N}. \tag{17.32}$$

A female spends time $(\lambda S \varrho_M)^{-1}$ to search for a male, where the overall proportion of searching males is

$$\varrho_M = p\varrho_H + (1-p)\varrho_N. \tag{17.33}$$

A male she encounters will be helpful with probability

$$r = \frac{p\varrho_H}{p\varrho_H + (1-p)\varrho_N}, \tag{17.34}$$

in which case, she spends time τ_H to asses him and with probability $(1 - \alpha)$ mates with him and spends an additional unit of time caring for the offspring. It is clear that in general $r \neq p$, so that the game does not have the polymorphic-monomorphic equivalence property (see Section 8.1) and so Selten's Theorem 8.4 does not hold, and so mixed strategy solutions are possible.

With probability $1 - r$, the male is not helpful, so she spends time τ_N assessing him and with probability β mates with him and spends an additional unit of time caring for the offspring. Thus we get

$$\varrho_F = \frac{(\lambda S \varrho_M)^{-1}}{(\lambda S \varrho_M)^{-1} + r\big(\tau_H + (1 - \alpha)\big) + (1 - r)(\tau_N + \beta)}, \tag{17.35}$$

and the overall fitness will be

$$\gamma_F = \frac{r(1 - \alpha)B_H + (1 - r)\beta B_N}{(\lambda S \varrho_M)^{-1} + r\big(\tau_H + (1 - \alpha)\big) + (1 - r)(\tau_N + \beta)}. \tag{17.36}$$

17.4.2.3   Determining the ESS

Equation (17.36) allows us to calculate the fitness of a female from the parameters B_N, B_H, λ, S, a given male strategy p and female strategies α, β. McNamara et al. (2009) performed numerical computations which suggest that, for any given fixed p, there is always a unique best strategy (α, β) which, in turn, determines a female's fitness at her optimal strategy, $\gamma_F(p)$. For this unique $\gamma_F(p)$, we can use (17.30) and (17.32) to determine the fitness of helpful males, $\gamma_H(p)$, and non-helpful males, $\gamma_N(p)$, and consequently the advantage of helpful males, $D(p) = \gamma_H(p) - \gamma_N(p)$. By further calculations, we can look for values $p = p^*$ such that

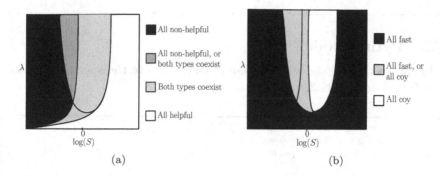

FIGURE 17.9: Qualitative outcomes of the coyness game of Example 17.5; (a) males, (b) females. A description of the parameters is in the text.

(i) If $p^* = 1$, then $D(1) > 0$ (i.e. helpful is an ESS),

(ii) If $p^* = 0$, then $D(0) < 0$ (i.e. non-helpful is an ESS),

(iii) If $0 < p^* < 1$, then $D(p) = 0$ and $D'(p^*) < 0$.

McNamara et al. (2009) found four qualitatively distinct outcomes; see also Figure 17.9.

1. All males are helpful and all females are fast, i.e. accept any male ($\alpha = 0, \beta = 1$ and, consequently $\tau_H = \tau_N = 0$); this happens when the sex ratio S is high or the encounter rate very low.

2. All males are non-helpful and all females are fast; this happens when the sex ratio is very low.

3. Both male types coexist and all females are fast; this happens either when the sex ratio is around 1 and the encounter rate medium to large or when the sex ratio is low and the encounter rate is low.

4. Both male types coexist and all females are coy (i.e. $\alpha > 0$, $\beta < 1$ and females inspect the males); this happens when the sex ratio takes an intermediate value and the encounter rate is not low.

In fact, in the case when females are coy, the authors find $\tau_H < \tau_N$ because if a male is accepted quickly then there is a time saving which partly makes up for the risk associated with early acceptance. On the other hand if a male is (wrongly) rejected quickly there is a cost to pay in the time it takes to find another male, but potentially there is a second cost as the new male might be one which should be rejected. Thus there is an asymmetry in costs which means females must be more certain before rejecting, as opposed to accepting, a male.

17.5   Python code

In this section we show how to use Python to solve the sperm allocation game from Example 17.3.

```
1   """ Sperm competition contest
2   Males with partial knowledge
3   The script solves a system of equations and
4   plots the ESS allocations to virgin and non-virgin females
5   """
6
7   ## Define parameters
8   E = 3        # Energy available for reproduction
9   D = 1        # Production cost of a unit of sperm
10  r = 0.2      # Discount factor for a second mating
11  e = 0.001    # Probability of female to stay unfertilized
12
13  ## Import basic packages
14  import numpy as np
15  from scipy.optimize import fsolve
16  import matplotlib.pyplot as plt
17
18  ## Initialize variables
19  # All possible probabilities of double mating q
20  all_q = np.linspace(0,1,101)
21
22  # Sperm allocation to virgin female
23  S01 = np.empty(len(all_q))
24
25  # Sperm allocation to non-virgin female
26  S2 = np.empty(len(all_q))
27
28  ## Define equations to solve
29  def eqs(variables, *Params) :
30      " returns the equations to be solved "
31      # Unpack the variables
32      (s01, s2, lbda) = variables
33
34      # Unpack the parameters
35      E, D, r, e, q = Params
36
37      # Set the equations
38      eq1 = (1-q)*e/((s01 + e)**2) + \
39                q*(r*s2 + e)/((s01 + r*s2 + e)**2) - lbda*D
40      eq2 = r*(s01 + e)/((s01 + r*s2 + e)**2) - lbda*D
41      eq3 = (s01*(1-q)/(s01 + e) + q*s01/(s01 + r*s2 + e) + \
42                q*r*s2/(s01 + r*s2 + e)) - lbda*E
43      return [eq1, eq2, eq3]
44
45  # For every q (and a corresponding index j)
46  for j, q in enumerate (all_q):
47      # Solve the equations
```

```
48      S01[j], S2[j], multiplier = fsolve(eqs, (0.1,0.5,0.5),
49                                  args = (E, D, r, e, q))
50
51  ## Plot the solutions
52  plt.plot(all_q, S01, label = '$s_{0,1}^*$, virgin female')
53  plt.plot(all_q, S2, label = '$s_2^*$, non-virgin female')
54  plt.xlabel('Probability of double mating, $q$')
55  plt.ylabel('Sperm allocation')
56  plt.legend()
57  plt.show()
```

17.6   Further Reading

For direct conflicts regarding mating, see the original classical works of Maynard Smith and Price (1973), Maynard Smith and Parker (1976) and Maynard Smith (1982), in addition to Houston and McNamara (1991) mentioned above. Broom and Rychtář (2018a) develop such a game as a sequential conflict. Laan et al. (2018) investigate such contests in zebrafish.

For a book on sperm competition see Birkhead and Møller (1998). Parker (1970); Simmons (2001) give an overview of sperm competition in insects; see Stockley et al. (1996); Young et al. (2013) for sperm competition in fish, Birkhead and Møller (1992) in birds and Shirokawa and Shimada (2020) in eukaryotes. Sperm competition has been modelled by a series of papers by Parker (1982); Parker et al. (1997); Ball and Parker (1997, 1998, 2000); Parker and Ball (2005); Ball and Parker (2007). Alternative models were developed in Fryer et al. (1999a,b), in Engqvist and Reinhold (2006, 2007); Engqvist (2012); Engqvist and Taborsky (2017), Mesterton-Gibbons (1999b) and Turnell et al. (2018).

For the battle of the sexes see Dawkins (1976). The dynamical analysis of the game with no ESS is done in Schuster and Sigmund (1981) (see also Mylius, 1999). Cressman and Křivan (2020) revisit the game where costs are given in terms of time lost (in the spirit of the original description of the problem). For the coyness game, see McNamara et al. (2009). For similar models involving the search for mates see Bergstrom and Real (2000); Collins et al. (2006); Janetos (1980); Ramsey (2011, 2012); Real (1990, 1991). For mating games similar to the coyness game, where males give costly but worthless gifts, see Sozou and Seymour (2005); Seymour and Sozou (2009). See Chapman et al. (2003) for a review on sexual conflict; see also Kokko and Johnstone (2002). Kokko et al. (2006a) suggest a unifying framework for sexual selection.

17.7   Exercises

Exercise 17.1 (Houston and McNamara, 1991). Show that the repeated Hawk-Dove game from Example 17.1 with only one contest per season (with payoffs given by (17.5)) can be analysed as a standard Hawk-Dove matrix game with payoff matrix given by

$$
\begin{array}{cc}
 & \begin{array}{cc} H & \quad\quad D \end{array} \\
\begin{array}{c} H \\ D \end{array} &
\left(\begin{array}{cc} R + \frac{V - zR}{2} & V + R \\ R & R + \frac{V}{2} \end{array} \right).
\end{array}
\tag{17.37}
$$

Find the ESS.

Exercise 17.2. Consider a repeated Hawk-Dove game as described in Example 17.1; but rather than assuming a monomorphic population (as done in Section 17.2.1), consider a polymorphic mixture where at the beginning of the season a proportion, p, of individuals play Hawk in every contest and a proportion $1 - p$ will always play Dove. Find an expression for the Population Hawk Probability in round $i + 1$ in terms of that in round i. Similarly, find an expression for the probability of survival to round $i + 1$ in terms of the probability of survival to round i. What is the payoff to Doves and to Hawks in this population if the season has $K_{\max}$ rounds?

Exercise 17.3 (Houston and McNamara, 1991). Show that the payoff function in the game described in Example 17.2 is given by (17.9). Show that for a fixed q, it is a monotone function of p.

Hint. Based on Figure 17.4, a male mates with probability $(1 - \theta) + \theta w$ after which it receives a payoff V. A male survives the whole cycle with probability $s(1 - \theta\delta)$. The expected lifetime payoff, E, is thus given by

$$
E = \big((1 - \theta) + \theta w\big)V + s(1 - \theta\delta)E.
\tag{17.38}
$$

Exercise 17.4. For the sperm competition game of Section 17.3.2, show that in the special case with $\varepsilon = 0$, the optimal sperm allocation strategy is given by (17.20).

Exercise 17.5. The Python code from Section 17.5 solves the sperm competition game in which the males have partial knowledge (Section 17.3.2). Adapt the code for the case when males have no knowledge.

Exercise 17.6. Considering the battle of the sexes game from Section 17.4.1. When there are no ESSs to this game find the internal equilibrium and comment on the long term behaviour in this case.

Hint. Use the results for bi-matrix games from Section 8.2.

Exercise 17.7. Analyse the battle of the sexes game from Section 17.4.1 under the assumption that it is a sequential game, where (a) the male chooses first, (b) the female chooses first.

Exercise 17.8. In the game from Section 17.4.2, assuming that all females are fast, find the evolutionarily stable male strategy as a function of the model parameters, i.e. show when (a) all males are helpful, (b) all males are non-helpful and (c) there is a mixture of the two.

Chapter 18

Signalling games

18.1 The theory of signalling games

In this chapter we will discuss the concept of biological signalling, and look at some of the important different categories. A biological signal is a behaviour or physical aspect by a transmitting individual, the *Sender*, which gives information to a recipient, the *Receiver*, which may affect the actions of the Receiver. In particular, signals evolve and are effective, because of the effect that they have on the Receiver's behaviour (Maynard Smith and Harper, 2003). The basic signalling game is that of Lewis (1969). Here, the Sender possesses some information unknown to the Receiver and sends a signal. The Receiver then takes some action in response to the signal, and this then gives a payoff both to the Sender and to the Receiver; the payoffs depend upon the action selected and the information (e.g. the quality of the Sender Skyrms, 2008).

We can distinguish signalling systems in a number of ways. Perhaps the most fundamental is the distinction between situations where the interests of the Signaller and Receiver are identical, and those where there is some, at least partial, conflict between the two (Kane and Zollman, 2015). Where there is common interest, such as dancing in honey bees (Kirchner et al., 1988) (an intraspecific example) or communication between the greater honeyguide (*Indicator indicator*) and humans (Isack and Reyer, 1989) (an interspecific example), both Sender and Receiver benefit from the signal communicating the underlying reality. Thus evolution favours accurate, and honest, signals.

When there is some potential conflict, however, it may be that dishonest signalling is beneficial for some senders. We note that a signal is only useful if it gives the Receiver (sufficiently accurate) information upon which it can act (Schaefer and Ruxton, 2011). for example, in the signalling game of (Zollman et al., 2013), honest signalling allows receivers to distinguish between different types of signallers (this is a very common use of signals). It is this more interesting situation, where there is the potential for dishonesty, that we will consider.

There are two distinctions that we make between signals of this type. Firstly, between fakable or unfakable signals (Holman, 2012; Biernaskie et al., 2014). If a signal is unfakable (also referred to as an *index*), it means that even though dishonesty might be preferable on the part of the Signaller, it is simply not possible (Maynard Smith and Harper, 1995). Thus Thomson's gazelles leap high into the air, *stotting*, when faced with some predators (such as wild dogs) to show that they are in good condition and so not worth chasing; those that are not in good condition are physically incapable of jumping as high as those that are (Fitzgibbon and Fanshawe, 1988). We address such signals in Section 18.3.1. If a signal is fakable, then dishonesty is possible, and we must consider the circumstances when it can or cannot evolve. There is a particular interest in honest signals, and in Section 18.2 we consider the classical example of the handicap principle (Zahavi, 1975). We also note that this neat distinction into two classes is not universally agreed upon, and in section 18.3.2 we consider a game which considers signals which have aspects of both types.

The second distinction is between costly and costless signals. In all of the examples above, there is some cost associated with the signal, and this is the subject of Sections 18.2 and 18.3 as a whole. Dishonesty is often discouraged because it is excessively costly, for example if a signal indicates that the Sender is physically strong, it may be more costly for a weak individual to fake than for a strong individual to honestly indicate. But what if there is no cost to a signal; how can the honesty of a signal be relied upon then? We consider such a scenario in Section 18.4.

We begin, however, with perhaps the most well-known signalling idea in evolutionary biology, that of the handicap principle.

18.2   Selecting mates: signalling and the handicap principle

In this section we consider sexual selection, and in particular, an important mechanism for it based upon signalling. Thus there is a natural link between this game and the mating models of Chapter 17; indeed in the first edition of this book, this section was part of that chapter, before we separated it out and added more signalling models to form its own chapter.

Sexual selection can affect the sexes to a different extent, based upon the level of parental investment (although this can also, in turn, interact with other fundamental parameters). In a population where parental investment is roughly equal, sexual selection is likely to be equal between the sexes, but if it is highly unequal, e.g. if females invest much more than males, then sexual selection amongst the sex providing the least investment should be stronger (Trivers, 1972). In mating systems which are highly polygynous, the male role in rearing young can be non-existent and sexual selection will act more strongly on males, which will increase the female tendency to be selective of which male(s) to mate with. How should females select the best mates, and how should males try to ensure it is they who are selected? This often leads to the development of impressive but non-functional ornaments, the classic example being the peacock's tail, or elaborate demonstrations of male quality in physical displays, or in the case of bowerbirds (Rowland, 2008) physical displays and the development of elaborate constructions which take many hours to produce, but which serve no function other than to attract a mate.

Fisher (1915) suggested that sexual selection might occur due to the self-reinforcement of female preferences (the sexy son hypothesis, see Section 5.4.6). Suppose that a certain type of male becomes preferred by females in the population for whatever reason. Provided that the preferred trait can be inherited by sons from their fathers and there is positive correlation between the preferences of females in successive generations, the fact of current preference makes mating with these preferred males beneficial, since any sons that the female has with these males will also be preferred and so the female's fitness will be higher than if she mated with other males.

We look at the most widely accepted explanation of such a selection mechanism, the classical *handicap principle* of Zahavi (1975, 1977). Zahavi states that these ornaments are signals of an individual's underlying quality. The more elaborate the ornament, the more costly it is to bear, and the greater the underlying quality of the individual which bears it, since it is able to do so and inferior individuals are not (at least except at prohibitive cost). Central to this argument is that for any signal of underlying quality to be honest, and so reliably inform the Receiver of the signal about the Sender's quality, there must be a cost associated with the signal. Initially considered as an alternative mechanism to the Fisher process, it is now arguably recognised as the more important of the two. We shall focus on the model of Grafen (1990a,b) which provides a mathematical modelling framework for the handicap principle.

Example 18.1 (Handicap principle, Grafen, 1990b). There is a population of males of varying quality. Quality cannot be directly observed by females, but males give a signal of their quality. Females receive the signals, and make the decision whether to mate with the male or not based upon their inference of the male's quality from the signal. Set up the model and analyse the game to see how males should signal and how females should interpret the signals.

18.2.1 Setting up the model

Assume a male of quality q gives a signal $a = A(q)$ of their quality. The male's strategy is the signalling level it chooses based upon its quality, specified by the function $A(q)$. When a female receives a signal a, she interprets it as a perceived quality $p = P(a)$. The female's strategy is the quality level that she allocates to a signal, specified by the function $P(a)$. Thus we have an asymmetric game with sequential decisions (see Section 8.4.2.2), but where there is only a single type of female, and an infinite number of types of male.

The fitness of the male depends upon his true quality, the selected level of advertising, and the female perception of his quality through his advertising level $w(a, p, q)$. The fitness of a female will depend upon the difference between her perception of a male's quality and his true quality, and she pays a cost $D(q, p)$. This may depend upon the male encountered, so it will also depend upon the likelihood of meeting each different quality of male. Assuming meeting any male is equally likely and the p.d.f. of males of quality q is $g(q)$, then the female payoff is

$$E_f = - \int D(q, p) g(q) dq, \qquad (18.1)$$

where we remember that p depends upon a which, in turn, depends upon q. An evolutionarily stable pair $A^*(q), P^*(a)$ satisfies the condition that if almost all individuals play these strategies, any mutant playing an alternative will do no better. This means that,

1) for every q, there is no benefit to a male of quality q from changing his advertising level from $A^*(q)$ to some other a, and also

2) any female which changes her perception function will do no better, averaged over all of the males that she might meet.

Thus we require

$$w\Big(A^*(q), P^*(A^*(q)), q\Big) \geq w\Big(a, P^*(a), q\Big) \text{ for all } a, q, \qquad (18.2)$$

and

$$\int D\Big(q, P^*(A^*(q))\Big) g(q) dq \leq \int D\Big(q, P(A^*(q))\Big) g(q) dq \text{ for all } P(a), \quad (18.3)$$

where the equalities happen only if $a = A^*$ and $P = P^*$.

18.2.2   Assumptions about the game parameters

We see that the ESS depends upon three functional forms: w, D and g. A number of assumptions are made by Grafen (1990b) about these functions. As seen below, the assumptions on g and D are relatively mild, but the assumptions on the male's payoff need to be detailed.

We will assume that g is non-zero on an interval, so there are no distinct groupings of male qualities with gaps inbetween where there are no males of intermediate quality. In particular, there is a minimum male quality q_{min}. As all individuals are from a single species, and a reasonable level of mixing in terms of mating can be assumed, then these assumptions are certainly reasonable.

Note that there is also a minimum level of advertising allowed a_{min}. This can be thought of as giving no signal, or not possessing the trait in question at all.

We will assume that

$$D(q,p) = 0 \quad q = p,$$
$$D(q,p) > 0 \quad q \neq p. \tag{18.4}$$

This means that there is no penalty to the female for assessing the male's quality correctly, and some penalty for not assessing it correctly. In much of the theory, the precise functional form of D does not matter as completely honest signals are the focus, and any penalty for mistaking the quality may be sufficient. A typical penalty function is of the form

$$D(q,p) = c|p - q|. \tag{18.5}$$

It should be noted that it is not always reasonable to assume that all misinterpretations carry such a penalty. We see in Section 18.2.6 below that in situations when there are binary choices (e.g. mate or not mate) errors in perception are only costly if this involves the wrong decision being made, so that many errors have no cost.

The conditions on the male payoff w are more detailed. The key assumptions are as follows.

- $w(a, p, q)$ is continuous. More than that, we need all of the first partial derivatives to exist. Grafen (1990b) uses the terms w_1, w_2, w_3 for the partial derivatives of w with respect to a, p and q, respectively.

- Increasing advertising increases cost, so $w_1 < 0$.

- In any reasonable system it is better to be thought of as high quality than low, so $w_2 > 0$.

- A higher advertising level is more beneficial to a good male than a poor one. This means that

$$\frac{w_1(a, p, q)}{w_2(a, p, q)} \tag{18.6}$$

is strictly increasing with q, i.e. the better the male, the smaller the ratio of the marginal cost to the marginal benefit for an increase in the advertising level (costs being negative rewards).

If the second derivatives of w with respect to a and q, w_{13}, and with respect to p and q, w_{23}, exist, then sufficient conditions for (18.6) to be increasing in q are

$$w_{13} > 0, \tag{18.7}$$
$$w_{23} \geq 0. \tag{18.8}$$

Inequality (18.7) means that the marginal cost of advertising is less for good quality males than for bad, whereas inequality (18.8) implies that an increase in a male's perceived quality by the females is at least as beneficial to a good male as a bad one.

18.2.3   ESSs

Grafen (1990b) shows that under the assumptions above, an evolutionarily stable pair of strategies P^*, A^* exists and satisfies the following conditions:

$$P^*(a_{\min}) = q_{\min}, \qquad (18.9)$$

$$\frac{dP^*(a)}{da} = -\frac{w_1\big(a, P^*(a), P^*(a)\big)}{w_2\big(a, P^*(a), P^*(a)\big)}, \qquad (18.10)$$

$$P^*\big(A^*(q)\big) = q \ \text{ for all } q. \qquad (18.11)$$

First, note that there is a unique solution to (18.9)–(18.11). The conditions (18.9)–(18.10) unambiguously define P^*, which will be monotone increasing, over the whole interval, since w_1 and w_2 exist and are of opposite signs. This then allows us to find an A^* which satisfies the third condition. Thus we have a unique pair (A^*, P^*) that satisfy the conditions.

Now we will show that such a pair is indeed an ESS. Consider a function $a \mapsto w\big(a, P^*(a), q\big)$ that corresponds to the value of advertising for a male with quality q. To show that (18.2) holds, we will show that the function is maximised at $a = A(q)$. Indeed, the marginal value of advertising is

$$\frac{d}{da} w(a, P^*(a), q) = w_1(a, P^*(a), q) + \frac{dP^*(a)}{da} w_2(a, P^*(a), q). \qquad (18.12)$$

Using (18.10) above and dividing by (the positive value) $w_2(a, P^*(a), q)$, we obtain that the marginal value has the same sign as

$$\frac{w_1\big(a, P^*(a), q\big)}{w_2\big(a, P^*(a), q\big)} - \frac{w_1\big(a, P^*(a), P^*(a)\big)}{w_2\big(a, P^*(a), P^*(a)\big)}. \qquad (18.13)$$

By (18.6), the first term of (18.13) is increasing with q. Thus the expression in (18.13) is negative if $q < P^*(a)$ and positive if $q > P^*(a)$. $P^*(a)$ is an increasing function of a, with inverse function A^*, so the marginal value of advertising is positive if $a < A^*(q)$ and negative if $a > A^*(q)$. Hence $A^*(q)$ is the (globally) optimal strategy.

The ESS condition (18.3) also holds since the minimum possible value of the integral (18.1) is zero and this minimum is achieved at $P^*\big(A^*(q)\big) = q$.

18.2.4   A numerical example

Example 18.2 (Grafen, 1990b). Consider a specific male payoff function

$$w(a, p, q) = p^r q^a = p^r e^{a \ln q}, \qquad (18.14)$$

with qualities q in the range $q_0 \le q < 1$ and advertising levels $a \ge a_0$. Here the payoff represents the probability of a male's survival to adulthood (the viability) q^a, multiplied by the probability of a successful mating given survival, p^r. Find the ESS.

We can see that

$$\frac{w_1}{w_2} = \frac{\ln(q)p^r e^{a\ln q}}{rp^{r-1}e^{a\ln q}} = \frac{p\ln q}{r}. \tag{18.15}$$

From (18.10) and (18.11),

$$\frac{dP}{da}\big(A(q)\big) = -\frac{P\big(A(q)\big)\ln\big(P\big(A(q)\big)\big)}{r} = -\frac{q\ln q}{r}. \tag{18.16}$$

Also, by (18.11),

$$\frac{dA}{dq}(q) = \frac{1}{\frac{dP}{da}\big(A(q)\big)} = -\frac{r}{q\ln q}. \tag{18.17}$$

This yields

$$A(q) = -r\ln\big(\ln(q)\big) + C, \tag{18.18}$$

where, since $A(q_0) = a_0$, the constant C is given by

$$C = a_0 + r\ln\big(\ln(q_0)\big). \tag{18.19}$$

Thus

$$A(q) = a_0 - r\ln\left(\frac{\ln(q)}{\ln(q_0)}\right), \tag{18.20}$$

which implies that

$$\frac{\ln(q)}{\ln(q_0)} = \exp\left(\frac{a_0 - A(q)}{r}\right). \tag{18.21}$$

Finally, since $P\big(A(q)\big) = q$, we obtain

$$P(a) = q_0^{\exp(-(a-a_0)/r)}. \tag{18.22}$$

These functions are shown in Figure 18.1. We see that the viability of intermediate males is low in comparison to both low and high quality males. The fitness $w(a, p, q)$ of the intermediate males is still higher than lower quality males, however, as they have a much higher chance of being mated with by females than low quality males.

18.2.5 Properties of the ESS—honest signalling

The property (18.11) of the ESS is particularly interesting, because it states that the female perception of the advertising level of the male is precisely the quality of the male for all male qualities. Every male picks a unique advertising level which can be used to pinpoint their quality, with the higher the quality the higher the level of advertising. Thus the signalling system is completely honest. It is the fact that signals are costly, and in particular, more intense signals are relatively more expensive to poorer quality males than better males (condition (18.6) above) which allows this to happen. The better

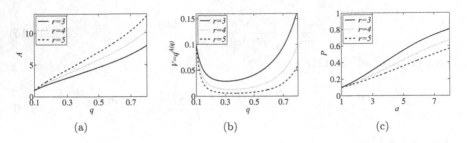

FIGURE 18.1: ESSs in the handicap principle from Example 18.2 with male payoff function $w = p^r q^a$ and parameters $q_0 = 0.1, a_0 = 1, r = 3, 4, 5$. (a) The ESS advertising level $A(q)$, (b) the ESS viability level $V = q^{A(q)}$, (c) the ESS perception level $P(a)$.

males produce expensive signals (or handicaps) because they are able to do so and poor males cannot, so any individual seen with a large handicap must be of good quality.

Thus honest signals relating to the handicap principle are one possible solution of such systems. Remarkably, Grafen (1990b) also shows that if (P^*, A^*) is an ESS, $w_2 > 0$ and the signalling level $A^*(q)$ is increasing, then

$$P^*\big(A^*(q)\big) = q \text{ for all } q, \tag{18.23}$$

$$w_1 < 0, \tag{18.24}$$

and the function from (18.6) is strictly increasing in q in the vicinity of the curve $(A^*(q), q, q)$.

Thus, effectively, in any system where the level of advertising increases with quality, signalling must be honest and follow the handicap principle. Following Grafen (1990b), this can be explained as follows. Suppose that different quality males give distinct signals, so that signals are reliable, in the sense that their correct quality can be inferred from them. Grafen (1990b) contends that such signals must be reliable or females would not use them, but see Johnstone (1994) and Broom and Ruxton (2011). Since females can infer exact quality, they can set their assessment rules to attribute the correct quality to any signal, so the signals are honest. To be stable it is clear that higher costs must be associated with signals of higher quality, otherwise an individual could pay the same or less for a more favourable outcome. So better males advertise more; for this to be stable, the marginal cost of advertising must be more for worse males.

Thus the handicap principle is enough to explain ornaments such as the peacock's tail, and the Fisher process above is not necessary (this is, of course, not to say that it does not play a role).

We have seen from above that under reasonable assumptions if all errors in assessment are costly, and that signals are transmitted with complete accuracy, then a completely honest signalling system results with a continuous range of signals used (unless there is no signalling at all). However, it is reasonable to assume that signals will not generally be transmitted without error, and that sometimes the signal received by the Receiver may not be the same as the one transmitted by the signaller. This scenario was investigated by Johnstone (1994) and he found that the existence of such errors can have significant consequences.

When there are errors in the transmission process, rather than a continuous range of signals, the signalling population splits into two groups, those which signal high and those which signal low. This result is particularly interesting, because the outcomes in real populations (Morris, 1957; Cullen, 1966) are often of this "all or nothing" variety. Thus it seems that perceptual error may be an important driving force for signalling mechanisms.

18.2.6 Limited options

We mentioned above that the existence of perceptual errors could lead to all or nothing signalling. Another way in which this can arise is when the number of choices of the Receiver is limited. It was assumed in Grafen (1990b) that any misclassification of quality would have a cost. However, it may be that there is no cost to misclassification unless it leads to a change in decision. In many scenarios the Receiver is limited to a small number of (often two) choices; e.g. attack or not attack, mate or not mate. Typically, the strategies adopted by receivers will be of the form attack if and only if the level of signal is sufficiently small, or mate if and only if it is sufficiently high. Broom and Ruxton (2011) considered this in general, and used the following simple example for illustration.

Example 18.3 (Signalling game, Broom and Ruxton, 2011). Males signal their quality to females; signals are costly (a better signal costs more; a male of lesser quality also pays more). Females can choose either to mate with a specific male or not. A female gets no benefits and pays no cost if she does not mate. If she mates, she pays a fixed cost for the mating itself and gets a benefit based on the mate's quality. Males get a benefit of 1 if mated with, and 0 otherwise. Set up and analyse the game.

Let q be the quality of a male. The cost of a signal a can be modelled as a/q. Thus the reward for a male is given by

$$E_m(a, q) = \begin{cases} -\frac{a}{q} & \text{if he did not mate,} \\ 1 - \frac{a}{q} & \text{if he mated.} \end{cases} \tag{18.25}$$

Let α be the fixed cost for a female to mate. We may assume that the female's benefits of mating with a male of quality q is just q and thus the reward function for a female is given by

$$E_f(q) = \begin{cases} 0 & \text{if she did not mate,} \\ q - \alpha & \text{if she mated.} \end{cases} \tag{18.26}$$

Broom and Ruxton (2011) show that there is an "honest" signalling system with all males above quality α giving a signal of level α and all other males giving a signal of 0, with the females mating only if they receive a signal of α or higher. To show that this system is stable against any change in strategy by either males or females is Exercise 18.2. In particular, it is clear that it makes no sense to give anything but the minimum signal that will elicit any particular response.

We note that the system described here is not fully consistent with the conditions on $w(a, p, q)$ of Grafen (1990b) from Section 18.2.2, since the benefits to the individuals are allocated by discrete choices. A male's payoff

$$w(a, p, q) = p - \frac{a}{q} \tag{18.27}$$

would satisfy all of the conditions from Section 18.2.2, but this requires females being able to make a different choice for every different perceived fitness.

Similarly, the condition on $D(p, q)$ is not satisfied. In effect in the Broom and Ruxton (2011) model, the penalty paid by the Receiver will be 0 unless $p < \alpha < q$ or $q < \alpha < p$, so that misclassification leads to the wrong decision.

18.3   Alternative models of costly honest signalling

18.3.1   Index signals

In this section we discuss *index signals*, as described in Maynard Smith and Harper (1995). These are unfakable signals which are directly linked to the property of the animal that is being signalled about. Thus, for example, the stotting gazelles that we mentioned in Section 18.1 are signalling their physical prowess (in running) by a highly correlated physical activity (jumping); although we should note that while these two activities are highly correlated, the signal is probably not completely unfakable.

Maynard Smith and Harper (1995) give two further examples of this situation. Funnel web spiders signal their size by vibrations they send out on their webs, and clearly smaller spiders concede (Riechert, 1978). A smaller loser can be turned into a winner just by adding a weight to its back, indicating that the spider is not making a distinct strategic decision when it sends out vibrations; it just vibrates in the usual way, but the extra weight is faithfully reported to its opponent. Thapar (1986) showed tigers signalling their size by leaving marks on trees, as high as they could reach, which is similarly a signal that a smaller animal cannot make.

In this section we will not consider a model of signalling as an index. This is simply because such a model is not really needed if we are convinced that an index is at work. Low quality signallers cannot signal, and high quality signallers can send the signal, which will be positively responded to by receivers. They will thus send the signal if and only if the benefit outweighs the cost. We can see an appropriate model of index signals in Section 18.3.2 below, (under the assumption that $s_H = 1$ and $s_L = 0$, where these terms are defined therein, see Exercise 18.5).

Maynard Smith and Harper (1995) also discuss the example of swallows *Hirundo rustica* (Moller, 1993) as a case where the tails of male birds provide distinct forms of signal, one aspect an index and the other a handicap. Females prefer long and symmetric tails; a long tail makes flying more difficult, and so is a handicap, but a symmetric tail makes flying easier, and so is an index (asymmetric tails as well as reducing a bird's ability to fly also reduces its ability to get a mate and is thus likely simply a feature that the bird cannot avoid having).

18.3.2 The Pygmalion game: signalling with both costs and constraints

In Section 18.2 and Section 18.3.1 we have considered models where signals are classified as either handicaps (fakable signals) or indices (unfakable signals). Huttegger et al. (2015) proposed that these should not be considered as distinct alternatives, and proposed a new game, the Pygmalion game (see also Safley et al. (2020) for further developments on this game), which includes features of both models; in particular, examples of each are extreme special cases of the Pygmalion game.

Huttegger et al. (2015) initially consider a simple model which exhibits the key features of the handicap principle (see also Zollman et al., 2013). Here we have two types of sender, either high quality or low quality. The Sender either sends a signal to indicate it is of high quality, or it does not. The Receiver can take one of two actions, A or B. Both types of sender prefer A (obtaining a return of 1 if this is chosen, as opposed to 0 if B is chosen). The Receiver only wants to choose A for a high quality sender, but wants to choose B for a low quality sender. It gets a return of 1 if it makes the right choice, but 0 otherwise. Finally, the Sender pays a cost if they send the signal; this is c_H for a high quality sender, but c_L for a low quality sender.

The payoffs for individuals are then the returns as described above, reduced by the corresponding cost for the senders, if this is incurred. We summarise these in Table 18.1.

The game is illustrated in Figure 18.2, which shows the game in extensive form. It is clear (see Exercise 18.4) that, provided that $c_L > 1 > c_H$, so that the cost of the signal to the low quality type is in excess of its potential

TABLE 18.1: Strategies and payoffs for the Pygmalion game.

Sender quality	Signal	Receiver choice	Sender payoff	Receiver payoff
High	Yes	A	$1 - c_H$	1
High	Yes	B	$-c_H$	0
High	No	A	1	1
High	No	B	0	0
Low	Yes	A	$1 - c_L$	0
Low	Yes	B	$-c_L$	1
Low	No	A	1	0
Low	No	B	0	1

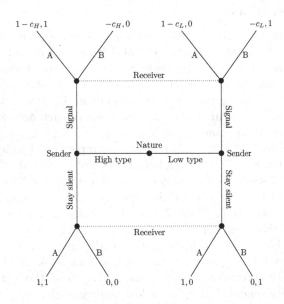

FIGURE 18.2: Extensive form representation of a "differential cost" model of the handicap principle, adapted from Huttegger et al. (2015).

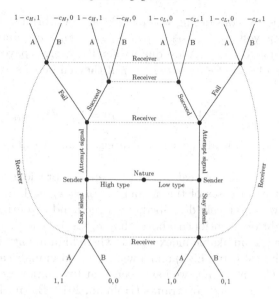

FIGURE 18.3: Pygmalion game. Conflict-of-interest game that includes uncertainty as to whether the signal will be successfully sent, adapted from Huttegger et al. (2015).

reward but the cost to the high quality type is less than the potential reward, there is an equilibrium where the high quality senders send the signal, the low quality senders do not, and the Receiver responds to the signal by choosing A and the absence of the signal by choosing B. This is an example of the handicap principle at work, where the differential costs of signals lead to honest signalling.

Huttegger et al. (2015) then generalized the above game to give the Pygmalion game of our section title. Here there are the same types, costs and rewards as in the previous game and shown in Table 18.1. The proportion of high quality senders in the population is p (see Exercise 18.6). The different types can still send signals, but this time it is not certain that they succeed in communicating with the Receiver. High (low) quality individuals succeed in communicating the signal with probability s_H (s_L). The Receiver then either sees a signal, which has occurred if the Sender (of whichever type) has attempted to signal and this has then succeeded, or does not see a signal, if the Sender did not attempt to send a signal or did attempt to, but failed.

Thus we have the new game which can be summarised in Figure 18.3. It was shown that an honest signalling system (high quality individuals attempt to signal, low quality ones do not, Receivers choose A if a signal is received and B if it is not) occurs if and only if

$$s_H > c_H, s_L < c_L, p < \frac{1}{2 - s_H}. \tag{18.28}$$

We can see this by considering the payoffs at equilibrium. A high quality sender receives payoff $s_H(1 - c_H) - c_H(1 - s_H) = s_H - c_H$, and would receive 0 if they chose not to signal. A low quality sender receives payoff 0, and would receive $s_L(1 - c_L) - c_L(1 - s_L) = s_L - c_L$ if they chose to signal. A Receiver receives

$$E_R = ps_H + 0 \times p(1 - s_H) + (1 - p) = 1 - p(1 - s_H) \qquad (18.29)$$

and would receive p if it always chose strategy A and $1 - p$ if it always chose strategy B.

We can see that the classical signalling scenarios previously discussed represent two (extreme) cases of this model. If $s_H = s_L = 1$ then we have the situation that we previously described, with the handicap principle (with the additional condition mentioned above that $c_L > 1 > c_H$). If $s_H = 1$ and $s_L = 0$ we have the unfakable index signal, where high quality individuals signal (unless their cost is too high) and low quality individuals cannot signal (so their cost is irrelevant). Thus we have seen that two major factors for honest signalling, signal costs and constraints (Holman, 2012; Biernaskie et al., 2014) are combined in the Pygmalion game. In general, we would assume that

$$1 \geq s_H \geq s_L \geq 0, \qquad (18.30)$$

so that the higher quality individuals have at least as good a chance to succeed in giving the signal as the lower quality individuals. For intermediate cases, we might expect that the inequalities in (18.30) become strict, so that each type may either succeed or fail to give the signal, but that the higher quality individuals have a higher chance of success.

We note that there is another interesting solution here, which does not have a counterpart in the original handicap principle theory. Here both types of sender send the signal, and the Receiver chooses A if and only if the signal is received. The conditions for this to occur are

$$s_H > c_H, s_H \geq s_L > c_L, p(s_H + s_L - 1) > 2s_L - 1. \qquad (18.31)$$

To show this is Exercise 18.7.

18.3.3   Screening games

An alternative scenario was considered by Archetti et al. (2011) (see also e.g. Archetti, 2019). Here there are two types of individual, termed by Archetti et al. (2011) as Principal and Agent. All agents want to interact with principals but principals have specific requirements for agents. In the simplest scenario there are two types of agent only, those of high quality q_H and those of low quality q_L, and principals want to interact with q_H agents but not with q_Ls. This is exactly as described in Section 18.3.2, where principals correspond to receivers and agents to signallers (see also Section 18.2.6, where this would

there were only two quality types). Other aspects of the scenario
however, as we describe below.

ing games, the Sender chooses a signal, and then the Receiver
ce based upon it (in Section 18.2.6 to accept or reject the sig-
reening games there is no signal, rather it is the Principal who
They set a task with a specific cost and the Agent then must
er to meet that cost or not. The Principal will accept any agent
ie cost.

rincipal to achieve its desired outcome it should set a cost that
y q_H individual is willing to pay and a low quality q_L individual
hen leads to a fixed cost task met by q_Hs who thus pay a fixed
selected, but not met by q_Ls who thus pay a zero cost and are

ome is thus very similar to the games in Sections 18.3.2 and 18.2.6,
specifics (and the order) of the choices made are different. Given
y we have omitted adding the theoretical calculations, though
e distinctive features, see Archetti et al. (2011) and Exercise 18.9.
ig games while distinct from the previously described signalling
ot actually involving any signalling, are perhaps best thought of
s family.

nalling without cost

ental feature of the handicap principle, and the models described
t signalling has to be costly to be effective. Without such a cost,
hoose whichever signal maximises their return from the interac-
eans that signals are not reliable indicators of quality, and so in
rs ignore the signal. Bergstrom and Lachmann (1998) came up
guing way that signals can be (partially) honest when they do
associated cost, based upon the Sir Philip Sidney game (May-
1991; Johnstone and Grafen, 1992).

3.4 (Sir Philip Sidney game, Bergstrom and Lachmann, 1998).
individuals, donor and signaller with relatedness k. The signaller
ost) signal of need based upon its fitness $x \leq 1$, and the donor
nate or not based upon the level of the signal and its own fitness
signaller receives the food item its fitness increases to 1, otherwise
x; if the donor keeps the food item its fitness becomes 1, and
nains at y. Find the ESS of the game.

that the inclusive fitness of the donor is $y + k$ if it donates, and
oes not. Similarly, the inclusive fitness of the signaller is $ky + 1$ if

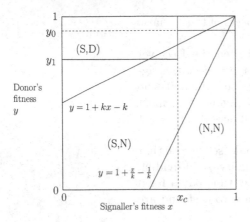

18.4: Optimal choices for the two individuals in the signalling
atives game from Example 18.4. (S,D)—signal and donate, (S,N)—
lo not donate, (N,N)—do not signal and do not donate. We note
will not be played in practice, as the donor does not know the
e signaller, only whether it signals or not. The signaller signals if
donor donates if $y > y_1$ and a signal is received, or $y > y_0$ and no
eived.

onates, and $k + x$ if it does not. We thus see that there are three
sibilities (see Figure 18.4).

$k + kx < y$, then it is in the interests of the donor to donate and
gnaller to receive.

$1/k + x/k < y < 1 - k + kx$, then it is not in the interests of the
to donate but it is in the interests of the signaller to receive.

$1 - 1/k + x/k$, then it is not in the interests of the donor to donate,
the interests of the signaller to receive.

ere are circumstances when both the donor would like to donate
eiver to receive (and circumstances when neither wishes the dona-
ade). If the donor knew the fitness of the signaller, then it should
ate if $1 - k + kx < y$. Bergstrom and Lachmann (1998) found an
system of "pooled equilibria" where:

gnaller signals high need if it has a sufficiently low fitness, below
critical threshold x_c and low need otherwise,

onor donates if it receives a signal and its own fitness is above a
l threshold y_1, or if it does not receive a signal and is above a
old $y_0 > y_1$, and does not donate otherwise.

he stability of this system has been brought into question, however

linator signalling games

tion we consider a patch of plants which produce a floral signal to
inating animal. The bigger the combined signal from the patch,
linators will visit (Moeller, 2004; Ghazoul, 2006). For instance
(2014) introduced a very conspicuous species into a meadow
th many different species, and found that it greatly enhanced
ness of the plants around it. This is known as the *magnet effect*
2; Molina-Montenegro et al., 2008). When a pollinator visits the
visit plants in the patch whether or not they are producing a
ık signal. Thus there is the potential to cheat by free-riding on
others.

al. (2018) showed that the plant *Moricandia moricandioides* was
ţe the level of investment it made in its floral display, depending
ı of neighbours it had in its patch, specifically whether plants were
t. Plants growing with kin would typically have larger displays
ıat were not. Torices et al. (2018) suggested that kin selection
ork here, and that this might have an influence on floral displays

(2021) developed the following mathematical model based upon
Torices et al. (2018). Whilst there is significant complexity in
analysis, the components of the model are simple. It considers
ı N plants, where the number of seeds that a plant produces
nds upon two factors, f the size of its floral display, and $P =$
f_N) the total number of pollinators attracted to the patch, which
ːurn, upon the floral displays of all of the patch members. The
increasing in each of the individual terms f_i due to the magnet

d note that whilst $S(f, P)$ is an increasing function of P, it is
ın increasing function of f. For low f few flowers attract few
ut for large f so much investment has gone into flower production
·e few remaining resources to produce seeds. A number of other
assumptions on the functions S and P and their derivatives were
ːr to prove some key mathematical results (we will omit the full

ortion of viable seeds produced within a patch could potentially
ıng function of the average relatedness r within the patch, due
ţ depression (Liao et al., 2000). This led to the number of viable
ed by a plant to be given as

$$\bar{S}(f, P) = \big(1 - \delta(r)\big)S(f, P), \qquad (18.32)$$

s the proportion of unviable seeds, and a non-decreasing func-

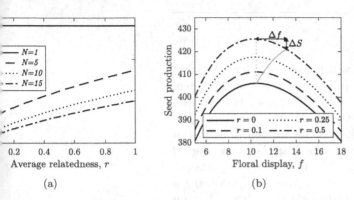

18.5: (a) Nash equilibrium floral display as a function of the av-
dness coefficient for different patch sizes. (b) An individual plant
in terms of decreasing its own seed production (by ΔS) as it in-
loral display (by Δf) to attract pollinators for the whole patch.
0 and $N = 15$. The vertical grey dashed line represents the flo-
it which the plant achieves the maximum S. The short grey solid
the seed production if the focal plant display is at the equilibrium
.

s accrued from all of the other plants, is then given by

$$\ldots, f_N) = \bar{S}\big(f_i, P(f_1, \ldots, f_N)\big) + r \sum_{j=1, j\neq i}^{N} \bar{S}\big(f_j, P(f_1, \ldots, f_N)\big).$$

(18.33)

l. (2021) showed a number of results for this system. In particular,
l that the optimal floral display is positively correlated with the
coefficient, as illustrated in Figure 18.5a. Secondly, in the model,
a cost in terms of decreasing their individual seed production in
crease their total payoff. They do this by increasing their floral
attracting pollinators to the whole patch, so helping their relatives
re seeds. This is illustrated in Figure 18.5b. We can see this as
o the cost from Hamilton's rule as discussed in Chapter 15 and
ection 15.1. We should note that the model here uses relatedness in
rward way as part of a game-theoretical model. This book similarly
n the game theory, and so we do not consider the significant and
debate around inclusive fitness (Nowak et al., 2010; Abbot et al.,

lso shown that, if there is no inbreeding suppression and assuming
nts are using the equilibrium level of floral display, the individual
production is an increasing function of relatedness r. It was noted

ɔdel of Sun et al. (2021) assumed that a plant's attractiveness
 the likelihood of pollinators visiting a patch, and once there
 ɪsit all plants with equal probability. As they noted, this might
 ɔe. Pollinators can distinguish between the individual plants and
 more likely to visit those with a larger floral display (Biernaskie
 7). Sun et al. (2021) considered this possibility and showed that
 ɔould not be affected when variability in floral display was small,
 variability there could be significant differences, for example the
 ɪsplay could become decreasing in r.

ɔhon code

ction we show how to use Python to solve the screening game
18.3.3.

```
l solution of the Screening game from Archeti et al (2011)
rs are set as in their Fig 1


t basic packages
umpy as np

e inputs/variables

sts of interaction with a principal for agents q and Q
= 0.6
= 0.1

sts of interactions with q and Q for principal
.3


sts of not interacting
.2  # Cost to Principal for not interacting with Agent
= 0.2 # Cost to Agent for not interacting with Principal

st of being fully demanding
05

equency of Q Agents


l possible efforts of being demanding
not demanding at all
fully demanding
e 2 by 100000 to get continuous rather than discrete case
```

```
 calculations
late Principal's cost of interaction with q
eracts only if 1-alpha_q*effort > 1-alpha_0 and
Principal pays pi_q.
wise, q does not interact and Principal pays pi_0.
i_q*(alpha_q*effort<alpha_0) + \
i_0*(alpha_q*effort≥alpha_0)


late Principal's cost of interaction with Q
i_Q*(alpha_Q*effort<alpha_0) + \
i_0*(alpha_Q*effort≥alpha_0)

late full Principal's payoff
1 minus dlt*effort minus costs for interaction with
 Q  (weighted by frequency of interactions)
  dlt*effort - (1-f)*c_q-f*c_Q

optimal effort level where payoff is maximal
Optimal demanding effort  is ', effort[P==max(P)])
```

ırther Reading

.rical test of the sexy son hypothesis (Fisher process) is described in
and Qvarnström (2006). See Huk and Winkel (2008) for a review.
ɔrk on the handicap principle can be found in Zahavi (1975, 1977),
)0a,b), see also Johnstone (1994); for a review see Cotton et al.
a book full of good pictures of bird mating rituals, see Léveillé

ɪg systems are common in biology; examples include a diverse range
n signalling in bacteria (Bassler, 1999), signalling among relatives
and Lachmann, 1997; Lachmann and Bergstrom, 1998), signalling
 mates (Zahavi, 1975), signalling between plants and pollinators
2018), between predators and potential prey (Ruxton et al., 2009;
ıl., 2006) and human language (Skyrms, 2008). For a two prey
ey model of signalling, see Ramesh and Mitchell (2018). These
ɤe partial conflicts of interest, for example in mate selection in
ɔecilia reticulata) (Kodric-Brown, 1985).
e on the Sir Phillip Sydney game, see Whitmeyer (2021). Coordina-
ap talk was discussed in Lewis (2008). Possible noise transmission
 in Crawford and Sobel (1982). For more on the handicap principle
ó (2011); Számadó and Penn (2015); Penn and Számadó (2020).
2020) for a recent review on signalling games.

18.8 Exercises

Exercise 18.1. For Example 18.2 on the handicap principle, confirm the mathematical solutions given. In the case of large r, find a linear approximation for $P(a)$, and interpret these results.

Exercise 18.2. Show that for the signalling game with two receiver responses of Broom and Ruxton (2011) described in Section 18.2.6, the strategy of all males with quality above α giving signal level α, all other males giving signal level 0 and females only mating with males that give a signal of α (or more) is stable against all alternative strategies by males or females.

Exercise 18.3. Show that the alternative reward function given by (18.27) satisfies the conditions of Grafen (1990b) for the handicap principle from Section 18.2.2.

Exercise 18.4. Consider the Pygmalion model from Section 18.3.2. Show that there is an equilibrium where high (low) quality senders (do not) send the signal and the Receiver responds to the (absence of the) signal by choosing A (B), provided that $c_L > 1 > c_H$.

Exercise 18.5. (i) Show that the Pygmalion game of Section 18.3.2 reduces to the Index signal model of Section 18.3.1 when $s_H = 1$ and $S_L = 0$.
(ii) Formulate the Index signal model in the extensive form representation used in Figure 18.3.

Exercise 18.6. (i) Explain why for the Pygmalion game we need to know the proportion of high quality senders p and why for the Index signal model this is irrelevant.
(ii) Suppose that each type of individual pays the same signalling cost c but that high quality individuals have twice the probability of signalling success of low quality individuals ($2s$ as compared to s). If $2/3$ of individuals are high quality, what values of s and c can lead to honest signalling?

Exercise 18.7. Show that if the conditions from inequality (18.31) hold, there is a solution to the Pygmalion game where both types of sender signal, and the Receiver chooses A if and only if the signal is received.

Exercise 18.8. Set up the game from Section 18.3.3 formally, writing it in the extensive form described in Figure 18.2 and then discuss your model and its potential solutions in comparison to earlier described signalling games.

Exercise 18.9. Set up and solve the following game. Consider a population of principals and agents, similarly as described in Section 18.3.3. Agents are high quality or low quality. Principals prefer to interact only with high quality agents and they can exert a varying level of effort to screen the agents. The more effort the Principal puts into the screening, the higher the cost is for any agent that interacts with the Principal. At the same time, the high quality agents incur smaller cost than the low quality agents. Find the optimal screening effort.

Hint. A possible setup and a numerical solution is provided in the Python code in Section 18.6.

Exercise 18.10. Show that for the Sir Philip Sidney game from Section 18.4, the payoffs are as described, and hence give the circumstances when it is optimal for the donor to donate and the Receiver to receive the reward.

Exercise 18.11. Consider a community of two plant species and two pollinator species (such a scenario is described in Metelmann et al., 2020) and set up and analyze the following game. The players are the plant species. Their strategy is to decide whether they allow visits by the first, the second, or both (or neither) of the pollinator species. Once the decision is made, the payoff is evaluated as follows. Each individual pollinator visits exactly two plants of any of the species that allow the visit. Pollen uptake happens at the first visit and pollination at the second plant (if of the same species as the first plant). What should be the optimal strategies of the plant species?

Chapter 19

Food competition

19.1 Introduction

When animals forage for food, they often do so in groups. There are many potential benefits to foraging in groups, but also many disadvantages, as we saw in Chapter 16. In this chapter we shall ignore all external influences such as the risk of predation, and concentrate on the effect of the presence of group-mates on foraging success. Later in the chapter we will focus on direct interference from other individuals, such as kleptoparasitism, where others actively try to steal food from those who have found it first. Before we do that, however, we will investigate indirect interference, where the presence of others simply means that there is less food available for any individual. We focus on the important idea of the *Ideal Free Distribution* (IFD) that we first met in Section 7.2.2.1 as Parker's matching principle.

19.2 Ideal Free Distribution for a single species

Fretwell and Lucas (1969) introduced the idea of the ideal free distribution.

Example 19.1 (IFD game, Fretwell and Lucas, 1969). Individuals have a number n of potential habitats or patches that they can forage on. The quality of these patches are different, but all individuals are essentially identical, and within each patch all individuals forage at an equal rate and so have identical payoffs, with this payoff determined by both the patch and the density of other individuals occupying the patch. Individuals can choose which patch they will forage on. Determine the ESSs.

19.2.1 The model

We will model this as a playing the field game; see Section 7.2.2. An individual strategy will be given by $\mathbf{y} = (y_i)$, where y_i is the proportion of the time spent on patch i.

DOI: 10.1201/9781003024682-19

To determine the payoff of an individual on a patch we assume that a patch i has a *basic suitability* B_i. Without loss of generality we can list the patches in decreasing order of quality, so that $B_1 > B_2 > \ldots > B_n$ (we may assume strict inequalities corresponding to the idea of generic games, see Section 2.1.3.3). If the density of individuals on patch i is d_i, then the payoff R_i to all individuals on the patch is given by

$$R_i(d_i) = B_i - f_i(d_i), \tag{19.1}$$

where f_i is a non-negative, increasing and continuous function, with $f_i(0) = 0$. By a suitable rescaling of the functions f_i, we can assume that the entire population always being on one patch has density 1, so that in the working that follows d_i will represent the (expected) proportion of individuals on patch i.

Individuals are able to judge precisely which patch will be best for them in relation to survival and reproduction, and hence are *ideal* individuals, which make perfect judgements. In other words they act as rational individuals with perfect information. They are also *free* to move between patches without restriction or cost. Note that if such movement costs do exist, then the results can be significantly different, see Charnov (1976); DeAngelis et al. (2011). The *distribution* of individuals over the patches will be governed by these two facts, combined with the properties of the different patches (B_i, f_i).

We suppose that we have an essentially infinite population of individuals playing strategy $\mathbf{p} = (p_i)$ against a small invading population of fraction u playing $\mathbf{q} = (q_i)$, so that

$$d_i = (1 - u)p_i + uq_i = p_i + u(p_i - q_i). \tag{19.2}$$

This gives us the payoff

$$\mathcal{E}[\mathbf{q}; (1 - u)\delta_{\mathbf{p}} + u\delta_{\mathbf{q}}] = \sum_{i=1}^{n} q_i \Big(B_i - f_i\big(p_i + u(q_i - p_i)\big) \Big), \tag{19.3}$$

$$\tag{19.4}$$

which implies, when setting $R_i = B_i - f_i(p_i)$, we obtain the incentive function

$$h_{\mathbf{p},\mathbf{q},0} = \mathcal{E}[\mathbf{p}; \delta_{\mathbf{p}}] - \mathcal{E}[\mathbf{q}; \delta_{\mathbf{p}}] = \sum_{i=1}^{n} (p_i - q_i)R_i. \tag{19.5}$$

We are now ready to state and prove the main result about the IFD game.

Theorem 19.2 (Cressman et al., 2004). *A strategy $\mathbf{p} = (p_i)$ is an ESS of the IFD game if and only if there is $l \leq n$ such that the following are true:*

(i) $p_i > 0$ if and only if $i \leq l$, and

(ii) $R_1 = R_2 = \ldots = R_l \geq B_{l+1}$.

This result is analogous to the result for matrix games requiring the payoff to all pure strategies within the support of an ESS to be equal, and the payoff of those outside the support to be no larger; see Lemma 6.7 and also Theorem 7.5.

It should be noted that although the IFD solution of the single species patch-foraging problem has often been (correctly) claimed to be an ESS, the first formal proof of this fact for some special circumstance appears in Cressman et al. (2004) and the full proof then appears in Cressman and Křivan (2006).

Proof. Assume that $\mathbf{p} = (p_i)$ is an ESS. To see (i), assume the contrary that there are $j < k \leq n$ such that $p_j = 0$ and $p_k > 0$. Then

$$R_j = B_j - f_j(p_j) = B_j \geq B_k > B_k - f_k(p_k) = R_k, \qquad (19.6)$$

so that when the strategy $\mathbf{q} = (q_i)$ is defined by

$$q_i = \begin{cases} p_i & i \neq j, k, \\ p_k & i = j, \\ 0 & i = k, \end{cases} \qquad (19.7)$$

we get, by (19.5),

$$h_{\mathbf{p},\mathbf{q},0} = p_k(R_k - R_j) < 0, \qquad (19.8)$$

which contradicts the necessary condition of Theorem 7.3 for $\mathbf{p} = (p_i)$ to be an ESS (note that formally to apply Theorem 7.3 we require the functions R_i to be differentiable, but the above argument works more generally). Moreover, if there are $j, k \leq l$ such that $R_j > R_k$, then for $\mathbf{q}$ defined similarly as in (19.7) by

$$q_i = \begin{cases} p_i & i \neq j, k, \\ p_j + p_k & i = j, \\ 0 & i = k, \end{cases} \qquad (19.9)$$

we again get (19.8), contradicting the ESS property. Hence,

$$R_1 = R_2 = \ldots = R_l. \qquad (19.10)$$

The fact that, in an ESS, $R_l \geq B_{l+1}$ follows in a similar manner as above and is left as Exercise 19.1.

We will now prove the other implication. Assume that there is l such that $\mathbf{p}$ satisfies conditions (i) and (ii). We will assume that $l = n$, the case $l \neq n$ is left as Exercise 19.2.

Because of condition (ii), we have that for any $\mathbf{q}$,

$$\sum_{i=1}^{n}(p_i - q_i)R_i - \sum_{i=1}^{n}(p_i - q_i)R_1 - R_1\sum_{i=1}^{n}(p_i - q_i) = 0. \qquad (19.11)$$

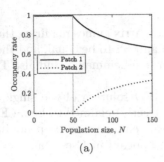 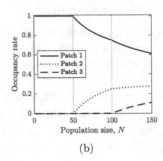

(a)   (b)

FIGURE 19.1: The Ideal Free Distribution for one species on (a) two patches and (b) three patches. The parameter values are $B_1 = 10$, $B_2 = 5$, $B_3 = 2.5$, $f_1(d) = f_2(d) = f_3(d) = \frac{N}{10}d$ where N is the population size. B_3 and $f_3(d)$ are only relevant for case (b).

Thus

$$\mathcal{E}[\mathbf{p}; \delta_\mathbf{q}] - \mathcal{E}[\mathbf{q}; \delta_\mathbf{q}] = \sum_{i=1}^{n}(p_i - q_i)\big(B_i - f_i(q_i)\big) \tag{19.12}$$

$$= \sum_{i=1}^{n}(p_i - q_i)\Big(\big(B_i - f_i(q_i)\big) - R_i\Big) \tag{19.13}$$

$$= \sum_{i=1}^{n}(p_i - q_i)\big(f_i(p_i) - f_i(q_i)\big) > 0, \tag{19.14}$$

where the last inequality holds because f_i are increasing functions. Consequently, $\mathbf{p}$ is locally superior, and thus it is an ESS by Theorem 7.8.   □

Example 19.3. Suppose that we have two patches 1 and 2, with the basic suitabilities satisfying $B_1 > B_2$. This means that patch 1 must be occupied. All individuals will occupy patch 1 provided that patch 2 is sufficiently poor, i.e.

$$B_1 - f_1(1) \geq B_2. \tag{19.15}$$

Otherwise both patches will be occupied, with equal payoffs on each patch, i.e. the ESS $\mathbf{p} = (p_1, p_2)$ will satisfy

$$B_1 - f_1(p_1) = B_2 - f_2(1 - p_1). \tag{19.16}$$

The situation is illustrated in Figure 19.1.

Example 19.4. Suppose that we have three patches $1, 2$ and 3, with the basic suitabilities satisfying $B_1 \geq B_2 \geq B_3$. This means that patch 1 must be occupied. As before, all individuals will occupy patch 1 provided that patch 2 is sufficiently poor, i.e.

$$B_1 - f_1(1) \geq B_2. \tag{19.17}$$

Otherwise patches 1 and 2 will be occupied. If there exists p_1 such that

$$B_1 - f_1(p_1) = B_2 - f_2(1 - p_1), \tag{19.18}$$
$$B_1 - f_1(p_1) \geq B_3, \tag{19.19}$$

then only patches 1 and 2 will be occupied. Otherwise, all three patches will be occupied and the ESS strategy $\mathbf{p} = (p_1, p_2, p_3)$ will satisfy the following equations

$$B_1 - f_1(p_1) = B_2 - f_2(p_2), \tag{19.20}$$
$$B_1 - f_1(p_1) = B_3 - f_1(p_3), \tag{19.21}$$
$$p_1 + p_2 + p_3 = 1. \tag{19.22}$$

See Figure 19.1.

19.3   Ideal Free Distribution for multiple species

We now suppose that there are (at least) two species which forage on our food patches (e.g. see Sutherland and Parker, 1985) and extend Example 19.1 to multiple species. There are a number of ways that the species can differ from each other which will affect how they will be distributed upon the patches, and we shall see that, in general, there will be many ways that the IFD conditions can be satisfied. As we see in Section 19.4, it is also possible that non-IFD related effects will decide which distributions of individuals are more likely to occur.

Example 19.5 (Křivan et al., 2008). Consider two species that have two habitats available to forage on. The habitats differ in quality and the species differ in their foraging rate. Determine the ESS distribution of individuals.

19.3.1   The model

Let the payoff from patch i to individuals of species 1 (species 2) be denoted by V_i (W_i). We will model the payoffs as Parker (1978) and as we saw in Section 7.2.2.1. We could use more general payoffs such as in Exercise 19.9 without changing the analysis in any substantial way.

Let us assume fixed population sizes N and M of species 1 and species 2, respectively, and let species 1 (species 2) have foraging rate λ (Λ). If species

1 uses strategy $\mathbf{p} = (p_1, p_2)$ and species 2 uses strategy $\mathbf{q} = (q_1, q_2)$, then we set

$$V_i(p_i M, q_i N) = \lambda R_i^*, \qquad (19.23)$$
$$W_i(p_i M, q_i N) = \Lambda R_i^*, \qquad (19.24)$$

where

$$R_i^* = \frac{r_i}{\lambda p_i M + \Lambda q_i N} \qquad (19.25)$$

is the ratio of the resources available and the total foraging rate on patch i, and so would be the total resource obtained by an individual foraging at unit rate. The payoffs to an individual in species 1 (2) using strategy $\mathbf{p}' = (p_1', p_2')$ (strategy $\mathbf{q}' = (q_1', q_2')$) in a population of $\mathbf{p}$ and $\mathbf{q}$ is given by

$$V(\mathbf{p}'; \delta_{(\mathbf{p}, \mathbf{q})}) = p_1' \lambda R_1^* + p_2' \lambda R_2^*, \qquad (19.26)$$
$$W(\mathbf{q}'; \delta_{(\mathbf{p}, \mathbf{q})}) = q_1' \Lambda R_1^* + q_2' \Lambda R_2^*. \qquad (19.27)$$

For the analysis, there are a number of possible cases that we need to distinguish.

19.3.2 Both patches occupied by both species

Firstly consider the case where both species occupy both patches. Here, in the notation from Section 19.2, we have $l = n = 2$ and thus by property (ii) we need $R_1^* = R_2^*$. This implies that $1/R_1^* = 1/R_2^*$, so that

$$\frac{\lambda p_1 M + \Lambda q_1 N}{r_1} = \frac{\lambda(1 - p_1)M + \Lambda(1 - q_1)N}{r_2}. \qquad (19.28)$$

Solving (19.28) for q_1, for a given value of p_1, we obtain

$$q_1 = \frac{r_1(\lambda M + \Lambda N)}{\Lambda N(r_1 + r_2)} - \frac{\lambda M}{\Lambda N} p_1. \qquad (19.29)$$

Thus we do not have a unique solution, and any distribution satisfying the above equation is possible. This strategy is not an ESS as such but provides a line of equilibria, which form an evolutionarily stable set ESSet (see Section 3.2.3; Thomas, 1985), where invasion by any strategy from outside the set is not possible, but invasion from inside the set is. Similar phenomena occur, for example, when extensive form games are reduced to normal form, and there are many equivalent strategies (see Chapter 10).

19.3.3 One patch occupied by one species, another by both

Now consider the case where one species occupies only one patch (suppose that this is species 1), and the other species occupies both patches. As both patches are occupied, they must give identical payoffs to the individuals of species 2 foraging on each patch, so that again $R_1^* = R_2^*$ and (19.29) holds. Now suppose that without loss of generality that patch 2 is the patch occupied by species 1. This means that $p_1 = 0$, which provides a unique value of q_1 which satisfies the IFD conditions for species 2, i.e.

$$q_1 = \frac{r_1(\lambda M + \Lambda N)}{\Lambda N(r_1 + r_2)}. \tag{19.30}$$

In general, there are four cases to consider here (see for example Exercise 19.8), where it could be species 1 or 2 which has no individuals on a patch, and this patch is patch 1 or 2.

19.3.4 Species on different patches

Now suppose that both species only occupy one patch, but that these are different patches. This means that again $R_1^* = R_2^*$ (otherwise it is beneficial for some individuals of one species to move) so that, if patch 1 (2) is occupied by species 1 (2) i.e. $p_1 = 1, q_1 = 0$, we obtain

$$\frac{r_1}{M} = \frac{r_2}{N}. \tag{19.31}$$

This precise coincidence of parameters is a non-generic case, and can effectively be ignored.

19.3.5 Species on the same patch

Finally, we consider the case where both species only occupy the same patch. In this case it is clear that under the Parker payoffs such a situation cannot be stable, as the empty patch would have an infinite available reward.

19.4 Distributions at and deviations from the Ideal Free Distribution

We saw above in Section 19.3.2 that where there is more than one species dividing its foraging between two or more patches, then there are a number of ways in which the ideal free distribution can be realised. But which of these distributions is more likely in practice? This question was addressed by Houston and McNamara (1988).

Let us suppose that we have a two-species and two-patch model as in Example 19.5. With M individuals of type 1 and N individuals of type 2, the number of ways of allocating m_i type 1 individuals and n_i type 2 individuals

TABLE 19.1: Four possible pairs of $m_1 = p_1 M$ and $n_1 = q_1 N$ that satisfy (19.33). There are a total of 10215 combinations. The probabilities in the last column assume a uniform distribution over all of the combinations.

m_1	n_1	Combinations	Probability
0	4	126	0.012
3	3	7056	0.691
6	2	3024	0.296
9	1	9	0.001

to patch i is given by

$$\left(\frac{M!}{m_1! m_2!}\right)\left(\frac{N!}{n_1! n_2!}\right). \tag{19.32}$$

Thus if we assume that every possible combination of individuals which results in the IFD is equally likely to occur, allocations with a large number of combinations are far more likely to occur than those with very few.

Consider the situation in Example 19.5 with $M = N = 9$, so there are a total of 18 individuals, $r_1 = 1, r_2 = 2, \lambda = 1$ and $\Lambda = 3$. As shown in Exercise 19.8, to have both species on both patches, we need to satisfy

$$q_1 = \frac{4}{9} - \frac{1}{3} p_1. \tag{19.33}$$

There are four possible pairs $(n_1, m_1) = (p_1 M, q_1 N)$ with integer values (m_1, n_1) that satisfy (19.33) and thus having an ideal free distribution of individuals over the two sites. These pairs of values are $(0, 4), (3, 3), (6, 2), (9, 1)$. We represent these in Table 19.1, together with the number of ways of achieving them, and the corresponding probability of getting this division within an IFD population. Thus we can see that the second pair $(3, 3)$ is three orders of magnitude more likely than the fourth $(9, 1)$.

We have assumed in the above that each of the different combinations of individuals is equally likely. That is not necessarily true, and the effect of non-IFD movements caused by disturbances, e.g. from predators, was investigated in Jackson et al. (2004) and Yates and Broom (2005), modelling foraging as a Markov chain. In the model of Yates and Broom (2005), each individual would leave its patch for one of two reasons. Firstly due to random disturbance (non-IFD movements), each individual on patch i would leave at constant rate α_i. Secondly, because the foraging rate at the other patch would be greater if it moved there (IFD movements); individuals of type i moved from patch j at rate λ_{ij}, the foraging advantage, assuming this was positive (otherwise there was no movement). The foraging advantage for an individual of type 1 moving from patch 1 to patch 2, in the terminology of Křivan et al. (2008) is

$$\lambda_{11} = \max\{V_2(M_2 + 1, N_2) - V_1(M_1, N_1), 0\}. \tag{19.34}$$

TABLE 19.2: Transition rates between patches when there are non-IFD movements (α's) as well as IFD movements (λ's).

Type	Movement	Rate
1	$1 \to 2$	$\alpha_1 + \lambda_{11}$
1	$2 \to 1$	$\alpha_2 + \lambda_{12}$
2	$1 \to 2$	$\alpha_1 + \lambda_{21}$
2	$2 \to 1$	$\alpha_2 + \lambda_{22}$

Thus we have a Markov process with transition rates given in Table 19.2. A general result from the model of Yates and Broom (2005) (and of Jackson et al., 2004) is that undermatching occurs, i.e. fewer foragers than expected occupy the superior patch, as non-IFD movements are more likely from the more heavily occupied patch.

19.5 Compartmental models of kleptoparasitism

We now look at a range of models where there is direct interference by some individuals with the foraging of others, which reduces their foraging rates. This can be through aggressive contests for food resources or through fights between individuals which do not necessarily result in food changing hands, but which nevertheless hamper individuals' foraging. We start by looking at the idea of compartmental models, and in particular, their application to direct food stealing, or kleptoparasitism.

Kleptoparasitism is a very common behaviour in nature and is practiced by a very diverse collection of species. Examples include insects (Jeanne, 1972), fish (Grimm and Klinge, 1996) and mammals such as hyenas (Kruuk, 1972). It is perhaps most common, or at least most visible, in birds (see Brockmann and Barnard, 1979, for a review). In particular, we associate it with seabirds in spectacular aerial contests for fish (Steele and Hockey, 1995; Triplet et al., 1999; Spear et al., 1999), and it is this scenario that was the inspiration for the models that we discuss, although they are more general than that. A good review paper covering this is Iyengar (2008).

Compartmental models divide a population of individuals up into various states or "compartments". Individuals occupy one compartment at any time, and generally move between different compartments following a (usually Markov) random process. Such models are common in epidemic models, for instance the Susceptible-Infectious-Removed (SIR) model, and we look at such models in Chapter 21. This process may be entirely mechanistic, in the sense that no strategies are involved such as the model of Ruxton and Moody (1997). Here we will describe a more general model of Broom and Ruxton (1998a), which for $p = 1$ reduces to the model of Ruxton and Moody (1997).

TABLE 19.3: Transitions between the states of the kleptoparasitism game.

Transition	Meaning	Rate
$S \to H$	a searcher finds food	$\nu_f f S$
$H \to S$	a handler eats food	$t_h^{-1} H$
$S + H \to G_2$	a searcher attacks a handler	$p\nu_h SH$
$G_2 \to S + H$	a fight is resolved	$t_c^{-1} G_2$

Example 19.6 (Kleptoparasitism game—to steal or not to steal, Broom and Ruxton, 1998a). Consider a population of individuals that feeds on a resource that cannot be immediately consumed, but takes some handling time (e.g. cracking a shell). Individuals can thus find not only a food item, but also another individual already handling a food item. Should a searcher challenge the handling individual and attempt to steal the food item?

19.5.1   The model

Let the population consist of three basic behavioural states: handlers (H), searchers (S) and those involved in a aggressive contest (G_2), each represented by a compartment. Handlers are individuals handling a food item (such as cracking its shell). Searchers are individuals searching for food and/or a handler. If a searcher finds a handler, then with a probability p the searcher challenges the handler and the pair engages in an aggressive contest and will move to the state G_2. The probability to challenge a handler, p, is the strategy of an individual. Our task is to find its ESS value.

Transitions between the states follow a Markov process and are given in Table 19.3; see also Figure 19.2. Each handler finds food at rate $\nu_f f$, the product of its searching rate and the food density, and this is multiplied by the searcher density to give the searcher transition rate. Each searcher finds a handler and challenges it at rate $p\nu_h H$, the product of its searching rate for handlers, the handler density and the probability to challenge; and this is again multiplied by the searcher density.

Fights last an exponential amount of time with mean t_c, and so the overall rate is again the reciprocal of the mean scaled by the density G_2. We assume that time is the only currency in this model, i.e. the only cost of the fight is its duration. At the end of the fight, one individual emerges as a winner (and becomes a handler, handling the contested item) and the other one becomes a searcher. We assume that a searcher challenging a handler will win with probability $\alpha \in [0, 1]$.

There are many ways how the food handling can be modelled. Broom and Ruxton (2003) considered two types of food, labelled "oranges" which took a constant amount of time to handle, the entire item being consumed at the end of the handling period, and "apples" which again took a constant amount of time to handle but were continuously consumed throughout the

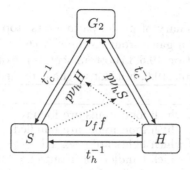

FIGURE 19.2: Transition diagram for the kleptoparasitism game from Example 19.6. Individuals are in one of the three states: searchers (S), handlers (H), fighting pairs (G_2). Full arrows represent the transitions between states, expressions by the arrows are the per capita rates. Dotted arrows show how a particular state can affect transitions from another state. Parameters are explained in Table 19.3; see also Table 19.4.

period. Here we stick to the analogy of cracking a shell. With this analogy, we assume that it takes an exponential amount of time with mean t_h to handle a food item, and that this handling time remains the same regardless of the history (i.e. how long and by whom the food item was already handled). The corresponding handling rate is thus the reciprocal of this mean (again scaled by the density H). When the food item is consumed, the individual resumes a search for another food item (and/or handler). The advantage of this assumption is that when a fight is over, the winner of the fight will, from the modelling perspective, face exactly the same handling scenario as a searcher with a newly discovered food item. This significantly reduces the number of compartments in this model and thus simplifies the analysis. See Table 19.4 for a summary of the model parameters.

The natural payoff in this scenario is the food consumption rate, i.e. the reciprocal of the expected time needed to find and consume (possibly including defence of) a food item.

19.5.2 Analysis

The total population density is P.

$$P = S + H + 2G_2, \tag{19.35}$$

since G_2 represents how many fighting pairs we have, so the density of fighting *individuals* is $2G_2$.

TABLE 19.4: A summary of model parameters (top section) and notation for the kleptoparasitism game from Example 19.6 (middle section) and also its extension from Section 19.5.3 (bottom section). Note that parameter α is relevant only to Example 19.6, but not to the extended model.

	Meaning
$\nu_f f$	the rate at which food items are found
$1/t_h$	the rate at which food items are handled
$1/t_c$	the rate at which fights are resolved
ν_h	the rate at which handlers or groups are found
α	the probability a challenging searcher will win the fight
S	the density of searchers in the population
H	the density of handlers in the population
G_2	the density of fighting pairs in the population
p	the probability a handler is challenged if found
G_i	the density of fighting groups of size i in the population
p_i	the probability a group of i individuals is challenged if found
$\mathbf{p}$	the challenging strategy, $\mathbf{p} = (p_1, p_2, p_3, \ldots)$
V_k	the strategy to challenge groups of size $< k$ only

The model introduced above leads to the following differential equations:

$$\frac{dS}{dt} = \frac{H}{t_h} - \nu_f S f - p\nu_H S H + \frac{G_2}{t_c}, \tag{19.36}$$

$$\frac{dH}{dt} = -\frac{H}{t_h} + \nu_f S f - p\nu_H S H + \frac{G_2}{t_c}, \tag{19.37}$$

$$\frac{dG_2}{dt} = p\nu_H S H - \frac{G_2}{t_c}. \tag{19.38}$$

We can see that (19.38) follows immediately from (19.35), (19.36) and (19.37).

Assuming a particular choice of strategy p in the population, we find the equilibrium state of the population by setting the differential equations (19.36), (19.37) and (19.35) to be equal to zero; solving the system is left as Exercise 19.12. This is only sensible if it can be assumed that the population spends most of its time at or near the equilibrium; that convergence happens sufficiently fast for this model was shown in Luther and Broom (2004). The food consumption rate of a general member of the population is then directly proportional to the handling ratio H/P, since individuals feed at a constant rate $1/t_h$ when handling, and not otherwise. Thus, see also Exercise 19.14,

$$\mathcal{E}[p; \delta_p] = \frac{H}{Pt_h}, \tag{19.39}$$

where H is the solution of

$$\left(\frac{H}{P}\right)^2 2t_c p\nu_h P + \frac{H}{P}(1 + t_h\nu_f f) - \nu_f f t_h = 0. \tag{19.40}$$

Can a single individual, playing alternative strategy q, improve its consumption rate when in such a population? If so, invasion occurs; if this cannot happen for any other q, then p is an ESS. The consumption rate of the focal individual is given by

$$\mathcal{E}[q; \delta_p] = \frac{1}{T_S(q,p) + T_H(q,p)}, \tag{19.41}$$

where $T_S = T_S(q,p)$ is the time to become a handler for the first time from the start of searching, and $T_H = T_H(q,p)$ is the time to finish eating a food item from the time of becoming a handler, where T_S may involve time spent fighting and T_H may involve time spent fighting and even time spent back in the searching state if a fight is lost. We note that the models in this section are related to those from Křivan and Cressman (2017) and Garay et al. (2017), see Section 7.4, with fitness functions of the form of the ratio of the expected value of an item (here just 1) and the time to acquire that item.

To determine T_S and T_H, we follow the diagram in Figure 19.3.

It follows that

$$T_S = \frac{1}{\nu_f f + \nu_h H} + \frac{\nu_h H}{\nu_f f + \nu_h H} \cdot ((1-q)T_S + q(t_c + \alpha \cdot 0 + (1-\alpha) \cdot T_S)), \tag{19.42}$$

$$T_H = \frac{1}{t_h^{-1} + p\nu_h S} + \frac{p\nu_h S}{t_h^{-1} + p\nu_h S}(t_c + (1-\alpha)T_H + (\alpha)(T_S + T_H)). \tag{19.43}$$

The analysis can then be completed by manipulation with formulae (19.41) and (19.42)–(19.43) and is left as Exercise 19.13. Here we present a simpler solution from Broom and Ruxton (1998a). They observed that to maximise the consumption rate (19.41) it is enough to maximise the instantaneous rate at which a searcher becomes a handler directly after the critical decision point where it has just found a handler. If a searcher challenges a handler, it enters a fight for a duration of t_c and wins it with probability α. Thus the rate in this case is αt_c^{-1}. If it does not challenge, it remains a searcher and the rate to become a handler is thus $\nu_f f$. Hence, $p = 1$ (always challenge) is an ESS if

$$\nu_f f t_c < \alpha, \tag{19.44}$$

and otherwise it is best never to challenge, $p = 0$.

This leads to an interesting step-change of behaviour, as parameters cross a threshold. Plotting the food uptake rate versus the food density f (see Figure 19.4), as f varies, we see that challenging is optimal if the food availability is sufficiently low. Challenging reduces food consumption even further by the costly fights that ensue (this is another example of the best individual behaviour leading to a bad result for the population, as in Section 4.2). Thus supposing that we have a high level of food availability, there is no incentive to fight. If this availability drops, then foraging success decreases gradually until

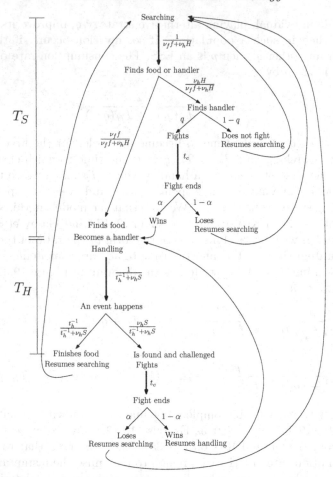

FIGURE 19.3: State transitions of an individual playing the kleptopara-sitism game from Example 19.6. Expressions by bold vertical lines represent the actual times taken for a particular process (see Exercise 19.10), expressions by angled lines represent probabilities of an event. The densities of handlers, H, and searchers, S, depend on the population strategy p. The strategy of the focal individual is q. Other parameters are explained in Table 19.4.

as food availability reduces below the threshold value, consumption falls suddenly. While the reduction in food is the indirect cause of this drop through the behaviour change it causes, it is this change which has the most significant effect on the population. For example some birds will engage in kleptoparasitism primarily in the winter when food is scarce (e.g. Olrog's gull, see Delhey et al., 2001).

As mentioned above, the models considered here lead to payoffs similar in character to the time-constraint models of Section 7.4. In both cases

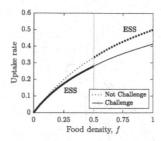

FIGURE 19.4: Step change in the uptake rate for the ESS in the game from Example 19.6. Parameters are $t_h = 1, \nu_h = 1, t_c = 1, \nu_f = 1, \alpha = 1/2$.

individuals have choices of actions which (can) lead to rewards but take time, with the aim of maximising their intake per unit time (here, however, strategies control movement between compartments and contests between individuals are simple).

19.5.3 Extensions of the model

The model of Broom and Ruxton (1998a) has been developed in a number of ways. Many kleptoparasitism models, starting with Broom et al. (2004b), allow individuals to give up their food instead of fighting. Broom et al. (2008a) modelled a population where individuals could either challenge or not, and resist challenges or not, giving four basic types of individuals.

Here we will focus on another extension. Broom and Rychtář (2011) developed the original model by allowing multiple fights ("mêlées"), where defenders must still always resist. Thus now, when a handler finds a group of two (or more) contesting a resource it may challenge, and so potentially very large fighting groups can occur. This is realistic, as such multiple fights are common in some populations (Steele and Hockey, 1995). The strategy of the population now becomes an infinite vector, with the element p_i representing the probability that a challenger should challenge a group of size i when encountered. The set of possible transitions is more complex, as we see in Figure 19.5. Nevertheless the same principle leads to a unique equilibrium distribution of individuals over the different states, given the strategy of the population. Broom and Rychtář (2011) showed that, under reasonable conditions on the parameters, the only possible ESSs must be of the form that all groups of size less than K are always challenged, and all larger groups are never challenged. This strategy was labelled V_K.

As food becomes more scarce, individuals should be prepared to challenge larger and larger groups; see Figure 19.6. When the population is not dense, then large groups rarely form even if individuals would be prepared to form them, since there will not be enough contact between individuals. However, for very dense populations, if food becomes scarce then such groups can form. In

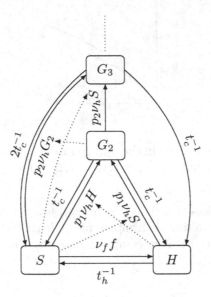

FIGURE 19.5: Transition diagram for kleptoparasitic mêlées. Individuals are in one of the potentially infinite number of states: searchers (S), handlers (H), fighting n-tuples ($G_n, n \geq 2$). Full arrows represent the transitions between states, expressions by the arrows are the per capita rates. Dotted arrows show how a particular state can affect transitions from another state.

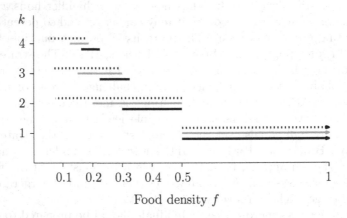

FIGURE 19.6: Regions of the food density where V_k is an ESS for kleptoparasitic mêlées. Parameter values are $\nu_f = t_c = t_h = 1$. Full line $P = 1$, grey line $P = 10$, dotted line $P = 100$.

these circumstances it is also possible to see a large number of simultaneous ESSs. We note however that in the most extreme cases food consumption practically collapses to zero, so such cases are clearly not viable for anything other than short periods. A situation like this may occur if food occurs in clumps which attract a number of foragers, e.g. a large dead fish on a beach. At the point where the food has almost all been consumed, there will be both a large number of foragers and limited food, so that fighting might become optimal (however see Section 19.7 for a model based upon large divisible food items).

At this point it is worth discussing the nature of these strategies. An ESS is one that persists over time, so how can we have one that is a solution over a short time interval? Here we think of conditional strategies, rather than fixed ones. Thus if an individual can recognise the level of food availability and change its strategy accordingly, an ESS may well comprise a number of different challenging strategies, each chosen when the food availability falls into a given range, as in the case of the behaviour of Olrog's gull described above (Delhey et al., 2001). It is then possible to imagine desperate strategies being employed when food availability is low. In reality, this kind of approach needs to be taken whenever parameters fluctuate on a short timescale, e.g. annually, as we saw in Chapter 11.

19.6   Compartmental models of interference

A related set of models of the more general behaviour of interference have been considered (e.g. Beddington, 1975; DeAngelis et al., 1975; Ruxton et al., 1992). Interference is the reduction of foraging efficiency by negative actions of other individuals, for example aggressive territory defence. It can potentially include the direct stealing of kleptoparasitism but is more usually thought of as interrupting foraging without stealing (but it is more than reducing the foraging of another individual merely by finding the food first and eating it). More recently, a series of papers (e.g. van der Meer and Ens, 1997; van der Meer and Smallegange, 2009; Smallegange and van der Meer, 2009) have involved a more complex sequence of states modelling foraging and interference behaviour crabs, and so the model of Ruxton and Moody (1997) can be thought of as effectively a special case of such interference models. There has as yet been no strategic element involved in these models, so these are not game-theoretic, but they are worth mentioning here because of their relation to the above models, their intrinsic interest and because game-theoretic versions could be developed, as we discuss at the end of the section.

A population of individuals is foraging for food. In addition, individuals display interference behaviour; often if they get sufficiently close to other individuals, although no food-stealing takes place, this nevertheless prevents

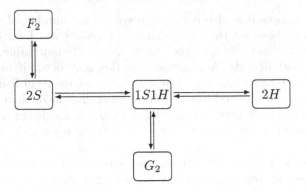

FIGURE 19.7: Transition diagram for the interference model of van der Meer and Smallegange (2009).

individuals from either searching or handling. Thus searchers and handlers can interfere with each other, as can searchers with other searchers. There are four states, with searchers under interference F and handlers under interference G, as well as S and H.

van der Meer and Smallegange (2009) considered a stochastic version of this model and used the detailed balance conditions (see e.g. Kelly, 1979) to find solutions for this interference system, giving explicit forms for small populations, and general expressions for arbitrarily large populations. They compared their models to data from pairs of crabs feeding on mussels (as in Broom et al., 2010a, this smallest non-trivial population was where predictions differed most from the infinite population case). They were able to get estimates for the key model parameters, and showed a good fit to the data in this case.

We can consider transitions in this system as follows:

$$S \to H, \tag{19.45}$$
$$H \to S, \tag{19.46}$$
$$2S \to F_2, \tag{19.47}$$
$$F_2 \to 2S, \tag{19.48}$$
$$S + H \to G_2, \tag{19.49}$$
$$G_2 \to S + H. \tag{19.50}$$

The transition diagram in the population with two individuals is shown in Figure 19.7.

In a related paper van der Meer and Smallegange (2009) considered two types of individuals, dominants and subordinants, foraging on two distinct patches, which they could move between (thus this resembles the IFD models of Section 19.2) and they compared their results to such models. The most significant interference was between dominants, and this made the population distribute more evenly on the patches than would be suggested by standard IFD theory.

Unlike in kleptoparasitism, interference is a by-product of some other activity. As no food is taken, no direct benefit is made by the individuals who interfere with others in the models of van der Meer and Smallegange (2009), Smallegange and van der Meer (2009). Thus the obvious strategic element, to interfere or not interfere, is not there, as all other things being equal, non-interference is best, and interference is just caused by the proximity of conspecifics. However, individuals do make choices, for example to move from one patch to another or not. Thus parameters which are deemed fixed, and estimated from data, may be thought of as strategic choices in future models. The size of the individuals can also be thought of as a life history strategy, and an evolutionarily stable mixture of individual sizes may be sought, on the assumption that there is an extra cost to growing large.

19.7 Producer-scrounger models

In the above models of kleptoparasitism or interference, food is discovered as single items, without the possibility of division. Therefore, although animals can potentially steal from each other, they cannot share their food (although see Hadjichrysanthou and Broom, 2012). What if food is discovered in clumps which are divisible? An individual which discovers the food can start eating straight away, generally with little or no handling time, but it will take some time to consume all of the food. If one or more individuals find the food before the Finder has consumed it all, then they may try to steal some of it, and division between the individuals can theoretically happen in many ways. This is the scenario in the producer-scrounger game developed by Barnard and Sibly (1981). A number of variants of this model have been developed to consider different real factors and take on board different assumptions (see for example Caraco and Giraldeau, 1991; Vickery et al., 1991; Dubois and Giraldeau, 2005). Here we look at the model of Dubois et al. (2003).

Example 19.7 (Finder-Joiner game, Dubois et al., 2003). Consider the following conflict between two individuals. One individual (the Finder) finds a patch of food and starts consuming it. Some time later, before the food has been completely consumed, a second individual (the Joiner) appears and attempts to feed at the patch. Model their interaction as a Hawk-Dove game and determine the ESS.

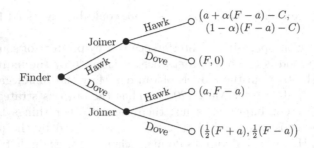

FIGURE 19.8: The sequential Finder-Joiner game from Example 19.7 in extensive form.

19.7.1 The Finder-Joiner game—the sequential version with complete information

We will follow the first model of Dubois et al. (2003) and assume the interaction between the two individuals to be a sequential Hawk-Dove game with the Finder choosing its strategy first. See also Exercises 19.15 and 19.16 for slightly modified versions of this game.

19.7.1.1 The model

Let the total value of the patch be F (possibly a number of distinct items), of which value a, the *Finder's share*, has already been consumed by the Finder before the Joiner arrives. We assume that both the Finder and the Joiner know the value of a. The sequential Hawk-Dove contest is modelled as a game in extensive form as in Figure 19.8.

When two Doves meet, they share the remaining resource, each trying to eat as much as they can (scramble competition), but it is assumed that it is eventually divided equally. When a Hawk meets a Dove, the Dove retreats and the Hawk consumes the entire remaining resource. When two Hawks meet, they fight and both pay an energetic cost of value C. The loser retreats and the winner keeps the entire resource, the probability of the Finder winning the contest being denoted by α.

19.7.1.2 Analysis

As a sequential game with each individual making just a single choice, there is a unique ESS in this game (see Section 10.1.2 and Exercise 10.1).

Following the diagram in Figure 19.8, we see that if the Finder plays Dove, the Joiner should play Hawk when $F - a > \frac{F-a}{2}$, which is always satisfied since $a < F$. Consequently, if the Finder plays Dove, it receives a payoff of a and the Joiner receives a payoff $F - a$.

If the Finder plays Hawk, the Joiner should also play Hawk if

$$(1 - \alpha)(F - a) - C > 0, \tag{19.51}$$

TABLE 19.5: Summary of the ESSs in the sequential Finder-Joiner game from Figure 19.8.

ESS	Finder's action	Joiner's action	conditions
(H,D)	Escalate	Withdraw	$a > F - \frac{C}{1-\alpha}$
(H,H)	Escalate	Escalate	$a < F - \frac{C}{\alpha}$ and $a < F - \frac{C}{1-\alpha}$
(D,H)	Withdraw	Escalate	$F - \frac{C}{\alpha} < a < F - \frac{C}{1-\alpha}$

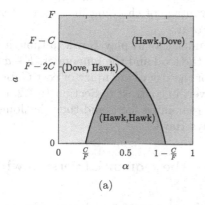

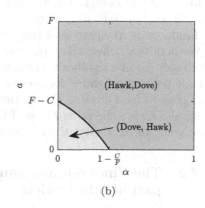

(a)
(b)

FIGURE 19.9: ESSs in the Finder-Joiner game; (a) there are three possible ESSs when $C < \frac{F}{2}$, (b) there are only two possible ESS when $\frac{F}{2} < C < F$. There is only one ESS (Hawk, Dove) when $C > F$, and this simple case is not shown.

and in this case, the Finder gets $a + \alpha(F - a) - C$. Thus if

$$a < F - \frac{C}{1 - \alpha}, \tag{19.52}$$

the Joiner should play Hawk irrespective of the strategy of the Finder. Consequently, if (19.52) holds, the Finder should play Hawk if

$$a + \alpha(F - a) - C > a, \tag{19.53}$$

i.e. if

$$a < F - \frac{C}{\alpha}. \tag{19.54}$$

If (19.52) does not hold then the Joiner will play Hawk if the Finder plays Dove, but will play Dove if the Finder plays Hawk. Thus the Finder should play Hawk, since $F > a$. The results are summarised in Table 19.5 and also in Figure 19.9.

19.7.1.3 Discussion

We can see from Figure 19.9, that for small values of the Finder's advantage a there can be a significant range of intermediate values of α where a fight for resources occurs, with the more extreme values of α leading to victory for the stronger individual. As the Finder's advantage increases, fights become less likely, until when $a = F - 2C$ there can no longer be fights, and the stronger animal wins. As the Finder's advantage increases further, there remains a division where low α leads to the Joiner winning, and higher α to the Finder winning, but the boundary between the regions is at lower and lower values, until we reach $a = F - C$, when the advantage of choosing first means that the Finder wins irrespective of the value of α.

We note that when a is large, the Finder should play Hawk, although it has already eaten a significant portion of the food and the remainder, $F - a$ is actually not worth fighting for (because of the risk of an injury of cost C for a small piece of food of value $F - a$). However, in contrast to Section 19.7.2, the Joiner also knows the value $F - a$ of the remaining food, and thus the Joiner is not going to fight when the Finder plays Hawk.

19.7.2 The Finder-Joiner game—the sequential version with partial information

We will now re-examine the model from Section 19.7.1. Assume this time that only the Finder has the information about the amount of food already eaten a. The Finder's strategy will thus depend on F, C, α, a, while the Joiner's strategy will depend only on F, C, α and the choice of the Finder. For a fixed F, C, α we are interested for which values of a the Finder will play Hawk, and for the corresponding response of the Joiner. Let

$$A = A(F, C, \alpha) = \{a; \text{Finder plays Hawk given it has eaten } a\}, \qquad (19.55)$$

and let $\langle q \rangle_A$, respectively $\langle q \rangle_{\neg A}$, be the expected value of some quantity q, given than the Finder played Hawk, respectively Dove. Following the same analysis as above, we get that when the Finder plays Dove, the Joiner will play Hawk because

$$\langle F - a \rangle_{\neg A} > \left\langle \frac{F - a}{2} \right\rangle_{\neg A}. \qquad (19.56)$$

Hence, the Joiner will play Hawk if the Finder has played Hawk, if

$$(1 - \alpha) \langle F - a \rangle_A - C > \langle 0 \rangle_A \Leftrightarrow \qquad (19.57)$$

$$\langle a \rangle_A < F - \frac{C}{1 - \alpha}, \qquad (19.58)$$

and will play Dove otherwise. Hence, if (19.58) does not hold, then the Finder will play Hawk. Also, the Finder will play Hawk when (19.58) holds and if

$$a < F - \frac{C}{\alpha}. \qquad (19.59)$$

It thus follows that the Finder plays Hawk whenever (19.59) holds. Condition (19.59) means that the Finder plays Hawk when there is sufficient food left, i.e. when a is small. But this implies that $\langle a \rangle_A$ cannot be too large. If we assume that a is uniformly distributed over $[0, F]$, we find the following bound for $\langle a \rangle_A$,

$$\langle a \rangle_A \leq \frac{F - C/\alpha}{2} \left(1 - \frac{C/\alpha}{F} \right) + F \frac{C/\alpha}{F} = \frac{F}{2} + \frac{C^2}{2\alpha^2 F}, \tag{19.60}$$

since $\frac{F-C/\alpha}{2}$ is the expected value of a, given $a \in [0, F - C/\alpha]$, and F is the maximal value that a can take.

Hence, when C/F is small enough to satisfy

$$\left(\frac{C}{F} \right)^2 \frac{1}{\alpha^2} + \frac{C}{F} \frac{2}{1 - \alpha} < 1, \tag{19.61}$$

we get

$$\langle a \rangle_A < F - \frac{C}{1 - \alpha}, \tag{19.62}$$

and so (19.58) is satisfied. Consequently, when $a > F - C/\alpha$, the Finder has to play Dove. We note that when (19.61) does not hold, the situation becomes a lot more complicated, and we shall not discuss this here.

This should be compared to the variant of the game with complete information when, e.g. in the case when $\alpha = 1/2$ and $a > F - 2C$, the Finder plays Hawk, since then the Joiner must play Dove, see Figure 19.9a. Now, when the Joiner does not know the true value of a, the Finder playing Hawk is an indication that there may be a significant amount of food remaining that is worth fighting over (since the Finder has to fight in such a situation), and thus the Joiner could calculate that fighting is beneficial in this case. Thus although the Finder knows more than the Joiner, this is disadvantageous to the Finder (it gets a lower payoff than in the situation when both the Finder and the Joiner both know the value of a).

19.8   Python code

In this section we show how to use Python to solve the IFD game from Example 19.1.

```
1   """  Ideal Free distribution
2   Payoffs of patches defined inside of  this  function
3   """
4
5   ## Import basic packages
```

```python
6   import numpy as np
7   from scipy.optimize import fsolve
8   import matplotlib.pyplot as plt
9
10  ## Give payoffs in the form 'Constant - f(x)'
11  # Constants must be in decreasing order
12  # Can be as many patches as needed
13  patch = [ lambda x: 6 - x*5,     # payoff on the patch 0
14            lambda x: 5 - x*5,     # payoff on the patch 1
15            lambda x: 4 - x*5,
16            lambda x: 3 - x*5]     # payoff on the last patch
17
18  num_of_patches = len(patch)  # Total number of patches
19
20  ## Define equations to solve
21  def eqs(x) :
22      " returns the equations to solve (for roots)"
23      # x[p] is occupancy on patch p
24      # Patch 0 will always be occupied
25      # If patch p>0 is occupied, the payoff there should be same
26      #     as on the patch 0
27      list_of_eqs  = [patch[0](x[0]) - patch[p](x[p])
28                      for p in range(1,LastPatchOccupied+1)]
29
30      # If unocuppied, the patch p will have x[p] = 0
31      for p in range(LastPatchOccupied+1, num_of_patches):
32          list_of_eqs.append(x[p])
33
34      # Add the condition that occupancy has to sum to 1
35      list_of_eqs.append(sum(x)-1)
36      return list_of_eqs
37
38  # Make initial guess of the occupancy
39  # All but the 0th patch will be unoccupied
40  guess = np.zeros(num_of_patches)
41  # 0th patch has everybody on it
42  guess[0] = 1
43
44  # Initialize the index of the last occupied patch
45  LastPatchOccupied = 0
46
47  # While not all patches are occupied and
48  # the individuals on the first patch do worst than
49  # a small number of them would do on the first unoccupied patch,
50  # increase the number of occupied patches and solve again
51  while (LastPatchOccupied<num_of_patches-1) and (
52          patch[0](guess[0]) < patch[LastPatchOccupied+1](0)):
53      # Print out what is going on
54      print('Current guess is')
55      print(guess)
56      print('Last occupied patch is ' + str(LastPatchOccupied))
57      for p in range(num_of_patches):
58          print('paoyff on patch ' + str(p) + ' is : '
59                  + str(patch[p](guess[p])))
60      print('Not the IFD.  Trying to occupy one more patch')
61
62      # Try to occupy one more patch
```

```
63      LastPatchOccupied += 1
64
65      # Solve the eqs again with the last guess as starting position
66      guess = fsolve(eqs, guess )
67
68  # We arrived to the final solution. Print it out
69  print('*** Ideal Free Distribution is ***')
70  print(guess)
71  # Print out the payoffs on individual patches
72  for p in range(num_of_patches):
73      print('Paoyff on patch ' + str(p) + ' is : '
74              + str(patch[p](guess[p])))
```

19.9   Further reading

The classical work on IFD theory is Fretwell and Lucas (1969). The fact that the IFD strategy is an ESS was proved in Cressman and Křivan (2006). Křivan and Sirot (2002) developed the IFD model to two species in two patches; see also Cressman et al. (2004). For population dynamics IFD models see Abrams et al. (2007), Křivan and Cressman (2009) and Cressman and Křivan (2010). For an overview of IFD theory up to its publication, see Křivan et al. (2008). For the IFD will allee effects see Křivan (2014); Cressman and Tran (2015), for the IFD on a structured population see Broom and Rychtář (2018b); see also Reding et al. (2016). The IFD in non-ideal populations was considered in Street et al. (2018). Miller and Coll (2010) show deviation from the IFD in spatially distributed populations. The situation where there is a cost of moving between patches (and the existence of distinct patches implies some geographical distance and consequent loss of time through travel) is considered, for example, in Charnov (1976) and DeAngelis et al. (2011). For studies of the IFD in real populations see Tyler and Gilliam (1995) for minnows, Bautista et al. (1995) for cranes , Griffen (2009) for crabs (this paper, in particular, also considers developments on the IFD model) and tuna Mariani et al. (2016). For the IFD in plants see McNickle and Brown (2014); McNickle (2020).

See Crowe et al. (2009b) for a review of some models of kleptoparasitism, and Vahl (2006) for an overview of a wider class of models. Broom and Ruxton (2003) consider kleptoparasitism models with different prey types where handling was not memoryless, and Broom and Rychtář (2009) look at such models with incomplete information. Crowe et al. (2009a) consider the specific example of kleptoparasitism in dung beetles and Spencer and Broom (2018) consider kleptoparasitism amongst urban gulls. Recent theoretical models are in Chowdhury et al. (2020); Garay et al. (2020). Stochastic versions of compartmental models were investigated in Yates and Broom (2007). Broom et al.

(2010a) consider the compartmental model of Ruxton and Moody (1997) for a finite number of states.

For more recent work on producer scrounger models see Afshar and Giraldeau (2014); Afshar et al. (2015); Dubois and Richard-Dionne (2020). A variant of producer-scrounger models where individuals can choose their level of fighting effort is given in Broom et al. (2015a, 2018).

A good book on social foraging theory is Giraldeau and Caraco (2000); see also Stephens and Krebs (1986) and Giraldeau and Livoreil (1998). For work on interference competition see Stillman et al. (1997), Vahl et al. (2005) and Smallegange (2007). Sirot (2000) gives a game-theoretical model of aggressiveness in feeding groups.

19.10　Exercises

Exercise 19.1 (Fretwell and Lucas, 1969). Assume **p** is an ESS of the IFD game of Example 19.1. Complete the proof of Theorem 19.2 that an ESS in this game must satisfy conditions (i) and (ii) (i.e. show that if individuals occupy only $l < n$ patches, then R $\geq B_{l+1}$).

Exercise 19.2 (Cressman and Křivan, 2006). Show that if **p** satisfies conditions (i) and (ii) of Theorem 19.2, then it is an ESS of the IFD game of Example 19.1 (including in the case when $l < n$).

Exercise 19.3. Show that there is exactly one $p_1 > 0$ satisfying (19.16) if and only if (19.17) does not hold (and otherwise there is no such p_1), and consequently that there is only one ESS for a single species IFD game on two patches.

Exercise 19.4. As in Exercise 19.3, show that there is exactly one ESS for a single species IFD game on n patches for general n.

Exercise 19.5. As in Example 19.4, determine the ESS for a single species IFD game on 3 patches, with parameters as in Figure 19.1.

Exercise 19.6. Compare the Ideal Free Distribution from Section 19.2 with Parker's matching principle from Section 7.2.2.1. Discuss the similarities and differences in these two models. Explain why the IFD model can have an empty patch but Parker's model cannot.

Exercise 19.7. Suppose that we have two patches 1 and 2 following the Fretwell and Lucas (1969) model with $B_1 = 1+\alpha, B_2 = 1, f_A(x) = x, f_B(x) = x^2$ for $\alpha > 0$. Find the IFD for (a) $\alpha = 1/4$, (b) general α.

Exercise 19.8. Suppose that in the two species two patches game described in Example 19.5, we have $M = N, r_1 = 1, r_2 = 2, \lambda = 1$ and $\Lambda = 3$. Show that both species can live on both patches with arbitrary $p_1 \in (0,1)$. Also, show that since species 2 is the more resource-intensive species, it is impossible to have an equilibrium with all of species 2 on either patch. Show that if all of species 1 is on patch 1, then $p_1 = 1$ and so $q_1 = 1/9$; if all of species 1 is on patch 2 then $p_1 = 0$ and so $q_1 = 4/9$.

Exercise 19.9 (Křivan, 2003, Křivan et al., 2008). Consider a modification of the two species two patches game from Example 19.5 where the resource level is restored depending upon the current level, i.e. the total level of resource on patch i is given by

$$\frac{dR_i}{dt} = r_i(R_i) - f_i(R_i)p_iM - g_i(R_i)q_iN \quad i = 1, 2, \tag{19.63}$$

where individuals of type 1 consume the resource at rate $f_i(R_i)$, individuals of type 2 consume it at rate $g_i(R_i)$ and the resource is replenished at rate $r_i(R_i)$. Note that this is equivalent to the previous analysis from Example 19.5 if $r_i(R_i) = r_i$, a constant, and $f_i(R_i) = \lambda R_i$, $g_i(R_i) = \Lambda R_i$ are linear functions where each species has the same constant term for both patches. Assume that the patches are in resource equilibrium, so that the differential equations (19.63) are equal to zero and there is no change in the available resources. Find the IFDs.

Exercise 19.10. Assume that there are two independent Markov processes M_1 and M_2, happening with rates v_1 and v_2. Show that the expected time for one of the processes to end is $(v_1 + v_2)^{-1}$ and that the probability the process M_1 will end first is $\frac{v_1}{v_1 + v_2}$.

Exercise 19.11 (Broom and Rychtář, 2007). Simplify formulae (19.42) and (19.43) to find expressions for the handling times T_S and T_H. Hence find an expression for the payoff function (19.41).

Exercise 19.12 (Broom and Rychtář, 2007). Given the parameters $P, t_h, t_c, \nu_f f, \nu_h, \alpha$, and the population strategy p, using the system (17.35)–(17.38), show (17.40).

Exercise 19.13. Using the results of Exercises 19.11 and 19.12, find the individual strategy q that for given population strategy p maximises (19.41). Use it to identify the ESS.

Exercise 19.14 (Broom and Rychtář, 2007). Substituting p for q in formula (19.41) and using the results of Exercises 19.11 and 19.12, prove the formula (19.39).

Exercise 19.15. Consider a variant of the Finder-Joinder game from Example 19.7 where the Joiner makes the decision first. Find the ESS. Also, for $\alpha = 1/2$, compare this with the variant when both individuals have full information and when only the Finder knows a.

Exercise 19.16. Consider a variant of the Finder-Joiner game from Example 19.7 where the individuals play simultaneously rather than a sequential Hawk-Dove game. Find the ESS.

Hint. This is an asymmetric game, as seen in Chapter 8, and the payoffs can be represented by a bimatrix.

Chapter 20

Predator-prey and host-parasite interactions

In this chapter we look at a class of interactions which are among the most fundamental in nature, that between two species, where one species exploits the other as a resource in some way. This can be between a predator and its prey, for instance between lynx and hare; a host and a parasite, for instance dogs and fleas; or between a host and a parasitoid, which can perhaps be thought of as an intermediate relationship between the first two, since the parasitoid lives within its host like the parasite but kills the host like the predator kills the prey.

Predator-prey systems, in particular, are the subject of a vast effort of modelling, generally of the differential equation type, following the classical models originating with Lotka and Volterra (Lotka, 1925; Volterra, 1926; Hofbauer and Sigmund, 1998). These models were used to explain the counter-intuitive result that the disruption of fishing in the Adriatic caused by the First World War had benefitted predatory fish more than the prey fish which were the prime focus of the fishing effort. Since then such models have proliferated and we will not discuss them here, since they are generally not game-theoretic.

In the following section we consider game-theoretical models of the classical predator-prey encounter, as well as related host-parasite models. We then go on to consider some more specific interactions and how they have been modelled using game theory.

20.1 Game-theoretical predator-prey models

The classical predator-prey interaction is often given in the form of differential equations (Hofbauer and Sigmund, 1988). Brown and Vincent (1992) consider a discrete version of this system, where the difference equations for

DOI: 10.1201/9781003024682-20

predators and prey are

$$N_{t+1} = N_t \left(1 + r_1 \left(\frac{K - N_t}{K} \right) - b P_t \right), \tag{20.1}$$

$$P_{t+1} = P_t \left(1 + r_2 \left(1 - \frac{P_t}{cbN_t} \right) \right), \tag{20.2}$$

where N_t (P_t) is the number of prey (predators) at time t, r_1 and r_2 are the intrinsic growth rates of the prey and predator, respectively, c represents the efficiency factor for converting captured prey into additional predators and b is the probability that a predator will capture a prey during a time step.

Below is a simplified version of the model of Brown and Vincent (1992).

Example 20.1 (Brown and Vincent, 1992). Consider a population of predators and prey. Both prey and predators use one out of a number of strategies, the choice of which influences their success. The prey's strategy influences their carrying capacity and their capture susceptibility by predators. Predators possess a strategy that influences the capture success of prey. Construct a model and determine the ESSs.

20.1.1 The model

Brown and Vincent (1992) generalise (20.1) and (20.2) to

$$N_i(t+1) = N_i(t) G_1(u, v, N, P), \tag{20.3}$$
$$P_j(t+1) = P_j(t) G_2(u, v, N, P), \tag{20.4}$$

where

$$G_1(u, v, N, P) = 1 + r_1 \left(\frac{K(u) - N}{K(u)} \right) - b(u, v) P, \tag{20.5}$$

$$G_2(u, v, N, P) = 1 + r_2 \left(1 - \frac{P}{cb(u, v)N} \right), \tag{20.6}$$

are the per capita growth rates of a prey (predator) playing strategy $u(v)$ in a population composed of population strategies and sizes u, v, N, P. The functions G_1 and G_2 play the role of our standard fitness functions $\mathcal{E}$.

Following Brown and Vincent (1992), we consider the following functions to model the influence of strategies:

$$K(u) = K_{\max} \exp \left(-\frac{u^2}{s_k^2} \right), \tag{20.7}$$

$$b(u, v) = b_{\max} \exp \left(-\frac{(u - v)^2}{s_p^2} \right), \tag{20.8}$$

for some parameter values $K_{\max}, s_k, b_{\max}$ and s_p. Equation (20.7) indicates that the carrying capacity is maximal at $u = 0$ and declines following the shape of the bell-shaped density function of the normal distribution. The larger the value of s_k, the slower the rate of decline, so that large s_k indicates a large range of potentially good prey strategies. Equation (20.8) assumes that a potential re-scaling of the predator strategy took place so that predators are most successful at catching prey if they use the same matching strategy $u = v$. The parameter s_p indicates the niche breadth of the predators; the higher the value of s_p, the greater the variety of prey a predator can catch.

20.1.2 Analysis

In the terminology of Chapter 8 we have a game with two roles, where every individual is always in the same role, either predator or prey, so that $\rho = 0$ or 1, and the payoffs are given by

$$\mathcal{E}_1[u; \delta_v] = G_1(u, v, N, P), \tag{20.9}$$

$$\mathcal{E}_2[v; \delta_u] = G_2(u, v, N, P). \tag{20.10}$$

The difference equation formulation requires any ESS (u^*, v^*, N^*, P^*) to be normalised by

$$G_1(u^*, v^*, N^*, P^*) = 1, \tag{20.11}$$

$$G_2(u^*, v^*, N^*, P^*) = 1. \tag{20.12}$$

In order for a strategy to be non-invadable, we require

$$G_1(u^*, v^*, N^*, P^*) > G_1(u, v^*, N^*, P^*), \tag{20.13}$$

$$G_2(u^*, v^*, N^*, P^*) > G_2(u^*, v, N^*, P^*), \tag{20.14}$$

for all $u \neq u^*, v \neq v^*$. Note that technically we can conceive of solutions where either or both of the above inequalities are replaced by equalities, but that given the form of the payoff functions G_1 and G_2 such cases would be non-generic. This yields two additional equations that our ESS must satisfy

$$\left.\frac{\partial}{\partial u} G_1\right|_{u=u^*} (u, v^*, N^*, P^*) = 0, \tag{20.15}$$

$$\left.\frac{\partial}{\partial v} G_2\right|_{v=v^*} (u^*, v, N^*, P^*) = 0. \tag{20.16}$$

The equations (20.11)–(20.12) and (20.15)–(20.16) give four equations for four unknowns and the solutions are candidates to be an ESS. However, we need to keep in mind that (20.15)–(20.16) are only necessary not sufficient conditions for an ESS. In order for (20.13) and (20.14) to hold, we also need

$$\left.\frac{\partial^2}{\partial u^2} G_1\right|_{u=u^*} (u, v^*, N^*, P^*) < 0, \tag{20.17}$$

$$\left.\frac{\partial^2}{\partial v^2} G_2\right|_{v=v^*} (u^*, v, N^*, P^*) < 0. \tag{20.18}$$

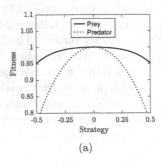 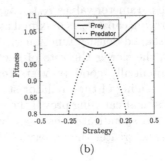

(a)   (b)

FIGURE 20.1: Plots of $G_1(u, v^*, N^*, P^*)$ for a prey individual and $G_2(u^*, v, N^*, P^*)$ for a predator. The parameter values are set to $K_{max} = 100, b_{max} = 0.1, s_k = 1, r_1 = r_2 = 1$ and thus $N^* = 50, P^* = 5$ and also $u^* = v^* = 0$. (a) $s_p = 1.1$ and (u^*, v^*, N^*, P^*) is an ESS, (b) $s_p = 0.6$ and (u^*, v^*, N^*, P^*) is not an ESS.

20.1.3   Results

It is left to the reader as Exercise 20.1 that (20.16) is satisfied only if $v^* = u^*$, i.e. if predators match the prey perfectly. Consequently, (20.15) is satisfied if $u^* = 0$, i.e. if prey maximises its carrying capacity. However, as observed in Brown and Vincent (1992), $u^* = v^* = 0$ is an ESS only for sufficiently large values of s_p (i.e. the predator has a sufficiently broad niche) because for small values of s_p, an inequality opposite to (20.17) holds (so that G_1 has a local minimum for $u^* = v^* = 0$). The situation is illustrated in Figure 20.1. Brown and Vincent (1992) conclude that for small s_p there is a polymorphism, where there are two prey species rather than one, with a single predator species. It should be noted that Brown and Vincent (1992) developed a more general model than the one presented here and that their model is able to handle the scenario of multiple predator and prey species.

20.2   The evolution of defence and signalling

Prey animals possess a variety of defences against predators. Some have defensive weapons such as sharp spines, others are able to move rapidly to evade predators. Many species of animals use toxins to make them unpalatable or worse, so that predators avoid them. One problem associated with this type of defence is that, unlike horns or rapid movement, the defence may not be apparent to predators until the prey has actually been eaten, which is of no help to the prey animal concerned. Toxic prey have thus often developed bright colouration to advertise the fact that they are defended. Thus unlike undefended animals which may use camouflage to hide from predators, these individuals make themselves clearly visible, for example the poison dart frogs of Central and South America. Predators, in turn, associate bright colouration with toxicity and avoid prey with such an appearance.

One question of interest is, how do predators come to associate bright colours with high levels of defence? Are they born with a natural aversion, or do they learn by attacking some brightly coloured individuals when young and finding them inedible? It is usually assumed that the latter is the case, and that some learning mechanism is required.

Example 20.2 (Signalling toxicity, Leimar et al., 1986). Consider a population of prey who possess varying levels of unobservable defence (which we shall call toxins) and of colouration. The prey is faced by a predator who is able to learn to avoid certain prey. The prey can advertise its toxicity with bright colouration, or alternatively aim to be as invisible as possible. Develop a model for this situation.

20.2.1   The model

We will follow the model developed in Leimar et al. (1986), but note that the symbols for the parameters are not the original ones used by Leimar et al. (1986), but chosen for consistency with subsequent working in this section.

Assume that a prey individual has two properties, its level of toxic defence t and its colouration r, with high t meaning high defence and high r high brightness and so visibility. The toxicity is assumed to be fixed and we consider r to be a trait or a strategy that can be evolved. For a given t, we would like to find an ESS value $r = r(t)$.

The colouration r will determine how often an individual prey will interact with the predator and, together with t, what will be the outcome of the interaction.

20.2.1.1   Interaction of prey with a predator

The interaction can be divided into three distinct phases.

1. The predator discovers the prey.

2. The predator observes the prey and decides whether or not to attack it.

3. If attacked, the prey defends itself.

Modelling the first and last phase is simple. The prey is discovered by the predator at rate $D(r)$; Leimar et al. (1986) simply set $D(r) = r$. Individuals with minimum $r = r_c \geq 0$ are the least visible and are termed (maximally) cryptic. The defence mechanism is modelled via a function $K(t)$ which denotes the probability that the prey is killed during an attack and is assumed to be a decreasing function of defence t.

Modelling the observation is more complex. The individual is attacked with probability Q that depends upon a number of factors, namely the prey colouration r and the predator's own experiences with past individuals (and thus upon the strategy played by the rest of the population).

Suppose that the population only consists of (t_1, r_1) individuals, except for our focal individual which is (t, r), and that a predator that has observed our focal individual has had n encounters with (t_1, r_1) individuals previously. Here, $n = D(r_1)T$ where T is the age of the predator. Leimar et al. (1986) express Q as follows:

$$Q(r; r_1, t_1, T) = e(r)\big(1 - h(r; r_1, t_1)\big)^n. \qquad (20.19)$$

Without any other information, a naive predator will attack an individual with appearance r with probability $e(r)$, where $e(r)$ may decrease with r, as there may be a natural wariness for unusual looking prey. Leimar et al. (1986) used

$$e(r) = 0.8 * \exp\left(-\left(\frac{r}{0.4}\right)^2\right) + 0.2. \qquad (20.20)$$

We suppose that our predator has a history of interacting with n previous prey. The worse these experiences, the more likely the individual will be to avoid similar prey in future (it is assumed that the probability of subsequent interactions leading to attack cannot be increased through earlier interactions). The higher h, the more the probability of subsequent interactions leading to attack is reduced. In particular, they used the form

$$h(r; r_1, t_1) = w(r, r_1)v(r_1)t_1, \qquad (20.21)$$

where $v(r_1)$ is an increasing function of r_1 and $w(r, r_1)$ is a unimodal function with a maximum at $r = r_1$. Thus $h(r; r_1, t_1)$ is increasing with t_1, with r_1 and also with the similarity of r and r_1. In an example, Leimar et al. (1986) use

$$v(r_1) = 1 - \exp\left(-\frac{1}{2}\left(\frac{r_1}{0.4}\right)^2\right), \qquad (20.22)$$

$$w(r, r_1) = 1 - \exp\left(-4\left(\frac{r - r_1}{r_1}\right)^2\right). \qquad (20.23)$$

The functions h and Q for various parameters are illustrated in Figure 20.2.

20.2.1.2   Payoff to an individual prey

We see that this game belongs to the class of games called playing the field, see Section 7.2, as the individual prey do not really play directly against themselves but are playing a predator (whose actions are influenced by the whole population of individuals).

The payoff to an individual prey using (r, t) in a population using (r_1, t_1) is defined as the probability it can survive some predetermined time interval in a population observed by N_p predators. The exact formula for the payoff is derived in Leimar et al. (1986) but is complex, so we omit it here.

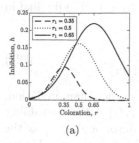 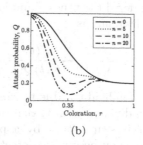 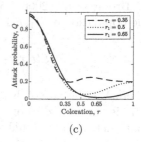

(a) (b) (c)

FIGURE 20.2: The attack probability as a function of an individual's colouration. In all figures, the population toxicity is $t_1 = 0.3$. (a) Function $h(r, r_1, t_1)$ given in (20.21). (b) The attack probability Q from (20.19) as a function of individual colouration r for various numbers of previous encounters n; here $r_1 = 0.35$. (c) The attack probability Q from (20.19) as a function of individual colouration r for various colourations of the population; here $n = 10$.

20.2.2 Analysis and results

The models involved are quite complex, and it was not possible to give an exhaustive analysis of possible results. However Leimar et al. (1986) used examples to show a number of interesting results. For instance, they found circumstances where there were two ESSs, a maximally cryptic one and an aposematic, i.e. brightly coloured, one. They observed two conditions for the existence of an aposematic solution, relating to how the value of r affects learning. If learning increases with r, or if the maximum of the generalisation gradient $h(r; r_1, t_1)$ increases with r_1, then such a solution can be stable. Similarly, two factors were identified for the evolution of increased defence t, in the model variant where t was variable and r fixed; the rate of avoidance learning increases with increasing t, and there is an increase in the chance of survival with increasing t.

20.2.3 An alternative model

The above model of Leimar et al. (1986) effectively considered a group of naive predators emerging at the start of every season (such as wasps) and replacing the experienced predators from the end of last season. Broom et al. (2006) developed a related model where the predator population was considered at equilibrium (some old and some new predators) and so the effect of learning, given the parameters and strategies, was constant, and did not change throughout the season. This is reasonable for longer-lived predators like birds, where there will be a mix of young and older individuals.

Broom et al. (2006) considered the fitness function to a (t, r) individual in the population of (t_1, r_1) individuals as

$$\mathcal{E}(t, r; t_1, r_1) = \frac{F(t)}{\lambda + D(r)K(t)Q(t, r; t_1, r_1)}, \qquad (20.24)$$

where λ is the rate of death of individuals due to causes other than predation, $F(t)$ is the fecundity of an individual with toxicity t, $K(t)$ is the probability that such an individual survives an attack by a predator and $D(r)$ is the rate that individuals of conspicuousness r are detected by predators. Q is the probability that a predator will attack a given individual. Q is considered a non-increasing function of

$$I = (1 - a)H(t_1)D(r_1)S(r, r_1) + aH(t)D(r)S(r, r), \qquad (20.25)$$

where $H(t)$ is a measure of the aversiveness of an individual of toxicity t (toxicity t reduces subsequent attacks if and only if $t > t_c$) and $S(r, r_1)$ is the similarity function between r and r_1 and is decreasing with $|r - r_1|$. The parameter a is the effect of local clustering, e.g. if a colony of insects plays the mutant strategy, then this may affect the predation of mutants because of high local mutant density, even if global density is negligible. See Exercise 20.3 for an example of a specific fitness function.

Broom et al. (2006) have shown the following possible results (depending on the parameter values).

1. $(0, 0)$ is an ESS (no investment in toxicity, maximally cryptic);

2. $(t_1, 0)$ for $0 < t_1 < t_c$ is an ESS (some investment in toxicity, but insufficient to be aversive, maximally cryptic);

3. $(t_1, 0)$ for $t_c < t_1$ is an ESS (high investment in toxicity to be aversive, still maximally cryptic);

4. $(t, 0)$ is not an ESS for any value of t (no ESS with maximum crypsis).

There can also be ESSs involving $r_1 > 0$. When these occur, for each r_1 there is a unique value t_1 associated with it, such that (t_1, r_1) is an ESS. However, for reasonable conditions, it is possible that such a pair exists for all $r_1 > r_c$ for some critical value r_c. Thus for such aposematic strategies, any colouration that is sufficiently bright is sufficient to deter predators.

The presence or absence of such a set of aposematic ESSs can occur with each of the four solutions above, yielding eight possible cases, including one with no ESSs of the above form. We note that Broom et al. (2006) also discussed the possibility of ESSs involving poorly defended individuals and a mixture of different values of r (it is good for poorly defended individuals to be cryptic, but it is also good for them not to look like others which predators wish to eat), which could be stable in such circumstances.

The Broom et al. (2006) solutions are co-evolutionary ESSs, whereas the Leimar et al. (1986) ones are not (either r or t was fixed), but it is likely that their model could also produce a continuum of aposematic ESSs. In both models an increased survivability of toxic individuals during an attack was important to allow non-minimal defence to evolve. There are other important features of the models which we will not discuss, one of which is the shape of the similarity function (S in Broom et al., 2006 and $w(r, r_1)$ in Leimar et al., 1986). In Broom et al. (2006) this function has a sharp (e.g. a Laplacian) peak at 0 and the Leimar et al. (1986) model has a flat (e.g. a Gaussian) peak. Which of these is used has important consequences, and this has been discussed at length in Ruxton et al. (2008).

20.2.4 Cheating

Another question is, how is cheating prevented? It is reasonable to assume that some cost is associated with being toxic, for example the cost of production or acquisition from the environment. This cost was explicitly incorporated in the original Leimar et al. (1986) and the Broom et al. (2006) model also implements the cost via the function $F(t)$ in (20.24). Thus if an individual had the bright colouration indicating that it was defended, but did not actually possess the defence, it would be able to deter the predators in the same way as defended animals, but not bear the cost of the defence. To prevent such cheating requires a sufficiently large value of the clustering coefficient a, when cheats have enough of a negative effect on their nest-mates to prevent invasion.

Such cheating may be prevented within a species, but entire species of cheats exist. A distinctive colouration used to deter predators can be common to a number of species. This can involve two defended species who reinforce each other's defence (*Mullerian mimicry*). But it can also involve one defended and one undefended species, where the undefended species effectively parasitises the first species (*Batesian mimicry*), for example hoverflies mimic the black and yellow striped patterning of wasps. The fact that the first species is defended makes predators avoid the individuals of both species, but the lack of defence of the second species reduces the effectiveness of the colouration, as predators may discover defenceless individuals with that patterning. See Ruxton et al. (2004) for a discussion.

We finally note that aposematism can be regarded as a signal in a similar way to the signals used in mating (see Chapter 17), where the signalling individual tries to communicate some information about itself to the receiver (I am of high quality, I am highly toxic) to persuade the receiver to take a particular action (mate with him, not attack it).

20.3   Brood parasitism

Various bird species do not raise their own chicks, but rather lay an egg in the nest of another bird and trick it into raising the chick, which is called brood parasitism. The two most well-known types of brood parasite are the cowbirds of the Americas and the cuckoos of Europe and Asia. Cuckoos are the more destructive parasites, as their chicks will completely destroy the host clutch, leaving just the cuckoo chick for the host bird to raise. Cowbird chicks will typically not destroy the clutch, but will outcompete the host chicks, thus significantly increasing mortality, but allowing some real chance of survival.

Following Planque et al. (2002), we consider a model of the interaction between cuckoos and their hosts. A remarkable feature of the interaction is that defence against cuckoos is not more evident. One can think of defence happening broadly in one of two stages; prior to hatching (egg-rejection) and after hatching (chick-rejection). In the Common Cuckoo, in particular, there is often egg-rejection behaviour by the host, but rarely chick rejection. Why does this not occur? This seems an obvious strategy, and cuckoo chicks appear very different to their hosts (as opposed to the eggs, which can often be very similar, as the cuckoos attempt to defeat egg rejection). This was the problem addressed by Planque et al. (2002).

Example 20.3 (Cuckoo-host interaction, Planque et al., 2002). Consider the interaction between cuckoos and their hosts, where a host defence happens in one of two stages, egg-rejection and chick-rejection. Find the ESSs.

20.3.1   The model

We will follow the model of Planque et al. (2002). They consider four strategies for the host: all acceptors, egg rejectors, chick rejectors and all-rejectors, the frequencies of each being denoted by h_t^a, h_t^e, h_t^c and h_t^k, respectively. Ignoring any costs in being able to carry out more complex behaviours, they concentrate instead on the costs associated with errors.

Suppose that cuckoo search for nests follows a Poisson process, leading to the probability of any host nest escaping parasitism being given by e^{-aP_t}, where P_t is the density of cuckoos in year t (May and Robinson, 1985). The expression for the density of parasites is

$$P_{t+1} = s_P P_t + (1 - e^{-aP_t})H_t(h_t^e + q_e h_t^e + q_c h_t^c + q_k h_t^k)G, \qquad (20.26)$$

where s_P is the survival probability of adult parasites to the next season, G is the probability that a cuckoo chick will survive to the next breeding season in the absence of any host defence, $H_t = h_t^a + h_t^e + h_t^c + h_t^k$ is the density of hosts and the q parameters relate to errors by the host in defence. Specifically, let q_e be the probability of an egg rejector accepting a cuckoo egg by mistake,

q_c be the probability of accepting a cuckoo chick by mistake and q_k be the probability of an all-rejector accepting a cuckoo by mistake after both of its defences.

The density of hosts, H_t, is assumed to follow

$$H_{t+1} = \frac{H_t}{1 + H_t/K}(s_H + f_a h_t^a + f_e h_t^e + f_c h_t^c + f_k h_t^k), \qquad (20.27)$$

where K is included to take into account limitations of natural resources and within-host competition, s_H is the probability of an adult host surviving to the next season and the f terms are fitnesses represented by the number of surviving offspring at the end of the season. These are given by

$$f_a = f e^{-aP_t}, \qquad (20.28)$$

$$f_e = e_1 f e^{-aP_t} + e_2 f(1 - e^{-aP_t}), \qquad (20.29)$$

$$f_c = c_1 f e^{-aP_t} + c_2 f(1 - e^{-aP_t}), \qquad (20.30)$$

$$f_k = k_1 f e^{-aP_t} + k_2 f(1 - e^{-aP_t}), \qquad (20.31)$$

where f is the number of offspring per annum raised by an all-accepting host pair that is not parasitized, $e_1 f$ and $e_2 f$ are the expected number of offspring raised by unparasitised and parasitised egg rejectors, and correspondingly for the similar parameters for the chick rejectors and all rejectors.

Let p_e be the probability that an egg rejector rejects one of its own eggs by mistake when there is no parasite present, and b_e be the payoff for raising the resultant brood with one chick less. Let p_c and b_c be the corresponding values for a chick-rejector.

This leads to the following values in equations (20.28)–(20.31) above:

$$e_1 = (1 - p_e) + p_e b_e, \qquad (20.32)$$

$$c_1 = (1 - p_c) + p_c b_c, \qquad (20.33)$$

$$k_1 = (1 - p_e)(1 - p_c) + b_e p_e(1 - p_c) + b_c p_c(1 - p_e) + b_k p_e p_c, \qquad (20.34)$$

$$e_2 = (1 - q_e)b_e, \qquad (20.35)$$

$$c_2 = (1 - q_c)\gamma b_e, \qquad (20.36)$$

$$k_2 = (1 - q_e)b_e + q_e(1 - q_c)\gamma b_e, \qquad (20.37)$$

where b_k is the cost of losing both an egg and a chick and γ is a measure of the damage done by the cuckoo chick before it is ejected (the chick will try to destroy all of the host brood). The derivation of the above formulae is left for the reader as Exercise 20.4.

20.3.2 Results

We note that for parasitism to be possible there must be the possibility of errors, and if there are no errors the cuckoos become extinct. Since there are

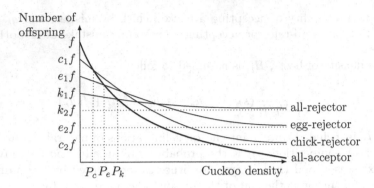

FIGURE 20.3: Host fitness as a function of cuckoo density for four possible host strategies: all-acceptor, egg-rejector, chick-rejector and all-rejector.

errors, and in particular, errors which result in host egg/chick loss irrespective of whether a parasite is present, at low cuckoo densities non-defending hosts do better. Similarly, at extremely high densities defence is best. See Figure 20.3 for a comparison of the host fitness for each of the four possible strategies as a function of the cuckoo density for a particular set of parameter values.

For many, but not all, parameter values, the populations eventually settled to equilibrium values of the number of hosts and parasites, and stable solutions occurred for three types of host strategy, a mixture of egg rejectors and all acceptors, a mixture of chick rejectors and all acceptors and a mixture of all rejectors and all acceptors.

The numbers of hosts and parasites were labelled H_e and P_e if the defence in the population was egg rejection (and similarly H_c, P_c or H_k, P_k for the other two cases).

Assuming there are errors, what defensive strategy will evolve, i.e. which of the three equilibria will be reached? Planque et al. (2002) found that the evolutionarily stable defensive strategy was precisely the one that gave the lowest equilibrium proportion of cuckoos in the mixture, so the problem reduced to comparing the sizes of P_c, P_e and P_k.

Egg ejection was worse than chick rejection if $P_e > P_c$ which is equivalent to

$$\frac{1 - e_1}{e_2} > \frac{1 - c_1}{c_2}, \tag{20.38}$$

and substitution of terms leads to

$$\frac{p_e}{1 - q_e} \frac{1 - q_c}{p_c} > \frac{1 - b_c}{1 - b_e} \frac{1}{\gamma} > 1. \tag{20.39}$$

We leave the derivation of (20.39) as Exercise 20.5. Since $b_c \leq b_e$ and $\gamma < 1$ (possibly γ is much less than 1), this means that the left-hand side of (20.39) must be sufficiently larger than 1. This corresponds to chick rejection having to be very much easier than egg rejection.

They also show that if chick rejectors cannot invade egg rejectors then nor can all rejectors. All rejectors use two layers of defence, the cost of each in terms of errors are unaffected by the presence of the other defence. But, for example, if the first defence is reasonably effective, when the second defence is employed many of the cuckoos will already have been removed. Thus it will be similar to not employing the first defence with a lower parasite level (and the smaller the number of parasites, the worse the defence is). As the authors note, this is a generalisable effect of multiple layers of defence; multiple defence is unlikely to be worthwhile, unless the component defences are unreliable.

We should mention that the game-theoretic explanation for the lack of defence to cuckoo parasitism is not the only one, and an alternative is that of evolutionary lag, i.e. that parasitism is sufficiently recent that defences have not evolved to meet it yet (see e.g. Soler et al., 1995; Zahavi, 1979).

20.4   Parasitic wasps and the asymmetric war of attrition

Parasitic wasps lay their eggs within the larvae of other species such as moth or butterfly caterpillars. These insects are parasitoids since they eventually kill the host larvae, although the larvae live for some time before this occurs, which allows any given larva to be parasitised by more than one egg (Godfray, 1994). This is called superparasitism and was modelled in Haccou et al. (2003). In such parasitic species it is often the case that females leave a pheromone mark on hosts that they have parasitised, and so females can recognise hosts that are unparasitised, parasitised by themselves or parasitised by conspecifics (van Lenteren, 1981). Thus they are able to distinguish four different types of host; those unparasitised, those parasitised by themselves but not others, those parasitised by others but not themselves and those parasitised by both themselves and others.

Example 20.4. Consider a patch of host larvae where one female parasitoid has just arrived and a second one may arrive sometime in the future. Whilst a female remains in the patch, she will encounter the hosts which will fall into one of four types (those unparasitised, those parasitised by herself but not the other, those parasitised by the other but not herself and those parasitised by both herself and the other). Parasitism of a host can only result in one surviving parasitoid larva, so the presence of other parasitoids brings costs. It is assumed that a host which only contains a single parasitoid will lead to a surviving offspring with probability 1.

Parasitoids will always attack unparasitised hosts (this leads to the maximum gain that they can receive) and never attack those only parasitised by themselves (this will entail additional costs but no extra payoff). They may or may not choose to attack hosts parasitised at least once by others; whether to do so is a strategic decision. The parasitoids aim to maximise their rate of production of future offspring, and there is a fixed rate of production in the environment which can be achieved when a patch is left. The female must also find the right time to leave a patch.

20.4.1   The model

In the following analysis we follow Haccou et al. (2003). Suppose that two parasitising wasps F_v and F_w are present at a patch of host larvae. For simplicity (and the analysis is still not simple) we shall assume that no other females will visit the patch subsequently. The first female enters a patch of non-parasitised hosts, the second entering the patch at a random time later (uniformly distributed from 0 to the time that a female would leave the patch if she were alone). Each host can be in one of five states:

(1) unparasitised,

(2) parasitised only by F_v,

(3) parasitised only by F_w,

(4) parasitised by both, and currently will yield a surviving offspring to F_v,

(5) parasitised by both, and currently will yield a surviving offspring to F_w.

Females know if the host is in state 1,2 or 3 and also if it is in 4 or 5 (but cannot distinguish between 4 and 5).

It is also assumed that females begin by parasitising only unparasitised eggs (a non-superparasitising female), and switch at some point (which may in practice be never) to both parasitising and superparasitising (just termed a superparasitising female), when the supply of parasitised hosts becomes sufficiently large compared to the supply of non-parasitised hosts.

Non-superparasitising females encounter a given unparasitised host (and parasitise it) at rate λ. Superparasitising females encounter both non-parasitised and parasitised hosts to be superparasitised at rate $\mu < \lambda$ (the lower rate accounts for the loss due to the proportion of effort now allocated to superparasitism). If they are playing the superparasitism strategy, they will superparasitise a host if it contains at least one egg of another individual, and then there is a probability σ that this latest egg will be the one to yield an offspring.

We thus have a Markov chain where each host larva is in one of the five states. There are four possibilities for the two females; neither superparasitises, only F_v superparasitises, only F_w superparasitises, or both superparasitise. The situation is summarised in Figure 20.4.

It can be seen from Figure 20.4 that states 2 and 4 (3 and 5) can be lumped together into a single state V (W). Let U denote state 1 (see e.g. Haccou and Meelis, 1992).

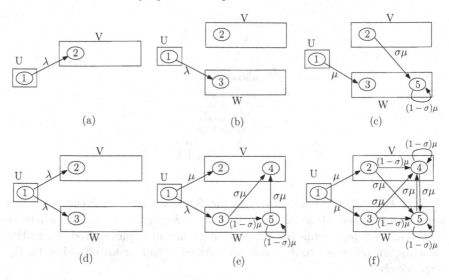

FIGURE 20.4: Markov chain diagrams for the model of parasitic wasps by Haccou et al. (2003). (a) Female F_v is alone on the patch. (b) F_w is alone on the patch (F_v has left) but has not started to superparasitize yet. (c) F_w is alone on the patch (F_v has left) and has started to superparasitize. If F_v superparasitized before leaving, there would also be an arrow from 4 to 5 at rate $\sigma\mu$. (d) F_v and F_w are both on the patch, neither superparasitising. (e) F_v and F_w are on the patch, with only F_v superparasitising. The case when only F_w superparasitises is analogous. (f) F_v and F_w are both on the patch and both are superparasitising.

The complete state of the patch is thus denoted by the proportions in the different states u, v and w. The states can thus be represented by our familiar triangle figure analogous to Figure 2.1. The strategy of any female will be a choice of what to do (leave, parasitise or superparasitise) for every possible scenario, i.e. every point in the triangle, as either first or second female. For example, Figure 20.5 shows an example of the strategy of a female (when she is the second female). We have an asymmetric game similar to those described in Chapter 8, where the different roles are indicated by whether a female is first or second, but also by the time elapsed between the arrival of the two females. An individual's role is thus the length of time t, positive or negative, which she arrives before the other female.

20.4.2 Analysis—evaluating the payoffs

Each female aims to maximise her number of offspring, and we can think of the payoff to each female pursuing a given strategy as the rate of increase in the number of hosts that will lead to her offspring. The payoff to F_v (F_w)

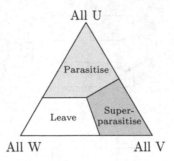

FIGURE 20.5: Optimal strategies for F_w (leaving, parasitising or super-parasitising) when alone on the patch after F_v has already left, where the point (u, v, w) represents the proportion of larvae unparasitised, parasitised (yielding an offspring to F_v) and parasitised (yielding an offspring to F_w), respectively.

is the derivative of $v(w)$ with respect to t. When there is only a single non-superparasitising female F_w present, transitions follow the differential equations

$$\frac{du}{dt} = -\lambda u, \frac{dw}{dt} = \lambda u, \frac{dv}{dt} = 0. \tag{20.40}$$

When there is only a single superparasitising female F_w present, the transitions follow the equations

$$\frac{du}{dt} = -\mu u, \frac{dw}{dt} = \mu u + \sigma \mu v, \frac{dv}{dt} = -\sigma \mu v. \tag{20.41}$$

The derivation of equations (20.40) and (20.41) is left for the reader as Exercise 20.7.

Assume that the environmental gain rate is γ. The female wishes to choose the strategy that maximises her gain rate, which in equations like (20.40) or (20.41) is simply the value of the derivative of v.

The payoff of a female F_w when still on the patch is simply the rate of change of w. Hence, when F_w is already alone on the patch, the payoff is

$$\frac{dw}{dt} = \begin{cases} \lambda u & F_w \text{ has not started superparasitising yet,} \\ \mu u + \sigma \mu v & F_w \text{ is superparasitising already.} \end{cases} \tag{20.42}$$

By comparing λu with $\mu u + \sigma \mu v$, the female F_w can thus determine whether she should superparasitise or not (when alone on the patch). Assuming that the environmental payoff (the expected rate of gain after leaving the patch) is γ, one can also determine when it is a good time to leave the patch. Note that once F_v has left, then (by our assumption) no other female will enter the patch after F_w leaves. Hence, the optimal strategies for F_w when already alone on the patch are as follows:

$$\text{leave if } \gamma > \lambda u \qquad \qquad \gamma > \mu u + \sigma \mu v, \qquad (20.43)$$

$$\text{parasitise if } \gamma < \lambda u \qquad \qquad \lambda u > \mu u + \sigma \mu v, \qquad (20.44)$$

$$\text{superparasitise if } \lambda u < \mu u + \sigma \mu v \qquad \gamma < \mu u + \sigma \mu v. \qquad (20.45)$$

This is summarised in Figure 20.5, which shows the most interesting case where the parameter values allow all three behaviours.

The situation when neither female has yet left is more complicated because even after leaving a patch, the payoff may not be fixed and will depend upon whether the other female is still on the patch (or when she will arrive), if she will superparasitise and how long she will stay. Let us assume that both females are on the patch and so F_v must choose whether to leave or not. Since the best strategy for F_w when alone on the patch is a straightforward optimisation problem with solution as above, it is possible to calculate the reward for F_v when leaving before F_w. This reward will be $V_\infty = V(U_0, V_0, W_0)$, the number of hosts that will be in the state V after F_w leaves the patch, conditional on F_v leaving the patch as the first female when the state was (U_0, V_0, W_0). The gain rate for F_v will then be V_∞ / T, where T is the total time spent on the patch. A similar reasoning works for F_w. Thus at any given point, we see that the two females choose a strategy based upon how long they will wait, and when one leaves the reward to the other will suddenly increase (equivalent to receiving a positive reward at that point) and F_v and F_w can be thought of as playing in a (generalised) war of attrition (see Bishop and Cannings, 1978).

20.4.3   Discussion

In general, within a patch the number of hosts available to parasitise decreases, so pure parasitism is less effective over time within a patch. In particular, for a patch only visited by a single female F_v, $w(t) = 0$, and the female will parasitise until $u(t) = \gamma / \lambda$, when she will leave (it is assumed that there is no later arrival). When there are two females, the number of hosts available to superparasitise increases initially, and then decreases (at some point after when she starts superparasitising, if she does). Depending upon when a female arrives and on the parameters, she may: leave immediately; parasitise, and then leave; parasitise, superparasitise and then leave; superparasitise and then leave.

We note that the solution of the model relies on the assumption that the background fitness γ does not depend upon the strategies of the wasps themselves. It is possible that if the population strategy changed, e.g. a switch to never superparasitising, then the background fitness could be changed, and so it might be possible to have more than one ESS (see the kleptoparasitism models in Section 19.5.3 for a similar situation). The assumption of constant γ is reasonable as long as the population density is sufficiently low (again see

Section 19.5.3). In fact, the assumption of not more than two females meeting on a patch also effectively relies on this.

It is also (reasonably) assumed that the role of the female (the time t she arrives before the other female) is independent of her strategy, i.e. the game is strategy-role independent (see Chapter 8).

The asymmetry of the model means that the female who arrives first finds a better quality patch than the second arrival, since then it is completely unparasitised. This led Haccou et al. (2003) to predict that the first female would spend longer in the patch than the second, a prediction empirically verified by Le Lann et al. (2011) in experiments with the aphid parisitoid *Aphidius ervi Haliday* in a patch of grain aphid *Sitobion avenae Fabricius* larvae.

20.5   Complex parasite lifecycles

Helminths are parasitic worms, for example tapeworms. They exhibit a variety of life stages and host types, and it is common for there to be a number of host stages, three levels of host in a food chain being common.

Example 20.5 (Tapeworms' lifecycle). A human infected with a tapeworm *Taenia solium* may pass eggs or segments of the adult tapeworm into soil. Each segment contains thousands of microscopic tapeworm eggs that can be ingested via food contaminated with the faeces. Once ingested by a pig, eggs develop into larvae. Larvae can migrate out of the intestines and form cysts in other tissues. If a raw or undercooked pork meat with cysts is ingested by a human, the larvae then develop into adult tapeworms in the intestines. On the other hand, the dwarf tapeworm *Hymenolepiasis hana* can complete its entire life cycle, egg to larva to adult tapeworm, in one host, rat or human.

The question is how can a parasite develop a behaviour where the parasite evolves to using two levels of host, from having a single host. This has been addressed by Parker et al. (2003). They model two types of mechanism:

- downwards incorporation, where a new intermediate host is added down the food chain (which will be preyed upon by the original host);

- upwards incorporation, where the new host is a predator of the first, and this is what we consider here.

20.5.1   A model of upwards incorporation

We consider three different stages of the development of the complex life cycle. Initially, a parasite has a lifecycle that involves a single host. Each

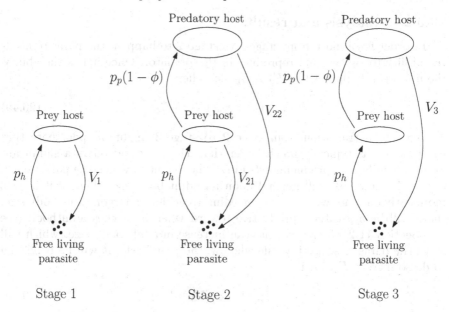

FIGURE 20.6: The development of a complex helminth lifecycle in three stages. Stage 1: single host. Stage 2: two hosts, with both the prey host and the predator host producing the parasite. Stage 3: two hosts, with only the predator host producing the parasite.

parasite individual enters the host with probability p_h, which then generates V_1 new free living parasites. As an intermediate stage, each initial host is eaten by a predator with probability p_p and the parasite can reproduce in both hosts, prey and predator, producing V_{21} and V_{22} free living parasites, respectively. The third and final stage is that reproduction in the initial prey host can become suppressed, so that reproduction only occurs in the predator host, resulting in V_3 free living parasites. This is summarised in Figure 20.6.

The question is, when can this sequence evolve, so that the parasites obtain an extra layer of host?

It is assumed that there is an extra cost to surviving in the predator host (the parasites now have to be able to resist the defences of both types of host), so that the fitness of any parasite within the higher level host is multiplied by $1 - \phi$. We give a simplified version of the argument in Parker et al. (2003) as an illustration (in particular, the time-based analysis which is a key feature of their argument is completely omitted here). We denote the payoff to a parasite playing any of the three strategies above as E_1, E_2 and E_3, and these are

$$E_1 = p_h V_1, \tag{20.46}$$

$$E_2 = p_h(V_{21} + p_p(1 - \phi)V_{22}), \tag{20.47}$$

$$E_3 = p_h p_p(1 - \psi)V_3. \tag{20.48}$$

20.5.2   Analysis and results

In order for a move from stage 1 to stage 2 to happen, the parasite needs the ability to survive and reproduce in the predator. Once it has the ability, the move can happen when $E_2 > E_1$, i.e. when

$$V_1 - V_{21} < p_p(1 - \phi)V_{22}. \tag{20.49}$$

To see how a transition from stage 2 to stage 3 might happen, note that the predator is typically much larger than the prey and offers a significant extra possibility for producing offspring. The extent to which the parasite can exploit its new host will depend upon its adult body size. Thus if it invests more early on in growth to delay breeding to achieve a larger adult body size, there will be a greater reproductive reward when it reaches adulthood (see also Section 11.2). If the parasite can increase in adult body size (which will leave the parasite still a juvenile when in the prey host), it will be beneficial to do so if $E_3 > E_2$, i.e. if

$$V_{21} < p_p(1 - \phi)(V_3 - V_{22}). \tag{20.50}$$

In simple terms this means that the benefit of each change must exceed its cost. Clearly this requires a sufficiently high predation rate of the prey host by the predator host p_p for this upwards incorporation to occur.

We see that this model is one of simple optimisation as stated. There is no competition either between host and parasite or between different parasites, so each parasite is finding the best solution to an environmental problem independent of the strategies of others. We thus note that not all models involving strategies are game-theoretical, and that "simple" models can also be illuminating. The model would become game-theoretic if either there was competition by different types of parasite for resources within a single host (see Exercise 20.9), and/or if the parasites had a direct effect on the fitness of the host (we shall observe a related situation in Section 21.2).

20.6   Search games involving predators and prey

A predator searches for a prey individual who is at, or soon to appear at, one of a fixed number of locations, for example a polar bear hunting for a seal which needs to use one of a number of air holes in the ice (Alpern et al., 2019). Which hole(s) should the predator guard, and which should the prey visit? This is a common type of scenario which is played out in many predator-prey interactions. Below we consider how to model situations like this one using game theory.

20.6.1 Search games

Search games are a class of game not traditionally associated with the biological world, but rather involving human actors. In the classical scenario there are two players, a Searcher and a Hider, where the Searcher wishes to capture the Hider, and the Hider wishes to avoid capture. Thus, for example, the Searcher could be a policeman and the Hider a criminal. There is also a related class, Rendezvous games, where the two players both wish to meet.

In common with many classical areas of mathematics, search games often feature questions that are easy to ask, i.e. with a simple formulation, but which are difficult to answer. For a detailed treatment of this type of problem, see Alpern and Gal (2003). It is clear that the predator-prey scenario that we described at the start of this section maps onto this search game problem well, with the predator as the Searcher and the prey as the Hider, and in the following we follow the model developed by Gal and Casas (2014) and Gal et al. (2015) for this case. In keeping with the spirit of this book, we shall use the terms predator and prey throughout, even though Searcher and Hider were used in the original works.

20.6.2 The model of Gal and Casas

In Gal and Casas (2014) a prey chooses one of n locations $\{1, 2, \ldots, n\}$ and the predator can search k of them, capturing the prey with probability p_i if it was hiding at location i. The prey is not allowed to move once the initial location is chosen, and so the probability that the predator captures the prey is simply

$$R(S, i) = \begin{cases} p_i & \text{if } i \in S, \\ 0 & \text{if } i \notin S, \end{cases} \tag{20.51}$$

where S is the set of locations that are searched.

The prey's strategy is just its selection of hiding location, and the predator's strategy is its selection of the set S, out of all sets with cardinality k, the order of search not being important in this game. Thus we can write the prey's strategy as the mixed strategy $\mathbf{h} = (h_i)$, where h_i is the probability that i is the hiding place, and the predator's strategy $\mathbf{r} = (r_i)$, where r_i is the probability that i is searched. Note that wheras $\sum_{i=1}^{n} h_i = 1$ as usual, we have $\sum_{i=1}^{n} r_i = k$, as a total of k places are searched.

The payoff to the predator is the capture probability and the payoff to the prey is the escape probability, i.e. one minus the predator's payoff. Thus we have a game of constant total payoff, equivalent to a zero-sum game.

Let us assume that the probabilities of predator success p_i increase (or at least do not decrease) with i, so that p_1 is the smallest of these. Define λ as follows:

$$\lambda = \frac{1}{\sum_{i=1}^{n} 1/p_i}. \tag{20.52}$$

Gal and Casas (2014) showed that there were two distinct solutions. If $k < p_1/\lambda$, then the optimal prey and predator strategies are to hide in or search location i with probabilities

$$h_i = \frac{\lambda}{p_i}, \quad r_i = \frac{k\lambda}{p_i}, \tag{20.53}$$

respectively. Both individuals choose a mixed strategy where the prey chooses every place with some non-zero probability and the predator visits (and does not visit) every place with some non-zero probability. Here the predator searches sufficiently few places that its success probability (which is $k\lambda$) is below the threshold value of p_1 (the success probability of visiting place 1 if the prey is known to be hiding there).

If $k \geq p_1/\lambda$ on the other hand, the prey should hide in place 1, and the predator should search place 1 with probability 1 as well as searching $r_i \geq \min(k\lambda/p_i)$ for $i > 1$. Here the predator searches so many places that it can guarantee a success rate of at least p_1 and provided it chooses the searching strategy above, the prey can keep it down to this level by always choosing the place where the predator finds it hardest to catch it. Note that if the predator deviates from the above strategy, the prey will have a location with a better result than choosing place 1.

20.6.3   The repeated game

In Gal et al. (2015) a version of the above game was developed to include a sequence of interactions. Each interaction includes the predator exploring k locations, and the following distinct events could happen after each interaction.

1. The predator does not find the prey in any of its k choices and the search stops. Here the prey is not caught and so "wins" the game, the predator (prey) getting payoff 0 (1).

2. The predator finds the prey and captures it. Here the predator (prey) gets payoff 1 (0).

3. The predator finds the prey but fails to catch it (at any of the choices). Here the prey interaction stops and after its escape the prey hides in another patch (perhaps the same one) and the game continues.

The game was also generalised to include a discount factor β, so that if capture happens after j interactions, the payoff to the predator is multiplied by β^{j-1} (the longer the prey avoids capture the better, and so the discount factor leads to an increase in the prey payoff).

Note that, just like the original game of Gal and Casas (2014), the searching order within each interaction is not important, as predator and prey encounter each other at most once, after which the interaction stops, regardless of its position in the order.

They defined the *equally attractive* hiding strategy by

$$h_i = \frac{v/k}{p_i + (1 - p_i)v},$$ (20.54)

where v is the value of the game, the payoff to the predator (for the undiscounted game v is just the probability of the predator capturing the prey). The value v is the unique solution to the equation

$$\sum_{i=1}^{n} \frac{v}{p_i + (1 - p_i)\beta v} = k.$$ (20.55)

Then just as in Section 20.6.2 we have two cases. If the equation

$$\sum_{i=1}^{n} \frac{p_1}{p_i(1 - \beta_k) + p_1\beta_k} = k$$ (20.56)

has a solution this will be unique and then for $\beta < \beta_k$ the prey should hide at place 1 with probability 1, and otherwise it should use all places, following the "equally attractive" strategy defined in (20.54).

20.6.4 Capture can occur in transit

The work above was further developed in Alpern et al. (2019) who introduced an extra factor to the game. In particular, if the predator finds the prey and the prey escapes, there is some chance that the predator catches it before it can reach a new location, and this probability α_{ij} depends upon both the original location i and its destination j. For example this might be greater if the two places are far apart, or if i is on difficult terrain for the prey.

Whilst we shall not consider analysis of their model in detail here, we briefly mention some interesting results from it. Playing the optimal strategy can reduce the prey's probability of capture very considerably (for the particular case considered 37%) over hiding randomly, and this emphasises the difference between considering the full model and simpler predator only optimisation, as used in many models. Increasing the probability of capture during relocation can paradoxically increase the probability that the prey chooses to relocate under the optimal strategy. If the participants can learn, a more variable capture probability over the locations (so that information and strategy is likely more important) is favourable to the predator if such probabilities are generally high, and favours the prey otherwise. If the patch is disrupted so that these probabilities are reset and so needing to be learnt again, then this is favourable to the prey. Thus, in general, such more realistic models can give results which depart significantly from the simpler examples.

20.7   Python code

In this section we show how to use Python to implement the Gillespie stochastic simulation algorithm to simulate two superparasiting wasps from Example 20.4. There are two files in this Section. The first one is a general implementation of the Gillespie algorithm.

```
1   """ Gillespie algorithm
2   this script provides a function gillespie_ssa
3   that can be used for stochastic simulations
4
5   Adapted from
6   http://be150.caltech.edu/2016/handouts/gillespie_simulation.html
7
8   gillespie_ssa(params, propensity_func, update, population_0,
9                   time_points):
10      Parameters:
11      params : arbitrary
12          The set of parameters to be passed to propensity_func.
13      propensity_func : function
14          Function of the form f(params, population) that takes
15          the current population of particle counts and return
16          an array of propensities for each reaction.
17      update : ndarray, shape (num_reactions, num_chemical_species)
18          Entry i, j gives the change in particle counts of
19          species j for chemical reaction i.
20      population_0 : array_like, shape (num_chemical_species)
21          Array of initial populations of all chemical species.
22      time_points : array_like, shape (num_time_points,)
23          Array of points in time for which to sample the
24          probability distribution.
25
26      Returns:
27      sample : ndarray, shape (num_time_points, num_species)
28          Entry i, j is the count of species j at time
29          time_points[i].
30      """
31
32   # Import basic package
33   import numpy as np
34
35   # Define auxiliary function to randomly sample given distribution
36   def sample(probs):
37       " Randomly sample an index with probability given by probs."
38
39       # Generate random number
40       q = np.random.rand()
41
42       # Find index
43       i = 0
44       p_sum = 0.0
45       while p_sum < q:
```

```
46          p_sum += probs[i]
47          i += 1
48      return i - 1
49
50  # Function to draw time interval and choice of reaction
51  def gillespie_draw(params, propensity_func, population):
52      " Draws a reaction and the time it took to do that reaction."
53
54      # Get propensities
55      props = propensity_func(params, population)
56
57      # Sum the propensities
58      props_sum = props.sum()
59
60      # Get time of the next event
61      time = np.random.exponential(1.0 / props_sum)
62
63      # Compute discrete probabilities of each reaction
64      rxn_probs = props / props_sum
65
66      # Draw reaction from this distribution
67      rxn = sample(rxn_probs)
68
69      return rxn, time
70
71
72  def gillespie_ssa(params, propensity_func, update, population_0,
73                    time_points):
74      "Main function, params and return described above"
75
76      # Initialize output
77      pop_out = np.empty((len(time_points), update.shape[1]),
78                         dtype=np.int)
79
80      # Initialize and perform simulation
81      i_time = 1
82      i = 0
83      t = time_points[0]
84      population = population_0.copy()
85      pop_out[0,:] = population
86      while i < len(time_points):
87          while t < time_points[i_time]:
88              # draw the event and time step
89              event, dt = gillespie_draw(params, propensity_func,
90                                         population)
91
92              # Update the population
93              pop_previous = population.copy()
94              population += update[event,:]
95
96              # Increment time
97              t += dt
98
99          # Update the index
100         i = np.searchsorted(time_points > t, True)
101
102         # Update the population
```

```
103          pop_out[i_time:min(i,len(time_points))] = pop_previous
104
105          # Increment index
106          i_time = i
107
108      return pop_out
```

The second file shows how to use the Gillespie algorithm for MCMC simulations.

```
1   """ Superparasitisng wasps
2   Simulating two superparasitising wasps on a patch of hosts
3   Uses the Gillespie algorithm to plot the history of the system
4
5   Parameters:
6   mu .. rate at which wasp find hosts
7   sigma .. probability of successful superparasitism
8
9   Outputs:
10  U .. Number of unparasitised hosts
11  V .. Number of hosts parasitised by wasp Fv
12  W .. Number of hosts parasitised by wasp Fw
13  """
14
15  ## Import basic packages
16  import numpy as np
17  import scipy.stats as st
18  import matplotlib.pyplot as plt
19
20  # Import Gillespie algorithm
21  from gillespie import gillespie_ssa
22
23  # Specify parameters for calculation
24  params = np.array([1, 0.5])      # mu, sigma
25  hosts_0 = np.array([10, 0, 0])   # Initial U, V, W numbers
26  time_points = np.linspace(0, 10, 21)  # when to track U, V, W
27  n_simulations = 100      # How many times to run the simulation
28
29  # Set up the single step update of host population
30  # Column 0 is change in U, 1 is change in V, 2 is change in W
31  update = np.array([[-1, 1, 0],   # U --> V, parasitized by Fv
32                     [-1, 0, 1],   # U --> W, parasitized by Fw
33                     [ 0,-1, 1],   # V --> W, superparasitized by Fw
34                     [ 0, 1,-1]],  # W --> V, superparasitized by Fv
35                     dtype=np.int)
36
37  # Set the propensities for the changes
38  def propensity(params, hosts):
39      """
40      Input: set of parameters and distribution of hosts
41      Output: an array of propensities
42      """
43      # Unpack parameters
44      mu, sigma = params
45
```

```
46      # Unpack population
47      U, V, W = hosts
48
49      return np.array([mu*(U>0),        # U --> V speed mu
50                       mu*(U>0),         # U --> W speed mu
51                       sigma* mu*(V>0),  # V --> W speed sigma*mu
52                       sigma* mu*(W>0)]) # W --> V speed sigma*mu
53
54  # Seed random number generator for reproducibility
55  np.random.seed(1)
56
57  # Initialize output array
58  pops = np.empty((n_simulations, len(time_points), 3))
59
60  # Run the simulations
61  for i in range(n_simulations):
62      pops[i,:,:] = gillespie_ssa(params, propensity, update,
63                                  hosts_0, time_points)
64
65  ## Plot
66  # Set up subplots to graph U, V, W in separate graphs
67  fig, axs = plt.subplots(1, 3, figsize=(10, 3))
68
69  for graph, ylabel, Color in zip(range(3), ['U', 'V', 'W'],
70                                  ["blue", "green", "brown"]):
71      # Plot the actual outcomes
72      for i in range(n_simulations):
73          axs[graph].plot(time_points, pops[i,:,graph], '-',
74                          lw=0.3, alpha=0.05, color=Color)
75      # Plot the means
76      axs[graph].plot(time_points, pops[:,:,graph].mean(axis=0),
77                      '-', lw=1, color="red")
78      # Label the axis
79      axs[graph].set_xlabel('Time')
80      axs[graph].set_ylabel('number of ' + ylabel)
81
82  plt.tight_layout()
83  plt.show()
```

20.8   Further reading

For game-theoretical models of predator-prey interactions see Brown and Vincent (1987), Brown and Vincent (1992) and Vincent and Brown (2005). For more recent work on the G-function methodology see Bukkuri and Brown (2021), including Morris et al. (2016) for its application to plant evolution. Křivan and Cressman (2009) model the evolutionary stability of the classical predator-prey model. Křivan (2011) considers a variant of the predator-prey model where there is a refuge for the prey which prevents foraging when the prey population gets sufficiently small. Berec (2010) studies the effect of cooperative hunting on predator-prey dynamics. Berryman et al. (1995) look at a large number of predator-prey models. Pintor et al. (2011) use evolutionary game theory to model competing species, as opposed to the predator-prey case.

The classic model of the evolution of defence and signalling is Leimar et al. (1986); see also Broom et al. (2006); Summers et al. (2015); Sasmal and Takeuchi (2021); Holen and Svennungsen (2012); Gamberale and Tullberg (1996). As well as the brood parasitism model of Planque et al. (2002), an important series of models of brood parasitism including explicit genetic elements is due to Takasu (see for example Takasu, 1998a,b; Takasu et al., 1993); see also Yamauchi (1995); Baran and Reeve (2015). For brood parasitism in dung beetles, see Barker et al. (2012); Crowe et al. (2009a). For an extensive form game with a more complex sequence of interaction between host and parasite, where both players make choices, see Harrison and Broom (2009). The complex parasitic lifecycles of helminth parasities are addressed in Chubb et al. (2010), as well as Parker et al. (2003).

For games involving searching behaviour, see Alpern and Gal (2003); Alpern (2010) for games in economics, and for a predator-prey search see Broom and Ruxton (2005). See also Ydenberg and Dill (1986) for a related case where prey must find the optimal time to flee predators. For a more recent series of search game papers involving predators, see Alpern et al. (2011); Gal et al. (2015); Alpern et al. (2019). Rael et al. (2009) is a life history-based game-theoretical model.

20.9 Exercises

Exercise 20.1. For the predator prey model of Section 20.1 with $c = 1$ substitute (20.7) and (20.8) into (20.5) and (20.6) and find a pair (u^*, v^*) that solves (20.15) and (20.16). From that, derive that $P^* = b_{\max} N^*$ and $N^* = r_1/(b_{\max}^2 + r_1/K_{\max})$.

Exercise 20.2. Also for the predator prey model of Section 20.1, show that (20.17) holds for sufficiently large s_p but fails for sufficiently small s_p.

Exercise 20.3 (Broom et al., 2008b). Consider the model from Section 20.2.3 with fitness functions given by (20.24) and (20.25) with the following parameters and model functions: $\lambda = 0$, $F(t) = e^{-ft}$, with $f > 1/2 K(t) = \frac{1}{1+t}$, $H(t) = t - t_c = t - 1$, $a = 1/2$, $S(r, r_1) = 1 - v|r - r_1|$, $Q = \min(1, q_0 e^{-I})$ for small q_0 and $D(r) = 1 + d_m(1 - e^{-r/d_m})$ where $d_m > 1$. Show that this gives:

1. a cryptic solution with minimal defence if $v < 3, 3/2 < f$,

2. a cryptic solution with non-aversive but not minimal defence if $v < (4f - 3)/(2f - 2)/1 < f < 3/2$,

3. a cryptic solution with aversive defence if $u > (4f - 3)/(2f - 2)$, $1/2 < f < 1$, and

4. no cryptic solution otherwise.

Exercise 20.4 (Planque et al., 2002). Derive and explain equations (20.28)–(20.31) and (20.32)–(20.37).

Exercise 20.5 (Planque et al., 2002). Show that (20.38) is equivalent to (20.39).

Exercise 20.6. The model of cuckoo-host interaction presented in Section 20.3 has a sequence of two choices for the host, the first to decide whether to reject an egg, the second whether to eject the chick(s). Write this as an extensive form game and comment on information sets.

Exercise 20.7 (Haccou et al., 2003). Use Figure 20.4 to derive equations (20.40) for the transitions between the states U,V,W when only a non-superparasitising F_v is present and (20.41) for transitions when only a superparasitising F_v is present. Derive the transition rates for the remaining cases as well.

Exercise 20.8. Use Figure 20.5 and the differential equations (20.40) and (20.41) to investigate the possible sequence of behaviours of the second female F_w once F_v has already left, for all of the possible starting scenarios that F_w can find herself in.

Exercise 20.9. Consider a version of the model from Section 20.5 where the predator is expected to eat K infected prey, and the surviving parasites on average share the rewards associated with being in the predator depending upon their strategy, so that if there are k_1 playing strategy S_1, k_2 playing strategy S_2 and $k_3 = K - k_1 - k_2$ playing strategy S_3, the strategy S_i players receive rewards F_i given by

$$F_1 = p_h V_1,$$ (20.57)

$$F_2 = p_h \left(V_{21} + p_p(1 - \phi)V_{22} \frac{V_{22}}{k_2 V_{22} + k_3 V_3} \right),$$ (20.58)

$$F_3 = p_h p_p (1 - \phi) V_3 \frac{V_3}{k_2 V_{22} + k_3 V_3}.$$ (20.59)

Analyse the game, finding the conditions when strategies S_1, S_2, S_3 or any mixture of the above are optimal.

Exercise 20.10. Consider the single shot search game of Section 20.6.2. Find the evolutionarily stable searching and hiding strategies in the following cases:
(i) $k = 2$ and $n = 3$ with $p_1 = 1/3$, $p_2 = 2/3$, $p_3 = 1$,
(ii) $k = 2$ and $n = 4$ with $p_1 = 2/7$, $p_2 = 3/7$, $p_3 = 4/7$, $p_4 = 5/7$,
(iii) General k and n with $p_i = a^{n-i+1}$ for $i = 1, \ldots, n$ where $a < 1$.

Chapter 21

Epidemic models

21.1 SIS and SIR models

The study of epidemics, and how to prevent them, has a long history and is as important today as ever. Historical epidemics, the most famous being the 14th century Black Death (the plague), have caused widespread death and played decisive roles in wars and in civilisation in general. The Black Death killed around a third of Europe's 85 million people.

Modern epidemics include malaria and AIDS, both of which have killed and continue to kill millions, and the current Covid-19 pandemic. Diseases such as Covid-19 and influenza spread rapidly around the world; worldwide influenza pandemics occur roughly every 20-30 years, though the time between pandemics varies significantly. Modern means of travel mean people travel further and faster than ever before, and coupled with increasing population sizes this means that dealing with epidemics can require quick action. Diseases can be infectious, passed between members of the population, or vector-borne, where there is an alternative source of the disease. For instance malaria is passed on to humans through the bite of infected mosquitos, and so is vector-borne. We shall only consider models of infectious diseases.

In an epidemic model of an infectious disease, an individual with the disease is called an *infective*. An individual without the disease, but who can catch the disease, is called a *susceptible*. It is assumed that initially there is at least one infective. When an infective comes into contact with a susceptible, they can pass the disease on. In the standard models, eventually each infective will recover from the disease, and in many diseases this makes them immune to further infection. An individual without the disease, but who cannot catch the disease, is called *removed*. These three states are the only ones that we shall consider, but we note that in more complex models, and real epidemics, there can be a fourth state. When an individual initially catches a disease she enters a *latent period* when she is unable to infect others and is termed *exposed*. In our models there is no latent period, and this class does not exist.

Epidemics are often modelled using compartmental models (see Chapter 19, and in particular Section 19.5). For the standard epidemic model, these compartments are labelled S (susceptible), I (infective) and R (removed). The original model for such a system is due to Kermack and McKendrick

DOI: 10.1201/9781003024682-21

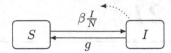

FIGURE 21.1: The SIS epidemic model. A susceptible individual becomes infected at the rate it encounters infected individuals which is $\beta\frac{I}{N}$, an infected individual recovers (and becomes susceptible again) at rate g. The dotted line highlights the fact that the number of individuals in compartment I affects the rate of movement out of compartment S.

(1927) (see also Kermack and McKendrick, 1991a), and we look at a number of variants here.

The initial models that we shall consider are not game-theoretical in character, but nonetheless it is useful to consider them before moving on to related game-theoretical models.

21.1.1   The SIS epidemic

Not all disease models have removed states. We shall start by considering the SIS epidemic, where individuals do not become immune, but become susceptible again when they recover from the disease. Many sexually transmitted diseases fall into this category, for instance.

Example 21.1 (Chlamydia epidemics). Chlamydia is the most commonly sexually transmitted disease, caused by the bacteria *Chlamydia trachomatis*. A susceptible individual becomes infected after sexual contact with an infected individual. An infected individual can recover (be cured by antibiotics) at rate g and after a recovery, (s)he becomes susceptible again. Describe and analyze the dynamics of the two types in the population.

21.1.1.1   The model

The SIS model is represented by Figure 21.1.

Let β be an expected number of individuals one interacts with per unit of time. We will assume that there is a fixed population of N individuals with no births and no deaths.

We note that if N was not constant, it is important to consider how interaction rates vary with population size. As formulated, the expected number of contacts per individual would be independent of N, and our model is what is referred to as frequency dependent. If β was replaced by a constant times N, then the number of contacts would increase linearly in population size, and we would have a model of the type referred to as density dependent. We could of course make the encounter rate depend upon N in other ways. However by choosing N to be constant, we avoid these issues here. Such an assumption is reasonable provided that the population size does not vary significantly, and that infection and recovery rates are sufficiently high compared to birth and death rates.

Let I be the number of infectives and S be the number of susceptibles. In reality I and S are natural numbers and so change in them cannot be governed by differential equations, but we assume that N is large and we will build a differential equation model that is a good approximation to the discrete process. Recall that we make a similar approximation for our models in general (see Section 3.1.1), but it is particularly important to remember for epidemic models, as we see below.

Over a time interval δt, a susceptible individual meets $\beta \delta t$ other individuals. Each one of those is infected with probability $\frac{I}{N-1} \approx \frac{I}{N}$. This yields

$$\frac{dS}{dt} = -\beta \frac{I}{N} S + gI, \tag{21.1}$$

$$\frac{dI}{dt} = \beta \frac{I}{N} S - gI. \tag{21.2}$$

Since we have $S + I = N$, choosing $v = I/N$, we obtain

$$\frac{dv}{dt} = v\big(\beta(1 - v) - g\big). \tag{21.3}$$

21.1.1.2 Analysis

The steady states of (21.3) are given by $v\big(\beta(1 - v) - g\big) = 0$. There are potentially two such steady states,

1) the *disease-free equilibrium*, $v = 0$, or

2) the *endemic equilibrium* with a non-zero disease state, $v^* = 1 - \frac{g}{\beta}$.

The endemic equilibrium occurs (and is stable) if and only if $\beta > g$.

Obviously if the disease is completely extinct it cannot spread, but often the disease will start close to the disease-free equilibrium, e.g. with a single new case in a large population. Considering the stability of the disease-free equilibrium is thus very important. For small $v \approx 0$, (21.3) becomes

$$\frac{dv}{dt} \approx (\beta - g)v. \tag{21.4}$$

Thus the disease-free equilibrium is stable, if and only if $\beta < g$.

The *basic reproduction rate* of an epidemic R_0 is the number of secondary infections produced by one primary infection in a wholly susceptible population. In this model, it is equal to the expected number of individuals one will meet before recovery, i.e. $R_0 = \beta \cdot \frac{1}{g}$. Note that, for simplicity, we have assumed that every contact between an infective and a susceptible results in transmission of the disease. The model is easily adapted to the case where the transmission probability is not 1, by replacing the rate β by the product of β and the transmission probability.

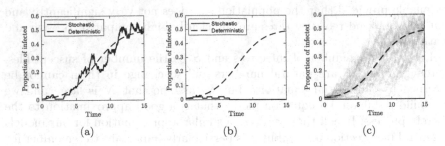

(a)   (b)   (c)

FIGURE 21.2: The SIS epidemic model. A comparison between the deterministic and stochastic model. In both figures, $\beta = 1$, $g = 1/2$. The deterministic model predicts an epidemic. However, although the simulation in (a) gives an epidemic similar to the deterministic case, as the simulation in (b) shows, the epidemic may not occur. (c) Results of 100 simulation runs.

21.1.1.3 Summary of results

1) If $R_0 = \frac{\beta}{g} < 1$ the disease-free equilibrium is stable against small perturbations (e.g. a new case of the disease) and the disease cannot spread.

2) If $R_0 = \frac{\beta}{g} > 1$ the disease-free equilibrium is unstable against small perturbations and a small perturbation will cause an epidemic, which leads to the endemic equilibrium state.

There is thus a unique stable equilibrium proportion of susceptibles, which is the reciprocal of R_0 if $R_0 > 1$ and is 1 otherwise. In simple terms, if any infection on average causes less than one subsequent infection the disease will die out; if it causes more than one, it will spread.

We note that stochastic models make different predictions in case 2. For stochastic models an epidemic *can* occur, but it is far from certain. Starting with a single individual in a large population, the number of infectives in a population at the start of any potential epidemic can be approximately modelled as a random walk, with the probability that the next change is an increase being

$$\frac{\beta I}{\beta I + gI} = \frac{R_0}{R_0 + 1}. \tag{21.5}$$

There is no epidemic if the number of infectives goes to 0 early in the process. The calculation for the probability that this does not occur is effectively the same as the probability of mutant fixation in the Moran process from Section 12.1.1 (which is also modelled using a random walk). Substituting R_0 for r and letting $N \to \infty$ in (12.11) yields the probability of an epidemic occurring as $1 - 1/R_0$, when $R_0 > 1$. For R_0 only a little above 1, this is small, so an epidemic is still unlikely. For a numerical simulation, see Section 21.5 and also Figure 21.2.

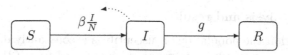

FIGURE 21.3: An SIR epidemic model. A susceptible individual becomes infected at the rate it encounters infected individuals which is $\beta\frac{I}{N}$, an infected individual recovers (and becomes immune) at rate g.

21.1.2 The SIR epidemic

We now consider a population model where removal can occur, in this case where individuals recover from the disease and are subsequently immune.

Example 21.2 (Influenza outbreak). Assume that there is a fixed population of N individuals with no births and no deaths (a reasonable assumption as influenza outbreaks typically occur in a short period of time). Let β be the expected number of individuals one interacts with per unit time. A susceptible individual becomes infected after interaction with an infected individual. An infected individual can recover at rate g and after recovery, it becomes immune to the disease. Describe and analyse the dynamics of the three types in the population.

21.1.2.1 The model

Let I again be the number of infectives and S the number of susceptibles, with R the number of removed (immune) individuals. Similarly to Example 21.1, we get the following equations

$$\frac{dS}{dt} = -\beta\frac{I}{N}S, \tag{21.6}$$

$$\frac{dI}{dt} = \beta\frac{I}{N}S - gI, \tag{21.7}$$

$$\frac{dR}{dt} = gI. \tag{21.8}$$

This model is shown in Figure 21.3. Here $S + I + R = N$, so we only need consider two variables. Thus using $v = I/N$ and $u = S/N$, (21.6) and (21.7) become

$$\frac{du}{dt} = -\beta uv, \tag{21.9}$$

$$\frac{dv}{dt} = (\beta u - g)v. \tag{21.10}$$

21.1.2.2   Analysis and results

It is clear that the population of susceptibles is constantly decreasing until $uv = 0$. Hence, in a steady state we need $uv = 0$ and from (21.10) it is clear that we also need $v = 0$.

It is left to the reader as Exercise 21.2 that an epidemic occurs (and the steady state $v = 0$ is not stable) if and only if

$$R_0 = \frac{\beta}{g} > 1. \tag{21.11}$$

If an epidemic occurs, we are interested in the size of the epidemic, i.e. the proportion of individuals that caught the disease. It follows from above that once started, the epidemics will continue until the proportion of individuals with the disease goes to zero. Hence, the final size of the epidemic is $1 - u(\infty)$.

It follows from (21.9) and (21.10) that

$$\frac{dv}{du} = \frac{dv}{dt} \div \frac{du}{dt} = \frac{(\beta u - g)v}{-\beta uv} = \frac{1}{R_0 u} - 1. \tag{21.12}$$

Separating variables gives

$$v(t) = -u(t) + \frac{1}{R_0} \ln\big(u(t)\big) + C_0. \tag{21.13}$$

At $t = 0$ using $u(0) = 1$ and $v(0) = 0$ we obtain $C_0 = 1$. As $t \to \infty$ $v(t) \to 0$ and so we have

$$1 - u(\infty) + \frac{1}{R_0} \ln\big(u(\infty)\big) = 0. \tag{21.14}$$

Consequently, the final size of the epidemic, w, solves

$$w = 1 - e^{-R_0 w}. \tag{21.15}$$

Figure 21.4 shows a plot of the final epidemic fraction w against R_0.

21.1.2.3   Some other models

The above two examples assume an epidemic where transmission and recovery are fast, without a significant mortality rate. Models are different for slower epidemics that occur on timescales where births and deaths cannot be neglected. In this case extra terms are added to the equations, for example in the SIS epidemic equations (21.1), (21.2) become

$$\frac{dS}{dt} = b(N) - \beta \frac{I}{N} S + gI - dS, \tag{21.16}$$

$$\frac{dI}{dt} = \beta \frac{I}{N} S - gI - dI, \tag{21.17}$$

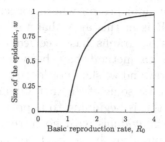

FIGURE 21.4: Final epidemic size in an SIR epidemic model, w, as a function of the basic reproduction rate, R_0.

on the assumption that each individual dies at a constant rate d but that new individuals that are born, at total rate $b(N)$, are initially all in the susceptible class (see Exercise 21.1). Models can involve the mortality rate of those with the disease being significantly different to those without it, so that the death rate of infectives exceeds that of susceptibles. Similarly, vaccination strategies can affect the development of epidemics greatly and in particular can prevent outbreaks if they are sufficiently efficient (Murray, 2002, 2003).

From the point of view of this book, we see that the above is completely free of game-theoretic analysis. Nevertheless, these are important models that provide the foundation for any such subsequent analysis. We shall return to these models to consider the evolution of virulence and superinfection, which are game-theoretic, in Section 21.2.

It should be noted that there are both deterministic and stochastic versions of SIS, SIR and similar models. Stochastic models are more realistic, but much harder to analyse; thus often some components need to be neglected, or analysis dispensed with altogether, and simulation used. Often stochastic and deterministic models of biological systems are equally good, so that the simpler deterministic ones can be used, but there are some circumstances when deterministic models can be seriously misleading; these occur commonly in the modelling of epidemics, in particular, when at some point the infective category contains a small number of individuals.

21.1.3 Epidemics on graphs

Models from Section 21.1 rely on the usual assumption of a well mixed population. In fact, real populations are generally not well-mixed, with some pairs of individuals far more likely to interact than others. As we saw in Section 12.3, well-mixed models need not be good approximations to such structured populations. In Section 12.3 we considered evolutionary processes on graphs, where the population structure was a central feature. The effect of population structure can be particularly important in the study of the spread of epidemics, where local effects can be crucial in the early stages of the disease and can have a significant influence on whether there is no epidemic at all or a severe one. In fact, it is in this area that the mathematical modelling of structured populations on graphs has received the most attention, and where a range of mathematical methods have been developed. These are again generally not game-theoretic, and we shall not look at any specific models here. However, there is potential for some of the methodology to transfer over into evolutionary models, and this is beginning to happen.

A fixed population of individuals with no births and deaths is often assumed. Similarly, models often involve a fixed population structure, so that individuals have connections with some members of the population but not others, and that the set of these connections does not change with time. Note that if there are any births and deaths in our model then the structure must necessarily evolve, so the populations we describe primarily relate to fast epidemics which spread and die out quickly. It is also assumed that the graph is connected, so there is a path between any pair of individuals.

Ames et al. (2011) were able to explain more than 98% of the variation in endemic disease levels in the stochastic simulations of epidemics on graphs using only three characteristics of the graphs:

(1) degree distribution,

(2) clustering coefficient,

(3) path length.

A sophisticated industry of mathematical modelling of diseases has developed which can involve vast computer simulations. Models are used to estimate the impact of, and combat, real diseases such as SARS. Contact networks between individuals are developed to model the spread of the disease. This is extremely difficult, as the data take time to collect and such diseases can spread rapidly. In the outbreak of foot and mouth disease amongst livestock such as sheep and cattle in the UK, models were developed where individual vertices of the graph were collections of animals, especially farms. Typically once a disease reached animals on a farm it spread rapidly, but transmission between farms was slower. A systematic policy of culling animals which were at risk of catching the virus was followed. The costs of making a wrong decision in such cases are large, and the use of this policy was controversial. Network models have been used to evaluate the effectiveness of the cull, see e.g. Tildesley et al. (2009).

While the methodology for modelling epidemics using graphs is quite straightforward and intuitive, it is often limited to individual-based stochastic simulations that can be time consuming to run, and the results generated may lack generality. Instead, methods to approximate the process using low-dimensional ordinary differential equation are used. These include pairwise approximation (e.g Matsuda et al., 1992; van Baalen and Rand, 1998; Eames and Keeling, 2002; House and Keeling, 2011) and the effective degree model

(Lindquist et al., 2011). We note that such models can also be applied to evolution on graphs more generally, as shown in Hadjichrysanthou et al. (2012).

21.2 The evolution of virulence

In this section we follow Nowak and May (1994). These authors developed models of the evolution of virulence in a number of papers, see e.g. Nowak et al. (1990, 1991); Nowak and May (1991, 1993), which are important to our understanding of how the virulence of parasites may change over time, including in response to human interventions.

21.2.1 An SI model for single epidemics with immigration and death

We now return to another variant of the epidemic model of Kermack and McKendrick (1927).

Example 21.3 (myxomatosis in rabbits). Assume that there is a population of N individuals with a constant immigration (perhaps birth) rate $b(N) = b$ and each individual has a death rate d. There are two types of individuals, susceptible (S) and infected (I). Newly arrived individuals come as susceptible. Susceptibles have a natural death rate d and become infected at rate τI; assuming that infection is passed on with probability 1 when a susceptible and an infective meet (as assumed in Section 21.1.1), this corresponds to τ being the rate of contact between any pair of individuals, so that $\tau = \beta/N$ in the terminology from Section 21.1.1. An infected individual has a death rate $d + \nu$ and can never recover from an infection. The parameter ν is called the level of virulence. Describe and analyse the dynamics of S and I.

21.2.1.1 Model and results

The model is shown in Figure 21.5. We note here that the situation modelled effectively involves the interaction between a host and its parasite, and so there is a connection with the models that we have discussed in Chapter 20. Similarly to Example 21.1 and ((21.1)–(21.2)) we have

$$\frac{dS}{dt} = b - Sd - \tau SI, \tag{21.18}$$

$$\frac{dI}{dt} = (\tau S - d - \nu)I. \tag{21.19}$$

It is left for the reader as Exercise 21.4 that R_0 for ((21.18)–(21.19)) is

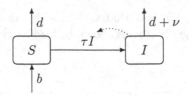

FIGURE 21.5: An SI epidemic model with births and deaths involving a disease with virulence ν, from Example 21.3.

given by

$$R_0 = \frac{\tau b}{d(d + \nu)}. \tag{21.20}$$

The disease spreads within the population if and only if $R_0 > 1$, as usual, and in this case there is an equilibrium distribution of individuals given by

$$S^* = \frac{d + \nu}{\tau}, \tag{21.21}$$

$$I^* = \frac{\tau b - d(d + \nu)}{\tau(d + \nu)}. \tag{21.22}$$

21.2.2　An SI model for two epidemics with immigration and death and no superinfection

Example 21.4. Consider a generalisation of Example 21.3. Suppose that there are two disease strains 1 and 2, which differ in both their transmission rate $\tau_1(\tau_2)$ and in the rate at which they kill their host $\nu_1(\nu_2)$. Assume that once infected by one strain it is not possible for a host to be infected by the other. Describe and analyse the dynamics of S, I_1 and I_2.

21.2.2.1　Model and results

The model is shown in Figure 21.6. Similarly to work done above, we have

$$\frac{dS}{dt} = b - Sd - \tau_1 S I_1 - \tau_2 S I_2, \tag{21.23}$$

$$\frac{dI_1}{dt} = (\tau_1 S - d - \nu_1) I_1, \tag{21.24}$$

$$\frac{dI_2}{dt} = (\tau_2 S - d - \nu_2) I_2. \tag{21.25}$$

Clearly, for generic parameter values, no equilibrium of (21.23)–(21.25) with $I_1, I_2 > 0$ is possible. Moreover, if both strains could invade a disease-free population (if they both have $R_0 > 1$), then parasite strain 1 outcompetes strain 2 if and only if

$$\frac{\tau_1}{d + \nu_1} > \frac{\tau_2}{d + \nu_2}. \tag{21.26}$$

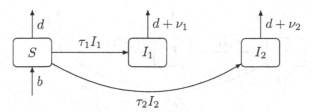

FIGURE 21.6: An SI epidemic model with two infections with different virulence levels but no superinfection from Example 21.4.

Clearly a high value of τ and a low value of ν will make a strain fitter. These will generally not be independent, however, and a high value of τ will correspond to a high value of ν. The more aggressive the strain (i.e. the higher its virulence ν), the greater damage it causes but also the greater rate of transmission it has.

For example, when the parasites can freely evolve any level of virulence ν and that the relationship between ν and τ is given by

$$\tau = \frac{a\nu}{c + \nu}, \tag{21.27}$$

the optimal level of virulence is given by $\nu_{opt} = \sqrt{cd}$; see Exercises 21.5 and 21.6.

21.2.3 Superinfection

Example 21.5. Consider a generalisation of Example 21.4. Suppose there are n disease strains $1, 2, \ldots, n$ that differ in both their transmission rate τ_i and in the rate at which they kill their host ν_i. Now, assume that if a more virulent strain comes into contact with an individual with a less virulent strain, the virulent strain replaces the other in that individual completely. This is called *superinfection*. Assume that superinfection occurs at rate μ. Describe and analyse the dynamics of S and I_i, $i = 1, \ldots, n$.

21.2.3.1 Model and results

Assume that $\nu_i > \nu_j$ if and only if $i > j$. The model for two infections is shown in Figure 21.7.

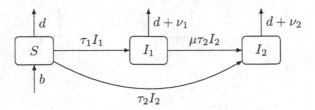

FIGURE 21.7: An SI epidemic model with two infections with different virulence levels and superinfection from Example 21.5.

$$\frac{dS}{dt} = k - dS - S \sum_{i=1}^{n} \tau_i I_i, \tag{21.28}$$

$$\frac{dI_i}{dt} = I_i \left(\tau_i S - d - \nu_i + \mu \tau_i \sum_{j=1}^{i-1} I_j - \mu \sum_{j=i+1}^{n} \tau_j I_j \right) \quad i = 1, \ldots, n. \tag{21.29}$$

If $\mu < 1$, it indicates that superinfection cannot occur at the same rate as an original infection (possibly due to defences of the host or from the existing parasite). It is possible to have $\mu > 1$, which would indicate that the original parasite makes the host more susceptible to superinfection.

Nowak and May (1994) assumed the relationship between virulence ν_i and infectivity τ_i was as given in (21.27). This led to a value of R_0 for strain i given by

$$R_{0,i} = \frac{ab\nu_i}{d(c + \nu_i)(d + \nu_i)}, \tag{21.30}$$

which is maximised by the virulence level $\nu_{opt} = \sqrt{cd}$, and this is best when $\mu = 0$, as we found above. Nowak and May (1994) found that with superinfection, suddenly a range of strains coexisted, between a minimum level ν_{min} and a maximum ν_{max} but where $\nu_{min} > \nu_{opt}$, so superinfection leads to a higher level of virulence. This higher level of virulence is caused by within-host competition. The strain with the biggest R_0 is not only not dominant, it is removed from the population completely. An interesting feature of the model is that the higher the rate of superinfection, the smaller the number of infected hosts.

21.3   Viruses and the Prisoner's Dilemma

Example 21.6 (Prisoner's Dilemma game in RNA viruses, Turner and Chao, 1999). A virus takes over the control of a bacterium's biomolecular machinery to manufacture proteins for its own reproduction. However, the manufactured proteins diffuse within the cell and this prevents an individual virus from having exclusive access to its own gene products. This creates a conflict situation whenever multiple viruses infect a single host.

21.3.1   The model

Assume that there are two types of virus in a cell. Viruses of type $i = 1, 2$ make the infected cell produce π_i units of protein per virus (per unit of time). When P units of protein are present, viruses of type i assemble $a_i \cdot P$ new viruses (per unit of time per virus). It is reasonable to assume that a_i and π_i are negatively correlated. We assume that there are some natural bounds, $a_{\min}$ and $a_{\max}$, for the speed of assembly a. This yields the following equation for the protein concentration

$$\frac{dP}{dt} = \pi_1 c_1 - a_1 P c_1 + \pi_2 c_2 - a_2 P c_2, \tag{21.31}$$

where c_i is the concentration of the virus of type i in the cell (note that it is assumed that the produced viruses go and infect other cells rather than stay in the originally infected one). When $c_2 = 0$ or $c_1 = 0$ (i.e. if only one type of virus is present), we see that when the situation stabilises, type one viruses produce exactly $\pi_1 c_1$ and type two exactly $\pi_2 c_2$ viruses per unit of time, i.e. π_1 or π_2 can be seen as a measure of fitness. When both types of virus are present, a total number $\pi_1 c_1 + \pi_2 c_2$ viruses will be produced per unit time, out of which $\frac{a_i c_i}{a_1 c_1 + a_2 c_2}$ will be of type i.

21.3.2   Results

Type 1 outperforms type 2 (i.e. has a higher per capita generation of new viruses) if

$$(\pi_1 c_1 + \pi_2 c_2)\frac{a_2}{a_1 c_1 + a_2 c_2} < (\pi_1 c_1 + \pi_2 c_2)\frac{a_1}{a_1 c_1 + a_2 c_2}, \tag{21.32}$$

which is equivalent to

$$a_2 < a_1. \tag{21.33}$$

It follows from (21.33) that any population of viruses that does not have the maximal assembly rate a is vulnerable to invasion by "defecting" mutants having larger a (and via negative correlation, a lower π). This means that the ESS strategy of the virus is to have maximal assembly rate a (and the minimal protein production rate π).

21.3.3   A real example

The situation was observed and studied in Turner and Chao (1999) (see also Turner and Chao, 2003) on the example of Bacteriophage $\Phi 6$. It is a double-stranded RNA virus of the family *Cystoviridae* that infects the bacterium *Pseudomonas phaseolicola*.

Turner and Chao (1999) modelled the interaction of phages as a Prisoner's Dilemma. Within each cell viruses both generate products and use the products of both themselves and others. If an individual both produces and uses products it can be termed a cooperator. If an individual concentrates solely on using products it can be termed a defector (such defectors are those that have lost most of their protein-coding sequences). If such an individual is within a cell with many that are generating products it can gain a significant advantage. When the level of infection is low, the virus may be alone in a cell, and it is important to be able to both generate and use products, so cooperators will flourish at low virus loads. If the virus load suddenly increases then initially there will be cooperators; when defectors appear they should rapidly spread through the population, but once they are too prevalent the production of products dries up, and the viruses are then unable to reproduce.

Using real data comparing the "defector" $\Phi H2$ and the "cooperator" $\Phi 6$ they obtained the following payoff matrix,

$$
\begin{array}{c}
\begin{array}{cc} \Phi 6 & \Phi H2 \end{array} \\
\begin{array}{c} \Phi 6 \\ \Phi H2 \end{array}
\begin{pmatrix} 1 & 0.65 \\ 1.99 & 0.83 \end{pmatrix}.
\end{array}
\tag{21.34}
$$

We note that fitnesses are scaled so that the baseline fitness of 1 is for $\Phi 6$ against a population of its own type. Also we see that $\Phi H2$ is able to survive in a population of its own type but is less fit than the $\Phi 6$ population. Thus using the four classical terms for the payoffs of a Prisoner's Dilemma R, S, T and P we have the values $R = 1, S = 0.65, T = 1.99$ and $P = 0.83$ so that $T > R > P > S$ as required.

21.4   Vaccination models

The recent Covid-19 epidemic has made the concept of vaccination one of general and widespread discussion as well as great political and social controversy. A number of recent papers from a game theoretical perspective have been produced, especially by Tanimoto and colleagues (see Section 21.6).

It is recognised that effective vaccination requires a sufficiently high uptake within the population. If vaccination is voluntary, and there are associated costs, either financial or in actual or perceived risk to health, then individuals may choose to stay unvaccinated and let others be vaccinated to provide the benefit to the population as a whole. This is similar in idea to the Prisoner's dilemma where individuals can cooperate (be vaccinated) or defect (not).

How should we model this type of scenario? This problem was explicitly considered by Bauch and Earn (2004), who developed a population game model which is the baseline for most recent and more disease-specific models. They presented the fundamental idea in an apparently straightforward model, which is nevertheless very adaptable to more complex scenarios.

Consider a population of individuals, who each have to strategically select their probability of being vaccinated. Let us assume that in the rest of the population the proportion of vaccinated individuals is p. For simplicity we just assume that there is negligible excess death of vaccinated or unvaccinated individuals, and this is simply the proportion that choose to be vaccinated, equivalent to if everyone played strategy p. Within such a population, the payoff to a single individual playing P is then $-r_v P - r_i \pi_p (1-P)$ where $r_v > 0$ is the cost associated with being vaccinated and $r_i > 0$ is the cost of acquiring the disease. Here π_p is the probability that an unvaccinated individual will become infected in a population that has vaccination coverage p, and we assume that the vaccine is 100% effective, so that the corresponding probability for vaccinated individuals is 0.

We can then divide the payoff function above by the constant r_i and denote $r = r_v/r_i$ as the relative cost of vaccination to obtain the following standardised payoff, which we will use as the payoff function for convenience,

$$\mathcal{E}[P, p] = -rP - \pi_p(1 - P) = -\pi_p + P(\pi_p - r). \tag{21.35}$$

Hence we can obtain the following incentive function for being vaccinated, the difference between the payoffs of vaccinated and unvaccinated individuals,

$$\mathcal{E}[1, p] - \mathcal{E}[0, p] = \pi_p - r. \tag{21.36}$$

Here in any plausible scenario, π_p is a decreasing function of p (and strictly decreasing until a high enough p is reached when the virus is completely eliminated).

We recall from Section 9.1 that a strategy p is an ESS if the incentive function takes value less than (greater than) 0 for $p = 0$ ($p = 1$), or is equal to 0 with negative derivative for any $0 < p < 1$. We can see from (21.36) that the incentive function is decreasing in p, and so there will be a unique ESS of this game. If $\pi_0 = 0$, so that the virus will not spread even in a completely unvaccinated population, or for any $\pi_0 \leq r$, this occurs at $p^* = 0$. If $\pi_1 \geq r$ so there is some (sufficiently large) risk to a single unvaccinated individual even in an otherwise fully vaccinated population, then $p^* = 1$. In all other cases, and these would include most diseases of concern, we have a unique $0 < p^* < 1$ which solves $\pi_{p^*} = r$.

This then is a simple model which requires only two pieces of information, the value of r and the function π_p over the range $0 \leq p \leq 1$. The latter,

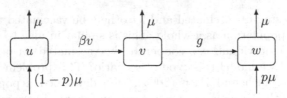

FIGURE 21.8: An SIR epidemic model with vaccination. Individual are born, and also die, at rate μ. A fraction p of them receives a vaccine and become immune to the disease (w). A susceptible individual (u) becomes infected at the rate it encounters infected individuals (v) which is βv, an infected individual recovers (and becomes immune) at rate g.

however, depends upon the disease dynamics, and to model this fully an appropriate epidemiological model is required. Bauch and Earn (2004) used a version of the SIR model (see Section 21.1.2) with births and deaths (see Section 21.1.2.3), where the birth and death rates were equal (denoted by μ) to ensure a fixed population size. The proportions of susceptible, infective and removed individuals are denoted by u, v and w, respectively and are given by

$$\frac{du}{dt} = (1 - p)\mu - \beta uv - \mu u, \tag{21.37}$$

$$\frac{dv}{dt} = \beta uv - gv - \mu v, \tag{21.38}$$

$$\frac{dw}{dt} = p\mu + gv - \mu w. \tag{21.39}$$

The dynamics is illustrated in Figure 21.8.

Clearly $u+v+w = 1$ from their definition, and as before the third equation is therefore redundant. Denoting $f = \mu/\lambda$ and $R_0 = \beta/(\gamma + \mu)$ (here R_0 has its usual meaning of the number of secondary infections per infective in an otherwise completely susceptible population), Bauch and Earn (2004) show (see Exercise 21.9) that the long-term proportion of susceptibles is $u = 1 - p$ and that p^* solves

$$p^* = 1 - \frac{1}{R_0(1 - r)}. \tag{21.40}$$

This p^* value is shown as a function of r for various R_0 values in Figure 21.9.

Bauch and Earn (2004) gave an interesting discussion of various consequences of their model. The most striking single point is that the ESS value p^* is always less than the critical threshold of p required for elimination of the disease. Thus in a population of well-informed and selfish individuals, a disease cannot be eradicated using a voluntary vaccination programme. In a recent review of game-theoretical models, Chang et al. (2020) found that perceived risk and perceived vaccination costs, as opposed to social benefits, are key factors in determining the level of vaccination within a population.

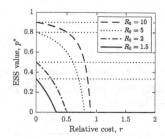

FIGURE 21.9: Vaccine coverage p^* at the ESS versus relative cost r, from (21.40), for various values of R_0. Dotted horizontal lines show the critical coverage level that eliminates the disease from the population.

Versions of the above model have then been applied by Rychtář and colleagues in a series of papers (Scheckelhoff et al., 2021; Acosta-Alonzo et al., 2020; Bankuru et al., 2020; Brettin et al., 2018; Cheng et al., 2020; Chouhan et al., 2020; Han et al., 2020; Sykes and Rychtář, 2015). They observed both the conceptual simplicity of the Bauch and Earn (2004) model, and the fact that a number of very good models of epidemics had been developed, but without the strategic element associated with individual choice which is potentially crucial to the effectiveness of a preventative strategy. They thus set about finding and extending the appropriate models, and addressed a wide variety of diseases as a result. They found that for many vector-born diseases, the optimal use of protection strategies such as sleeping under insecticide treated nests can significantly reduce disease incidence and often eliminate the disease as a public health concern (Broom et al., 2016; Crawford et al., 2015; Dorsett et al., 2016; Han et al., 2020; Klein et al., 2020; Fortunato et al., 2021; Angina et al., in press).

This type of model can be extended to consider more than one type of potential protection, and hence more that one strategic choice as done in Kobe et al. (2018). A model relevant to the Covid-19 epidemic was created by Choi and Shim (2020) who considered both vaccination and social distancing as strategies that individuals could adopt. Although individuals could use both strategies, they focused on the one that was the most effective for the given situation, and so in practice only one of the two was used, with the result that the disease is (usually) not eradicated. Thus again individual choice has led to a situation which leaves the whole population facing some level of risk.

21.5   Python code

In this section we show how to use Python to carry out stochastic simulation of the SIR epidemic with vaccination at birth presented as in Bauch and Earn (2004) discussed in Section 21.4.

```python
1   """ Vaccination game
2   Analysis of the simple SIR model with vaccination (V) at birth
3
4   This script has 3 parts
5   1) analytical: coding exact formulas for equilibrium states
6   2) numerical: numerical solutions of ODE system
7   3) stochastic: Gillespie algorithm
8   """
9
10  ## Import basic packages
11  import numpy as np
12  from scipy.integrate import odeint
13  import matplotlib.pyplot as plt
14
15  # Import the Gillespie algorithm
16  from gillespie import gillespie_ssa
17
18  ## Define input
19  beta = 5      # Disease transmission rate
20  mu = 0.5      # Natural birth and death rate
21  r = 1         # Recovery rate
22  p = 0.1       # Probability of being vaccinated
23  N = 100       # Population size
24
25  # Specify the time interval over which we will solve
26  t = np.linspace(0, 5, 51)
27
28  # Code the dynamics/ODE system
29  # Column 0 is change in S, 1 is change in I,
30  #          2 is change in R, 3 is change in V
31  update = np.array([[ 1, 0, 0, 0],    # S born
32                     [ 0, 0, 0, 1],    # V born
33                     [-1, 1, 0, 0],    # S --> I
34                     [ 0,-1, 1, 0],    # I --> R
35                     [-1, 0, 0, 0],    # S dies
36                     [ 0,-1, 0, 0],    # I dies
37                     [ 0, 0,-1, 0],    # R dies
38                     [ 0, 0, 0,-1]],   # V dies
39                     dtype=np.int)
40
41  def propensity(params, population):
42      """
43      Returns an array of propensities given a set of parameters
44      and an array of populations.
45      """
46      # Unpack parameters
```

```
47      beta, mu, r, p = params
48
49      # Unpack population
50      S, I, R, V = population
51
52      return np.array([mu*(1-p)*(S+I+R+V),   # S born,   N*mu*(1-p)
53                       mu*p*(S+I+R+V),        # V born,   N*mu*p
54                       beta*I*S/(S+I+R+V),    # S --> I,  beta*S*I/N
55                       r*I,                   # I --> R,  r*I
56                       mu*S,                  # S dies,   mu*S
57                       mu*I,                  # I dies,   mu*I
58                       mu*R,                  # R dies,   mu*R
59                       mu*V])                 # V dies,   mu*V
60
61  ###### Code analytical solution for equilibrium points
62
63  # Basic reproduction number
64  # If R0<1 we will have disease free equilibrium
65  # If R0>1, we will have endemic equilibrium
66  R0 = (1-p)*beta/(mu+r)
67
68  # Calculate equilibrium for S,
69  Sstar = N*(1-p)           * (R0<1) +\
70          N*(mu+r)/beta * (R0>1)      # Endemic equlibrium
71  # Calculate equilibria for I
72  Istar = 0 * (R0<1) + \
73          ((1-p)*mu*N - Sstar*mu)/(Sstar*beta/N) * (R0>1)
74  # Calculate equilibrium value of R
75  Rstar = Istar * r/mu
76  # Calculate equilibrium value of V
77  Vstar = N*p
78
79  #####  Numerical solutions of the ODE system
80  def SIRVeqs(Variables, t, *Params):
81      """ Provides equations for the SIR with vaccination model
82      uses the already defined propensities and 'update' table """
83
84      # Get the number of equations
85      n = update.shape[1]  # number of columns of `update'
86      # Initialize the array of equations
87      eqs = np.empty(n)
88      # Write the equations
89      for i in range(n):
90          eqs[i] = np.dot(propensity(Params, Variables),
91                          update[:,i])
92      return eqs
93
94
95  # Give initial proportion of S, I, R, V
96  # Start almost at the disease free equilibrium
97  Sols0 = [(1-p)*N-1, 1, 0, Vstar]
98
99  # Numerically solve the dynamics
100 Sols = odeint(SIRVeqs, Sols0, t, args = (beta, mu, r, p))
101
102 ##### Stochastic simulation
103
```

```
104   # Specify parameters for calculation
105   params = np.array([beta, mu, r, p])
106   population_0 = np.array([100*(1-p)-1, 1, 0, 100*p])
107   n_simulations = 100
108
109   # Seed random number generator for reproducibility
110   np.random.seed(1)
111
112   # Initialize output array
113   pops = np.empty((n_simulations, len(t), 4))
114
115
116   # Run the simulations
117   for i in range(n_simulations):
118       pops[i,:,:] = gillespie_ssa(params, propensity, update,
119                                   population_0, t)
120
121   # Get the overall mean over the simulations
122   Mean = pops.mean(axis=0)
123
124   # Get the mean given I persisted, i.e. there was I at the end
125   PopWithI = pops[pops[:, -1, 2]>0, :, :]
126   MeanWithI = PopWithI.mean(axis = 0)
127
128   # Get the mean given I disappeared
129   PopWithoutI = pops[pops[:, -1, 2]==0, :, :]
130   MeanWithoutI = PopWithoutI.mean(axis = 0)
131
132   ## Plot the outcomes
133   # Set up subplots to graph S, I, R V in separate graphs
134   fig, axs = plt.subplots(1, 4, figsize=(13, 3))
135
136   SolAn = [Sstar, Istar, Rstar, Vstar]
137
138   for ax, comp, Title, Color in zip(axs.flat, range(4),
139                               ['S', 'I', 'R',      'V'],
140                               ["blue","red","black","green"]):
141       # Plot the actual outcomes from all simulations
142       for i in range(n_simulations):
143           ax.plot(t, pops[i,:,comp], '-', lw=0.3,
144                   alpha=0.09, color=Color)
145
146       # Plot the analytical solutions as brown lines
147       ax.axhline(SolAn[comp], color = 'brown', alpha = 0.5)
148
149       # Plot the means
150       ax.plot(t, Mean[:, comp], '-', lw=1, color="black")
151       ax.plot(t, MeanWithI[:, comp], '-', lw=1, color="red")
152       ax.plot(t, MeanWithoutI[:, comp], '-', lw=1, color="blue")
153
154       # Plot the numerical solutions
155       ax.plot(t, Sols[:, comp], ':', lw=1, color="red")
156
157       # Label the axis
158       ax.set_xlabel('Time')
159       ax.set_ylabel('Count')
160       ax.set_title(Title)
```

```
161
162  plt.tight_layout()
163  plt.show()
```

21.6   Further reading

The original mathematical modelling of epidemic systems is due to Kermack and McKendrick (1927, 1932, 1933) (alternatively Kermack and McKendrick, 1991a,b,c). Two classic works on mathematical biology which include significant discussion of these and related models are Murray (2002, 2003). Diekmann et al. (1995) is an overview of some more models following the work of Kermack and McKendrick (1933); see also Diekmann and Heesterbeek (2000) and Daley and Gani (2001) for similar overviews of epidemic modelling. For epidemic modelling with a more explicit focus on genetics, see May and Anderson (1983).

For models of epidemics on graphs see Keeling and Eames (2005), which as well as a good general introduction, also discusses important pair approximation methods. Pastor-Satorras and Vespignani (2001) consider SIS epidemics on scale free networks and May and Lloyd (2001) consider SIR models, again on scale free networks. There has been quite a lot of more recent work on epidemic modelling on graphs, see for example Ball and House (2017); Pellis et al. (2015a,b). For models using exact equations see Sharkey et al. (2015); Sharkey and Wilkinson (2015); Simon and Kiss (05 2020). For a good book on this type of epidemic modelling see Kiss et al. (2017).

For the modelling of the evolution of virulence, see Nowak and May (1994), Nowak and May (1992a); May and Nowak (1994, 1995). For the use of game theory in epidemiological models, see for example Bauch and Earn (2004), Bhattacharyya and Bauch (2011) or Bauch and Bhattacharyya (2012), or the work by Rychtář and colleagues mentioned in Section 21.4.

Following Fu et al. (2011), many recent studies such as Iwamura and Tanimoto (2018); Kabir et al. (2019); Kabir and Tanimoto (2019); Kuga et al. (2019); Arefin et al. (2019, 2020); Huang et al. (2020) use multi-agent-simulation (MAS) methodology, thus allowing considerable flexibility and realism in the modelling approach, especially in prevention of seasonal influenza where individuals have to re-vaccinate every year.

For recent game-theoretical work on epidemics particularly relevant to the Covid-19 epidemic, see Martcheva et al. (2021); Kabir et al. (2021b); Kabir (2021); Glaubitz and Fu (2020); Kabir et al. (2021a); Alam et al. (2021). Amaral et al. (2021), see also Agusto et al. (in press), which considers a game with strategic use of quarantine; Xu and Cressman (2016) consider a game with voluntary vaccination. Reluga and Galvani (2011) provides a general population games approach and in particular also considers imperfect vaccination. For a review of game theoretical models in epidemics, see Chang et al. (2020).

21.7 Exercises

Exercise 21.1. Consider an SIS epidemic as described in Section 21.1.1. Suppose that each individual dies at rate d, and also gives birth at rate d, so that the population size is fixed. Assuming that all newborns are susceptible, find the steady state proportion of infectives in the population.

Exercise 21.2. Show that in the model of the influenza outbreak in Example 21.2, the epidemic occurs and the state $I = 0$ is not stable if and only if $R_0 = \beta/g > 1$.

Exercise 21.3 (Keeling, 1999). A key feature of a graph (or network) is the adjacency matrix. For a graph with N vertices, this is the $N \times N$ matrix $A = (a_{ij})$ where

$$a_{ij} = \begin{cases} 1 & \text{if } i \neq j \text{ and there is a link from vertex } i \text{ to vertex } j, \\ 0 & \text{otherwise.} \end{cases} \qquad (21.41)$$

The two most fundamental properties of a graph are perhaps the average number of edges per individual k and the clustering coefficient, ϕ, which is the ratio of 3 times the number of triangles to the number of connected triples in the population. Assuming that a graph is undirected, show that $A = A^T$ and that

$$k = \frac{1}{N} \sum_{i,j} a_{ij} = \frac{1}{N} \text{trace}(A^2). \qquad (21.42)$$

As a harder exercise, show that

$$\phi = \frac{\text{trace}(A^3)}{\sum_{ij}(A^2)_{ij} - \text{trace}(A^2)}. \qquad (21.43)$$

Exercise 21.4. For the virulence model of Section 21.2 show that R_0 for (21.18)–(21.19) is given by (21.20).

Exercise 21.5. In the context of Example 21.4, suppose that parasites can freely evolve any level of virulence ν, but that the relationship between ν and τ is given by (21.27). Show that the optimal level of virulence is given by $v_{opt} = \sqrt{cd}$. Hence show that the optimal level of virulence can be arbitrarily high, or arbitrarily close to 0, depending upon the parameter values.

Exercise 21.6. For Exercise 21.5, find the number of infectives associated with the optimal level of virulence. Hence comment on the possible combinations of parasite load and virulence that can occur.

Exercise 21.7. For the superinfection model of Example 21.5 with $k = 2$ and $d = 1.5$, assume that there are two disease strains where $\nu_1 = 1, \nu_2 = 2$. Find the level of evolutionarily stable virulence in the population for a general value of $\mu > 0$.

Exercise 21.8. Consider the Prisoner's Dilemma virus model of Section 21.3. By considering pure populations of the two strains of virus within a cell (possibly with a negligible proportion of the alternative strain), relate the payoffs from (21.34) with the parameters a_1, a_2, π_1 and π_2 and suggest an explanation for your results.

Exercise 21.9 (Bauch and Earn, 2004). For the vaccination model of Section 21.4 show that vaccine coverage at the ESS solves (21.40).

Exercise 21.10. Consider the vaccination model of Section 21.4. Suppose that the interaction rate for individuals is 2, the recovery rate is 0.1 the birth/death rate μ is 0.05 and the cost of being vaccinated is 1. Find the ESS value p^* as a function of the infection cost r_i. How high does this cost have to be for it to be worth having some level of vaccination in the population?

Chapter 22

Evolutionary cancer modelling

In Chapter 21 we saw that an epidemic could be treated as an evolutionary process, both in its overall spread within a human or animal population, but also involving competition between rival strains within individuals (see Section 21.2). Here the human/animal population can be considered as the environment, although there are games with human strategic options including with regard to vaccination or the use of various protection measures such as bed nets etc. as discussed in Section 21.4. The situation is, however, complicated by the fact that as well as there being competition within individuals, the whole population needs to be taken into account.

More recently, it has been realised that cancer can be modelled in a very similar way. Cancers are diseases which involve abnormal cell growth, so that affected cells proliferate within the body, in many cases leading to the death of the person or animal (for most cancers the cause of death is metastasis as the cancer spreads beyond its original site, rather than simply volume). They typically require a number of gene mutations before such dangerous proliferation can occur (Stratton et al., 2009; Hanahan and Weinberg, 2000; Marusyk et al., 2012; Dujon et al., 2021).

Tumours are not homogeneous, however, and contain cells of different types which respond to the body and treatments in different ways, and also compete with each other. Thus we can think of cancer as a population, living within an environment which is the human or animal body. Indeed this analogy is in some ways cleaner than that for epidemics, since cancers are in general non-transmissable (although there are exceptions, (e.g. Devil Facial Tumour Disease in the Tasmanian Devil Bender, 2010) and so a single organism is the whole environment, and every cancer is a new development unique to the patient (though distinct types of cancer also have many common features too).

The first paper that modelled cancer using evolutionary game theory was that of Tomlinson (1997). Notably, Vincent and Gatenby (2008) developed an eco-evolutionary model which we discuss below in Section 22.1. In Section 22.2 we consider a spatial model where the games involved are comparatively simple and, as in Chapter 12, the interesting behaviour comes from the spatial aspect.

With this understanding it has come the possibility to make use of this knowledge to help with the treatment of cancer, including with so-called "adaptive therapies" also called Darwinian, evolutionary, or evolutionarily

DOI: 10.1201/9781003024682-22

enlightened therapies (Stroh et al., 2014). In Sections 22.3 and 22.4 we consider important examples of such an evolutionary approach.

22.1 Modelling tumour growth — an ecological approach to cancer

Vincent and Gatenby (2008) developed a model of the evolution of cancer, by treating the cancer cells as an evolving population within the human environment. In particular, they used the G-function methodology that we discussed in Section 20.1 and considered the evolution of different cell types, and we follow their model below.

Proliferation of cells depends upon R, which is a level of resource available to the cells (such as glucose). We assume that there are n_s different cell types, where the strategy of type i is denoted as u_i and the population density of that type is given by x_i. Here the strategy $\mathbf{u}$ represents the density of hypoxia inducable factor (HIF) cells, which affects the number and type of membrane glucose transporters (GLUT) which, in turn, affects the uptake of glucose into the cell.

The fitness of a cell using strategy v is then given by

$$G(v, \mathbf{u}, \mathbf{x}, R) = B_n \left(1 - \frac{\sum_{j=1}^{n_s} a_j(v, \mathbf{u}) x_j}{K(v)} \right) \max\left(\frac{E(v)R^2}{R_0^2 + R^2} - m, 0 \right). \quad (22.1)$$

Equation (22.1) contains a number of terms which need to be defined and we look at these here. B_n is the intrinsic growth rate of the cells when not constrained. The first bracketed term is a Lotka-Volterra competition term (which is 1 in the absence of such competition). The second bracketed term represents regulation of the intrinsic growth term, as a function of excess uptake over and above the basal demand level m; here $E(v)$ represents the rate at which a v-strategist could consume an infinitely abundant resource, and R_0 governs how far below this maximum level is achieved for the given finite resource level R.

R is a variable resource and its level also evolves subject to the type and number of cells as follows:

$$\frac{dR}{dt} = r - \sum_{i=1}^{n_s} \frac{E(u_i)R^2}{R_0^2 + R^2} x_i, \quad (22.2)$$

where $r = \max\left(r_e m \sum_{i=1}^{n_s} x_i, r_{max}\right)$, and $r_e m$ represents the flow of re-source through vascular networks (r_e must exceed 1 to be sufficient for this to meet the demand), and there is an inherent maximum level for r to take, r_{max}.

The functions a and K relate to specific biological properties. a_j is a measure of competition between our focal cell and type j. This is the combination of an intrinsic property of j, a_{meanj}, which increases with the value of u_j, and the difference between our focal type and u_j; the more similar they are, the greater the level of competition, analogous to real animal competition. Specifically the function is

$$a_j(v, \mathbf{u}) = a_{meanj}\exp\left(-\frac{(v - u_j)^2}{2\sigma_a^2}\right). \tag{22.3}$$

The function K is the carrying capacity in the "environment". Tumour cells use glucose in two distinct ways, aerobic glycolysis and anaerobic glycolysis and the optimal value of u for each method is different. Thus this carrying capacity is represented by the function

$$K(v) = K_{mean1}\exp\left(-\frac{(v - u_{K1})^2}{2\sigma_K^2}\right) + K_{mean2}\exp\left(-\frac{(v - u_{K2})^2}{2\sigma_K^2}\right), \tag{22.4}$$

which is a bimodal function where the value of the first peak at u_{K1} represents the optimal value for the aerobic metabolism and the value at the second peak u_{K2} represents the optimal value for the anaerobic metabolism.

In the full system the densities x_i and the level of resources R change, but also so do the values of the strategies themselves, as the most successful strategies both proliferate and mutate, leading to the emergence of fitter nearby strategies. The system evolves following equations

$$\frac{dx_i}{dt} = x_i G(u_i, \mathbf{u}, \mathbf{x}, R), \tag{22.5}$$

$$\frac{du_i}{dt} = \sigma_i^2 \frac{\partial G(v, \mathbf{u}, \mathbf{x}, R)}{\partial v}\bigg|_{v=u_i}, \tag{22.6}$$

together with the equation for R, (22.2), above.

Following simulation of the above process, three distinct stages were observed. The first stage produced little change in the strategy of the tumour cell population. This is consistent with the initiation phase of experimental carcinogenesis and the Fearon and Vogelstein (1990) model, where the first step in carcinogenesis is tumour suppressor mutation. The second stage had more significant strategy changes, leading to growth of the number of tumour cells. This again is consistent with real cancer evolution and the Fearon-Vogelstein model. Finally, there was a third stage of high strategy change and rapid growth of the tumour results, representing an invasive cancer. The authors note that whilst this stage is consistent with real carcinogenesis, this does not appear in the Fearon-Vogelstein model.

22.2   A spatial model of cancer evolution

The model from the previous section had a number of complex elements, but space was not considered. As we have seen in Chapters 12 and 13, space can be very important in the analysis of evolutionary models, but adding it greatly complicates a model. Thus spatial models often have quite simple games at their heart. This approach to consider a spatial model of cancer was followed by Nanda and Durrett (2017) where matrix games represented interactions between cells, normal and cancer, played on a three dimensional integer lattice (so all individuals have six neighbours). In fact, this paper was more a general methodology paper, though it contained interesting cancer examples. For example, a three strategy model of multiple myeloma was considered, which was based upon a simple non-spatial model of Dingli et al. (2009).

Multiple myeloma is a bone disease which damages the health of bones by disrupting the balance between two classes of normal bone cell; osteoclasts (OC) and osteoblasts (OB). The key interactions were modelled in Dingli et al. (2009) using the three strategy matrix game below:

$$
\begin{array}{c}
\text{OC} \quad \text{OB} \quad \text{MM} \\
\begin{array}{c} \text{OC} \\ \text{OB} \\ \text{MM} \end{array}
\left(
\begin{array}{ccc}
0 & a & b \\
e & 0 & -d \\
c & 0 & 0
\end{array}
\right),
\end{array}
\qquad (22.7)
$$

where a, b, c, d and e are all non-negative.

Without MM, there is a mixed ESS for the OC and OB cells. A cell is only of one type, so this is a polymorphic rather than a monomorphic population. Following standard results from Section 6.4 if $c > e$ then MM can invade the OC and OB mixture. Since $d > 0$, this clearly cannot lead to an internal ESS, and indeed if $c > e$ then MM dominates OB, so OB must be eliminated from the population. This then leaves a stable mixed ESS of OC and MM. This is a bad outcome as it completely disrupts the natural balance of the cells and can lead, for example, to a significant risk of bone fracture.

An alternative scenario is when $c < e$ and $c/e + dc/be > 1$ which leads to an internal saddle point, with two stable pairwise solutions, a mixture of OB and OC (i.e. completely free of MM) and the OC and MM mixture mentioned above (if $c < e$ and $c/e + dc/be < 1$ we have the desired, but less interesting from the modelling situation, case where the OB and OC mixture is a global attractor).

We note that Dingli et al. (2009) considered the full dynamics over the three strategies, not just starting from the mixture with no MM, but this would seem the most pertinent, as where the real scenario would initiate. In such a case if we were in the scenario with the unique OC and MM mixture, then this is of course where the dynamics would end up, but in the case with

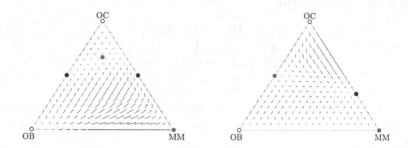

FIGURE 22.1: Evolutionary dynamics of osteoclasts (OC), osteoblasts (OB) and multiple myeloma (MM) cell types. Vertices mean that the cell population is monotypic. Bone homeostasis occurs in the absence of MM (the OC–OB line), remaining stable in the presence of MM if $c/e < 1$ (left, where $a = b = e = 1, c = 0.5, d = 2$). The fixed point along OC–MM is unstable whenever $c/e + cd/be < 1$. For $c/e > 1$ only the OC-MM co-existence equilibrium is stable (right, where $a = b = e = 1, c = 2, d = 0$). The black dots are stable points, grey dots saddle points and white dots are unstable points of the dynamics.

two pairwise solutions, we would remain at the OB and OC mixture as this is where we would begin.

However, this assumed fixed parameters, and this is perhaps unrealistic. Suppose that the parameters had initially been in the former case allowing MM to proliferate? If these then shift after MM has reached some significant level, then the precise mixture can be crucial for the final outcome. Indeed, if clinicians are able to affect the parameters in some way, with sufficient knowledge of the system, they can alter the speed of disease development, improve (without fundamentally changing) the outcome (change the precise location of a given mixture), or even change the solution completely to eradicate the disease.

The conclusion of Dingli et al. (2009) is one that shows the change in thinking that an evolutionary perspective can bring, even from such a simple model: "Instead of trying to kill all cancer cells, therapies should aim at reducing the fitness of malignant cells compared with normal cells, allowing natural selection to eradicate the tumour."

The more that a system is understood, the more chance that such a beneficial outcome can be brought about. Nanda and Durrett (2017) considered a development of the Dingli et al. (2009) model. They considered an updated version of the payoff matrix

$$
\begin{array}{c}
 \\
\text{OC} \\
\text{OB} \\
\text{MM}
\end{array}
\begin{array}{ccc}
\text{OC} & \text{OB} & \text{MM} \\
\left(\begin{array}{ccc}
0 & A & B \\
E & 0 & -D \\
C & F & 0
\end{array} \right),
\end{array}
\qquad (22.8)
$$

where $A = (1+\theta)a - \theta e, B = (1+\theta)b - \theta c, C = (1+\theta)c - \theta b, D = (1+\theta)d, E = (1+\theta)e - \theta a, F = \theta d$ for $\theta \geq 0$, so that when $\theta = 0$ the original version is recovered.

This is simply the original matrix where each entry not on the leading diagonal has a constant θ multiplied by the difference between itself and its reflection in the leading diagonal added to it. This is a natural perturbation which keeps important matrix properties intact (specifically the negative-definiteness condition from Haigh's procedure from condition 6.45).

The spatial game led to no significant differences to the non-spatial model when there was a unique stable boundary solution, but for the case with an unstable internal equilibrium and two alternative solutions these could be very different (Durrett, 2014), and so Nanda and Durrett (2017) expressed doubts about some of the conclusions of Dingli et al. (2009). In particular, all three strategies can coexist in the spatial game for a long time, whereas this does not occur for the replicator equation.

22.3   Cancer therapy as a game-theoretic scenario

As we have discussed previously, cancer is subject to evolutionary pressures; in particular, it can be considered as a heterogeneous ecology of competing strains. One consequence of this is that treatment is more successful against some strains than others, so that it changes the composition of the tumour, allowing resistance to evolve. Thus whilst some cancers can be successfully cured with a high rate of success, in other cases progress has proved very hard and mortality rates in metastatic cancer, in particular, have only improved a little over the last fifty years (Jemal et al., 2017; Yi et al., 2014).

Is it possible for us to use this understanding to develop ways of treating cancer? One possible way to do this is to exploit an *evolutionary double bind*, as modelled by Basanta et al. (2012). An evolutionary double bind is a concept from ecology, describing a situation where a prey species faces two different predators, and defence against each predator makes it more likely that the prey will fall victim to the other. For example gerbils in the Negev desert were observed to face precisely this problem when contending with both snakes and owls (Bouskila, 2001).

The cancer equivalent of this idea involves the prey as the cancer cells and the two predators as the components of a combination therapy, which each target the cancer in different ways, and so have different effects on different strains of the cancer. In particular, strains that are resistant to one therapy may not be resistant to the other. Often resistance is assumed to come at some cost, but see Staňková (2019); Kaznatcheev et al. (2019); Viossat and Noble (2021), so that strains which are not resistant are likely to be the fittest in the absence of treatment. A combination therapy is thus used in the hope of causing a significant fitness reduction to the cancer, without allowing the evolution of resistance. It is noted in Basanta et al. (2012) that this description is a somewhat idealised picture, and for example it is perfectly possible for resistance to develop to multiple treatments. It is nevertheless harder, and combination therapies are widely used because they are often more successful than individual therapies.

The model of Basanta et al. (2012) contains three strains of cancer cells and two treatments. Susceptible cells S are susceptible to both treatments, R_A cells are completely resistant to treatment A but susceptible to treatment B and R_B cells are completely resistant to treatment B but susceptible to treatment A. There are (fixed) costs to resistance to treatments A and B, c_A and c_B, respectively, irrespective of the level of treatment. The treatments also levy a cost to susceptible cells depending upon the level of treatment, and so this changes with time as the treatment levels are adjusted; these are denoted $d_A(t)$ and $d_B(t)$. It may be that resistant strains even face an extra cost from alternative treatments, and these are included in the fitnesses as multipliers α and β for strains A and B, respectively.

The proportions of the three cell types are denoted p_A, p_B and $1 - p_A - p_B$ for celltypes R_A, R_B and S, respectively. It is assumed that there is also direct cell competition, which will especially be true for larger tumours when resources are scarce. In particular, if treatment A (B) is being delivered, resistant cells have an advantage $X(t)$ $(Y(t))$ over non-resistant cells in their interactions with them. Non-resistant cells under treatments A and B are in proportion $1 - p_A$ and $1 - p_B$, respectively.

Putting all of the above together, the fitnesses of the three types of cell are as follows;

$$W(R_A) = 1 - c_A - \alpha d_B(t) + (1 - p_A)X(t), \qquad (22.9)$$

$$W(R_B) = 1 - c_B - \beta d_A(t) + (1 - p_B)Y(t), \qquad (22.10)$$

$$W(S) = 1 - d_A(t) - d_B(t). \qquad (22.11)$$

This then yields a mean fitness of

$$\overline{W} = p_A W(R_A) + p_B W(R_B) + (1 - p_A - p_B)W(S). \qquad (22.12)$$

We would thus follow the evolution of the proportions of the three cell types, the population evolving according to the discrete replicator dynamic from Section 9.1.1.1.

For any particular example situation you would just need to substitute in the appropriate payoff functions (or make suitable modifications if there were additional/distinct factors). In Basanta et al. (2012) they considered an example where the two treatments are a p53 vaccine and chemotherapy, following a clinical trial on lung cancer carried out by Antonia et al. (2006). In the clinical trial (and a subsequent experimental paper using mice Ramakrishnan et al., 2010) it was shown that a combination of the therapies was more effective than using single therapies; in particular, chemotherapy increased tumour susceptibility to the p53 treatment. Basanta et al. (2012) then explored this using a variant of their mathematical model, especially cases where the application of the treatment were given singly, in varying order.

22.4   Adaptive therapies

As mentioned in the previous section, mortality rates in some cancers have greatly reduced over recent years with more effective treatment, but in metastatic cancer this is not the case. The usual treatment regime for metastatic cancer is for the patient to take a drug or drug combination at the maximum tolerable dose, either continuously or cyclically. This process continues until the patient is cured (unlikely for metastatic cancer), or it is halted by excessive toxicity or tumour progression. The logic behind this approach is that it is best for the patient to kill cancer cells as quickly as possible (Evans, 2018; Gatenby, 2009). This procedure often appears to work initially, but becomes less successful as resistance is developed (Evans, 2018; Gatenby, 2009; Marusyk et al., 2012; Staňková, 2019; Staňková et al., 2019; Werner et al., 2011, 2016).

As discussed in Section 22.3, cancer can be modelled following an evolutionary process where the evolution of resistance can depend upon the therapies used. In this section we shall follow the model of Staňková et al. (2019) and consider *evolutionary therapies*, where the clinicians attempt to use understanding of the evolutionary process to choose therapies which control the evolution of resistance, steering evolution onto as favourable a path as possible (see also Staňková et al., 2019; Ledzewicz and Schaettler, 2016; Maley et al., 2004). Such therapies were employed at the Moffitt Cancer Center in clinical trials and enabled patients with metastatic cancers to have both a higher quality of life and much longer mean time to progression of the disease (e.g., see Zhang et al., 2017).

The model of Staňková et al. (2019) in particular follows a *Stackelberg Evolutionary Game*. The Stackelberg game is a classical game from Economics (von Stackelberg, 1952) which is a two-player sequential game of perfect information, involving a leader and a follower, where the follower responds to the strategies of the leader. In this evolutionary game, the clinician is the leader who acts rationally, and the followers are an infinite population of cancer cells which respond following evolution.

Let us assume that the rate of cancer growth follows a function based upon the treatment level $u_T \in [0,1]$, where 1 is the maximal dose toxicity of the treatment, and the strategy of the cancer $u_c \in [0,1]$, where higher u_c implies a higher resistance to treatment, but also a lower level of intrinsic growth. We have

$$\frac{dN}{dt} = N\left(r\frac{(1-u_c)K - N}{K} - \frac{u_T}{k + bu_c}\right), \tag{22.13}$$

where N is the number of cancer cells. Here r, b, k and K are positive parameters which govern the cancer's growth properties; for example, in the absence of treatment or resistance, the intrinsic rate of growth is r when cancer cells are rare, following logistic growth up to a carrying capacity K. For any given combination of strategies u_c and u_T, N will evolve to $N(u_T, u_c) = \max\{0, N^*\}$, where

$$N^* = K(1 - u_c) - \frac{u_T K}{r(k + bu_c)} \tag{22.14}$$

is an equilibrium value provided that $N^* > 0$ (0 is always an equilibrium, see Exercise 22.8). The best strategy for the cancer, treating it as a single entity, is the value of u_C (as a function of the treatment strategy u_T) which maximises the growth rate,

$$R_c(u_T) = \min\left(1, \max\left\{0, \sqrt{\frac{u_T}{rb}} - \frac{k}{b}\right\}\right). \tag{22.15}$$

We consider the patient and the clinician as the same player, so that the clinician's aim is to maximise the payoff for the patient. The patient does not simply want to minimise the above growth rate, but has a more complex payoff function based upon quality of life. This takes the form (note that the function we have chosen is different to that of Staňková et al. (2019) to help better illustrate the results)

$$J_T(u_T, u_c) = \alpha(D^2 - u_T^2) + (1 - \alpha)\frac{B - N(u_T, u_c)}{B}(1 + 4u_c) \tag{22.16}$$

whenever $0 \le N(u_T, u_c) \le B$, where B is the maximum tumour burden. It is typically assumed that $B < K$, as otherwise the cancer would be benign. Outside of this range when $N > B$, the payoff is 0 (the patient has died).

Here our clinician knows that the cancer employs the best response function from (22.15) and so can then strategically choose the value of u_T which maximises (22.16) when allowing for the substitution of $R_c(u_T)$ from (22.15) for u_c. This gives a better outcome for the patient than the game where the physician and the cancer effectively play simultaneously. In the former the clinician predicts what the cancer will do and reacts accordingly, whereas in the latter the physician chooses the strategy to maximise short term gain based upon the current cancer strategy. The difference between these two approaches is illustrated in Figure 22.2.

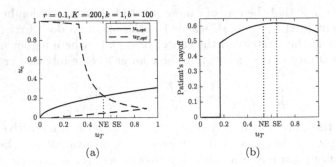

(a)   (b)

FIGURE 22.2: The difference between the Stackelberg and Nash equilibria. (a) The best response function for the cancer and the patient. The Nash equilibrium is at the intersection of the curves. Here, we assume $B = 150 < K$. (b) Patient's payoff $J_T(u_T, R_c(u_T))$. The maximum occurs at Stackelberg equilibrium.

It should be noted that there are of course factors that the above model has not taken into account. In Section 22.2 we emphasised the importance of spatial factors in cancer modelling and this is important when considering adaptive therapies too. Modelling by Bacevic et al. (2017) indicates that the use of such therapies may be more effective than non-spatial models indicate, as spatial factors constrain resistance. See also Noble et al. (2020) for a more detailed investigation into the effect of tumour composition and its effects.

22.5   Python code

In this section we show how to use Python for the model from Dingli et al. (2009).

```
1   """
2   Generates the vector field for myeloma example
3   """
4
5   ## Import basic packages
6   import numpy as np
7   import matplotlib.pyplot as plt
8
9   ## Define payoff matrix
10  A=[ [ 0,  1,   1],   # Payoffs to OC
11      [ 1,  0,  -2],   # Payoffs to OB
12      [.5,  0,   0]]   # Payoffs to MM
13
```

```
14  ## Define replicator dynamics
15  def dynamics(y,t):
16      " Defines the dynamics "
17      dydt = y*(np.dot(A,y) - np.dot(y,A).dot(y));
18      # dy/dt[i] = y[i] * ((A y^T)[i] - yAy^T)
19      return dydt
20
21  ## Define coordinate transformation to the triangle
22  OC = [0.5, 0.866]     # OC top vertex
23  OB = [0, 0]           # OB lower left
24  MM = [1, 0]           # MM lower right
25
26  ## Get uniform points in the population space
27  # Will be given as [b, c-b, 1-c]
28  mesh = 0.05  # Size of the mesh
29  for a in np.arange(0, 1, mesh):
30      for b in np.arange(0, a, mesh):
31          for c in np.arange(a, 1, mesh):
32              # Make the triple into triangle coords
33              x = b*OC[0] + (c-b)*OB[0] + (1-c)*MM[0]
34              y = b*OC[1] + (c-b)*OB[1] + (1-c)*MM[1]
35
36              # Get the direction of the vector
37              dV = dynamics([b, c-b, 1-c],[])
38
39              # Make the triple into triangle coords
40              u = dV[0]*OC[0] + dV[1]*OB[0] + dV[2]*MM[0]
41              v = dV[0]*OC[1] + dV[1]*OB[1] + dV[2]*MM[1]
42
43              # Plot the arrow from (x,y) to (x+u,y+v)
44              plt.arrow(x,y,u,v)
45
46  # Show the plot
47  plt.show()
```

22.6   Further reading

The following papers apply game theoretical methods to the treatment of cancer Vincent and Gatenby (2005); Axelrod et al. (2006); Gatenby (2009); Aktipis et al. (2015); Hurlbut et al. (2018); Archetti and Pienta (2019); Staňková (2019); Staňková et al. (2019). In particular, Staňková et al. (2019) includes a discussion of the methods for a non-mathematical audience. Less aggressive treatment strategies for containment of cancer are considered in Viossat and Noble (2021); Kuosmanen et al. (2021). Cunningham et al. (2020) considers an optimal control approach to metastatic prostate cancer. Tavakoli et al. (2020) considers a model of immunoediting.

Related mathematical models of cancer are given in Altrock et al. (2015) and Anderson and Chaplain (1998). Wu (2015) considers the emergence of resistance in depth. For more on the non-heterogeneity of cancer tumours and its relevance to treatments, including evolutionary ones, see Marusyk et al. (2012); Baker et al. (2017); Ledzewicz and Schaettler (2016). For mathematical models applied to cases involving data see Werner et al. (2011, 2016); Maley et al. (2004). For work on the transmissable cancer found in the Tasmanian devil, see Ujvari et al. (2016a,b).

The production of diffusible molecules in cancer cells has been studied as a nonlinear public goods game, see, for example, Archetti (2013a,b, 2015, 2016, 2018).

Mathematical modelling of cancer is a rapidly developing area, and whilst we refer the reader to Wölfl et al. (in press) for a recent review on the contribution of evolutionary game theory to cancer, we recognise that this section will be somewhat out of date even by the time of publication. The reader will have to do their own research!

22.7   Exercises

Exercise 22.1. Consider the model of Section 22.1. Assume that there is just a single cell type u. Using (22.1), describe the evolution of the density of the cell population and find an expression for its long-term value.

Exercise 22.2. Using the methods from Section 6.4, find the ESSs for the payoff matrix (22.7), thus confirming the results discussed in Section 22.2.

Exercise 22.3. Repeat Exercise 22.2 for the payoff matrix (22.8).

Exercise 22.4. Demonstrate how varying the value of θ in the matrix (22.8) affects the set of ESSs of the game; in particular show that satisfaction of the negative definiteness condition for an internal ESS from Section 22.2 is unaltered by varying θ, provided that θ does not change sign.

Exercise 22.5. Consider the equations (22.9)–(22.12) from Section 22.3. Assuming that $d_A(t), d_B(t), X(t)$ and $Y(t)$ are constant (the former two indicating a non-varying treatment regime) find the conditions for neither type of resistance to evolve. Can this tell you what the optimal fixed treatment strategy would be?

Exercise 22.6. For the model of Section 22.3, assume fixed treatment strategies $d_A(t) = d_A$ and $d_B(t) = d_B$ and associated fixed resistant cell advantages $X(t) = X, Y(t) = Y$. Find the conditions for a mixed equilibrium of all three cell types to exist (i.e. where all three can co-exist with equal fitnesses) and find the distribution proportions of the cell types at that equilibrium.

Exercise 22.7. Show that the optimal cancer strategy for the model of Section 22.4, as a function of the treatment strategy u_T, is described by (22.15).

Exercise 22.8. For the evolutionary process from Section 22.4, show that $N = 0$ and $N = N^*$ from (22.14) are equilibria (assuming the latter is positive). Further show that the dynamics from (22.13) converges to $N(u_t, u_c) = \max\{0, N^*\}$.

Exercise 22.9. Adapt the code from Section 22.5 to get trajectories of solutions in the multiple myeloma model from Dingli et al. (2009).

Hint. Use the function `odeint` from `scipy.integrate`.

Chapter 23

Conclusions

In this chapter we round up the book. We have considered a large range of mathematical techniques and applications, and we start by summarising the models that we have seen throughout the book. We then briefly consider how best to make a game-theoretical model, and what components make a good model. Finally we look at some potential future developments.

23.1 Types of evolutionary games used in biology

23.1.1 Classical games, linearity on the left and replicator dynamics

Many of the early classical games, such as the Hawk-Dove game and the Prisoner's Dilemma that we saw in Chapter 4, are matrix games and have a number of good properties, as we saw in Chapter 6, such as the existence of a uniform invasion barrier. A number of classic results from evolutionary game theory actually rely on the game being a matrix game. This includes the Bishop-Cannings Theorem 6.9, and the study of patterns of ESSs, which was only really worthwhile because the Bishop-Cannings theorem holds. Another example is Haigh's procedure for finding all of the ESSs for a matrix game which works for (generic) games with any number of pure strategies. It also includes many of the results from replicator dynamics, for instance that any internal ESS is a global attractor of the replicator dynamics. In particular, as we discuss in Chapter 7, matrix games are what we have termed linear on the left and linear on the right. This, in turn, means that they have the important polymorphic-monomorphic equivalence property, that all payoffs (and consequently all Nash equilibria) are the same if played in a population of mixed strategists or in the equivalent weighted mixture of pure strategists.

We should recall that even for matrix games, there are some points where the static and dynamic analyses will lead to qualitatively distinct solutions. An ESS is always an attractor of the replicator dynamics, but not every attractor of the dynamics is an ESS, even if it is an internal strategy. This comes about because of the different way invaders are considered in the two cases. Invaders play mixed strategies in the static case, but a polymorphic mixture

DOI: 10.1201/9781003024682-23

(the composition of which can and will change) in the dynamic analysis. A key feature of the replicator dynamics is the fact that the population is a composition of pure strategists, and the population state is a simple average of these, weighted by their frequency. The distinction between a population of mixed strategists and a polymorphic mixture of pure strategists can be an important. one, and we have discussed this in Chapter 2 and beyond.

Replicator dynamics do not require games to be matrix games, and there is a natural extension to games that are linear on the left. Linearity on the left actually covers a wide class of games, since the games from conventional evolutionary game models that are nonlinear are usually only nonlinear on the right, rather than on the left. This arises naturally out of the way we normally think about the strategies as mixed strategies, i.e. probabilistic mixtures of a combination of pure strategies. Thus payoffs are usually just a weighted average of the probability that a particular pure strategy is played. Games such as the sex ratio game and the foraging games which lead to the ideal free distribution are of this type. The population can still be thought of as being composed of pure strategists, and hence the replicator equations can easily be derived. We also note that this is true of the other types of conventional dynamics such as the discrete replicator dynamics, best response dynamics, imitation dynamics and the replicator-mutation dynamics, since all of these are also based upon a population of pure strategists.

Games that are linear on the left but not on the right do retain some important properties. Providing that the payoff functions are continuous there is still a uniform invasion barrier, and as we see above, the replicator equation can still be applied. We have seen that the Bishop-Cannings theorem will no longer hold in general, and in fact, there may be multiple ESSs with the same support. Similarly, there may be significant differences between monomorphic and polymorphic populations, in the sense that the equilibrium solutions in each case may be different. We saw this for the case of pairwise games with non-constant interaction rates in Chapter 7.

Games with multiple players are nonlinear, and in particular, multi-player matrix games are examples of games that are linear on the left. In these games some of the good properties of matrix games are preserved, e.g. no two ESSs have the same support for three-player games, and we have some idea of bounds on the number of potential ESSs in general. Other examples, such as the multi-player war of attrition and the knockout models, whilst being more complicated than their two-player equivalents, at least possess some structure to help analyse the game (we note that knockout contests and the multi-player war of attrition with strategy adjustments are not linear on the left).

We have also seen one other type of game that, like matrix games, is both linear on the left and linear on the right. The (continuous time) war of attrition, while not a matrix game as such because of its infinite (and indeed uncountable) number of pure strategies, nevertheless is both linear on the left and linear on the right (and the multi-player war of attrition without strategy adjustments is linear on the left but not on the right). One major consequence of the uncountable number of strategies is that the war of attrition does not readily extend to the standard replicator dynamics (except in special cases with restricted strategy sets), since they rely on a population composed of pure strategists, as we discuss above.

23.1.2 Strategies as a continuous trait and nonlinearity on the left

What if a game is not linear on the left? We saw some examples of this in Chapter 7, including a tree height game and a game of sperm competition. These examples considered a strategy as a single measurement, such as the height of a tree, rather than a probabilistic mixture of two pure strategies. When an individual's strategy is a trait value like this, there is no reason for the payoff to an individual with a trait halfway between two others to be simply the average of the payoffs of the other two. Such cases where a strategy is of this form and in general nonlinear on the left are common in biological models, as we discuss below in Section 23.1.6.

The theory of adaptive dynamics, in particular, considers the evolution of continuous traits and thus assumes nonlinearity on the left. The key solutions of adaptive dynamics models, Evolutionarily Singular Strategies, can exhibit a rich variety of behaviour based upon the signs and sizes of the second derivatives of payoffs with respect to the left and right strategies. In particular, evolutionary branching points can result in a population splitting into a polymorphic population of distinct traits. In fact, all games that are linear on the left are non-generic in some sense in the theory of adaptive dynamics, as we saw in Chapter 14.

Adaptive dynamics considers evolution where only very small changes are possible, and this is logical when the strategic choices are really fixed traits such as height or horn size. There are games, as we have seen for example in Chapter 17, where a strategy is the volume of sperm to allocate, in which strategies could change significantly and in short timescales. Such games tend to be generated to solve an individual problem and thus the form in which they appear is quite varied. Thus, as we discuss in Section 23.3, there is scope for a wider theory of games that are nonlinear on the left, which is so far lacking.

23.1.3 Departures from infinite, well-mixed populations of identical individuals

Often, and in the games discussed above, populations are assumed to contain an infinite number of individuals, where every pair of individuals is equally likely to meet and where every individual is exactly the same, except perhaps in the strategy that they employ. Here we consider departures from these assumptions.

A different type of evolutionary process occurs when finite populations are considered. Of course, all real populations are finite, but if the population is sufficiently small then random effects come into play. Firstly, changes in the population occur within a finite set of states, and which occur are governed by probabilities. The evolutionary dynamics that is most commonly used is the Moran process, which as we saw in Chapter 12 is close to the discrete replicator dynamics if the population is not too small. This, in turn, means that populations can never consist of more than one type of individual in the long term (without repeated mutation), and the key concept in finite populations is the fixation probability, the probability that the population will eventually consist solely of a given type. It turns out that fixation times are also important, and if the population would take a very long time to fixate, it is worth investigating quasi-stationary equilibria (i.e. equilibria conditional on fixation not occurring).

Although populations of identical individuals are often assumed, nature is full of types of asymmetry such as asymmetry in size, age, horn size or speed. Any asymmetry of this type is termed a correlated asymmetry. Naturally, to model populations mathematically with a number of the asymmetries is very difficult, and the logical approach using mathematical models is to allow the population to differ in one respect only and observe how this affects the game. An interesting asymmetry with surprisingly wide potential consequences is that of role; i.e. individuals are actually completely identical, but they can distinguish a particular role that they currently occupy, such as the owner of a territory and an intruder. This is what is termed an uncorrelated asymmetry. If individuals can base their chosen strategy on role, in the simplest cases mixed strategy ESSs become impossible, as we saw in Chapter 8. This occurs for bimatrix games, the two-role equivalent of matrix games, which are both linear on the left and linear on the right, but in general, it also applies for a wider class of games. The classic example of this is the Owner-Intruder game, a generalisation of the Hawk-Dove game where there are four potential strategies (play either Hawk or Dove in either of the two roles Owner and Intruder). We note that we can recover mixed strategy solutions in other asymmetric games, for instance in a slightly more realistic model of the Owner-Intruder game, where the role occupied is not independent of strategy.

An interesting result from the Owner-Intruder game is that asymmetry, including the common occurrence of intruders backing down against owners in the real world, does not require any asymmetry other than role. This is not to say that other asymmetries are not important, just that they are not necessary. In a number of models we saw that asymmetry is important. Mating games that involve the interaction of males and females are an example, for instance the game of brood care and desertion that we saw in Chapter 8 (among other chapters). A game which involves not only males and females in different roles but also different types (in fact, an infinite number of types) of males, although only a single type of female, is the signalling game related to the handicap principle from Chapter 18.

Heterogeneity can occur within a population even if the individuals are not inherently different. In Chapter 11 we discussed the idea of the state of an individual, which may involve permanent differences like age or sex, but can also include temporary ones like its level of hunger. Hungry individuals may be more willing to take chances than those that are not hungry, for example. Similarly, small juvenile individuals may need to take more risks than the adults they will hopefully grow into. There may also be a structure to the population, so even if individuals do not differ, the interaction rates of certain pairs of individual do. Such structure can significantly affect how evolution works, as we saw in Chapters 12 and 13. In fact, even evolutionary dynamics which are essentially the same for well-mixed populations can be radically different for a structured population, as we saw when we compared a number of dynamics which were variants of the Moran process for finite populations.

23.1.4 More complex interactions and other mathematical complications

In the sections above we discussed games where even if there might be some structure or difference between the individuals, every pair of individuals has a single interaction with a single decision (although in multi-player games this might be more complicated as we mention below). Games can occur where each individual has more than one choice to make, in a potential complex exchange. These can be modelled by extensive form games as we saw in Chapter 10 (which are similar in form to multi-player structures such as knockout contests). Complex populations like human ones involve repeated interactions, and as we saw in Chapter 15 the possibility of meeting an individual again, such as in an Iterated Prisoner's Dilemma, can have a big impact on what is the best strategy to play. Even if interactions do not occur between the same pair of individuals, social mechanisms like punishment or the use of reputation can generate cooperation when it would otherwise not occur (as can structured populations which allow cooperators to cluster).

A second way that cooperation can emerge is through cooperation between relatives, using inclusive fitness, as we saw in Chapter 15 (and elsewhere). Here the fitness of an individual includes a component for all of its relatives, as they share genetic material with it. In the simplest models, whether altruism can occur or not is governed by the classic Hamilton's rule, which relates the value of the degree of relatedness between the individuals and the benefit obtained and the cost borne by recipient and performer of the action respectively. We saw, however, a number of types of altruism.

Games in real populations of course contain errors, and we have seen the effect of the possibility of error in a number of places in the book. We have also seen the concept of the trembling hand in Chapter 3 where the possibility of error means that a good solution to a game must be robust to such errors. For example, the occurrence of errors makes the strategy Tit for Tat in the Iterated

Prisoner's Dilemma far less effective, and error has an effect on essentially all of the models of the evolution of cooperation.

We have also seen games involving a lack of information to some or all participants. With the disturbed games from Chapter 10, individuals are not fully aware of their opponent's payoff. Alternatively, individuals may not be aware of what choices their opponent has made, so that the position on the game tree is not known (imperfect information). It may be the case that one individual has more information than the other, for instance the finder of a food patch may know its value, but a challenger may not, as we saw in Chapter 19 (although this can paradoxically sometimes benefit the individual with the least information).

In general, we have looked for Evolutionarily Stable Strategies (ESSs) as the solutions to our games. We note that for very complex games it may be hard to show whether a strategy is an ESS, for example as in the multi-player knockout contests from Chapter 9, although often it is much easier to say whether a strategy satisfies the weaker condition of being a Nash equilibrium. We note that Nash equilibria that are not ESSs can still have important properties, e.g. may still be an attractor of the replicator dynamics. However, they may not be attracting in any sense, for example in two-player coordination games there is a mixed equilibrium which signals the boundary of the basin of attractions of the two pure ESSs.

Finally, we should mention an assumption about the type of games that we have allowed ourselves to consider, that has been running through the book. We assume that our games represent interactions from nature, and that underlying any payoffs there is some uncertainly which means that neat functional relationships between payoffs (non-generic games) are not allowed. One way to tell whether a game is a non-generic game is to perturb it by adding a different infinitesimal perturbation to each payoff/parameter. If this changes the character of the solution (rather than just slightly altering some vectors), then the game is non-generic. In fact, it is natural for people to choose simple payoffs when modelling, so we should be vigilant about this problem, although formally non-generic games are often used, which is fine as long as they pass this perturbation test.

23.1.5 Some biological issues

In Chapter 5 we looked at the underlying biology and a number of issues that complicated the idealised view of the games we have used elsewhere. An individual's fitness or payoff, for us usually a simple numerical measure, is actually a complex notion, and in particular, it may be that individual animals are not the right units of selection to consider at all. Selection at the level of the gene is important, as genes do not always simply prosper only if they enhance the fitness of their individual. There are, in fact, a number of levels at which selection can occur, including the cell and the group, and the idea of multi-level selection is growing in influence. We shall discuss this a little more in Section 23.3 below. Other modes of evolution, including cultural evolution and the spread of ideas, have also been considered.

For pairwise games between individuals, genetics can also play a role in determining which strategies are played. Genes can restrict the potential strategies that can be played (the repertoire), and the outcomes are influenced if, for example, a particular strategy is associated with a recessive allele. Similarly, if genes are linked to strategic choices made by a single sex, such as the sex ratio game where genes in the mother only can be thought to encode the sex ratio, evolution can be significantly complicated. The so-called "streetcar theory" maintains that problems with genetics may not be a fundamental objection to our idealised game-theoretical models. Genetics, however, is still a factor of importance, whose influence in evolutionary games is still rather under-explored.

A common assumption of most evolutionary game models is the separation of timescales. In real biological populations, behavioural changes usually occur more rapidly than population dynamical ones, and evolutionary changes are slower still. Models typically assume equilibrium is reached in the quicker timescale while effectively no change happens on the other timescales at all. As we mentioned in Chapter 3, when these processes occur on the same timescale, analysis is more complicated, and results can be different. A related problem occurs when considering the interactions between two populations using bimatrix games (see Chapter 8), where the dynamics effectively assumes fixed (relative) population sizes, but the fluctuations in the strategy frequencies lead, in turn, to fluctuations in mean payoffs which should logically affect the relative size of the populations of the two types.

23.1.6 Models of specific behaviours

We have considered a large number of models that relate to specific animal behaviours, and often the models are more detailed as a result. In most of these models the aim is to find the ESSs of the model, to interpret their biological meaning and to investigate the effect of changing key biological parameters. There is considerable variety in such models, as models are selected on the basis of the real problem, and these are very variable, with the common feature that there must be clearly recognisable players, strategies and payoffs for the games.

When modelling group behaviour, as we saw in Chapter 16, different forms of multi-player games are used, and the use of inclusive fitness and relatedness are common features. Another feature of such models is an asymmetry in the capacities of different individuals, whether real e.g. strength, or psychological e.g. due to a dominance relationship. Such situations also lend themselves to repeated games between individuals which continuously meet.

In Chapter 17 on mating games, we saw games with a sequential structure and also nonlinearity, in particular nonlinearity on the left as strategic choices were often a measure of some trait or quantity (sperm investment, signalling intensity, as considered in the handicap principle game from Section 18.2) rather than a probability. Here, roles are also important, as males and females must necessarily be distinct. This is particularly complicated for the handicap principle game, as although there is just a single class of female, there are an infinite number of male classes, which can all be thought of as different roles (games are also effectively sequential, as female strategies are responses to observed male signals).

The classical patch foraging game that we saw in Chapter 19 is a nice example of a game that is linear on the left, and although nonlinear on the right, has a relatively simple payoff function. The multiple species case effectively introduces a number of distinct roles. The kleptoparasitism games use compartmental models, where individuals move between different categories (effectively states) where behaviour is different, and there is a separation of timescales involving behavioural and evolutionary dynamics. Pairwise interactions are very simple but the nonlinear structure of the compartments make the models nevertheless quite complex. In this chapter we also saw examples of games with incomplete information.

We saw that compartmental models with a similar timescale separation as described above are also the basis of the models of Chapter 21, where individuals move between distinct disease states. Important game-theoretical models consider the competition between disease strains and the evolution of virulence, as well as the behaviour of a population under possible vaccination and its consequences for infection levels. We also saw a range of parasitic interactions in Chapter 20. This involves a game-theoretical version of the classical predator-prey model (again with a separation of timescales, this time between the population and evolutionary dynamics), and diverse games involving nonlinear payoffs, sequential games and in the case of the parasitic wasps model a nice application of a generalised asymmetric war of attrition, with effectively an infinite number of available roles.

Finally in Chapter 22 we saw some models of the development and treatment of cancer using game theory. In particular, this included models of adaptive therapy where cancer and its human host are considered as an evolving ecosystem, where potentially many variants of the cancer compete with each other. By understanding the different strains and the effects of different treatments, the physician can hopefully develop more sophisticated and effective treatment strategies. This young and exciting area of research has been developing fast.

23.2　What makes a good mathematical model?

When we observe a real population closely, it is clear that any accurate model would include a vast number of factors. For example, consider modelling a group of animals that are foraging, and at the same time must look out for predators as in Chapter 16. In addition to the factors already considered, every individual in the group will be different in inherent capabilities, in its immediate food requirements, and its position in the group relative to potential cover and the areas of greatest danger of attack. There may be a number of potential predators, but also other groups to be attacked. Important factors about flight capability, how the vision of predators and prey works and so on could also be considered.

If we included all such factors we would have a very complicated model, and we would be unlikely to be able to do much with it, at least mathematically. It is of course possible to simulate complex models with modern computers, as we mention below. Nevertheless, even when simulating it is best not to include everything that you can think of, but to be more selective about the components that are included.

Ideally, we would like our model to be one which involves only parameters which could be measured by biologists and so allow the model to be statistically testable, which contains all of the parameters that are both realistic and "important" in some sense, and which is mathematically soluble, or at least one where the mathematics gets us far enough to have a good insight. In most situations it is difficult or impossible to find a model which does all of this.

What makes a good model depends upon the situation that we are modelling, and what we are going to do with it. If we are going to try to fit it to real data, then it must include enough of the factors that are needed to explain the data so that the effects being tested are not dominated by random effects (including effects of other factors not explicitly included in the model). We want to be able to explain much of the differences in behaviour by corresponding differences in measurable parameters, and in what manner those parameters affect the differences. To be fittable, models generally have to be relatively simple. The alternative problem of overfitting, including too many parameters given the number of data points and predicting spurious influences of parameters, also needs to be avoided.

If we do not believe that we can fit our model to data, but are using it to show how some important biological effect can come about (as in many of the models in this book including the classical Hawk-Dove game), then we should include as many parameters as we need to account for what we wish to model, but try to leave everything else out. The simpler the model that can nevertheless demonstrate the effect, the clearer the explanation, the better the model. Thus we should include enough complexity to explain the central features, but no more.

Increasingly, models that involve a variety of different modelling skills will be required, and cooperation between biologists, mathematicians and computer scientists is important, in particular when modelling real complex situations. We have noticed this shift in focus strengthening since our first edition, and there is no reason to think that this will not continue. This is not to say that purely mathematical models will not continue to play a key role, and we have also seen many new important models of this type developed since the first edition, some of which we have included here. Insights found from quite simplified models analysed mathematically can tell us a lot. However, attacking problems from a number of directions is useful, and simulations, and more sophisticated computational models, can be complementary to mathematical modelling.

23.3   Future developments

In the first edition of our book we listed eight areas (theoretical areas, as opposed to areas of application) where we thought there was significant potential for future work. This was not close to an exhaustive list of possibilities. In the text that follows we expand on our original comments and add two more areas.

23.3.1   Agent-based modelling

We start with an area that we have entirely neglected in this book, agent-based modelling (ABM) and computer simulation more generally (see Railsback and Grimm, 2019). Simulation of complex biological systems and the application of agent-based modelling for evolutionary processes are valuable in investigating systems which are too complex mathematically. The work in this area has markedly increased since the first edition (see, for example, Gilbert, 2019; Izquierdo et al., 2019; Hindersin et al., 2019). As the computer capabilities increase, simulations and ABMs are applied to a wide range of increasingly complex and sophisticated applications not amenable to mathematical analysis (Bodine et al., 2020), as well as being a useful complement to theory in simpler cases (Gräbner et al., 2019).

Agent-based models, particularly with agents that can carry out complex realistic behaviour and learn, have great potential to model biological populations. At the same time we strongly believe that mathematics still has much to offer, and that the first response to any problem should not just be to resort to simulation without trying to use mathematical modelling. As pointed out in Adami et al. (2016b), mathematics and agent-based simulations should be used side-by-side by anyone who is interested in making progress in the fascinating field of evolutionary game theory (see also Adami et al., 2016a; Hilbe and Traulsen, 2016; Tarnita, 2016). The mathematical underpinning of agent-based modelling still needs to be properly formalised, however; for a significant consideration of this issue, see Evans (2015).

23.3.2 Multi-level selection

Evolutionary models typically operate with a single fixed level of selection, most commonly the individual but also the gene, or more controversially even the group. Recently, the idea of multi-level selection, where evolution acts on multiple levels has been suggested, see Chapter 15, and there is growing interest in this idea. The interaction between such different levels will provide an interesting mathematical challenge. This is still an area that requires significant development; see Hertler et al. (2020) and also Abs et al. (2020) for a discussion of these ideas.

The evolutionary modelling of cancer at the level of the cell as well as the patient has been discussed in our newly introduced Chapter 22. This is still a rapidly developing area with a lot of potential for further progress (Wölfl et al., in press).

23.3.3 Unifying timescales

Just as most evolutionary models focus on a single unit of selection, most models focus on a single process (e.g. evolutionary dynamics or population dynamics), or consider distinct timescales for different processes, so that for example population dynamics are assumed to be in equilibrium when considering strategic changes on an evolutionary timescale. This is of course a simplification, and just as evolution can occur at many levels which interact, the different processes, albeit operating at different rates, may have a more dynamic interaction. This has the potential for more static theory, but in particular for the development of dynamical systems. This area is also relatively unexplored. Timescale separation is still the norm for theoretical evolutionary models, as this often is a good approximation to reality that greatly simplifies analysis. An increasing number of models do attempt to consider a common timescale, and we have discussed this in Section 3.1.4. Further evolution can occur quickly when considering infectious diseases (Chapter 21) and cancer (Chapter 22), and eco-evolutionary models with no separation of timescales are used in these scenarios as discussed in Section 22.1. For a recent discussion of this subject, see Sigmund and Holt (2021).

23.3.4 Games in structured populations

This is an area that has already received a lot of interest, principally in the form of evolutionary games on graphs as discussed in Chapters 12 and 13. Evolutionary graph theory gives a nice way for structure to be modelled, where each pair of individuals may have very different chances of interacting (many pairs may have no chance of direct interaction). This led to more flexible, general models of population structure which can incorporate more important features such as multi-player games, and moving populations, as we discussed in Chapter 13. This area is developing quite quickly, but there is still much potential for development, for example in calculating the fixation probabilities (McAvoy and Allen, 2021), or absorption times (Allen et al., 2021; Monk and van Schaik, 2021) without the assumption of weak selection.

23.3.5 Nonlinear games

The dynamic and static theory of general non-linear games is discussed for example in Sandholm (2015) and our Chapter 7. There is significant potential to develop a more generalised theory of nonlinear games. There are many types of games that can be classified, and relatively little progress has been made in the general theory. It would be of particular interest to see the extension of the replicator dynamics to more general and complicated games that are nonlinear on the right, such as multi-player matrix games. A field that has received much recent attention, and one of the most significant developments of recent years, is the theory of adaptive dynamics, and there is still plenty of potential for further work in this area, as well as the theory of games which are nonlinear on the left more generally.

23.3.6 Asymmetries in populations

A number of key models in game theory make use of asymmetries between individuals to show why some important behaviours occur. However, understandably (and indeed as we would recommend, see Section 23.2) such asymmetries are generally missing from mathematical models unless they have a particular role to play. Age-structured populations, in particular, are of course widespread, and the development of a more game-theoretic theory of life histories (as we briefly looked at in Chapter 11) would be of considerable value. In general, a more systematic look at the fundamental asymmetries in biology from the perspective of evolutionary games would be useful.

23.3.7 What is a payoff?

Related to the above, if we consider the life history of every individual, what constitutes its payoff will be some combination of all of the rewards that it achieves throughout its life (early rewards might be more or less valuable, depending on whether the population is growing or shrinking, for instance). This may be hard to define. At different snapshots of an individual's life, what its payoff is may be very different. A simple cost and benefits analysis that is commonly employed in evolutionary game theory (including in this book) is often not appropriate for more realistic situations. In general, a more complex idea of fitness depending upon state, as we discuss in Chapter 11, has already been developed, but wider application and integration into game-theoretical modelling would be beneficial. This issue is still outstanding for more general game theoretical models, though a good approach to considering fitness is that from adaptive dynamics, see Chapter 14, see especially Metz et al. (1992), and this is also briefly discussed in McNamara and Leimar (2020).

23.3.8 A more unified approach to model applications

Every real behaviour has its own key features, and so the best way to model it will be different for different behaviours. Thus there is a wide range of model types, assumptions made and so on, and this is only to be expected, and encouraged. However, it often seems that models are invented without any regard for what already exists. The myriad definitions of different types of stability that are used is one example, and the almost random use of parameter terms is another (we plead guilty to this as much as anyone). Cataloguing the different modelling approaches, which are the most suitable, and coming up with a more unified approach would be a huge task, but of real benefit if done effectively. This is related to the more general issue of model realism discussed below in Section 23.3.9. As far as we know, not much progress has been made in this area, although, for example, Allen et al. (2017) and Allen and McAvoy (2019) develop a unified approach to modelling general population structures.

23.3.9 A more integrated understanding of the role of natural selection

Evolutionary game theory often takes a "functional approach", where a specific problem is identified first and then appropriate evolutionarily stable strategies are found (addressing the first of the four questions identified below). This book is full of such models. Models thus ignore potentially important features, often for reasons of practicality. McNamara (in press) (see also McNamara and Leimar, 2020) proposes a new approach, considering behaviours according to four questions originally given in Tinbergen (1963)(see also Bateson and Laland, 2013).

1. What is it for (how does the behaviour increase fitness)?

2. How did it evolve (how does its phylogenetic history affect the behaviour)?

3. How does it work (what mechanisms implement the behaviour)?

4. How did it develop (how did the behaviour emerge initially)?

We await with interest developments from this project.

23.3.10 Integrating player and strategy evolution into evolutionary dynamics

In most evolutionary dynamics models, such as the replicator dynamics, the composition of the population (player types, available strategies) and its environment remain the same, and only the frequencies change. In real populations, environments change, new strategies appear, the population size changes. How can we incorporate this into an appropriate dynamics?

The above is a fundamentally dynamic notion, and so with this book's emphasis on static methodology, we have not discussed this in too much detail. This is related to the question of unification of timescales, as discussed in Section 23.3.3. Strategy evolution and speciation (Gavrilets, 2014) also occur in some of the models we have considered, especially in Adaptive Dynamics (see Chapter 14) and the G-function approach from Section 20.1, though adaptive dynamics, in particular, typically involves change on a slow timescale with timescale separation. What is the appropriate way to model a generalised evolutionary dynamics capable of incorporating all relevant features?

Appendix A

Python

Python is an interpreted, object-oriented, high-level programming language with dynamic semantics. It has a simple, easy to learn syntax that emphasizes readability and reduces the cost of program maintenance. It can be downloaded for free from `https://www.python.org/`. We refer the reader to Linge and Langtangen (2020) for an introduction to Python, including more detailed advice on installation and basic programming.

While not originally designed for numerical computing, it has been attracting the attention of the scientific and engineering community. As a result, many libraries have been developed, including

- NumPy (`https://numpy.org/doc/stable/user/index.html`) that provides support for large, multi-dimensional arrays and matrices, along with a large collection of high-level mathematical functions to operate on these arrays,

- SciPy (`https://www.scipy.org/`) that provides many efficient routines for numerical integration, interpolation, optimisation, linear algebra, and statistics,

- Matplotlib (`https://matplotlib.org/`) for creating static, animated, and interactive visualisations,

- SymPy (`https://www.sympy.org/`) for symbolic mathematics with the aim to become a full-featured computer algebra system,

- Pandas (`https://pandas.pydata.org/`) for data analysis and manipulation.

The power and strength of Python comes with its support of modules and packages, which encourages program modularity and code reuse. With the packages listed above, Python is comparable to matlab. The package networkx (`https://networkx.org/`) used in Chapter 13 is an effective tool for the creation, manipulation, and study of the structure, dynamics, and functions of graphs and complex networks. While not explicitly used in this book, we also recommend readers to explore libraries like

- DyPy (Nande et al., 2020) that can perform evolutionary simulations for any matrix form game for three common evolutionary dynamics: Moran, Wright-Fisher and Replicator, or

DOI: 10.1201/9781003024682-A

- Gambit (`https://gambitproject.readthedocs.io/en/latest/index.html`) that provides tools for the construction and analysis of finite extensive and strategic games.

Another advantage of Python is the existence of a large and active community of programmers. This makes learning Python easy with plenty of resources such as

- `https://www.w3schools.com/python/default.asp`

- `https://docs.python.org/3/tutorial/`

- `https://www.python-course.eu/index.php`

readily available. There are also many online forums with most of the standard questions already answered.

As a final remark, we note that the scripts in this book were written in Python 2.7.17 and then slightly modified and tested on Python 3.8.0.

Bibliography

Abakuks, A. (1980). Conditions for evolutionarily stable strategies. *Journal of Applied Probability*, 17:559–562.

Abbot, P., Abe, J., Alcock, J., Alizon, S., Alpedrinha, J., Andersson, M., Andre, J.-B., Van Baalen, M., Balloux, F., and Balshine, S. (2011). Inclusive fitness theory and eusociality. *Nature*, 471(7339):E1–E4.

Abrams, P. A., Cressman, R., and Křivan, V. (2007). The role of behavioral dynamics in determining the patch distributions of interacting species. *The American Naturalist*, 169(4):505–518.

Abrams, P. A., Matsuda, H., and Harada, Y. (1993). Evolutionarily unstable fitness maxima and stable fitness minima of continuous traits. *Evolutionary Ecology*, 7(5):465–487.

Abs, E., Leman, H., and Ferrière, R. (2020). A multi-scale eco-evolutionary model of cooperation reveals how microbial adaptation influences soil decomposition. *Communications Biology*, 3(1):1–13.

Acosta-Alonzo, C., Erovenko, I. V., Lancaster, A., Oh, H., Rychtář, J., and Taylor, D. (2020). High endemic levels of typhoid fever in rural areas of Ghana may stem from optimal voluntary vaccination behavior. *Proceedings of the Royal Society A: Mathematical, Physical and Engineering Sciences*, 476:20200354.

Adami, C., Schossau, J., and Hintze, A. (2016a). Evolutionary game theory using agent-based methods. *Physics of Life Reviews*, 19:1–26.

Adami, C., Schossau, J., and Hintze, A. (2016b). The reasonable effectiveness of agent-based simulations in evolutionary game theory. *Physics of Life Reviews*, 19:38–42.

Adlam, B., Chatterjee, K., and Nowak, M. A. (2015). Amplifiers of selection. *Proceedings of the Royal Society A: Mathematical, Physical and Engineering Sciences*, 471(2181):20150114.

Afshar, M. and Giraldeau, L.-A. (2014). A unified modelling approach for producer–scrounger games in complex ecological conditions. *Animal Behaviour*, 96:167–176.

Afshar, M., Hall, C. L., and Giraldeau, L.-A. (2015). Zebra finches scrounge more when patches vary in quality: experimental support of the linear operator learning rule. *Animal Behaviour*, 105:181–186.

Agusto, F. B., Erovenko, I. V., Fulk, A., Abu-Saymeh, Q., Romero-Alvarez, D. D., Ponce, J., Sindi, S., Ortega, O., Saint Onge, J. M., and Peterson, A. T. (2022). To isolate or not to isolate: The impact of changing behavior on COVID-19 transmission. *BMC Public Health* 22, 138 (2022). https://doi.org/10.1186/s12889-021-12275-6.

Aizouk, R. and Broom, M. (2022). Modelling conflicting individual preference: Target sequences and graph realization. *Discrete and Continuous Dynamical Systems B.* - Series B, doi: 10.3934/dcdsb.2022001

Aktipis, C. (2004). Know when to walk away: contingent movement and the evolution of cooperation. *Journal of Theoretical Biology*, 231(2):249–260.

Aktipis, C. A., Boddy, A. M., Jansen, G., Hibner, U., Hochberg, M. E., Maley, C. C., and Wilkinson, G. S. (2015). Cancer across the tree of life: cooperation and cheating in multicellularity. *Philosophical Transactions of the Royal Society B: Biological Sciences*, 370(1673):20140219.

Alam, M., Ida, Y., and Tanimoto, J. (2021). Abrupt epidemic outbreak could be well tackled by multiple pre-emptive provisions — A game approach considering structured and unstructured populations. *Chaos, Solitons & Fractals*, 143:110584.

Alboszta, J. and Miękisz, J. (2004). Stability of evolutionarily stable strategies in discrete replicator dynamics with time delay. *Journal of Theoretical Biology*, 231(2):175–179.

Ale, S. B. and Brown, J. S. (2007). The contingencies of group size and vigilance. *Evolutionary Ecology Research*, 9(8):1263–1276.

Allen, B., Lippner, G., Chen, Y.-T., Fotouhi, B., Momeni, N., Yau, S.-T., and Nowak, M. A. (2017). Evolutionary dynamics on any population structure. *Nature*, 544(7649):227–230.

Allen, B. and McAvoy, A. (2019). A mathematical formalism for natural selection with arbitrary spatial and genetic structure. *Journal of Mathematical Biology*, 78(4):1147–1210.

Allen, B., Sample, C., Jencks, R., Withers, J., Steinhagen, P., Brizuela, L., Kolodny, J., Parke, D., Lippner, G., and Dementieva, Y. A. (2020). Transient amplifiers of selection and reducers of fixation for death-Birth updating on graphs. *PLOS Computational Biology*, 16(1):e1007529.

Allen, B., Sample, C., Steinhagen, P., Shapiro, J., King, M., Hedspeth, T., and Goncalves, M. (2021). Fixation probabilities in graph-structured populations under weak selection. *PLOS Computational Biology*, 17(2):e1008695.

Allen, B., Traulsen, A., Tarnita, C. E., and Nowak, M. A. (2012). How mutation affects evolutionary games on graphs. *Journal of Theoretical Biology*, 299:97–105.

Alpern, S. (2010). Search games on trees with asymmetric travel times. *SIAM Journal on Control and Optimization*, 48:5547–5563.

Alpern, S., Fokkink, R., Timmer, M., and Casas, J. (2011). Ambush frequency should increase over time during optimal predator search for prey. *Journal of the Royal Society Interface*, 8(64):1665–1672.

Alpern, S. and Gal, S. (2003). *The Theory of Search Games and Rendezvous*, volume 55. Springer-Verlag, Berlin, Heidelberg.

Alpern, S., Gal, S., Lee, V., and Casas, J. (2019). A stochastic game model of searching predators and hiding prey. *Journal of the Royal Society Interface*, 16(153):20190087.

Altrock, P. M., Liu, L. L., and Michor, F. (2015). The mathematics of cancer: integrating quantitative models. *Nature Reviews Cancer*, 15(12):730–745.

Altrock, P. M., Traulsen, A., and Nowak, M. A. (2017). Evolutionary games on cycles with strong selection. *Physical Review E*, 95(2):022407.

Amaral, M. A., de Oliveira, M. M., and Javarone, M. A. (2021). An epidemiological model with voluntary quarantine strategies governed by evolutionary game dynamics. *Chaos, Solitons & Fractals*, 143:110616.

Ames, G. M., George, D. B., Hampson, C. P., Kanarek, A. R., McBee, C. D., Lockwood, D. R., Achter, J. D., and Webb, C. T. (2011). Using network properties to predict disease dynamics on human contact networks. *Proceedings of the Royal Society B: Biological Sciences*, 278(1724):3544–3550.

Anderson, A. R. and Chaplain, M. A. J. (1998). Continuous and discrete mathematical models of tumor-induced angiogenesis. *Bulletin of Mathematical Biology*, 60(5):857–899.

Angina, J., Bachhu, A., Talati, E., Talati, R., Rychtář, J., and Taylor, D. (2022). Game-theoretical model of the voluntary use of insect repellents to prevent Zika fever. *Dynamic Games and Applications*.

Antal, T., Redner, S., and Sood, V. (2006). Evolutionary dynamics on degree-heterogeneous graphs. *Physical Review Letters*, 96(18):188104.

Antal, T. and Scheuring, I. (2006). Fixation of strategies for an evolutionary game in finite populations. *Bulletin of Mathematical Biology*, 68(8):1923–1944.

Antonia, S., Mirza, N., Fricke, I., Chiappori, A., Thompson, P., Williams, N., Bepler, G., Simon, G., Janssen, W., and Lee, J.-H. (2006). Combination of p53 cancer vaccine with chemotherapy in patients with extensive stage small cell lung cancer. *Clinical Cancer Research*, 12(3):878–887.

Apaloo, J. (1997). Revisiting strategic models of evolution: The concept of neighborhood invader strategies. *Theoretical Population Biology*, 52(1):52–71.

Apaloo, J. (2006). Revisiting matrix games: The concept of neighborhood invader strategies. *Theoretical Population Biology*, 69(3):235–242.

Apaloo, J., Brown, J. S., and L., V. T. (2009). Evolutionary game theory: ESS, convergence stability, and NIS. *Evolutionary Ecology Reseach*, 11:489–515.

Appleby, M. C. (1983). The probability of linearity in hierarchies. *Animal Behaviour*, 31(2):600–608.

Archetti, M. (2009a). Cooperation as a volunteer's dilemma and the strategy of conflict in public goods games. *Journal of Evolutionary Biology*, 22(11):2192–2200.

Archetti, M. (2009b). The volunteer's dilemma and the optimal size of a social group. *Journal of Theoretical Biology*, 261(3):475–480.

Archetti, M. (2013a). Dynamics of growth factor production in monolayers of cancer cells and evolution of resistance to anticancer therapies. *Evolutionary Applications*, 6(8):1146–1159.

Archetti, M. (2013b). Evolutionary game theory of growth factor production: implications for tumour heterogeneity and resistance to therapies. *British Journal of Cancer*, 109(4):1056–1062.

Archetti, M. (2015). Heterogeneity and proliferation of invasive cancer subclones in game theory models of the Warburg effect. *Cell Proliferation*, 48(2):259–269.

Archetti, M. (2016). Cooperation among cancer cells as public goods games on Voronoi networks. *Journal of Theoretical Biology*, 396:191–203.

Archetti, M. (2018). How to analyze models of nonlinear public goods. *Games*, 9(2):17.

Archetti, M. (2019). Maintenance of variation in mutualism by screening. *Evolution*, 73(10):2036–2043.

Archetti, M. and Pienta, K. J. (2019). Cooperation among cancer cells: Applying game theory to cancer. *Nature Reviews Cancer*, 19(2):110–117.

Archetti, M. and Scheuring, I. (2012). Game theory of public goods in one-shot social dilemmas without assortment. *Journal of Theoretical Biology*, 299:9–20.

Archetti, M., Úbeda, F., Fudenberg, D., Green, J., Pierce, N., and Yu, D. (2011). Let the right one in: A microeconomic approach to partner choice in mutualisms. *The American Naturalist*, 177(1):75–85.

Arefin, M., Kabir, K. M. A., and Tanimoto, J. (2020). A mean-field vaccination game scheme to analyze the effect of a single vaccination strategy on a two-strain epidemic spreading. *Journal of Statistical Mechanics: Theory and Experiment*, 2020(3):033501.

Arefin, M., Masaki, T., Kabir, K. M. A., and Tanimoto, J. (2019). Interplay between cost and effectiveness in influenza vaccine uptake: a vaccination game approach. *Proceedings of the Royal Society A: Mathematical, Physical and Engineering Science*, 475(2232):20190608.

Argasinski, K. (2006). Dynamic multipopulation and density dependent evolutionary games related to replicator dynamics. A metasimplex concept. *Mathematical Biosciences*, 202(1):88–114.

Argasinski, K. (2012). The dynamics of sex ratio evolution: Dynamics of global population parameters. *Journal of Theoretical Biology*, 309:134–146.

Argasinski, K. (2013). The dynamics of sex ratio evolution: from the gene perspective to multilevel selection. *PLOS One*, 8(4):e60405.

Argasinski, K. (2018). The dynamics of sex ratio evolution: the impact of males as passive gene carriers on multilevel selection. *Dynamic Games and Applications*, 8(4):671–695.

Argasinski, K. and Broom, M. (2013a). Ecological theatre and the evolutionary game: how environmental and demographic factors determine payoffs in evolutionary games. *Journal of Mathematical Biology*, 67(4):935–962.

Argasinski, K. and Broom, M. (2013b). The nest site lottery: How selectively neutral density dependent growth suppression induces frequency dependent selection. *Theoretical Population Biology*, 90:82–90.

Argasinski, K. and Broom, M. (2018a). Evolutionary stability under limited population growth: Eco-evolutionary feedbacks and replicator dynamics. *Ecological Complexity*, 34:198–212.

Argasinski, K. and Broom, M. (2018b). Interaction rates, vital rates, background fitness and replicator dynamics: How to embed evolutionary game structure into realistic population dynamics. *Theory in Biosciences*, 137(1):33–50.

Argasinski, K. and Broom, M. (2021). Towards a replicator dynamics model of age structured populations. *Journal of Mathematical Biology*, 82(5):1–39.

Argasinski, K. and Kozlowski, J. (2008). How can we model selectively neutral density dependence in evolutionary games. *Theoretical Population Biology*, 73(2):250–256.

Arthur, W. B. (1994). Inductive reasoning and bounded rationality. *American Economic Review*, 84(2):406–411.

Axelrod, R. (1981). The emergences of cooperation among egoists. *The American Political Science Review*, 75:306–318.

Axelrod, R. (1984). *The Evolution of Cooperation*. Basic Books.

Axelrod, R., Axelrod, D. E., and Pienta, K. J. (2006). Evolution of cooperation among tumor cells. *Proceedings of the National Academy of Sciences*, 103(36):13474–13479.

Axelrod, R. and Hamilton, W. D. (1981). The evolution of cooperation. *Science*, 211:1390–1396.

Bacevic, K., Noble, R., Soffar, A., Ammar, O., Boszonyik, B., Prieto, S., Vincent, C., Hochberg, M., Krasinska, L., and Fisher, D. (2017). Spatial competition constrains resistance to targeted cancer therapy. *Nature Communications*, 8(1):1–15.

Bach, L. A., Helvik, T., and Christiansen, F. B. (2006). The evolution of *n*-player cooperation–threshold games and ESS bifurcations. *Journal of Theoretical Biology*, 238(2):426–434.

Baker, A.-M., Huang, W., Wang, X.-M. M., Jansen, M., Ma, X.-J., Kim, J., Anderson, C. M., Wu, X., Pan, L., and Su, N. (2017). Robust RNA-based in situ mutation detection delineates colorectal cancer subclonal evolution. *Nature Communications*, 8(1):1–8.

Bala, V. and Goyal, S. (2000). A noncooperative model of network formation. *Econometrica*, 68(5):1181–1229.

Ball, F. and House, T. (2017). Heterogeneous network epidemics: real-time growth, variance and extinction of infection. *Journal of Mathematical Biology*, 75(3):577–619.

Ball, M. A. and Parker, G. A. (1997). Sperm competition games: inter- and intra-species results of a continuous external fertilization model. *Journal of Theoretical Biology*, 186(4):459–466.

Ball, M. A. and Parker, G. A. (1998). Sperm competition games: a general approach to risk assessment. *Journal of Theoretical Biology*, 194(2):251–262.

Ball, M. A. and Parker, G. A. (2000). Sperm competition games: a comparison of loaded raffle models and their biological implications. *Journal of Theoretical Biology*, 206(4):487–506.

Ball, M. A. and Parker, G. A. (2007). Sperm competition games: the risk model can generate higher sperm allocation to virgin females. *Journal of Evolutionary Biology*, 20(2):767–779.

Ballerini, M., Cabibbo, N., Candelier, R., Cavagna, A., Cisbani, E., Giardina, I., Lecomte, V., Orlandi, A., Parisi, G., Procaccini, A., Viale, M., and Zdravkovic, V. (2008). Interaction ruling animal collective behavior depends on topological rather than metric distance: Evidence from a field study. *Proceedings of the National Academy of Sciences*, 105(4):1232–1237.

Bankuru, S., Kossol, S., Hou, W., Mahmoudi, P., Rychtář, J., and Taylor, D. (2020). A game-theoretic model of Monkeypox to assess vaccination strategies. *PeerJ*, 8:e9272.

Barabási, A. L. and Albert, R. (1999). Emergence of scaling in random networks. *Science*, 286(5439):509–512.

Baran, N. M. and Reeve, H. K. (2015). Coevolution of parental care, parasitic, and resistance efforts in facultative parasitism. *The American Naturalist*, 186(5):594–609.

Barker, H. A., Broom, M., and Rychtář, J. (2012). A game theoretic model of kleptoparasitism with strategic arrivals and departures of beetles at dung pats. *Journal of Theoretical Biology*, 300:292–298.

Barnard, C. J. and Sibly, R. M. (1981). Producers and scroungers: a general model and its application to captive flocks of house sparrows. *Animal Behaviour*, 29(2):543–550.

Barreto, W. P., Marquitti, F. M., and de Aguiar, M. A. (2017). A genetic approach to the rock-paper-scissors game. *Journal of Theoretical Biology*, 421:146–152.

Barton, N. H. (2000). Genetic hitchhiking. *Philosophical Transactions: Biological Sciences*, 355(1403):1553–1562.

Basanta, D., Gatenby, R. A., and Anderson, A. R. (2012). Exploiting evolution to treat drug resistance: combination therapy and the double bind. *Molecular Pharmaceutics*, 9(4):914–921.

Bassler, B. L. (1999). How bacteria talk to each other: regulation of gene expression by quorum sensing. *Current Opinion in Microbiology*, 2(6):582–587.

Basu, K. (2007). The traveler's dilemma. *Scientific American Magazine*, 296(6):90–95.

Bateson, P. and Laland, K. N. (2013). Tinbergen's four questions: An appreciation and an update. *Trends in Ecology & Evolution*, 28(12):712–718.

Bauch, C. T. and Bhattacharyya, S. (2012). Evolutionary game theory and social learning can determine how vaccine scares unfold. *PLOS Computational Biology*, 8(4):e1002452.

Bauch, C. T. and Earn, D. J. D. (2004). Vaccination and the theory of games. *Proceedings of the National Academy of Sciences*, 101(36):13391–13394.

Bauer, J., Broom, M., and Alonso, E. (2019). The stabilization of equilibria in evolutionary game dynamics through mutation: Mutation limits in evolutionary games. *Proceedings of the Royal Society A: Mathematical, Physical and Engineering Science*, 475(2231):20190355.

Baum, L. E. and Eagon, J. A. (1967). An inequality with applications to statistical estimation for probabilistic functions of Markov processes and to a model for ecology. *Bulletin of the American Mathematical Society*, 73(360-363):212.

Bautista, L. M., Alonso, J. C., and Alonso, J. A. (1995). A field test of ideal free distribution in flock-feeding common cranes. *Journal of Animal Ecology*, 64(6):747–757.

Beddington, J. R. (1975). Mutual interference between parasites or predators and its effect on searching efficiency. *The Journal of Animal Ecology*, 44(1):331–340.

Bellomo, N. and Ha, S.-Y. (2017). A quest toward a mathematical theory of the dynamics of swarms. *Mathematical Models and Methods in Applied Sciences*, 27(04):745–770.

Bender, H. (2010). Devil facial tumour disease (DFTD): Using genetics and genomics to investigate infectious disease in an endangered marsupial. In J. E. Deakin, P. D. Waters, J. A. Marshall Graves, editors, *Marsupial Genetics and Genomics*, pages 499–515. Springer.

Berec, L. (2010). Impacts of foraging facilitation among predators on predator-prey dynamics. *Bulletin of Mathematical Biology*, 72(1):94–121.

Bergstrom, C. T. and Lachmann, M. (1997). Signalling among relatives. I. Is costly signalling too costly? *Philosophical Transactions of the Royal Society of London. Series B: Biological Sciences*, 352(1353):609–617.

Bergstrom, C. T. and Lachmann, M. (1998). Signaling among relatives. III. Talk is cheap. *Proceedings of the National Academy of Sciences*, 95(9):5100–5105.

Bergstrom, C. T. and Real, L. A. (2000). Toward a theory of mutual mate choice: Lessons from two-sided matching. *Evolutionary Ecology*, 2:493–508.

Berryman, A. A., Gutierrez, A. P., and Arditi, R. (1995). Credible, parsimonious and useful predator-prey models: a reply to Abrams, Gleeson and Sarnelle. *Ecology*, 76(6):1980–1985.

Bhattacharyya, S. and Bauch, C. T. (2011). "Wait and see" vaccinating behaviour during a pandemic: A game theoretic analysis. *Vaccine*, 29(33):5519–5525.

Biernaskie, J. and Elle, E. (2007). A theory for exaggerated secondary sexual traits in animal-pollinated plants. *Evolutionary Ecology*, 21(4):459–472.

Biernaskie, J. M., Grafen, A., and Perry, J. C. (2014). The evolution of index signals to avoid the cost of dishonesty. *Proceedings of the Royal Society B: Biological Sciences*, 281(1790):20140876.

Binmore, K. and Samuelson, L. (2001). Can mixed strategies be stable in asymmetric games? *Journal of Theoretical Biology*, 210(1):1–14.

Birkhead, T. R. and Møller, A. P. (1992). *Sperm Competition in Birds. Evolutionary Causes and Consequences.* Academic Press, San Diego, USA.

Birkhead, T. R. and Møller, A. P. (1998). *Sperm Competition and Sexual Selection.* Academic Press.

Bishop, D. T., Broom, M., and Southwell, R. (2020). Chris Cannings: A life in games. *Dynamic Games and Applications*, 10(3):591–617.

Bishop, D. T. and Cannings, C. (1976). Models of animal conflict. *Advances in Applied Probability*, 8:616–621.

Bishop, D. T. and Cannings, C. (1978). A generalized war of attrition. *Journal of Theoretical Biology*, 70:85–124.

Bodine, E. N., Panoff, R. M., Voit, E. O., and Weisstein, A. E. (2020). Agent-based modeling and simulation in mathematics and biology education. *Bulletin of Mathematical Biology*, 82(8):1–19.

Bodnar, M., Miękisz, J., and Vardanyan, R. (2020). Three-player games with strategy-dependent time delays. *Dynamic Games and Applications*, 10(3):664—675.

Boerlijst, M. C., Nowak, M. A., and Sigmund, K. (1997). The logic of contrition. *Journal of Theoretical Biology*, 185(3):281–293.

Bolnick, D. and Doebeli, M. (2003). Sexual dimorphism and adaptive speciation: two sides of the same ecological coin. *Evolution*, 57(11):2433–2449.

Bomze, I. M. (1986). Non-cooperative two-person games in biology: A classification. *International Journal of Game Theory*, 15(1):31–57.

Bomze, I. M. (1990). Dynamical aspects of evolutionary stability. *Monatshefte für Mathematik*, 110:189–206.

Bomze, I. M. (1991). Cross entropy minimization in uninvadable states of complex populations. *Journal of Mathematical Biology*, 30(1):73–87.

Bomze, I. M. (1992). Detecting all evolutionarily stable strategies. *Journal of Optimization Theory and Applications*, 75(2):313–329.

Bomze, I. M. (2002). Regularity versus degeneracy in dynamics, games, and optimization: a unified approach to different aspects. *SIAM Review*, 44(3):394–414.

Bomze, I. M. and Pötscher, B. M. (1989). *Game Theoretical Foundations of Evolutionary Stability*. Springer-Verlag, Berlin, Heidelberg.

Bomze, I. M. and Schachinger, W. (2020). Constructing patterns of (many) ESSs under support size control. *Dynamic Games and Applications*, 10(3):618–640.

Bomze, I. M., Schachinger, W., and Ullrich, R. (2018). The complexity of simple models—a study of worst and typical hard cases for the standard quadratic optimization problem. *Mathematics of Operations Research*, 43(2):651–674.

Bonabeau, E., Theraulaz, G., and Deneubourg, J. L. (1999). Dominance orders in animal societies: the self-organization hypothesis revisited. *Bulletin of Mathematical Biology*, 61(4):727–757.

Borgia, G. (1985). Bower destruction and sexual competition in the satin bowerbird (*Ptilonorhynchus violaceus*). *Behavioral Ecology and Sociobiology*, 18(2):91–100.

Borries, C. and Koenig, A. (2000). Infanticide in hanuman langurs: social organization, male migration and weaning age. In van Schaik, C. P. and Janson, C. H., editors, *Infanticide by Males and Its Implications*, pages 99–122. Cambridge University Press, Cambridge, UK.

Bouskila, A. (2001). A habitat selection game of interactions between rodents and their predators. *Annales Zoologici Fennici*, 38:55–70.

Boyd, R. (1989). Mistakes allow evolutionary stability in the repeated prisoner's dilemma game. *Journal of Theoretical Biology*, 136(1):47–56.

Boyd, R. and Lorberbaum, J. P. (1987). No pure strategy is evolutionarily stable in the repeated prisoner's dilemma game. *Nature*, 327:58–59.

Boyd, R. and Richerson, P. (1988). *Culture and the Evolutionary Process*. University of Chicago Press.

Boyd, R. and Richerson, P. J. (2002). Group beneficial norms can spread rapidly in a structured population. *Journal of Theoretical Biology*, 215(3):287–296.

Brännström, Å., Gross, T., Blasius, B., and Dieckmann, U. (2010). Consequences of fluctuating group size for the evolution of cooperation. *Journal of Mathematical Biology*, 63(2):1–19.

Brettin, A., Rossi-Goldthorpe, R., Weishaar, K., and Erovenko, I. V. (2018). Ebola could be eradicated through voluntary vaccination. *Royal Society Open Science*, 5(1):171591.

Brockmann, H. J. and Barnard, C. J. (1979). Kleptoparasitism in birds. *Animal Behaviour*, 27:487–514.

Broom, M. (2000a). Bounds on the number of ESSs of a matrix game. *Mathematical Biosciences*, 167:163–175.

Broom, M. (2000b). Patterns of ESSs: the maximal pattern conjecture revisited. *Journal of Mathematical Biology*, 40:406–412.

Broom, M., Borries, C., and Koenig, A. (2004a). Infanticide and infant defence by males–modelling the conditions in primate multi-male groups. *Journal of Theoretical Biology*, 231(2):261–270.

Broom, M. and Cannings, C. (2002). Modelling dominance hierarchy formation as a multi-player game. *Journal of Theoretical Biology*, 219(3):397–413.

Broom, M. and Cannings, C. (2013). A dynamic network population model with strategic link formation governed by individual preferences. *Journal of Theoretical Biology*, 335:160–168.

Broom, M. and Cannings, C. (2015). Graphic deviation. *Discrete Mathematics*, 338:701–711.

Broom, M. and Cannings, C. (2017). Game theoretical modelling of a dynamically evolving network I: general target sequences. *Journal of Dynamics and Games*, 4:285–318.

Broom, M., Cannings, C., and Vickers, G. T. (1993). On the number of local maxima of a constrained quadratic form. *Proceedings of the Royal Society of London A*, 443:573–584.

Broom, M., Cannings, C., and Vickers, G. T. (1994). Sequential methods for generating patterns of ESS's. *Journal of Mathematical Biology*, 32:597–615.

Broom, M., Cannings, C., and Vickers, G. T. (1996). Choosing a nest site: Contests and catalysts *American Naturalist*, 147(6):1108–1114.

Broom, M., Cannings, C., and Vickers, G. T. (1997a). A sequential-arrivals model of territory acquisition. *Journal of Theoretical Biology*, 189(3):257–272.

Broom, M., Cannings, C., and Vickers, G. T. (1997b). Multi-player matrix games. *Bulletin of Mathematical Biology*, 59(5):931–952.

Broom, M., Cannings, C., and Vickers, G. T. (2000a). Evolution in knock-out conflicts: The fixed strategy case. *Bulletin of Mathematical Biology*, 62(3):451–466.

Broom, M., Cannings, C., and Vickers, G. T. (2000b). A sequential-arrivals model of territory acquisition II. *Journal of Theoretical Biology*, 207(3):389–403.

Broom, M., Cannings, M., and Vickers, G. T. (2000c). Evolution in knockout contests: the variable strategy case. *Selection*, 1(1):5–22.

Broom, M., Cressman, R., and Křivan, V. (2019a). Revisiting the "fallacy of averages" in ecology: expected gain per unit time equals expected gain divided by expected time. *Journal of Theoretical Biology*, 483:109993.

Broom, M., Crowe, M. L., Fitzgerald, M. R., and Rychtář, J. (2010a). The stochastic modelling of kleptoparasitism using a Markov process. *Journal of Theoretical Biology*, 264(2):266–272.

Broom, M., Erovenko, I. V., Rowell, J. T., and Rychtář, J. (2020). Models and measures of animal aggregation and dispersal. *Journal of Theoretical Biology*, 484:110002.

Broom, M., Erovenko, I. V., and Rychtář, J. (2021). Modelling evolution in structured populations involving multiplayer interactions. *Dynamic Games and Applications*, 11(2):270–293.

Broom, M., Hadjichrysanthou, C., and Rychtář, J. (2010b). Evolutionary games on graphs and the speed of the evolutionary process. *Proceedings of the Royal Society A: Mathematical, Physical and Engineering Science*, 466(2117):1327–1346.

Broom, M., Hadjichrysanthou, C., Rychtář, J., and Stadler, B. T. (2010c). Two results on evolutionary processes on general non-directed graphs. *Proceedings of the Royal Society A: Mathematical, Physical and Engineering Science*, 466(2121):2795–2798.

Broom, M., Johanis, M., and Rychtář, J. (2015a). The effect of fight cost structure on fighting behaviour. *Journal of Mathematical Biology*, 71(4):979–996.

Broom, M., Johanis, M., and Rychtář, J. (2018). The effect of fight cost structure on fighting behaviour involving simultaneous decisions and variable investment levels. *Journal of Mathematical Biology*, 76(1):457–482.

Broom, M., Koenig, A., and Borries, C. (2009). Variation in dominance hierarchies among group-living animals: modeling stability and the likelihood of coalitions. *Behavioral Ecology*, 20(4):844–855.

Broom, M. and Křivan, V. (2020). Two-strategy games with time constraints on regular graphs. *Journal of Theoretical Biology*, 506:110426.

Broom, M., Lafaye, C., Pattni, K., and Rychtář, J. (2015b). A study of the dynamics of multi-player games on small networks using territorial interactions. *Journal of Mathematical Biology*, 71:1551–1574.

Broom, M., Luther, R. M., and Ruxton, G. D. (2004b). Resistance is useless? - extensions to the game theory of kleptoparasitism. *Bulletin of Mathematical Biology*, 66(6):1645–1658.

Broom, M., Luther, R. M., Ruxton, G. D., and Rychtář, J. (2008a). A game-theoretic model of kleptoparasitic behavior in polymorphic populations. *Journal of Theoretical Biology*, 255(1):81–91.

Broom, M., Pattni, K., and Rychtář, J. . (2019b). Generalized social dilemmas: The evolution of cooperation in populations with variable group size. *Bulletin of Mathematical Biology*, 81(11):4643–4674.

Broom, M. and Ruxton, G. D. (1998a). Evolutionarily stable stealing: game theory applied to kleptoparasitism. *Behavioral Ecology*, 9(4):397–403.

Broom, M. and Ruxton, G. D. (1998b). Modelling responses in vigilance rates to arrivals to and departures from a group of foragers. *Mathematical Medicine and Biology*, 15(4):387–400.

Broom, M. and Ruxton, G. D. (2003). Evolutionarily stable kleptoparasitism: consequences of different prey types. *Behavioral Ecology*, 14(1):23.

Broom, M. and Ruxton, G. D. (2005). You can run or you can hide: optimal strategies for cryptic prey against pursuit predators. *Behavioral Ecology*, 16(3):534–540.

Broom, M. and Ruxton, G. D. (2011). Some mistakes go unpunished: the evolution of "all or nothing" signalling. *Evolution*, 65(10):2743–2749.

Broom, M., Ruxton, G. D., and Speed, M. P. (2008b). Evolutionarily stable investment in anti-predatory defences and aposematic signalling. In Deutsch, A., Bravo de la Parra, R., de Boer, R., Diekmann, O., Jagers, P., Kisdi, É., Kretzschmar, M., Lansky, P., and Metz, H., editors, *In Mathematical Modeling of Biological Systems*, Volume II, pages 37–48. Springer, Birkhauser, Boston.

Broom, M. and Rychtář, J. (2007). The evolution of a kleptoparasitic system under adaptive dynamics. *Journal of Mathematical Biology*, 54(2).151 177.

Broom, M. and Rychtář, J. (2008). An analysis of the fixation probability of a mutant on special classes of non-directed graphs. *Proceedings of the Royal Society A: Mathematical, Physical and Engineering Science*, 464(2098):2609–2627.

Broom, M. and Rychtář, J. (2009). A game theoretical model of kleptoparasitism with incomplete information. *Journal of Mathematical Biology*, 59(5):631–649.

Broom, M. and Rychtář, J. (2011). Kleptoparasitic melees - modelling food stealing featuring contests with multiple individuals. *Bulletin of Mathematical Biology*, 73(3):683–699.

Broom, M. and Rychtář, J. (2012). A general framework for analyzing multiplayer games in networks using territorial interactions as a case study. *Journal of Theoretical Biology*, 302:70–80.

Broom, M. and Rychtář, J. (2014). Asymmetric games in monomorphic and polymorphic populations. *Dynamic Games and Applications*, 4(4):391–406.

Broom, M. and Rychtář, J. (2018a). Evolutionary games with sequential decisions and dollar auctions. *Dynamic Games and Applications*, 8(2):211–231.

Broom, M. and Rychtář, J. (2018b). Ideal cost-free distributions in structured populations for general payoff functions. *Dynamic Games and Applications*, 8:79–92.

Broom, M., Rychtář, J., and Stadler, B. T. (2011). Evolutionary dynamics on graphs - the effect of graph structure and initial placement on mutant spread. *Journal of Statistical Theory and Practice*, 5(3):369–381.

Broom, M., Rychtář, J., and Spears-Gill, T. (2016). The game-theoretical model of using insecticide-treated bed-nets to fight malaria. *Applied Mathematics*, 7(09):852–860.

Broom, M., Speed, M. P., and Ruxton, G. D. (2006). Evolutionarily stable defence and signalling of that defence. *Journal of Theoretical Biology*, 242(1):32–43.

Brown, J. S. (1999). Vigilance, patch use and habitat selection: foraging under predation risk. *Evolutionary Ecology Research*, 1(1):49–71.

Brown, J. S. (2016). Why Darwin would have loved evolutionary game theory. *Proceedings of the Royal Society B: Biological Sciences*, 283(1838):20160847.

Brown, J. S. and Vincent, T. L. (1987). A theory for the evolutionary game. *Theoretical Population Biology*, 31(1):140–166.

Brown, J. S. and Vincent, T. L. (1992). Organization of predator-prey communities as an evolutionary game. *Evolution*, 46(5):1269–1283.

Bukkuri, A. and Brown, J. S. (2021). Evolutionary game theory: Darwinian dynamics and the *g* function approach. *Games*, 12(4):72.

Bukowski, M. and Miękisz, J. (2004). Evolutionary and asymptotic stability in symmetric multi-player games. *International Journal of Game Theory*, 33(1):41–54.

Bulmer, M. (1994). *Theoretical Evolutionary Ecology*. Sinauer Associates Sunderland.

Burgess, J. W. (1976). Social spiders. *Scientific American*, 234(3):100–106.

Buss, L. W. (1987). *The Evolution of Individuality*. Princeton University Press, Princeton, NJ, USA.

Cannings, C. (1990). Topics in the theory of ESS's. In Lessard, S., editor, *Mathematical and Statistical Developments of Evolutionary Theory*, Lecture Notes in Mathematics, pages 95–119. Kluwer Academic Publishers.

Cannings, C. (2009). The majority game on regular and random networks. In *2009 International Conference on Game Theory for Networks*, pages xii–xiii. IEEE.

Cannings, C. (2015). Combinatorial aspects of Parker's model. *Dynamic Games and Applications*, 5(2):263–274.

Cannings, C. and Broom, M. (2020). Game theoretical modelling of a dynamically evolving network II: Target sequences of score 1. *Journal of Dynamics and Games*, 7(1):37–64.

Cannings, C., Tyrer, J. P., and Vickers, G. T. (1993). Routes to polymorphism. *Journal of Theoretical Biology*, 165(2):213–223.

Cannings, C. and Vickers, G. T. (1988). Patterns of ESS's II. *Journal of Theoretical Biology*, 132:409–420.

Cannings, C. and Vickers, G. T. (1990). Patterns and invasions of evolutionarily stable strategies. *Journal of Applied Mathematics and Computation*, 32:227–253.

Cannings, C. and Vickers, G. T. (1991). The Genealogy of Patterns of ESS's. In Basawa, I. V. and Taylor, R. L., editors, *Selected Proceedings of the Sheffield Symposium on Applied Probability: Held at the University of Sheffield, Sheffield, August 16–19, 1989*, pages 193–204. Institute of Mathematical Statistics, Hayward, CA.

Cannings, C. and Whittaker, J. C. (1995). The finite horizon war of attrition. *Games and Economic Behavior*, 11(2):193–236.

Cant, M. A., Nichols, H. J., Johnstone, R. A., and Hodge, S. J. (2014). Policing of reproduction by hidden threats in a cooperative mammal. *Proceedings of the National Academy of Sciences*, 111(1):326–330.

Caraco, T. and Giraldeau, L. A. (1991). Social foraging: Producing and scrounging in a stochastic environment. *Journal of Theoretical Biology*, 153(4):559–583.

Cardillo, A., Meloni, S., Gómez-Gardenes, J., and Moreno, Y. (2012). Velocity-enhanced cooperation of moving agents playing public goods games. *Physical Review E*, 85(6):067101.

Caswell, H. (2000). *Matrix Population Models*. Wiley Online Library.

Cazaubiel, A., Lütz, A. F., and Arenzon, J. J. (2017). Collective strategies and cyclic dominance in asymmetric predator-prey spatial games. *Journal of Theoretical Biology*, 430:45–52.

Chalub, F. and Rodrigues, J. F. (2011). *The Mathematics of Darwin's Legacy*. Birkhauser.

Chalub, F. A. (2016). Asymptotic expression for the fixation probability of a mutant in star graphs. *Journal of Dynamics and Games*, 3:217–223.

Chamberland, M. and Cressman, R. (2000). An example of dynamic (in)consistency in symmetric extensive form evolutionary games. *Games and Economic Behavior*, 30(2):319–326.

Chang, S., Piraveenan, M., Pattison, P., and Prokopenko, M. (2020). Game theoretic modelling of infectious disease dynamics and intervention methods: a review. *Journal of Biological Dynamics*, 14(1):57–89.

Chapman, T., Arnqvist, G., Bangham, J., and Rowe, L. (2003). Sexual conflict. *Trends in Ecology & Evolution*, 18(1):41–47.

Chappell, J. M., Iqbal, A., and Abbott, D. (2012). N-player quantum games in an EPR setting. *PLOS One*, 7(5):e36404.

Charlesworth, B. (2007). A hitch-hiking guide to the genome: a commentary on "The hitch-hiking effect of a favourable gene" by John Maynard Smith and John Haigh. *Genetics Research*, 89(5-6):389–390.

Charnov, E. L. (1976). Optimal foraging, the marginal value theorem. *Theoretical Population Biology*, 9(2):129–136.

Charnov, E. L. (1982). *Sex Allocation*. Princeton University Press, Princeton, NJ, USA.

Chase, I. D. (1980). Social process and hierarchy formation in small groups: a comparative perspective. *American Sociological Review*, 45(6):905–924.

Chase, I. D., Bartolomeo, C., and Dugatkin, L. A. (1994). Aggressive interactions and inter-contest interval: how long do winners keep winning? *Animal Behaviour*, 48(2):393–400.

Chase, I. D. and Rohwer, S. (1987). Two methods for quantifying the development of dominance hierarchies in large groups with applications to harris' sparrows. *Animal Behaviour*, 35(4):1113–1128.

Chase, I. D., Tovey, C., Spangler-Martin, D., and Manfredonia, W. (2002). Individual differences versus social dynamics in the formation of animal dominance hierarchies. *Proceedings of the National Academy of Sciences*, 99(8):5744–5749.

Cheng, E., Gambhirrao, N., Patel, R., Zhowandai, A., Rychtář, J., and Taylor, D. (2020). A game-theoretical analysis of Poliomyelitis vaccination. *Journal of Theoretical Biology*, 499:110298.

Choi, W. and Shim, E. (2020). Optimal strategies for vaccination and social distancing in a game-theoretic epidemiological model. *Journal of Theoretical Biology*, 505:110422.

Chopard, B. and Droz, M. (1998). *Cellular Automata Modeling of Physical Systems*. Cambridge University Press, Cambridge, UK.

Chouhan, A., Maiwand, S., Ngo, M., Putalapattu, V., Rychtář, J., and Taylor, D. (2020). Game-theoretical model of retroactive hepatitis B vaccination in China. *Bulletin of Mathematical Biology*, 82:80.

Chowdhury, N., Kentiba, K., Mirajkar, Y., Nasseri, M., Rychtář, J., and Taylor, D. (2020). Kleptoparasitic interactions modeling varying owner and intruder hunger awareness. *Theoretical Population Biology*, 136:31–40.

Chubb, J. C., Ball, M. A., and Parker, G. A. (2010). Living in intermediate hosts: evolutionary adaptations in larval helminths. *Trends in Parasitology*, 26(2):93–102.

Cloak, F. T. (1975). Is a cultural ethology possible? *Human Ecology*, 3(3):161–182.

Clutton-Brock, T. H. and Albon, S. D. (1979). The roaring of red deer and the evolution of honest advertising. *Behaviour*, 69:145–170.

Cohen, Y., Vincent, T. L., and Brown, J. S. (1999). A *G*-function approach to fitness minima, fitness maxima, evolutionarily stable strategies and adaptive landscapes. *Evolutionary Ecology Research*, 1(8):923–942.

Collet, P., Méléard, S., and Metz, J. A. J. (2013). A rigorous model study of the adaptive dynamics of mendelian diploids. *Journal of Mathematical Biology*, 67(3):569–607.

Collins, E. J., McNamara, J. M., and Ramsey, D. M. (2006). Learning rules for optimal selection in a varying environment: mate choice revisited. *Behavioral Ecology*, 17(5):799–809.

Conradt, L. and List, C. (2009). Group decisions in humans and animals: a survey. *Philosophical transactions of the Royal Society B: Biological Sciences*, 364(1518):719–742.

Constable, G. W. and McKane, A. J. (2014). Population genetics on islands connected by an arbitrary network: An analytic approach. *Journal of Theoretical Biology*, 358:149–165.

Cotton, S., Fowler, K., and Pomiankowski, A. (2004). Do sexual ornaments demonstrate heightened condition-dependent expression as predicted by the handicap hypothesis? *Proceedings of the Royal Society of London. Series B: Biological Sciences*, 271(1541):771–783.

Cox, J. T. (1989). Coalescing random walks and voter model consensus times on the torus in $\mathbb{Z}^d$. *The Annals of Probability*, 17(4):1333–1366.

Crawford, K., Lancaster, A., Oh, H., and Rychtář, J. (2015). A voluntary use of insecticide-treated cattle can eliminate African sleeping sickness. *Letters in Biomathematics*, 2(1):91–101.

Crawford, V. and Sobel, J. (1982). Strategic information transmission. *Econometrica: Journal of the Econometric Society*, 50(6):1431–1451.

Crawford, V. P. (1990a). Nash equilibrium and evolutionary stability in large- and finite- population 'playing the field' models. *Journal of Theoretical Biology*, 145(1):83–94.

Crawford, V. P. (1990b). On the definition of an evolutionarily stable strategy in the 'playing the field' model. *Journal of Theoretical Biology*, 143(2):269–273.

Cressman, R. (1988). Frequency-dependent viability selection (A single-locus, multi-phenotype model). *Journal of Theoretical Biology*, 130(2):147–165.

Cressman, R. (1990). Strong stability and density-dependent evolutionarily stable strategies. *Journal of Theoretical Biology*, 145(3):319–330.

Cressman, R. (1992). *The Stability Concept of Evolutionary Game Theory*. Springer-Verlag, Berlin.

Cressman, R. (2003). *Evolutionary Dynamics and Extensive Form Games*. MIT Press, Cambridge, MA, USA.

Cressman, R., Hofbauer, J., and Hines, W. G. S. (1996). Evolutionary stability in strategic models of single-locus frequency-dependent viability selection. *Journal of Mathematical Biology*, 34(7):707–733.

Cressman, R. and Křivan, V. (2006). Migration dynamics for the ideal free distribution. *The American Naturalist*, 168(3):384–397.

Cressman, R. and Křivan, V. (2010). The ideal free distribution as an evolutionarily stable state in density-dependent population games. *Oikos*, 119(8):1231–1242.

Cressman, R. and Křivan, V. (2019). Bimatrix games that include interaction times alter the evolutionary outcome: the Owner–Intruder game. *Journal of Theoretical Biology*, 460:262–273.

Cressman, R. and Křivan, V. (2020). Reducing courtship time promotes marital bliss: The battle of the sexes game revisited with costs measured as time lost. *Journal of Theoretical Biology*, 503:110382.

Cressman, R., Křivan, V., and Garay, J. (2004). Ideal free distributions, evolutionary games, and population dynamics in multiple-species environments. *The American Naturalist*, 164(4):473–489.

Cressman, R. and Tran, T. (2015). The ideal free distribution and evolutionary stability in habitat selection games with linear fitness and Allee effect. In *Interdisciplinary Topics in Applied Mathematics, Modeling and Computational Science*, pages 457–463. Springer.

Crockett, C. M. and Janson, C. H. (2000). Infanticide in red howlers: female group size, male membership and a possible link to folivory. In van Schaik, C. P. and Janson, C. H., editors, *Infanticide by Males and Its Implications*, pages 75–98. Cambridge University Press, Cambridge, UK.

Crow, J. F. (1979). Genes that violate Mendel's rules. *Scientific American*, 240(2):134–143.

Crowe, M., Fitzgerald, M., Remington, D. L., Ruxton, G. D., and Rychtář, J. (2009a). Game theoretic model of brood parasitism in a dung beetle *Onthophagus taurus*. *Evolutionary Ecology*, 23(5):765–776.

Crowe, M., Fitzgerald, M., Remington, D. L., and Rychtář, J. (2009b). On deterministic and stochastic models of kleptoparasitism. *Journal of Interdisciplinary Mathematics*, 12(2):161–180.

Cuesta, F. A., Guerberoff, G., and Rojo, Á. L. (2022). Bernoulli and binomial proliferation on evolutionary graphs. *Journal of Theoretical Biology*, 534:110942.

Cuesta, F. A., Sequeiros, P. G., and Rojo, Á. L. (2017). Suppressors of selection. *PLOS One*, 12(7):e0180549.

Cullen, J. M. (1966). Reduction of ambiguity through ritualization. *Philosophical Transactions of the Royal Society of London, Series B: Biological Sciences*, 251(772):363–374.

Cunningham, J., Thuijsman, F., Peeters, R., Viossat, Y., Brown, J. S., Gatenby, R. A., and Staňková, K. (2020). Optimal control to reach eco-evolutionary stability in metastatic castrate-resistant prostate cancer. *PLOS One*, 15(12):e0243386.

Daley, D. J. and Gani, J. (2001). *Epidemic Modelling: An Introduction*. Cambridge University Press, Cambridge, UK.

Darwin, C. (1859). *On the Origin of Species by Means of Natural Selection, or the Preservation of Favoured Races in the Struggle for Life*. London: John Murray.

Darwin, C. (1871). *The Descent of Man and Selection in Relation to Sex*. London: Murray.

Darwin, C. (1874). *The Descent of Man and Selection in Relation to Sex*. London: Murray.

Dawkins, R. (1976). *The Selfish Gene*. Oxford University Press, Oxford, UK.

Dawkins, R. (1999). *The Extended Phenotype: The Long Reach of the Gene*. Oxford University Press, New York, USA.

Dawkins, R. and Brockmann, H. J. (1980). Do digger wasps commit the concorde fallacy? *Animal Behaviour*, 28(3):892–896.

DeAngelis, D. L., Goldstein, R. A., and O'Neill, R. V. (1975). A model for trophic interaction. *Ecology*, 56(4):881–892.

DeAngelis, D. L., Wolkowicz, G. S. K., Lou, Y., Jiang, Y., Novak, M., Svanbäck, R., Araújo, M. S., Jo, Y. S., and Cleary, E. A. (2011). The effect of travel loss on evolutionarily stable distributions of populations in space. *The American Naturalist*, 178(1):15–29.

Débarre, F., Nuismer, S. L., and Doebeli, M. (2014). Multidimensional (co) evolutionary stability. *The American Naturalist*, 184(2):158–171.

DeGroot, M. H. (2004). *Optimal Statistical Decisions*, volume 82. John Wiley & Sons, New York, USA.

Delhey, J. K. V., Carrete, M., and Martínez, M. (2001). Diet and feeding behaviour of Olrog's gull *Larus atlanticus* in Bahía Blanca, Argentina. *Ardea*, 89(2):319–329.

Dercole, F. and Rinaldi, S. (2008). *Analysis of Evolutionary Processes: The Adaptive Dynamics Approach and Its Applications*. Princeton University Press, Princeton, NJ, USA.

Deutsch, A. and Dormann, S. (2005). *Cellular Automaton Modeling of Biological Pattern Formation: Characterization, Applications and Analysis*. Birkhauser.

Díaz, J., Goldberg, L. A., Mertzios, G. B., Richerby, D., Serna, M., and Spirakis, P. G. (2013). On the fixation probability of superstars. *Proceedings of the Royal Society A: Mathematical, Physical and Engineering Sciences*, 469(2156):20130193.

Díaz, J. and Mitsche, D. (2021). A survey of the modified moran process and evolutionary graph theory. *Computer Science Review*, 39:100347.

Dieckmann, U., Doebeli, M., Metz, J. A. J., and Tautz, D. (2012). *Adaptive Speciation*. Cambridge University Press, Cambridge, UK.

Dieckmann, U. and Law, R. (1996). The dynamical theory of coevolution: a derivation from stochastic ecological processes. *Journal of Mathematical Biology*, 34(5):579–612.

Diekmann, O. and Heesterbeek, J. A. P. (2000). *Mathematical Epidemiology of Infectious Diseases: Model Building, Analysis and Interpretation*, volume 5. John Wiley & Sons, New York, NY, USA.

Diekmann, O., Heesterbeek, J. A. P., and Metz, J. A. J. (1995). The legacy of kermack and mckendrick. In Mollison, D., editor, *Epidemic Models: Their Structure and Relation to Data*, pages 95–115. Cambridge University Press, Cambridge, UK.

Dingli, D., Chalub, F. A. d. C. C., Santos, F. C., Van Segbroeck, S., and Pacheco, J. M. (2009). Cancer phenotype as the outcome of an evolutionary game between normal and malignant cells. *British Journal of Cancer*, 101(7):1130–1136.

Doebeli, M. and Hauert, C. (2005). Models of cooperation based on the prisoner's dilemma and the snowdrift game. *Ecology Letters*, 8(7):748–766.

Doebeli, M. and Ispolatov, I. (2010). Complexity and diversity. *Science*, 328(5977):494–497.

Domenici, P., Batty, R. S., Simila, T., and Ogam, E. (2000). Killer whales (*Orcinus orca*) feeding on schooling herring (*Clupea harengus*) using underwater tail-slaps: kinematic analyses of field observations. *Journal of Experimental Biology*, 203(2):283–294.

Dong, R., Wang, X., and Chen, L. (2017). The controller design for promoting the evolution of cooperation in the prisoner's dilemma based on the non-uniform interaction rates. *Engineering Theory and Practice*, 37(10):2582–2591.

Dorsett, C., Oh, H., Paulemond, M. L., and Rychtář, J. (2016). Optimal repellent usage to combat dengue fever. *Bulletin of Mathematical Biology*, 78(5):916–922.

Drummond, H. and Canales, C. (1998). Dominance between booby nestlings involves winner and loser effects. *Animal Behaviour*, 55(6):1669–1676.

Dubois, F. and Giraldeau, L. A. (2005). Fighting for resources: the economics of defense and appropriation. *Ecology*, 86(1):3–11.

Dubois, F., Giraldeau, L. A., and Grant, J. W. A. (2003). Resource defense in a group-foraging context. *Behavioral Ecology*, 14(1):2.

Dubois, F. and Richard-Dionne, É. (2020). Consequences of multiple simultaneous opportunities to exploit others' efforts on free riding. *Ecology and Evolution*, 10(10):4343–4351.

Dugatkin, L. and Dugatkin, A. (2007). Extrinsic effects, estimating opponents' RHP, and the structure of dominance hierarchies. *Biology Letters*, 3(6):614–616.

Dugatkin, L. A. (1997a). *Cooperation among Animals: An Evolutionary Perspective*. Oxford University Press, New York, USA.

Dugatkin, L. A. (1997b). Winner and loser effects and the structure of dominance hierarchies. *Behavioral Ecology*, 8(6):583–587.

Dugatkin, L. A. (2009). *Principles of Animal Behavior*. New York.

Dugatkin, L. A. and Earley, R. L. (2003). Group fusion: the impact of winner, loser and bystander effects on hierarchy formation in large groups. *Behavioral Ecology*, 14(3):367–373.

Dujon, A. M., Aktipis, A., Alix-Panabières, C., Amend, S. R., Boddy, A. M., Brown, J. S., Capp, J.-p., DeGregori, J., Ewald, P., and Gatenby, R. A. (2021). Identifying key questions in the ecology and evolution of cancer. *Evolutionary Applications*, 14(4):877–892.

Duong, M. H. and Han, T. A. (2016). Analysis of the expected density of internal equilibria in random evolutionary multi-player multi-strategy games. *Journal of Mathematical Biology*, 73(6):1727–1760.

Durrett, R. (2014). Spatial evolutionary games with small selection coefficients. *Electronic Journal of Probability*, 19:1–64.

Düsing, C. (1884). On the regulation of the sex-ratio. *Theoretical Population Biology*, 58:255–257.

Dutta, B. and Jackson, M. O. (2000). The stability and efficiency of directed communication networks. *Review of Economic Design*, 5(3):251–272.

Eames, K. T. D. and Keeling, M. J. (2002). Modeling dynamic and network heterogeneities in the spread of sexually transmitted diseases. *Proceedings of the National Academy of Sciences*, 99(20):13330–13335.

Earley, R. L. and Dugatkin, L. A. (2002). Eavesdropping on visual cues in green swordtail (*xiphophorus helleri*) fights: a case for networking. *Proceedings of the Royal Society of London. Series B: Biological Sciences*, 269(1494):943–952.

Edwards, A. W. F. (2000). *Foundations of Mathematical Genetics*. Cambridge University Press, Cambridge, UK.

Eisert, J., Wilkens, M., and Lewenstein, M. (1999). Quantum games and quantum strategies. *Physical Review Letters*, 83(15):3077–3080.

Elias, D. O., Kasumovic, M. M., Punzalan, D., Andrade, M. C. B., and Mason, A. C. (2008). Assessment during aggressive contests between male jumping spiders. *Animal Behaviour*, 76(3):901–910.

Engqvist, L. (2012). Evolutionary modeling predicts a decrease in postcopulatory sperm viability as a response to increasing levels of sperm competition. *The American Naturalist*, 179(5):667–677.

Engqvist, L. and Reinhold, K. (2006). Theoretical influence of female mating status and remating propensity on male sperm allocation patterns. *Journal of Evolutionary Biology*, 19(5):1448–1458.

Engqvist, L. and Reinhold, K. (2007). Sperm competition games: optimal sperm allocation in response to the size of competing ejaculates. *Proceedings of the Royal Society B: Biological Sciences*, 274(1607):209.

Engqvist, L. and Taborsky, M. (2017). The evolution of strategic male mating effort in an information transfer framework. *Journal of Evolutionary Biology*, 30(6):1143–1152.

Enquist, M. and Leimar, O. (1983). Evolution of fighting behaviour: Decision rules and assessment of relative strength. *Journal of Theoretical Biology*, 102(3):387–410.

Ermentrout, G. B. and Edelstein-Keshet, L. (1993). Cellular automata approaches to biological modeling. *Journal of Theoretical Biology*, 160:97–97.

Erovenko, I. V. (2019). The evolution of cooperation in one-dimensional mobile populations with deterministic dispersal. *Games*, 10(1):2.

Erovenko, I. V., Bauer, J., Broom, M., Pattni, K., and Rychtář, J. (2019). The effect of network topology on optimal exploration strategies and the evolution of cooperation in a mobile population. *Proceedings of the Royal Society A: Mathematical, Physical and Engineering Sciences*, 475(2230):20190399.

Erovenko, I. V. and Rychtář, J. (2016). The evolution of cooperation in 1-dimensional mobile populations. *Far East Journal of Applied Mathematics*, 95(1):63.

Eshel, I. (1972). On the neighbor effect and the evolution of altruistic traits. *Theoretical Population Biology*, 3(3):258–277.

Eshel, I. (1982). Evolutionarily stable strategies and viability selection in Mendelian populations. *Theoretical Population Biology*, 22(2):204–217.

Eshel, I. (1983). Evolutionary and continuous stability. *Journal of Theoretical Biology*, 103(1):99–111.

Eshel, I. and Cavalli-Sforza, L. L. (1982). Assortment of encounters and evolution of cooperativeness. *Proceedings of the National Academy of Sciences*, 79(4):1331–1335.

Eshel, I. and Cohen, D. (1976). Cooperation, competition, and kin selection in populations. In Karlin, S. and Nevo, E., editors, *Population Genetics and Ecology*, pages 537–546. Academic Press, New York, USA.

Eshel, I. and Sansone, E. (1995). Owner-intruder conflict, Grafen effect and self-assessment. The Bourgeois principle re-examined. *Journal of Theoretical Biology*, 177(4):341–356.

Estes, R. (2012). *The Behavior Guide to African Mammals: Including Hoofed Mammals, Carnivores, Primates*. University of California Press.

Estrada, E. (2010). Quantifying network heterogeneity. *Physical Review E*, 82(6):066102.

Evans, C. P. (2018). Bipolar androgen therapy: an intriguing paradox. *The Lancet Oncology*, 19(1):8–10.

Evans, T. P. (2015). *Perspectives on the relationship between local interactions and global outcomes in spatially explicit models of systems of interacting individuals*. PhD thesis, University College London.

Fairbanks, D. J. and Rytting, B. (2001). Mendelian controversies: a botanical and historical review. *American Journal of Botany*, 88(5):737–752.

Falster, D. S. and Westoby, M. (2003). Plant height and evolutionary games. *Trends in Ecology & Evolution*, 18(7):337–343.

Farrell, J. and Ware, R. (1989). Evolutionary stability in the repeated prisoner's dilemma. *Theoretical Population Biology*, 36(2):161–166.

Fearon, E. and Vogelstein, B. (1990). A genetic model for colorectal tumorigenesis. *Cell*, 61(5):759–767.

Fehr, E. and Gachter, S. (2002). Altruistic punishment in humans. *Nature*, 415(6868):137–140.

Ferriere, R. and Michod, R. E. (1996). The evolution of cooperation in spatially heterogeneous populations. *The American Naturalist*, 147(5):692–717.

Fisher, L. (2008). *Rock, Paper, Scissors: Game Theory in Everyday Life.* Perseus Books Group.

Fisher, R. A. (1915). The evolution of sexual preference. *The Eugenics Review,* 7(3):184–192.

Fisher, R. A. (1930). *The Genetical Theory of Natural Selection.* Clarendon Press, Oxford, UK.

Fishman, M. (2020). Polymorphic evolutionary games and non-mendelian genetics. *Bulletin of Mathematical Biology,* 82(2):1–11.

Fishman, M. A. (2008). Asymmetric evolutionary games with non-linear pure strategy payoffs. *Games and Economic Behavior,* 63(1):77–90.

Fishman, M. A. (2016). Polymorphic evolutionary games. *Journal of Theoretical Biology,* 398:130–135.

Fishman, M. A. (2018). Animal conflicts in diploid populations with sexual reproduction. *Journal of Theoretical Biology,* 462:475–478.

Fitzgibbon, C. and Fanshawe, J. (1988). Stotting in Thomson's gazelles: an honest signal of condition. *Behavioral Ecology and Sociobiology,* 23(2):69–74.

Fletcher, J. A. and Doebeli, M. (2009). A simple and general explanation for the evolution of altruism. *Proceedings of the Royal Society B: Biological Sciences,* 276(1654):13–19.

Fletcher, J. A. and Zwick, M. (2006). Unifying the theories of inclusive fitness and reciprocal altruism. *The American Naturalist,* 168(2):252–262.

Fogarty, L. and Kandler, A. (2020). The fundamentals of cultural adaptation: implications for human adaptation. *Scientific Reports,* 10(1):1–11.

Fokker, A. D. (1914). The median energy of rotating electrical dipoles in radiation fields. *Annalen der Physik,* 43:810–820.

Forst, S. and Nealson, K. (1996). Molecular biology of the symbiotic-pathogenic bacteria *Xenorhabdus* spp. and *Photorhabdus* spp. *Microbiological Reviews,* 60(1):21–43.

Fortunato, A., Glasser, C., Watson, J., Lu, Y., Rychtář, J., and Taylor, D. (2021). Mathematical modeling of the use of insecticide treated nets for elimination of visceral leishmaniasis in Bihar, India. *Royal Society Open Science,* 8(6):201960.

Fournier, D. and Keller, L. (2001). Partitioning of reproduction among queens in the Argentine ant, *Linepithema humile. Animal Behaviour,* 62(6):1039–1045.

Fowler, J. H. (2005). Altruistic punishment and the origin of cooperation. *Proceedings of the National Academy of Sciences*, 102(19):7047–7049.

Fréchet, M. (1953). Emile Borel, initiator of the theory of psychological games and its application. *Econometrica (pre-1986)*, 21(1):95.

Fretwell, S. D. and Lucas, H. L. (1969). On territorial behavior and other factors influencing habitat distribution in birds. *Acta biotheoretica*, 19(1):16–36.

Friedman, D. and Sinervo, B. (2016). *Evolutionary Games in Natural, Social, and Virtual Worlds*. Oxford University Press.

Fryer, T., Cannings, C., and Vickers, G. T. (1999a). Sperm competition I: basic model, ESS and dynamics. *Journal of Theoretical Biology*, 196(1):81–100.

Fryer, T., Cannings, C., and Vickers, G. T. (1999b). Sperm competition II: post-copulatory guarding. *Journal of Theoretical Biology*, 197(3):343–360.

Fu, F., Nowak, M. A., and Hauert, C. (2010). Invasion and expansion of cooperators in lattice populations: Prisoner's dilemma vs. snowdrift games. *Journal of Theoretical Biology*, 266(3):358–366.

Fu, F., Rosenbloom, D., Wang, L., and Nowak, M. A. (2011). Imitation dynamics of vaccination behaviour on social networks. *Proceedings of the Royal Society B: Biological Sciences*, 278(1702):42–49.

Fu, F., Wang, L., Nowak, M. A., and Hauert, C. (2009). Evolutionary dynamics on graphs: Efficient method for weak selection. *Physical Review E*, 79(4):046707.

Gal, S., Alpern, S., and Casas, J. (2015). Prey should hide more randomly when a predator attacks more persistently. *Journal of the Royal Society Interface*, 12(113):20150861.

Gal, S. and Casas, J. (2014). Succession of hide–seek and pursuit–evasion at heterogeneous locations. *Journal of the Royal Society Interface*, 11(94):20140062.

Galanis, A., Göbel, A., Goldberg, L. A., Lapinskas, J., and Richerby, D. (2017). Amplifiers for the Moran process. *Journal of the ACM*, 64(1):1–90.

Gamberale, G. and Tullberg, B. S. (1996). Evidence for a peak-shift in predator generalization among aposematic prey. *Proceedings of the Royal Society of London. Series B: Biological Sciences*, 263(1375):1329–1334.

Garay, J., Cressman, R., Móri, T. F., and Varga, T. (2018). The ESS and replicator equation in matrix games under time constraints. *Journal of Mathematical Biology*, 76(7):1951–1973.

Garay, J., Cressman, R., Xu, F., Broom, M., Csiszár, V., and Móri, T. F. (2020). When optimal foragers meet in a game theoretical conflict: A model of kleptoparasitism. *Journal of Theoretical Biology*, 502:110306.

Garay, J., Csiszár, V., and Móri, T. F. (2017). Evolutionary stability for matrix games under time constraints. *Journal of Theoretical Biology*, 415:1–12.

García, J. and Traulsen, A. (2012). Leaving the loners alone: Evolution of cooperation in the presence of antisocial punishment. *Journal of Theoretical Biology*, 307:168–173.

Gardner, A. and West, S. A. (2004). Spite and the scale of competition. *Journal of Evolutionary Biology*, 17(6):1195–1203.

Gardner, A. and West, S. A. (2010). Greenbeards. *Evolution*, 64(1):25–38.

Gardner, M. (1970). Mathematical games: The fantastic combinations of John Conway's new solitaire game "life". *Scientific American*, 223(4):120–123.

Gatenby, R. A. (2009). A change of strategy in the war on cancer. *Nature*, 459(7246):508–509.

Gatenby, R. A., Brown, J. S., and Vincent, T. L. (2009). Lessons from applied ecology: Cancer control using an evolutionary double bind. *Cancer Research*, 69(19):7499.

Gatenby, R. A., Gillies, R. J., and Brown, J. S. (2010). The evolutionary dynamics of cancer prevention. *Nature Reviews Cancer*, 10(8):526–527.

Gatti, N., Panozzo, F., and Restelli, M. (2013). Efficient evolutionary dynamics with extensive-form games. In *Twenty-Seventh AAAI Conference on Artificial Intelligence*.

Gatti, N. and Restelli, M. (2016). Sequence-form and evolutionary dynamics: realization equivalence to agent form and logit dynamics. In *Proceedings of the AAAI Conference on Artificial Intelligence*, volume 30.

Gaunersdorfer, A., Hofbauer, J., and Sigmund, K. (1991). On the dynamics of asymmetric games. *Theoretical Population Biology*, 39(3):345–357.

Gavrilets, S. (2014). Models of speciation: Where are we now? *Journal of Heredity*, 105(S1):743–755.

Geritz, S. A. H., Kisdi, É., Meszéna, G., and Metz, J. A. J. (1998). Evolutionary singular strategies and the adaptive growth and branching of the evolutionary tree. *Evolutionary Ecology*, 12:35–57.

Geritz, S. A. H., Metz, J. A. J., and Rueffler, C. (2016). Mutual invadability near evolutionarily singular strategies for multivariate traits, with special reference to the strongly convergence stable case. *Journal of Mathematical Biology*, 72(4):1081–1099.

Ghazoul, J. (2006). Floral diversity and the facilitation of pollination. *Journal of Ecology*, 94(2):295–304.

Giakkoupis, G. (2016). Amplifiers and suppressors of selection for the Moran process on undirected graphs. arXiv preprint: 1611.01585.

Gilbert, N. (2019). *Agent-Based Models*, volume 153. Sage Publications.

Gintis, H. (2000). *Game Theory Evolving*. Princeton University Press, Princeton, NJ, USA.

Giraldeau, L. A. and Caraco, T. (2000). *Social Foraging Theory*. Princeton University Press, Princeton, NJ, USA.

Giraldeau, L. A. and Livoreil, B. (1998). Game theory and social foraging. In Dugatkin, L. A. and Reeve, H. K., editors, *Game Theory and Animal Behavior*, pages 16–37. Oxford University Press, New York, NY, USA.

Givnish, T. J. (1982). On the adaptive significance of leaf height in forest herbs. *The American Naturalist*, 120(3):353–381.

Glance, N. S. and Huberman, B. A. (1994). The dynamics of social dilemmas. *Scientific American*, 270(3):76–81.

Glaubitz, A. and Fu, F. (2020). Oscillatory dynamics in the dilemma of social distancing. *Proceedings of the Royal Society A: Mathematical, Physical and Engineering Science*, 476(2243):20200686.

Godfray, H. C. J. (1994). *Parasitoids: Behavioral and Evolutionary Ecology*. Princeton University Press, Princeton, NJ, USA.

Gokhale, C. S. and Traulsen, A. (2010). Evolutionary games in the multiverse. *Proceedings of the National Academy of Sciences*, 107(12):5500–5504.

Gokhale, C. S. and Traulsen, A. (2011). Strategy abundance in evolutionary many-player games with multiple strategies. *Journal of Theoretical Biology*, 283(1):180–191.

Gokhale, C. S. and Traulsen, A. (2014). Evolutionary multiplayer games. *Dynamic Games and Applications*, 4(4):468–488.

Grabbe, O. (2005). An introduction to quantum game theory. arXiv preprint https://arxiv.org/abs/quant-ph/0506219.

Gräbner, C., Bale, C. S., Furtado, B. A., Alvarez-Pereira, B., Gentile, J. E., Henderson, H., and Lipari, F. (2019). Getting the best of both worlds? Developing complementary equation-based and agent-based models. *Computational Economics*, 53(2):763–782.

Grafen, A. (1990a). Sexual selection unhandicapped by the Fisher process. *Journal of Theoretical Biology*, 144(4):473–516.

Grafen, A. (1990b). Biological signals as handicaps. *Journal of Theoretical Biology*, 144(4):517–546.

Grafen, A. (2007). An inclusive fitness analysis of altruism on a cyclical network. *Journal of Evolutionary Biology*, 20(6):2278–2283.

Grafen, A. (2009). Formalizing Darwinism and inclusive fitness theory. *Philosophical Transactions of the Royal Society B: Biological Sciences*, 364(1533):3135–3141.

Griffen, B. D. (2009). Consumers that are not ideal or free can still approach the ideal free distribution using simple patch-leaving rules. *Journal of Animal Ecology*, 78(5):919–927.

Grimm, M. P. and Klinge, M. (1996). Pike and some aspects of its dependence on vegetation. In J. F. Craig, editor, *Pike: Biology and Exploitation*, pages 125–156. Chapman & Hall.

Gustafsson, L. and Qvarnström, A. (2006). A test of the "sexy son" hypothesis: Sons of polygynous collared flycatchers do not inherit their fathers' mating status. *The American Naturalist*, 167(2):297–302.

Haccou, P., Glaizot, O., and Cannings, C. (2003). Patch leaving strategies and superparasitism: an asymmetric generalized war of attrition. *Journal of Theoretical Biology*, 225(1):77–89.

Haccou, P. and Meelis, E. (1992). *Statistical Analysis of Behavioural Data: An Approach Based on Time-Structured Models*. Oxford University Press, Oxford, UK.

Haccou, P. and van Alphen, J. J. M. (2008). Competition and asymmetric wars of attrition in insect parasitoids. In Wajnberg, E., Bernstein, C., and van Alphen, J., editors, *Behavioral Ecology of Insect Parasitoids*, pages 193–211. Blackwell Publishing Ltd.

Hadjichrysanthou, C. and Broom, M. (2012). When should animals share food? Game theory applied to kleptoparasitic populations with food sharing. *Behavioral Ecology*, 23(5):977–991.

Hadjichrysanthou, C., Broom, M., and Kiss, I. Z. (2012). Approximating evolutionary dynamics on networks using a neighbourhood configuration model. *Journal of Theoretical Biology*, 312:13–21.

Hadjichrysanthou, C., Broom, M., and Rychtář, J. (2011). Evolutionary games on star graphs under various updating rules. *Dynamic Games and Applications*, 1(3):386–407.

Haigh, J. (1975). Game theory and evolution. *Advances in Applied Probability*, 7:8–11.

Haigh, J. (1988). The distribution of evolutionarily stable strategies. *Journal of Applied Probability*, 25:233–246.

Haigh, J. (1989). How large is the support of an ESS? *Journal of Applied Probability*, 26:164–170.

Haigh, J. (2003). *Taking Chances: Winning with Probability*. Oxford University Press, New York, USA.

Haigh, J. and Cannings, C. (1989). The n-person war of attrition. *Acta Applicandae Mathematicae*, 14(1):59–74.

Halloway, A. H., Malone, M. A., and Brown, J. S. (2020). Unstable population dynamics in obligate co-operators. *Theoretical Population Biology*, 136:1–11.

Hamilton, W. D. (1964). The genetical evolution of social behaviour. I and II. *Journal of Theoretical Biology*, 7(1):17–52.

Hamilton, W. D. (1967). Extraordinary sex ratios. *Science*, 156:477–488.

Hamilton, W. D. (1970). Selfish and spiteful behaviour in an evolutionary model. *Nature*, 228:1218–1220.

Hammerstein, P. (1981). The role of asymmetries in animal contests. *Animal Behaviour*, 29(1):193–205.

Hammerstein, P. (1996). Darwinian adaptation, population genetics and the streetcar theory of evolution. *Journal of Mathematical Biology*, 34(5):511–532.

Hammerstein, P. and Noë, R. (2016). Biological trade and markets. *Philosophical Transactions of the Royal Society B: Biological Sciences*, 371(1687):20150101.

Hammerstein, P. and Parker, G. A. (1982). The asymmetric war of attrition. *Journal of Theoretical Biology*, 96:647–682.

Hammerstein, P. and Selten, R. (1994). Game Theory and Evolutionary Biology. In Aumann, R. J. and Hart, S., editors, *Handbook of Game Theory, Volume 2*. Elsevier Science.

Han, C. Y., Issa, H., Rychtář, J., Taylor, D., and Umana, N. (2020). A voluntary use of insecticide treated nets can stop the vector transmission of Chagas disease. *PLOS Neglected Tropical Diseases*, 14(11):e0008833.

Han, T. A., Traulsen, A., and Gokhale, C. S. (2012). On equilibrium properties of evolutionary multi-player games with random payoff matrices. *Theoretical Population Biology*, 81(4):264–272.

Hanahan, D. and Weinberg, R. A. (2000). The hallmarks of cancer. *Cell*, 100(1):57–70.

Hao, Y. and Wu, Z. (2018). Computation of sparse and dense equilibrium strategies of evolutionary games. *Games*, 9(3):46–60.

Harcourt, A. H., Harvey, P. H., Larson, S. G., and Short, R. V. (1981). Testis weight, body weight and breeding system in primates. *Nature*, 293(5827):55–57.

Hardin, G. (1998). Essays on Science and Society: Extensions of "The Tragedy of the Commons". *Science*, 280(5364):682–683.

Hardy, G. H. (1908). Mendelian proportions in a mixed population. *Science*, 28:49–50.

Hardy, I. (2002). *Sex Ratios: Concepts and Research Methods*. Cambridge University Press, Cambridge, UK.

Harrison, M. D. and Broom, M. (2009). A game-theoretic model of interspecific brood parasitism with sequential decisions. *Journal of Theoretical Biology*, 256(4):504–517.

Harsanyi, J. C. (1966). A general theory of rational behavior in game situations. *Econometrica: Journal of the Econometric Society*, 34(3):613–634.

Harsanyi, J. C. (1967). Games with incomplete information played by 'Bayesian' players. Part I. The basic model. *Management Science*, 14(3):159–182.

Harsanyi, J. C. (1968a). Games with incomplete information played by 'Bayesian' players. Part II. Bayesian equilibrium points. *Management Science*, 14(5):320–334.

Harsanyi, J. C. (1968b). Games with incomplete information played by 'Bayesian' players. Part III. The basic probability distribution of the game. *Management Science*, 14(7):486–502.

Harsanyi, J. C. (1973). Games with randomly disturbed payoffs: A new rationale for mixed-strategy equilibrium points. *International Journal of Game Theory*, 2(1):1–23.

Haslegrave, J. and Cannings, C. (2017). Majority dynamics with one nonconformist. *Discrete Applied Mathematics*, 219:32–39.

Hauert, C. and Doebeli, M. (2004). Spatial structure often inhibits the evolution of cooperation in the snowdrift game. *Nature*, 428(6983):643–646.

Hauert, C., Traulsen, A., Brandt, H., Nowak, M. A., and Sigmund, K. (2007). Via freedom to coercion: the emergence of costly punishment. *Science*, 316(5833):1905–1907.

Heather, B. D. and Robertson, H. A. (2000). *The Field Guide to the Birds of New Zealand*. Penguin Books, New Zealand.

Helbing, D. (1992). A mathematical model for behavioral changes by pair interactions and its relation to game theory. In Haag, G., Mueller, U., and Troitzsch, K. G., editors, *Economic Evolution and Demographic Change. Formal Models in Social Sciences*. Springer-Verlag, Berlin, Heidelberg.

Helgesson, P. and Wennberg, B. (2015). The n-player war of attrition in the limit of infinitely many players. *Dynamic Games and Applications*, 5(1):65–93.

Hertler, S., Figueredo, A., and Peñaherrera-Aguirre, M. (2020). *Multilevel Selection: Theoretical Foundations, Historical Examples, and Empirical Evidence*. Palgrave Macmillan.

Hilbe, C., Nowak, M. A., and Sigmund, K. (2013). Evolution of extortion in iterated prisoner's dilemma games. *Proceedings of the National Academy of Sciences*, 110(17):6913–6918.

Hilbe, C. and Traulsen, A. (2012). Emergence of responsible sanctions without second order free riders, antisocial punishment or spite. *Scientific Reports*, 2:458.

Hilbe, C. and Traulsen, A. (2016). Only the combination of mathematics and agent-based simulations can leverage the full potential of evolutionary modeling: Comment on "Evolutionary game theory using agent-based methods" by C. Adami, J. Schossau and A. Hintze. *Physics of Life Reviews*, 19:29–31.

Hildenbrandt, H., Carere, C., and Hemelrijk, C. K. (2010). Self-organized aerial displays of thousands of starlings: a model. *Behavioral Ecology*, 21(6):1349–1359.

Hindersin, L., Möller, M., Traulsen, A., and Bauer, B. (2016). Exact numerical calculation of fixation probability and time on graphs. *Biosystems*, 150:87–91.

Hindersin, L. and Traulsen, A. (2014). Counterintuitive properties of the fixation time in network-structured populations. *Journal of The Royal Society Interface*, 11(99):20140606.

Hindersin, L. and Traulsen, A. (2015). Most undirected random graphs are amplifiers of selection for birth-death dynamics, but suppressors of selection for death-birth dynamics. *PLOS Computational Biology*, 11(11):e1004437.

Hindersin, L., Wu, B., Traulsen, A., and García, J. (2019). Computation and simulation of evolutionary game dynamics in finite populations. *Scientific Reports*, 9(1):1–21.

Hines, W. G. S. (1980). Three characterizations of population strategy stability. *Journal of Applied Probability*, 17(2):333–340.

Hines, W. G. S. (1994). ESS modelling of diploid populations I: anatomy of one-locus allelic frequency simplices. *Advances in Applied Probability*, 26(2):341–360.

Hines, W. G. S. and Bishop, D. T. (1984). Can and will a sexual diploid population attain an evolutionary stable strategy? *Journal of Theoretical Biology*, 111(4):667–686.

Hock, K. and Huber, R. (2006). Modeling the acquisition of social rank in crayfish: winner and loser effects and self-structuring properties. *Behaviour*, 143(3):325–346.

Hofbauer, J. (1996). Evolutionary dynamics for bimatrix games: A Hamiltonian system? *Journal of Mathematical Biology*, 34(5):675–688.

Hofbauer, J. (2000). From Nash and Brown to Maynard Smith: Equilibria, dynamics and ESS. *Selection*, 1(1):81–88.

Hofbauer, J. and Hopkins, E. (2005). Learning in perturbed asymmetric games. *Games and Economic Behavior*, 52(1):133–152.

Hofbauer, J. and Sandholm, W. H. (2009). Stable games and their dynamics. *Journal of Economic theory*, 144(4):1665–1693.

Hofbauer, J., Schuster, P., and Sigmund, K. (1979). A note on evolutionary stable strategies and game dynamics. *Journal of Theoretical Biology*, 81:609–612.

Hofbauer, J. and Sigmund, K. (1988). *The Theory of Evolution and Dynamical Systems*. Cambridge University Press, Cambridge, UK.

Hofbauer, J. and Sigmund, K. (1990). Adaptive dynamics and evolutionary stability. *Applied Mathematics Letters*, 3:75–79.

Hofbauer, J. and Sigmund, K. (1998). *Evolutionary Games and Population Dynamics*. Cambridge University Press, Cambridge, UK.

Hofbauer, J. and Sigmund, K. (2003). Evolutionarily game dynamics. *Bulletin of the American Mathematical Society*, 40(4):479–519.

Hofmann, H. A. and Schildberger, K. (2001). Assessment of strength and willingness to fight during aggressive encounters in crickets. *Animal Behaviour*, 62(2):337–348.

Höglund, J. and Alatalo, R. V. (1995). *Leks*. Princeton University Press, Princeton, NJ, USA.

Holen, Ø. H. and Svennungsen, T. O. (2012). Aposematism and the handicap principle. *The American Naturalist*, 180(5):629–641.

Holman, L. (2012). Costs and constraints conspire to produce honest signaling: insights from an ant queen pheromone. *Evolution*, 66(7):2094–2105.

Houchmandzadeh, B. and Vallade, M. (2010). Alternative to the diffusion equation in population genetics. *Physical Review E*, 82(5):051913.

Houchmandzadeh, B. and Vallade, M. (2011). The fixation probability of a beneficial mutation in a geographically structured population. *New Journal of Physics*, 13:073020.

House, T. and Keeling, M. J. (2011). Insights from unifying modern approximations to infections on networks. *Journal of The Royal Society Interface*, 8(54):67–73.

Houston, A, . and McNamara, J. M. (1999). *Models of Adaptive Behaviour*. Cambridge University Press, Cambridge, UK.

Houston, A. I. and McNamara, J. M. (1988). The ideal free distribution when competitive abilities differ: an approach based on statistical mechanics. *Animal Behaviour*, 36(1):166–174.

Houston, A. I. and McNamara, J. M. (1991). Evolutionarily stable strategies in the repeated hawk–dove game. *Behavioral Ecology*, 2(3):219–227.

Huang, J., Wang, J., and Xia, C. (2020). Role of vaccine efficacy in the vaccination behavior under myopic update rule on complex networks. *Chaos, Solitons & Fractals*, 130:109425.

Huk, T. and Winkel, W. (2008). Testing the sexy son hypothesis - a research framework for empirical approaches. *Behavioral Ecology*, 19(2):456–461.

Hurlbut, E., Ortega, E., Erovenko, I. V., and Rowell, J. T. (2018). Game theoretical model of cancer dynamics with four cell phenotypes. *Games*, 9(3):61.

Hutson, V. C. L. and Vickers, G. T. (1995). The spatial struggle of tit-for-tat and defect. *Philosophical Transactions of the Royal Society of London. Series B: Biological Sciences*, 348(1326):393–404.

Huttegger, S. M., Bruner, J. P., and Zollman, K. J. S. (2015). The handicap principle is an artifact. *Philosophy of Science*, 82(5):997–1009.

Ibsen-Jensen, R., Chatterjee, K., and Nowak, M. A. (2015). Computational complexity of ecological and evolutionary spatial dynamics. *Proceedings of the National Academy of Sciences*, 112(51):15636–15641.

Imhof, L. and Nowak, M. A. (2010). Stochastic evolutionary dynamics of direct reciprocity. *Proceedings of the Royal Society B: Biological Sciences*, 277(1680):463–468.

Iqbal, A. and Cheon, T. (2008). Evolutionary stability in quantum games. In D. Abbott, P. C. W. D. and Pati, A. K., editors, *Quantum aspects of life*, pages 251–290. Imperial College Press London.

Iqbal, A., Cheon, T., and Abbott, D. (2008). Probabilistic analysis of three-player symmetric quantum games played using the Einstein-Podolsky-Rosen-Bohm setting. *Physics Letters, Section A: General, Atomic and Solid State Physics*, 372(44):6564–6577.

Iqbal, A. and Toor, A. H. (2001). Entanglement and dynamic stability of Nash equilibria in a symmetric quantum game. *Physics Letters, Section A: General, Atomic and Solid State Physics*, 286(4):245–250.

Isaac, R. M. and Walker, J. M. (1988). Group size effects in public goods provision: The voluntary contributions mechanism. *The Quarterly Journal of Economics*, 103(1):179–199.

Isack, H. A. and Reyer, H.-U. (1989). Honeyguides and honey gatherers: interspecific communication in a symbiotic relationship. *Science*, 243(4896):1343–1346.

Ito, H. and Sasaki, A. (2016). Evolutionary branching under multi-dimensional evolutionary constraints. *Journal of Theoretical Biology*, 407:409–428.

Iwamura, Y. and Tanimoto, J. (2018). Realistic decision-making processes in a vaccination game. *Physica A: Statistical Mechanics and its Applications*, 494:236–241.

Iyengar, E. V. (2008). Kleptoparasitic interactions throughout the animal kingdom and a re-evaluation, based on participant mobility, of the conditions promoting the evolution of kleptoparasitism. *Biological Journal of the Linnean Society*, 93(4):745–762.

Iyer, S. and Killingback, T. (2016). Evolutionary dynamics of a smoothed war of attrition game. *Journal of Theoretical Biology*, 396:25–41.

Izquierdo, L. R., Izquierdo, S. S., and Sandholm, W. H. (2019). An introduction to ABED: Agent-based simulation of evolutionary game dynamics. *Games and Economic Behavior*, 118:434–462.

Izquierdo, S. S., Izquierdo, L. R., and Vega-Redondo, F. (2010). The option to leave: Conditional dissociation in the evolution of cooperation. *Journal of Theoretical Biology*, 267(1):76–84.

Jackson, A. L., Beauchamp, G., Broom, M., and Ruxton, G. D. (2006). Evolution of anti-predator traits in response to a flexible targeting strategy by predators. *Proceedings of the Royal Society B: Biological Sciences*, 273(1590):1055–1062.

Jackson, A. L., Humphries, S., and Ruxton, G. D. (2004). Resolving the departures of observed results from the ideal free distribution with simple random movements. *Journal of Animal Ecology*, 73(4):612–622.

Jackson, A. L. and Ruxton, G. D. (2006). Toward an individual-level understanding of vigilance: the role of social information. *Behavioral Ecology*, 17(4):532–538.

Jackson, M. O. (2003). The stability and efficiency of economic and social networks. In M. R. Sertel and S. K., editors, *Advances in Economic Design*, pages 319–361. Springer-Verlag, Berlin, Heidelberg.

Jackson, M. O. (2010). *Social and Economic Networks*. Princeton University Press, Princeton, NJ, USA.

Jackson, M. O. (2011). An overview of social networks and economic applications. In J. Benhabib, A. B. and Jackson, M. O., editors, *Handbook of Social Economics*, volume 1, pages 511–585. Elsevier.

Jackson, M. O. and Wolinsky, A. (1996). A strategic model of social and economic networks. *Journal of Economic Theory*, 71(1):44–74.

Janetos, A. C. (1980). Strategies of female mate choice: a theoretical analysis. *Behavioral Ecology and Sociobiology*, 7(2):107–112.

Jansen, V. A. A. and van Baalen, M. (2006). Altruism through beard chromodynamics. *Nature*, 440(7084):663–666.

Jeanne, R. L. (1972). Social biology of the neotropical wasp *Mischocyttarus drewseni*. *Bulletin of the Museum of Comparative Zoology*, 144:63–150.

Jemal, A., Ward, E. M., Johnson, C. J., Cronin, K. A., Ma, J., Ryerson, A. B., Mariotto, A., Lake, A. J., Wilson, R., and Sherman, R. L. (2017). Annual report to the nation on the status of cancer, 1975–2014, featuring survival. *Journal of the National Cancer Institute*, 109(9):djx030.

Jiang, L.-L., Wang, W.-X., Lai, Y.-C., and Wang, B.-H. (2010). Role of adaptive migration in promoting cooperation in spatial games. *Physical Review E*, 81(3):036108.

Johnstone, R. A. (1994). Honest signalling, perceptual error and the evolution of 'all-or-nothing' displays. *Proceedings of the Royal Society B: Biological Sciences*, 256:169–175.

Johnstone, R. A. (2000). Models of reproductive skew: a review and synthesis. *Ethology*, 106(1):5–26.

Johnstone, R. A. and Grafen, A. (1992). The continuous Sir Philip Sidney game: a simple model of biological signalling. *Journal of Theoretical Biology*, 156(2):215–234.

Joyce, D., Kennison, J., Densmore, O., Guerin, S., Barr, S., Charles, E., and Thompson, N. S. (2006). My way or the highway: a more naturalistic model of altruism tested in an iterative prisoners' dilemma. *Journal of Artificial Societies and Social Simulation*, 9(2):4.

Kabir, K. M. A. (2021). How evolutionary game could solve the human vaccine dilemma. *Chaos, Solitons & Fractals*, 152:111459.

Kabir, K. M. A., Chowdhury, A., and Tanimoto, J. (2021a). An evolutionary game modeling to assess the effect of border enforcement measures and socio-economic cost: Export-importation epidemic dynamics. *Chaos, Solitons & Fractals*, 146:110918.

Kabir, K. M. A., Jusup, M., and Tanimoto, J. (2019). Behavioral incentives in a vaccination-dilemma setting with optional treatment. *Physical Review E*, 100(6):062402.

Kabir, K. M. A., Risa, T., and Tanimoto, J. (2021b). Prosocial behavior of wearing a mask during an epidemic: an evolutionary explanation. *Scientific Reports*, 11(1):1–14.

Kabir, K. M. A. and Tanimoto, J. (2019). Modelling and analysing the coexistence of dual dilemmas in the proactive vaccination game and retroactive treatment game in epidemic viral dynamics. *Proceedings of the Royal Society A: Mathematical, Physical and Engineering Science*, 475(2232):20190484.

Kalmár, L. (1928). Zur Theorie der abstrakten Spiele. *Acta Sci. Math. Univ. Szeged*, 4:65–85.

Kamiński, D., Miękisz, J., and Zaborowski, M. (2005). Stochastic stability in three-player games. *Bulletin of Mathematical Biology*, 67(6):1195–1205.

Kandori, M. (1997). Evolutionary game theory in economics. In Kreps, D. M. and Wallis, K. F., editors, *Advances in Economics and Econometrics: Theory and Applications: Seventh World Congress*. Volume I, Econometric Society Monographs, pages 243–277. Cambridge University Press, Cambridge, UK.

Kane, P. and Zollman, K. J. S. (2015). An evolutionary comparison of the handicap principle and hybrid equilibrium theories of signaling. *PLOS One*, 10(9):e0137271.

Karlin, S. and Lessard, S. (1986). *Theoretical Studies on Sex Ratio Evolution*. Princeton University Press, Princeton, NJ, USA.

Karlin, S. and Taylor, H. M. (1975). *A First Course in Stochastic Processes*. Academic press.

Kaznatcheev, A., Peacock, J., Basanta, D., Marusyk, A., and Scott, J. G. (2019). Fibroblasts and alectinib switch the evolutionary games played by non-small cell lung cancer. *Nature Ecology & Evolution*, 3(3):450–456.

Keeling, M. J. (1999). The effects of local spatial structure on epidemiological invasions. *Proceedings of the Royal Society of London. Series B: Biological Sciences*, 266(1421):859–867.

Keeling, M. J. and Eames, K. T. D. (2005). Networks and epidemic models. *Journal of the Royal Society Interface*, 2(4):295–307.

Kelly, F. P. (1979). *Reversibility and Stochastic Networks*, volume 40. John Wiley & Sons, New York, NY, USA.

Kermack, W. O. and McKendrick, A. G. (1927). Contributions to the mathematical theory of epidemics I. *Proceedings of the Royal Society of London. Series A*, 115A:700–721.

Kermack, W. O. and McKendrick, A. G. (1932). Contributions to the mathematical theory of epidemics II. The problem of endemicity. *Proceedings of the Royal Society of London. Series A*, 138(834):55–83.

Kermack, W. O. and McKendrick, A. G. (1933). Contributions to the mathematical theory of epidemics. III. Further studies of the problem of endemicity. *Proceedings of the Royal Society of London. Series A*, 141(843):94–122.

Kermack, W. O. and McKendrick, A. G. (1991a). Contributions to the mathematical theory of epidemics I. *Bulletin of Mathematical Biology*, 53(1):33–55.

Kermack, W. O. and McKendrick, A. G. (1991b). Contributions to the mathematical theory of epidemics II. The problem of endemicity. *Bulletin of Mathematical Biology*, 53(1):57–87.

Kermack, W. O. and McKendrick, A. G. (1991c). Contributions to the mathematical theory of epidemics III. Further studies of the problem of endemicity. *Bulletin of Mathematical Biology*, 53(1):89–118.

Kerr, B., Riley, M. A., Feldman, M. W., and Bohannan, B. J. M. (2002). Local dispersal promotes biodiversity in a real-life game of rock–paper–scissors. *Nature*, 418(6894):171–174.

Kimura, M. (1964). Diffusion models in population genetics. *Journal of Applied Probability*, 1(2):177–232.

Kimura, M. and Crow, J. F. (1963). The measurement of effective population number. *Evolution*, 17(3):279–288.

Kingman, J. F. C. (1982). The coalescent. *Stochastic Processes and Their Applications*, 13(3):235–248.

Kirchner, W. H., Lindauer, M., and Michelsen, A. (1988). Honeybee dance communication. *Naturwissenschaften*, 75(12):629–630.

Kisdi, É. (2020). TPB and the invasion of adaptive dynamics. *Theoretical Population Biology*, 133:52–55.

Kisdi, É. and Meszéna, G. (1993). Density dependent life history evolution in fluctuating environments. *Lecture Notes in Biomathematics*, 98:26–62.

Kiss, I. Z., Miller, J. C., and Simon, P. L. (2017). *Mathematics of Epidemics on Networks: From Exact to Approximate Models*. Springer.

Klein, S., Foster, A., Feagins, D., Rowell, J., and Erovenko, I. V. (2020). Optimal voluntary and mandatory insect repellent usage and emigration strategies to control the chikungunya outbreak on Reunion Island. *PeerJ*, 8:e10151.

Klopfer, P. H. (1973). *Behavioral Aspects of Ecology*. Englewood Cliffs, NJ: Prentice-Hall, Inc.

Kobe, J., Pritchard, N., Short, Z., Erovenko, I. V., Rychtář, J., and Rowell, J. T. (2018). A game-theoretic model of cholera with optimal personal protection strategies. *Bulletin of Mathematical Biology*, 80(10):2580–2599.

Koch, G. W., Sillett, S. C., Jennings, G. M., and Davis, S. D. (2004). The limits to tree height. *Nature*, 428(6985):851–854.

Kodric-Brown, A. (1985). Female preference and sexual selection for male coloration in the guppy (*Poecilia reticulata*). *Behavioral Ecology and Sociobiology*, 17(3):199–205.

Kokko, H. (2003). Are reproductive skew models evolutionarily stable? *Proceedings of the Royal Society of London. Series B: Biological Sciences*, 270(1512):265–270.

Kokko, H. (2007). *Modelling for Field Biologists and Other Interesting People*. Cambridge University Press, Cambridge, UK.

Kokko, H., Jennions, M. D., and Brooks, R. (2006a). Unifying and testing models of sexual selection. *Annu. Rev. Ecol. Evol. Syst.*, 37:43–66.

Kokko, H. and Johnstone, R. A. (2002). Why is mutual mate choice not the norm? operational sex ratios, sex roles and the evolution of sexually dimorphic and monomorphic signalling. *Philosophical Transactions of the Royal Society of London. Series B: Biological Sciences*, 357(1419):319–330.

Kokko, H., López-Sepulcre, A., and Morrell, L. J. (2006b). From hawks and doves to self-consistent games of territorial behavior. *The American Naturalist*, 167(6):901–912.

Kolmogoroff, A. (1931). Über die analytischen Methoden in der Wahrscheinlichkeitsrechnung. *Mathematische Annalen*, 104(1):415–458.

König, D. (1927). *Über eine Schlussweise aus dem Endlichen ins Unendliche*, volume 3. Egyet. Barátainak Egyes.

Kraines, D. P. and Kraines, V. Y. (1989). Pavlov and the prisoner's dilemma. *Theory and Decision*, 26(1):47–79.

Kraines, D. P. and Kraines, V. Y. (2000). Natural selection of memory-one strategies for the iterated prisoner's dilemma. *Journal of Theoretical Biology*, 203(4):335–355.

Krakauer, D. C. (1995). Groups confuse predators by exploiting perceptual bottlenecks: a connectionist model of the confusion effect. *Behavioral Ecology and Sociobiology*, 36(6):421–429.

Krause, J. and Ruxton, G. D. (2002). *Living in Groups*. Oxford University Press, New York, USA.

Kreps, D. M. and Wilson, R. (1982). Reputation and imperfect information. *Journal of Economic Theory*, 27(2):253–279.

Křivan, V. (2003). Ideal free distributions when resources undergo population dynamics. *Theoretical Population Biology*, 64(1):25–38.

Křivan, V. (2009). Evolutionary games and population dynamics. In Drábek, P., editor, *Proceedings of Seminar in Differential Equations, Kamenice nad Lipou*, volume II of *Lecture Notes in Mathematics*.

Křivan, V. (2011). On the Gause predator-prey model with a refuge: A fresh look at the history. *Journal of Theoretical Biology*, 274:67–73.

Křivan, V. (2014). The Allee-type ideal free distribution. *Journal of Mathematical Biology*, 69(6):1497–1513.

Křivan, V. and Cressman, R. (2009). On evolutionary stability in predator-prey models with fast behavioural dynamics. *Evolutionary Ecology Research*, 11(2):227–251.

Křivan, V. and Cressman, R. (2017). Interaction times change evolutionary outcomes: Two-player matrix games. *Journal of Theoretical Biology*, 416:199–207.

Křivan, V., Cressman, R., and Schneider, C. (2008). The ideal free distribution: a review and synthesis of the game-theoretic perspective. *Theoretical Population Biology*, 73(3):403–425.

Křivan, V. and Sirot, E. (2002). Habitat selection by two competing species in a two-habitat environment. *The American Naturalist*, 160(2):214–234.

Kruuk, H. (1972). *The Spotted Hyena: A Study of Predation and Social Behavior*. University of Chicago Press, Chicago.

Kuga, K., Tanimoto, J., and Jusup, M. (2019). To vaccinate or not to vaccinate: A comprehensive study of vaccination-subsidizing policies with multi-agent simulations and mean-field modeling. *Journal of Theoretical Biology*, 469:107–126.

Kuhn, H. W. (1953). Extensive games and the problem of information. *Contributions to the Theory of Games*, 2(28):193–216.

Kuosmanen, T., Cairns, J., Noble, R., Beerenwinkel, N., Mononen, T., and Mustonen, V. (2021). Drug-induced resistance evolution necessitates less aggressive treatment. *PLOS Computational Biology*, 17(9):e1009418.

Kura, K., Broom, M., and Kandler, A. (2015). Modelling dominance hierarchies under winner and loser effects. *Bulletin of Mathematical Biology*, 77(6):927–952.

Kura, K., Broom, M., and Kandler, A. (2016). A game-theoretical winner and loser model of dominance hierarchy formation. *Bulletin of Mathematical Biology*, 78(6):1259–1290.

Kurokawa, S. and Ihara, Y. (2009). Emergence of cooperation in public goods games. *Proceedings of the Royal Society B: Biological Sciences*, 276(1660):1379–1384.

Kuzmics, C. and Rodenburger, D. (2020). A case of evolutionarily stable attainable equilibrium in the laboratory. *Economic Theory*, 70(3):685–721.

Laan, A., Iglesias-Julios, M., and de Polavieja, G. G. (2018). Zebrafish aggression on the sub-second time scale: evidence for mutual motor coordination and multi-functional attack manoeuvres. *Royal Society Open Science*, 5(8):180679.

Lachmann, M. and Bergstrom, C. T. (1998). Signalling among relatives: II. Beyond the tower of babel. *Theoretical Population Biology*, 54(2):146–160.

Landau, H. G. (1951a). On dominance relations and the structure of animal societies: I. Effect of inherent characteristics. *Bulletin of Mathematical Biology*, 13(1):1–19.

Landau, H. G. (1951b). On dominance relations and the structure of animal societies: II. Some effects of possible social factors. *Bulletin of Mathematical Biology*, 13(4):245–262.

Landauer, R. and Buttiker, M. (1987). Diffusive traversal time: Effective area in magnetically induced interference. *Physical Review B*, 36(12):6255.

Laraki, R., Renault, J., and Sorin, S. (2019). *Mathematical Foundations of Game Theory*. Springer.

Laverty, T. (1992). Plant interactions for pollinator visits: a test of the magnet species effect. *Oecologia*, 89(4):502–508.

Le Lann, C., Outreman, Y., van Alphen, J. J. M., and van Baaren, J. (2011). First in, last out: Asymmetric competition influences patch exploitation of a parasitoid. *Behavioral Ecology*, 22(1):101–107.

le Roux, A., Cherry, M. I., Gygax, L., and Manser, M. B. (2009). Vigilance behaviour and fitness consequences: comparing a solitary foraging and an obligate group-foraging mammal. *Behavioral Ecology and Sociobiology*, 63(8):1097–1107.

Ledzewicz, U. and Schaettler, H. (2016). Optimizing chemotherapeutic anticancer treatment and the tumor microenvironment: an analysis of mathematical models. In Rejniak, K., editor, *Systems Biology of Tumor Microenvironment. Advances in Experimental Medicine and Biology*. Volume 936, pages 209–223. Springer-Verlag, Berlin, Heidelberg.

Lehmann, L., Keller, L., and Sumpter, D. J. T. (2007). The evolution of helping and harming on graphs: the return of the inclusive fitness effect. *Journal of Evolutionary Biology*, 20(6):2284–2295.

Leimar, O. (1997). Repeated games: a state space approach. *Journal of Theoretical Biology*, 184(4):471–498.

Leimar, O. (2009). Multidimensional convergence stability. *Evolutionary Ecology Research*, 11(2):191–208.

Leimar, O. and Enquist, M. (1984). Effects of asymmetries in owner-intruder conflicts. *Journal of Theoretical Biology*, 111(3):475–491.

Leimar, O., Enquist, M., and Sillen-Tullberg, B. (1986). Evolutionary stability of aposematic coloration and prey unprofitability: a theoretical analysis. *The American Naturalist*, 128(4):469–490.

Leimar, O. and Hammerstein, P. (2001). Evolution of cooperation through indirect reciprocity. *Proceedings of the Royal Society of London. Series B: Biological Sciences*, 268(1468):745–753.

Lessard, S. (2011). On the robustness of the extension of the one-third law of evolution to the multi-player game. *Dynamic Games and Applications*, 1(3):410–418.

Léveillé, J. (2007). *Birds in Love: The Secret Courting & Mating Rituals of Extraordinary Birds*. Voyageur Press.

Lewis, D. (2008). *Convention: A Philosophical Study*. John Wiley & Sons, New York, USA.

Lewis, D. K. (1969). *Convention*. Harvard University Press; Cambridge, MA, USA.

Lewontin, R. C. (1961). Evolution and the theory of games. *Journal of Theoretical Biology*, 1(3):382–403.

Li, A., Broom, M., Du, J., and Wang, L. (2016). Evolutionary dynamics of general group interactions in structured populations. *Physical Review E*, 93(2):022407.

Li, A., Wu, B., and Wang, L. (2014). Cooperation with both synergistic and local interactions can be worse than each alone. *Scientific Reports*, 4:5536.

Liao, W.-J., Hu, Y., Zhu, B.-R., Zhao, X.-Q., Zeng, Y.-F., and Zhang, D.-Y. (2009). Female reproductive success decreases with display size in monkshood, *Aconitum kusnezoffii* (Ranunculaceae). *Annals of Botany*, 104(7):1405–1412.

Lieberman, E., Hauert, C., and Nowak, M. A. (2005). Evolutionary dynamics on graphs. *Nature*, 433(7023):312–316.

Liebert, A. E. and Starks, P. T. (2006). Taming of the skew: transactional models fail to predict reproductive partitioning in the paper wasp *Polistes dominulus*. *Animal Behaviour*, 71(4):913–923.

Liggett, T. M. (2012). *Interacting Particle Systems*, volume 276. Springer-Verlag, Berlin, Heidelberg.

Lindquist, J., Ma, J., van den Driessche, P., and Willeboordse, F. H. (2011). Effective degree network disease models. *Journal of Mathematical Biology*, 62(2):143–164.

Lindquist, W. B. and Chase, I. D. (2009). Data-based analysis of winner-loser models of hierarchy formation in animals. *Bulletin of Mathematical Biology*, 71(3):556–584.

Lindquist, W. B. and Chase, I. D. (2016). The fragility of individual-based explanations of social hierarchies: A test using animal pecking orders. *PLOS One*, 11(7):e0158900.

Linge, S. and Langtangen, H. P. (2020). *Programming for Computations-Python: A Gentle Introduction to Numerical Simulations with Python 3. 6.* Springer Open.

Liu, L., Chen, X., and Szolnoki, A. (2017). Competitions between prosocial exclusions and punishments in finite populations. *Scientific Reports*, 7(1):1–8.

Liu, L., Wang, S., Chen, X., and Perc, M. (2018). Evolutionary dynamics in the public goods games with switching between punishment and exclusion. *Chaos: An Interdisciplinary Journal of Nonlinear Science*, 28(10):103105.

Lloyd, D. G. (1977). Genetic and phenotypic models of natural selection. *Journal of Theoretical Biology*, 69(3):543–560.

Lorberbaum, J. P. (1994). No strategy is evolutionarily stable in the repeated prisoner's dilemma. *Journal of Theoretical Biology*, 168(2):117–130.

Lorberbaum, J. P., Bohning, D. E., Shastri, A., and Sine, L. E. (2002). Are there really no evolutionarily stable strategies in the iterated prisoner's dilemma? *Journal of Theoretical Biology*, 214(2):155–169.

Lotka, A. J. (1925). *Elements of Physical Biology*. Williams & Wilkins company.

Luenberger, D. G. and Ye, Y. (2008). *Linear and Nonlinear Programming*, volume 116. Springer-Verlag, Berlin, Heidelberg.

Luther, R. M. and Broom, M. (2004). Rapid convergence to an equilibrium state in kleptoparasitic populations. *Journal of Mathematical Biology*, 48(3):325–339.

Lyon, J. E., Pandit, S. A., van Schaik, C. P., and Pradhan, G. R. (2011). Mating strategies in primates: A game theoretical approach to infanticide. *Journal of Theoretical Biology*, 274(1):103–108.

Maley, C. C., Reid, B. J., and Forrest, S. (2004). Cancer prevention strategies that address the evolutionary dynamics of neoplastic cells: simulating benign cell boosters and selection for chemosensitivity. *Cancer Epidemiology and Prevention Biomarkers*, 13(8):1375–1384.

Mariani, P., Křivan, V., MacKenzie, B. R., and Mullon, C. (2016). The migration game in habitat network: the case of tuna. *Theoretical ecology*, 9(2):219–232.

Marrow, P., Law, R., and Cannings, C. (1992). The coevolution of predator–prey interactions: ESSs and red queen dynamics. *Proceedings of the Royal Society of London. Series B: Biological Sciences*, 250(1328):133–141.

Marshall, J. A. R. (2011). Ultimate causes and the evolution of altruism. *Behavioral Ecology and Sociobiology*, 65(3):503–512.

Martcheva, M., Tuncer, N., and Ngonghala, C. (2021). Effects of social-distancing on infectious disease dynamics: an evolutionary game theory and economic perspective. *Journal of Biological Dynamics*, 15(1):342–366.

Marusyk, A., Almendro, V., and Polyak, K. (2012). Intra-tumour heterogeneity: a looking glass for cancer? *Nature Reviews Cancer*, 12(5):323–334.

Masuda, N. (2009). Directionality of contact networks suppresses selection pressure in evolutionary dynamics. *Journal of Theoretical Biology*, 258(2):323–334.

Masuda, N. and Ohtsuki, H. (2009). Evolutionary dynamics and fixation probabilities in directed networks. *New Journal of Physics*, 11:033012.

Matsuda, H., Ogita, N., Sasaki, A., and Sato, K. (1992). Statistical mechanics of population: The lattice Lotka-Volterra model. *Progress of Theoretical Physics*, 88(6):1035–1049.

Matsui, A. (1992). Best response dynamics and socially stable strategies. *Journal of Economic Theory*, 57(2):343–362.

Matsumura, S. and Kobayashi, T. (1998). A game model for dominance relations among group-living animals. *Behavioral Ecology and Sociobiology*, 42(2):77–84.

May, R. M. and Anderson, R. M. (1983). Epidemiology and genetics in the coevolution of parasites and hosts. *Proceedings of the Royal society of London. Series B. Biological Sciences*, 219(1216):281–313.

May, R. M. and Lloyd, A. L. (2001). Infection dynamics on scale-free networks. *Physical Review E*, 64(6):066112.

May, R. M. and Nowak, M. A. (1994). Superinfection, metapopulation dynamics, and the evolution of diversity. *Journal of Theoretical Biology*, 170(1):95–114.

May, R. M. and Nowak, M. A. (1995). Coinfection and the evolution of parasite virulence. *Proceedings of the Royal Society of London. Series B: Biological Sciences*, 261(1361):209–215.

May, R. M. and Robinson, S. K. (1985). Population dynamics of avian brood parasitism. *The American Naturalist*, 126(4):475–494.

Maynard Smith, J. (1964). Group selection and kin selection. *Nature*, 201:1145–1147.

Maynard Smith, J. (1974). The theory of games and the evolution of animal conflicts. *Journal of Theoretical Biology*, 47(1):209–221.

Maynard Smith, J. (1982). *Evolution and the Theory of Games*. Cambridge University Press, Cambridge, UK.

Maynard Smith, J. (1989). *Evolutionary Genetics*. Oxford University Press, Oxford, UK.

Maynard Smith, J. (1991). Honest signalling: The Philip Sidney game. *Animal Behaviour*, 42:1034–1035.

Maynard Smith, J. and Haigh, J. (1974). The hitch-hiking effect of a favourable gene. *Genetics Research*, 23(01):23–35.

Maynard Smith, J. and Harper, D. G. (1995). Animal signals: models and terminology. *Journal of Theoretical Biology*, 177(3):305–311.

Maynard Smith, J. and Harper, D. G. (2003). *Animal Signals*. Oxford University Press, New York, USA.

Maynard Smith, J. and Parker, G. A. (1976). The logic of asymmetric contests. *Animal Behaviour*, 24:159–175.

Maynard Smith, J. and Price, G. R. (1973). The logic of animal conflict. *Nature*, 246:15–18.

McAvoy, A. and Allen, B. (2021). Fixation probabilities in evolutionary dynamics under weak selection. *Journal of Mathematical Biology*, 82(3):1–41.

McAvoy, A. and Hauert, C. (2015). Structure coefficients and strategy selection in multiplayer games. *Journal of Mathematical Biology*, 72:203–238.

McCann, T. S. (1981). Aggression and sexual activity of male southern elephant seals, *Mirounga leonina*. *Journal of Zoology*, 195(3):295–310.

McKane, A. J. and Waxman, D. (2007). Singular solutions of the diffusion equation of population genetics. *Journal of Theoretical Biology*, 247(4):849–858.

McNamara, J. M. (2022). Game theory in biology: Moving beyond functional accounts. *The American Naturalist*.

McNamara, J. M., Barta, Z., and Houston, A. I. (2004). Variation in behaviour promotes cooperation in the prisoner's dilemma game. *Nature*, 428:745–748.

McNamara, J. M., Fromhage, L., Barta, Z., and Houston, A. I. (2009). The optimal coyness game. *Proceedings of the Royal Society B: Biological Sciences*, 276(1658):953.

McNamara, J. M. and Houston, A. I. (1992). Evolutionarily stable levels of vigilance as a function of group size. *Animal Behaviour*, 43:641–658.

McNamara, J. M., Houston, A. I., and Lima, S. L. (1994). Foraging routines of small birds in winter: a theoretical investigation. *Journal of Avian Biology*, 25(4):287–302.

McNamara, J. M. and Leimar, O. (2020). *Game Theory in Biology: Concepts and Frontiers*. Oxford University Press, USA.

McNickle, G. G. (2020). Interpreting plant root responses to nutrients, neighbours and pot volume depends on researchers' assumptions. *Functional Ecology*, 34(10):2199–2209.

McNickle, G. G. and Brown, J. S. (2014). An ideal free distribution explains the root production of plants that do not engage in a tragedy of the commons game. *Journal of Ecology*, 102(4):963–971.

Mendel, G. (1866). Versuche über Pflanzen-Hybriden. Verhandlungen des naturforschenden Vereines. *Abhandlungen, Brüünn*, 4:3–47.

Mesterton-Gibbons, M. (1994). The Hawk-Dove game revisited: Effects of continuous variation in resource-holding potential on the frequency of escalation. *Evolutionary Ecology*, 8(3):230–247.

Mesterton-Gibbons, M. (1999a). On sperm competition games: incomplete fertilization risk and the equity paradox. *Proceedings of the Royal Society of London. Series B: Biological Sciences*, 266(1416):269–274.

Mesterton-Gibbons, M. (1999b). On sperm competition games: raffles and roles revisited. *Journal of Mathematical Biology*, 39(2):91–108.

Mesterton-Gibbons, M. (1999c). On the evolution of pure winner and loser effects: a game-theoretic model. *Bulletin of Mathematical Biology*, 61(6):1151–1186.

Mesterton-Gibbons, M. (2000). *An Introduction to Game-Theoretic Modelling*. American Mathematical Society.

Mesterton-Gibbons, M. and Dugatkin, L. A. (1995). Toward a theory of dominance hierarchies: effects of assessment, group size and variation in fighting ability. *Behavioral Ecology*, 6(4):416–423.

Mesterton-Gibbons, M., Karabiyik, T., and Sherratt, T. N. (2014). The iterated Hawk-Dove game revisited: The effect of ownership uncertainty on Bourgeois as a pure convention. *Dynamic Games and Applications*, 4(4):407–431.

Mesterton-Gibbons, M., Karabiyik, T., and Sherratt, T. N. (2016). On the evolution of partial respect for ownership. *Dynamic Games and Applications*, 6(3):359–395.

Mesterton-Gibbons, M. and Sherratt, T. N. (2011). Information, variance and cooperation: minimal models. *Dynamic Games and Applications*, 1(3):419–439.

Mesterton-Gibbons, M. and Sherratt, T. N. (2014). Bourgeois versus anti-Bourgeois: a model of infinite regress. *Animal Behaviour*, 89:171–183.

Meszéna, G., Kisdi, É., Dieckmann, U., Geritz, S. A. H., and Metz, J. A. J. (2001). Evolutionary optimisation models and matrix games in the unified perspective of adaptive dynamics. *Selection*, 2(1):193–220.

Metelmann, S., Sakai, S., Kondoh, M., and Telschow, A. (2020). Evolutionary stability of plant–pollinator networks: efficient communities and a pollination dilemma. *Ecology Letters*, 23(12):1747–1755.

Metz, J. A. J. (2008). Fitness. In rgensen, S. E. J. and Fath, B. D., editors, *Encyclopedia of Ecology*, pages 1599–1612. Elsevier.

Metz, J. A. J. (2011). Thoughts on the geometry of meso-evolution: collecting mathematical elements for a postmodern synthesis. In Chalub, F. and Rodrigues, J., editors, *The Mathematics of Darwin's Legacy*, pages 193–231. Springer-Verlag, New York, NY, USA.

Metz, J. A. J. (2012). Adaptive dynamics. In Hastings, A. and Gross, L., editors, *Encyclopedia of Theoretical Ecology*. California University Press.

Metz, J. A. J., Geritz, S. A. H., Meszéna, G., Jacobs, F. J. A., and van Heerwaarden, J. S. (1996). Adaptive dynamics, a geometrical study of the consequences of nearly faithful reproduction. In van Strien, S. J. and Lunel, S. M. V., editors, *Stochastic and Spatial Structures of Dynamical Systems*, Lecture Notes in Mathematics. North Holland, Amsterdam.

Metz, J. A. J., Nisbet, R. M., and Geritz, S. A. H. (1992). How should we define 'fitness' for general ecological scenarios? *Trends in Ecology & Evolution*, 7(6):198–202.

Meyer, C. D. (2001). *Matrix Analysis and Applied Linear Algebra*. Society for Industrial and Applied Mathematics.

Miękisz, J. (2004). Stochastic stability in spatial three-player games. *Physica A: Statistical Mechanics and its Applications*, 343:175–184.

Miękisz, J. (2008). Evolutionary game theory and population dynamics. In Capasso, V. and Lachowicz, M., editors, *Multiscale Problems in the Life Sciences. Lecture Notes in Mathematics*, vol 1940, pages 269–316. Springer-Verlag, Berlin, Heidelberg.

Miękisz, J., Matuszak, M., and Poleszczuk, J. (2014). Stochastic stability in three-player games with time delays. *Dynamic Games and Applications*, 4(4):489–498.

Milinski, M., Semmann, D., Bakker, T., and Krambeck, H. J. (2001). Cooperation through indirect reciprocity: Image scoring or standing strategy? *Proceedings of the Royal Society of London. Series B: Biological Sciences*, 268(1484):2495–2501.

Miller, E. and Coll, M. (2010). Spatial distribution and deviations from the ifd when animals forage over large resource patches. *Behavioral Ecology*, 21(5):927–935.

Mirmirani, M. and Oster, G. (1978). Competition, kin selection, and evolutionary stable strategies. *Theoretical Population Biology*, 13:304–339.

Missakian, E. A. (1972). Genealogical and cross-genealogical dominance relations in a group of free-ranging rhesus monkeys (*Macaca mulatta*) on Cayo Santiago. *Primates*, 13(2):169–180.

Moeller, D. (2004). Facilitative interactions among plants via shared pollinators. *Ecology*, 85(12):3289–3301.

Molander, P. (1985). The optimal level of generosity in a selfish, uncertain environment. *The Journal of Conflict Resolution*, 29(4):611–618.

Molina-Montenegro, M., Badano, E., and Cavieres, L. (2008). Positive interactions among plant species for pollinator service: assessing the 'magnet species' concept with invasive species. *Oikos*, 117(12):1833–1839.

Moller, A. (1993). Sexual selection in the barn swallow *Hirundo rustica*. III. Female tail ornaments. *Evolution*, 47(2):417–431.

Möller, M., Hindersin, L., and Traulsen, A. (2019). Exploring and mapping the universe of evolutionary graphs identifies structural properties affecting fixation probability and time. *Communications Biology*, 2(1):137.

Monk, T. (2018). Martingales and the fixation probability of high-dimensional evolutionary graphs. *Journal of Theoretical Biology*, 451:10–18.

Monk, T., Green, P., and Paulin, M. (2014). Martingales and fixation probabilities of evolutionary graphs. *Proceedings of the Royal Society A: Mathematical, Physical and Engineering Sciences*, 470(2165):20130730.

Monk, T. and van Schaik, A. (2021). Martingales and the characteristic functions of absorption time on bipartite graphs. *Royal Society Open Science*, 8(10):210657.

Moran, P. A. P. (1958). Random processes in genetics. *Mathematical Proceedings of the Cambridge Philosophical Society*, 54(01):60–71.

Moran, P. A. P. (1962). *The Statistical Processes of Evolutionary Theory*. Clarendon Press; Oxford University Press.

Morris, D. (1957). "typical intensity" and its relation to the problem of ritualisation. *Behaviour*, 11(1):1–12.

Morris, D. W., Lundberg, P., and Brown, J. S. (2016). On strategies of plant behaviour: evolutionary games of habitat selection, defence, and foraging. *Evolutionary Ecology Research*, 17(5):619–636.

Murray, J. D. (2002). *Mathematical Biology I: An Introduction.* Springer-Verlag, New York, NY, USA.

Murray, J. D. (2003). *Mathematical Biology II: Spatial Models and Biomedical Applications.* Springer-Verlag, New York, NY, USA.

Myerson, R. B. (1978). Refinements of the nash equilibrium concept. *International Journal of Game Theory*, 15:133–154.

Mylius, S. D. (1999). What pair formation can do to the battle of the sexes: towards more realistic game dynamics. *Journal of Theoretical Biology*, 197(4):469–485.

Nanda, M. and Durrett, R. (2017). Spatial evolutionary games with weak selection. *Proceedings of the National Academy of Sciences*, 114(23):6046–6051.

Nande, A., Ferdowsian, A., Lubin, E., Yoeli, E., and Nowak, M. A. (2020). DyPy: A Python library for simulating matrix-form games. arXiv preprint: 2007.13815.

Nasar, S. (1998). *A Beautiful Mind.* New York. Touchstone Book.

Nåsell, I. (1991). On the quasi-stationary distribution of the Ross malaria model. *Mathematical Biosciences*, 107(2):187–207.

Nåsell, I. (1996). The quasi-stationary distribution of the closed endemic SIS model. *Advances in Applied Probability*, 28(3):895–932.

Nash, J. F. (1950). Equilibrium points in n-person games. *Proceedings of the National Academy of Sciences*, 36(1):48–49.

Nash, J. F. (1951). Non-cooperative games. *The Annals of Mathematics*, 54(2):286–295.

Neel, J. V. (1949). The inheritance of sickle cell anemia. *Science*, 110(2846):64–66.

Neill, D. B. (2001). Optimality under noise: Higher memory strategies for the alternating prisoner's dilemma. *Journal of Theoretical Biology*, 211(2):159–180.

Neill, S. R. J. and Cullen, J. M. (1974). Experiments on whether schooling by their prey affects the hunting behaviour of cephalopods and fish predators. *Journal of Zoology*, 172(4):549–569.

Nessler, S. H., Uhl, G., and Schneider, J. M. (2007). Genital damage in the orb-web spider *Argiope bruennichi* (*Araneae: Araneidae*) increases paternity success. *Behavioral Ecology*, 18(1):174–181.

Noble, R., Burley, J., Le Sueur, C., and Hochberg, M. (2020). When, why and how tumour clonal diversity predicts survival. *Evolutionary Applications*, 13(7):1558–1568.

Noë, R. (2001). Biological markets: Partner choice as the driving force behind the evolution of cooperation. In *Economics in nature. Social dilemmas, mate choice and biological markets*, pages 93–118. Cambridge University Press.

Noë, R. and Hammerstein, P. (1994). Biological markets: supply and demand determine the effect of partner choice in cooperation, mutualism and mating. *Behavioral Ecology and Sociobiology*, 35(1):1–11.

Noë, R. and Hammerstein, P. (1995). Biological markets. *Trends in Ecology & Evolution*, 10(8):336–339.

Nonacs, P., Liebert, A. E., and Starks, P. T. (2006). Transactional skew and assured fitness return models fail to predict patterns of cooperation in wasps. *The American Naturalist*, 167(4):467–480.

Nowak, M. A. (1990a). An evolutionary stable strategy may be inaccessible. *Journal of Theoretical Biology*, 142:237–241.

Nowak, M. A. (1990b). Stochastic strategies in the prisoner's dilemma. *Theoretical Population Biology*, 38(1):93–112.

Nowak, M. A. (2006a). *Evolutionary Dynamics, Exploring the Equations of Life*. Harvard University Press, Cambridge, MA, USA.

Nowak, M. A. (2006b). Five rules for the evolution of cooperation. *Science*, 314(5805):1560–1563.

Nowak, M. A. (2012). Evolving cooperation. *Journal of Theoretical Biology*, 299:1–8.

Nowak, M. A., Anderson, R. M., McLean, A. R., Wolfs, T. F., Goudsmit, J., and May, R. M. (1991). Antigenic diversity thresholds and the development of AIDS. *Science*, 254(5034):963–969.

Nowak, M. A. and May, R. M. (1991). Mathematical biology of HIV infections: antigenic variation and diversity threshold. *Mathematical Biosciences*, 106(1):1–21.

Nowak, M. A. and May, R. M. (1992a). Coexistence and competition in HIV infections. *Journal of Theoretical Biology*, 159(3):329–342.

Nowak, M. A. and May, R. M. (1992b). Evolutionary games and spatial chaos. *Nature*, 359(6398):826–829.

Nowak, M. A. and May, R. M. (1993). AIDS pathogenesis: mathematical models of HIV and SIV infections. *Aids*, 7:S3.

Nowak, M. A. and May, R. M. (1994). Superinfection and the evolution of parasite virulence. *Proceedings of the Royal Society of London. Series B: Biological Sciences*, 255(1342):81–89.

Nowak, M. A., May, R. M., and Anderson, R. M. (1990). The evolutionary dynamics of HIV-1 quasispecies and the development of immunodeficiency disease. *Aids*, 4(11):1095–1103.

Nowak, M. A., Sasaki, A., Taylor, C., and Fudenberg, D. (2004). Emergence of cooperation and evolutionary stability in finite populations. *Nature*, 428(6983):646–650.

Nowak, M. A. and Sigmund, K. (1990). The evolution of stochastic strategies in the prisoner's dilemma. *Acta Applicandae Mathematicae*, 20(3):247–265.

Nowak, M. A. and Sigmund, K. (1993). A strategy of win-stay, lose-shift that outperforms tit-for-tat in the prisoner's dilemma game. *Nature*, 364(6432):56–58.

Nowak, M. A. and Sigmund, K. (1998a). The dynamics of indirect reciprocity. *Journal of Theoretical Biology*, 194(4):561–574.

Nowak, M. A. and Sigmund, K. (1998b). Evolution of indirect reciprocity by image scoring. *Nature*, 393(6685):573–577.

Nowak, M. A. and Sigmund, K. (2004). Evolutionary dynamics of biological games. *Science*, 303:793–799.

Nowak, M. A. and Sigmund, K. (2005). Evolution of indirect reciprocity. *Nature*, 7063:1291–1298.

Nowak, M. A., Sigmund, K., and El-Sedy, E. (1995). Automata, repeated games and noise. *Journal of Mathematical Biology*, 33(7):703–722.

Nowak, M. A., Tarnita, C. E., and Wilson, E. O. (2010). The evolution of eusociality. *Nature*, 466(7310):1057–1062.

Oechssler, J. and Riedel, F. (2002). On the dynamic foundation of evolutionary stability in continuous models. *Journal of Economic Theory*, 107(2):223–252.

Ohta, T. and Kimura, M. (1969a). Linkage disequilibrium at steady state determined by random genetic drift and recurrent mutation. *Genetics*, 63(1):229–238.

Ohta, T. and Kimura, M. (1969b). Linkage disequilibrium due to random genetic drift. *Genetics Research*, 13(01):47–55.

Ohtsuki, H. (2004). Reactive strategies in indirect reciprocity. *Journal of Theoretical Biology*, 227(3):299–314.

Ohtsuki, H., Hauert, C., Lieberman, E., and Nowak, M. A. (2006). A simple rule for the evolution of cooperation on graphs and social networks. *Nature*, 441(7092):502–505.

Ohtsuki, H. and Iwasa, Y. (2004). How should we define goodness?–reputation dynamics in indirect reciprocity. *Journal of Theoretical Biology*, 231(1):107–120.

Ohtsuki, H. and Iwasa, Y. (2006). The leading eight: social norms that can maintain cooperation by indirect reciprocity. *Journal of Theoretical Biology*, 239(4):435–444.

Ohtsuki, H. and Nowak, M. A. (2006). The replicator equation on graphs. *Journal of Theoretical Biology*, 243(1):86–97.

Ohtsuki, H. and Nowak, M. A. (2008). Evolutionary stability on graphs. *Journal of Theoretical Biology*, 251(4):698–707.

Ohtsuki, H., Nowak, M. A., and Pacheco, J. M. (2007). Breaking the symmetry between interaction and replacement in evolutionary dynamics on graphs. *Physical Review Letters*, 98(10):108106.

Osborne, M. J. and Rubinstein, A. (1994). *A Course in Game Theory*. The MIT Press.

Overton, C. E., Broom, M., Hadjichrysanthou, C., and Sharkey, K. J. (2019). Methods for approximating stochastic evolutionary dynamics on graphs. *Journal of Theoretical Biology*, 468:45–59.

Pacheco, J. M., Santos, F. C., Souza, M. O., and Skyrms, B. (2009). Evolutionary dynamics of collective action in *n*-person stag hunt dilemmas. *Proceedings of the Royal Society B: Biological Sciences*, 276(1655):315–321.

Pacheco, J. M., Santos, F. C., Souza, M. O., and Skyrms, B. (2011). Evolutionary dynamics of collective action. In Chalub, F. and Rodrigues, J., editors, *The Mathematics of Darwin's Legacy*, pages 119–138. Springer-Verlag, Berlin, Heidelberg.

Pacheco, J. M., Traulsen, A., and Nowak, M. A. (2006a). Active linking in evolutionary games. *Journal of Theoretical Biology*, 243(3):437–443.

Pacheco, J. M., Traulsen, A., and Nowak, M. A. (2006b). Coevolution of strategy and structure in complex networks with dynamical linking. *Physical Review Letters*, 97(25):258103.

Page, K. M. and Nowak, M. A. (2002). Unifying evolutionary dynamics. *Journal of Theoretical Biology*, 219:93–98.

Palm, G. (1984). Evolutionary stable strategies and game dynamics for n-person games. *Journal of Mathematical Biology*, 19(3):329–334.

Panchanathan, K. and Boyd, R. (2003). A tale of two defectors: the importance of standing for evolution of indirect reciprocity. *Journal of Theoretical Biology*, 224(1):115–126.

Panov, E. N., Pavlova, E. Y., and Nepomnyashchikh, V. A. (2010). Signal behavior in cranes (the Siberian crane *Sarcogeranus leucogeranus*, the white-naped crane *Grus vipio*, and the red-crowned crane *Grus japonensis*) in the context of the ritualization hypothesis. *Biology Bulletin*, 37(9):915–940.

Parker, G. A. (1970). Sperm competition and its evolutionary consequences in the insects. *Biological Reviews*, 45(4):525–567.

Parker, G. A. (1978). Searching for mates. In Krebbs, J. and Davies, N., editors, *Behavioural Ecology: An Evolutionary Approach*, pages 214–244. Blackwell, Oxford.

Parker, G. A. (1979). Sexual selection and sexual conflict. In Blum, M. S. and Blum, N. A., editors, *Sexual Selection and Reproductive Competition in Insects*, pages 123–166. Academic Press, New York.

Parker, G. A. (1982). Why are there so many tiny sperm? Sperm competition and the maintenance of two sexes. *Journal of Theoretical Biology*, 96(2):281–294.

Parker, G. A. and Ball, M. A. (2005). Sperm competition, mating rate and the evolution of testis and ejaculate sizes: a population model. *Biology Letters*, 1(2):235–238.

Parker, G. A., Ball, M. A., Stockley, P., and Gage, M. J. G. (1997). Sperm competition games: a prospective analysis of risk assessment. *Proceedings of the Royal Society of London. Series B: Biological Sciences*, 264(1389):1793–1802.

Parker, G. A., Chubb, J. C., Ball, M. A., and Roberts, G. N. (2003). Evolution of complex life cycles in helminth parasites. *Nature*, 425(6957):480–484.

Parker, G. A. and Rubenstein, D. I. (1981). Role assessment, reserve strategy and acquisition of information in asymmetric animal conflicts. *Animal Behaviour*, 29:221–240.

Parker, G. A. and Thompson, E. A. (1980). Dung fly struggles: a test of the war of attrition. *Behavioral Ecology and Sociobiology*, 7(1):37–44.

Parmigiani, S., Palanza, P., Mainardi, D., and Brain, P. F. (1994). Infanticide and protection of young in house mice (*Mus domesticus*): female and male strategies. In Parmigiami, S. and vom Saal, F. S., editors, *Infanticide and Parental Care*, pages 341–363. Harwood Academic Publishers, London, UK.

Parmigiani, S. and vom Saal, F. S. (1994). *Infanticide and Parental Care*. Harwood Academic Publishers, London, UK.

Pastor-Satorras, R. and Vespignani, A. (2001). Epidemic spreading in scale-free networks. *Physical Review Letters*, 86(14):3200–3203.

Pattni, K., Broom, M., and Rychtář, J. (2017). Evolutionary dynamics and the evolution of multiplayer cooperation in a subdivided population. *Journal of Theoretical Biology*, 429:105–115.

Pattni, K., Broom, M., and Rychtář, J. (2018). Evolving multiplayer networks: modelling the evolution of cooperation in a mobile population. *Discrete and Continuous Dynamical Systems B*, 23:1975–2004.

Pattni, K., Broom, M., Rychtář, J., and Silvers, L. J. (2015). Evolutionary graph theory revisited: when is an evolutionary process equivalent to the moran process? *Proceedings of the Royal Society A: Mathematical, Physical and Engineering Sciences*, 471(2182):20150334.

Pattni, K., Overton, C. E., and Sharkey, K. J. (2021). Evolutionary graph theory derived from eco-evolutionary dynamics. *Journal of Theoretical Biology*, 519:110648.

Pavlogiannis, A., Tkadlec, J., Chatterjee, K., and Nowak, M. A. (2017). Amplification on undirected population structures: Comets beat stars. *Scientific Reports*, 7(1):1–8.

Pavlogiannis, A., Tkadlec, J., Chatterjee, K., and Nowak, M. A. (2018). Construction of arbitrarily strong amplifiers of natural selection using evolutionary graph theory. *Communications Biology*, 1(1):1–8.

Pellis, L., Ball, F., Bansal, S., Eames, K., House, T., Isham, V., and Trapman, P. (2015a). Eight challenges for network epidemic models. *Epidemics*, 10:58–62.

Pellis, L., House, T., and Keeling, M. J. (2015b). Exact and approximate moment closures for non-Markovian network epidemics. *Journal of Theoretical Biology*, 382:160–177.

Pen, I. and Weissing, F. (2000). Sexual selection and the sex ratio: an ESS analysis. *Selection*, 1:111–121.

Pen, I., Weissing, F. J., and Daan, S. (1999). Seasonal sex ratio trend in the European kestrel: An ESS analysis. *The American Naturalist*, 153:384–397.

Peña, J., Lehmann, L., and Nöldeke, G. (2014). Gains from switching and evolutionary stability in multi-player matrix games. *Journal of Theoretical Biology*, 346:23–33.

Peña, J. and Nöldeke, G. (2016). Variability in group size and the evolution of collective action. *Journal of Theoretical Biology*, 389:72–82.

Peña, J. and Nöldeke, G. (2018). Group size effects in social evolution. *Journal of Theoretical Biology*, 457:211–220.

Peña, J., Pestelacci, E., Berchtold, A., and Tomassini, M. (2011). Participation costs can suppress the evolution of upstream reciprocity. *Journal of Theoretical Biology*, 273(1):197–206.

Peña, J., Wu, B., Arranz, J., and Traulsen, A. (2016a). Evolutionary games of multiplayer cooperation on graphs. *PLOS Computational Biology*, 12(8):e1005059.

Peña, J., Wu, B., and Traulsen, A. (2016b). Ordering structured populations in multiplayer cooperation games. *Journal of the Royal Society Interface*, 13:20150881.

Penn, D. and Számadó, S. (2020). The handicap principle: how an erroneous hypothesis became a scientific principle. *Biological Reviews*, 95(1):267–290.

Pepper, J. W. and Smuts, B. B. (2002). A mechanism for the evolution of altruism among nonkin: positive assortment through environmental feedback. *The American Naturalist*, 160(2):205–213.

Perc, M., Gómez-Gardenes, J., Szolnoki, A., Floría, L. M., and Moreno, Y. (2013). Evolutionary dynamics of group interactions on structured populations: a review. *Journal of the Royal Society Interface*, 10(80):20120997.

Perc, M. and Szolnoki, A. (2010). Coevolutionary games–A mini review. *BioSystems*, 99(2):109–125.

Pintor, L. M., Brown, J. S., and Vincent, T. L. (2011). Evolutionary game theory as a framework for studying biological invasions. *The American Naturalist*, 177(4):410–423.

Pitchford, J. W. and Brindley, J. (2001). Prey patchiness, predator survival and fish recruitment. *Bulletin of Mathematical Biology*, 63(3):527–546.

Pitchford, J. W., James, A., and Brindley, J. (2005). Quantifying the effects of individual and environmental variability in fish recruitment. *Fisheries Oceanography*, 14(2):156–160.

Planck, M. (1917). Über einen Satz der statistischen dynamik und seine Erweiterung in der quantentheorie Sitz. *Preuschen AkaD. Wissen*, 24:324–341.

Plank, M. (1997). Some qualitative differences between the replicator dynamics of two player and n player games. In *Proceedings of the Second World Congress on Nonlinear Analysts: part 3*, pages 1411–1417. Pergamon Press, InC.

Planque, R., Britton, N. F., Franks, N. R., and Peletier, M. A. (2002). The adaptiveness of defence strategies against cuckoo parasitism. *Bulletin of Mathematical Biology*, 64(6):1045–1068.

Płatkowski, T. (2004). Evolution of populations playing mixed multiplayer games. *Mathematical and Computer Modelling*, 39(9-10):981–989.

Płatkowski, T. (2016). Evolutionary coalitional games. *Dynamic Games and Applications*, 6(3):396–408.

Płatkowski, T. and Bujnowski, P. (2009). Cooperation in aspiration-based N-person prisoner's dilemmas. *Physical Review E*, 79(3):036103.

Płatkowski, T. and Stachowska-Pietka, J. (2005). ESSs in n-player mixed games. *Applied Mathematics and Computation*, 167(1):592–606.

Pohley, H. J. and Thomas, B. (1983). Non-linear ESS-models and frequency dependent selection. *Biosystems*, 16(2):87–100.

Poundstone, W. (1992). *Prisoner's Dilemma: John von Neumann, Game Theory and the Puzzle of the Bomb*. Anchor Books.

Press, W. H. and Dyson, F. J. (2012). Iterated Prisoner's Dilemma contains strategies that dominate any evolutionary opponent. *Proceedings of the National Academy of Sciences*, 109(26):10409–10413.

Proctor, C. J. and Broom, M. (2000). A spatial model of antipredator vigilance. *IMA Journal of Mathematical Control and Information*, 17(1):75–93.

Proctor, C. J., Broom, M., and Ruxton, G. D. (2001). Modelling antipredator vigilance and flight response in group foragers when warning signals are ambiguous. *Journal of Theoretical Biology*, 211(4):409–417.

Proctor, C. J., Broom, M., and Ruxton, G. D. (2003). A communication-based spatial model of antipredator vigilance. *Journal of Theoretical Biology*, 220(1):123–137.

Proctor, C. J., Broom, M., and Ruxton, G. D. (2006). Antipredator vigilance in birds: Modelling the 'edge' effect. *Mathematical Biosciences*, 199(1):79–96.

Pulliam, H. R., Pyke, G. H., and Caraco, T. (1982). The scanning behavior of juncos: a game-theoretical approach. *Journal of Theoretical Biology*, 95(1):89–103.

Pusey, A. E. and Packer, C. (1994). Infanticide in lions: consequences and counterstrategies. In Parmigiami, S. and vom Saal, F. S., editors, *Infanticide and Parental Care*, pages 277–299. Harwood Academic Publishers, London, UK.

Qin, J. (2017). *Biased Sampling, Over-identified Parameter Problems and Beyond*. Springer-Verlag, Berlin, Heidelberg.

Queller, D. C. (1984). Kin selection and frequency dependence: a game theoretic approach. *Biological Journal of the Linnean Society*, 23(2-3):133–143.

Rael, R. C., Costantino, R. F., Cushing, J. M., and Vincent, T. L. (2009). Using stage-structured evolutionary game theory to model the experimentally observed evolution of a genetic polymorphism. *Evolutionary Ecology Research*, 11(2):141–151.

Railsback, S. F. and Grimm, V. (2019). *Agent-Based and Individual-Based Modeling: A Practical Introduction*. Princeton University Press.

Ramakrishnan, R., Assudani, D., Nagaraj, S., Hunter, T., Cho, H.-I., Antonia, S., Altiok, S., Celis, E., and Gabrilovich, D. (2010). Chemotherapy enhances tumor cell susceptibility to CTL-mediated killing during cancer immunotherapy in mice. *Journal of Clinical Investigation*, 120(4):1111–1124.

Ramesh, D. and Mitchell, W. A. (2018). Evolution of signalling through pursuit deterrence in a two-prey model using game theory. *Animal Behaviour*, 146:155–163.

Ramsey, D. M. (2011). Mutual mate choice with multiple criteria. *Advances in Dynamic Games*, 11:337–355.

Ramsey, D. M. (2012). Partnership formation based on multiple traits. *European Journal of Operational Research*, 216:624–637.

Rand, D. G. and Nowak, M. A. (2011). The evolution of antisocial punishment in optional public goods games. *Nature Communications*, 2:434.

Real, L. A. (1990). Search theory and mate choice. I. Models of single-sex discrimination. *The American Naturalist*, 136(3):376–405.

Real, L. A. (1991). Search theory and mate choice. II. Mutual interaction, assortative mating, and equilibrium variation in male and female fitness. *The American Naturalist*, 138(4):901–917.

Reding, I., Kelley, M., Rowell, J. T., and Rychtář, J. (2016). A continuous ideal free distribution approach to the dynamics of selfish, cooperative and kleptoparasitic populations. *Royal Society Open Science*, 3(11):160788.

Reeve, H. K. (2000). A transactional theory of within-group conflict. *The American Naturalist*, 155(3):365–382.

Reeve, H. K. and Emlen, S. T. (2000). Reproductive skew and group size: an n-person staying incentive model. *Behavioral Ecology*, 11(6):640–647.

Reeve, H. K., Emlen, S. T., and Keller, L. (1998). Reproductive sharing in animal societies: reproductive incentives or incomplete control by dominant breeders? *Behavioral Ecology*, 9(3):267–278.

Reeve, H. K. and Ratnieks, F. L. W. (1993). Queen-queen conflicts in polygynous societies: mutual tolerance and reproductive skew. In Keller, L., editor, *Queen Number and Sociality in Insects*, pages 45–85. Oxford University Press, Oxford, UK.

Reluga, T. and Galvani, A. (2011). A general approach for population games with application to vaccination. *Mathematical Biosciences*, 230(2):67–78.

Richter, H. (2016). Analyzing coevolutionary games with dynamic fitness landscapes. In *2016 IEEE Congress on Evolutionary Computation (CEC)*, pages 609–616. IEEE.

Richter, H. (2017). Dynamic landscape models of coevolutionary games. *BioSystems*, 153:26–44.

Richter, H. (2019). Properties of network structures, structure coefficients, and benefit–to–cost ratios. *Biosystems*, 180:88–100.

Riechert, S. (1978). Games spiders play: behavioral variability in territorial disputes. *Behavioral Ecology and Sociobiology*, 3(2):135–162.

Riolo, R. L., Cohen, M. D., and Axelrod, R. (2001). Evolution of cooperation without reciprocity. *Nature*, 414(6862):441–443.

Roberson, B. (2006). The colonel blotto game. *Economic Theory*, 29(1):1–24.

Roff, D. A. (1992). *The Evolution of Life Histories: Theory and Analysis*. Springer-Verlag, Berlin, Heidelberg.

Rowland, P. (2008). *Bowerbirds*. Csiro.

Rutte, C., Taborsky, M., and Brinkhof, M. W. G. (2006). What sets the odds of winning and losing? *Trends in Ecology & Evolution*, 21(1):16–21.

Ruxton, G. D., Franks, D. W., Balogh, A. C. V., and Leimar, O. (2008). Evolutionary implications of the form of predator generalization for aposematic signals and mimicry in prey. *Evolution*, 62(11):2913–2921.

Ruxton, G. D., Gurney, W. S. C., and de Roos, A. M. (1992). Interference and generation cycles. *Theoretical Population Biology*, 42(3):235–253.

Ruxton, G. D. and Moody, A. L. (1997). The ideal free distribution with kleptoparasitism. *Journal of Theoretical Biology*, 186(4):449–458.

Ruxton, G. D., Sherratt, T. N., and Speed, M. P. (2004). *Avoiding Attack: The Evolutionary Ecology of Crypsis, Warning Signals, and Mimicry*. Oxford University Press, Oxford, UK.

Ruxton, G. D., Speed, M. P., and Broom, M. (2009). Identifying the ecological conditions that select for intermediate levels of aposematic signalling. *Evolutionary Ecology*, 23(4):491–501.

Sachs, J. L., Mueller, U. G., Wilcox, T. P., and Bull, J. J. (2004). The evolution of cooperation. *Quarterly Review of Biology*, 79(2):135–160.

Safley, J., Sun, S., and Rychtář, J. (2020). Dishonest signalling in a variant of Pygmalion game. *Dynamic Games and Applications*, 10:719–731.

Samuelson, L. (1997). *Evolutionary Games and Equilibrium Selection*. MIT Press, Cambridge, MA, USA.

Samuelson, L. and Zhang, J. (1992). Evolutionary stability in asymmetric games. *Journal of Economic Theory*, 57(2):363–391.

Sandholm, W. H. (2010). *Population Games and Evolutionary Dynamics*. MIT press.

Sandholm, W. H. (2015). Population games and deterministic evolutionary dynamics. In *Handbook of game theory with economic applications*, volume 4, pages 703–778. Elsevier.

Santos, F. C. and Pacheco, J. M. (2005). Scale-free networks provide a unifying framework for the emergence of cooperation. *Physical Review Letters*, 95(9):98104.

Santos, F. C. and Pacheco, J. M. (2006). A new route to the evolution of cooperation. *Journal of Evolutionary Biology*, 19(3):726–733.

Santos, F. C., Pacheco, J. M., and Lenaerts, T. (2006a). Evolutionary dynamics of social dilemmas in structured heterogeneous populations. *Proceedings of the National Academy of Sciences*, 103(9):3490–3494.

Santos, F. C., Pinheiro, F. L., Lenaerts, T., and Pacheco, J. M. (2012). The role of diversity in the evolution of cooperation. *Journal of Theoretical Biology*, 299:88–96.

Santos, F. C., Rodrigues, J. F., and Pacheco, J. M. (2006b). Graph topology plays a determinant role in the evolution of cooperation. *Proceedings of the Royal Society B: Biological Sciences*, 273(1582):51.

Santos, F. C., Santos, M. D., and Pacheco, J. M. (2008). Social diversity promotes the emergence of cooperation in public goods games. *Nature*, 454(7201):213–216.

Sasaki, T. and Uchida, S. (2013). The evolution of cooperation by social exclusion. *Proceedings of the Royal Society B: Biological Sciences*, 280(1752):20122498.

Sasmal, S. K. and Takeuchi, Y. (2021). Evolutionary dynamics of single species model with Allee effects and aposematism. *Nonlinear Analysis: Real World Applications*, 58:103233.

Schaefer, H. M. and Ruxton, G. D. (2011). *Plant-Animal Communication*. Oxford University Press, Oxford, UK.

Scheckelhoff, K., Ejaz, A., Erovenko, I. V., Rychtář, J., and Taylor, D. (2021). Optimal voluntary vaccination of adults and adolescents can help eradicate hepatitis B in China. *Games*, 12(4):82.

Schiff, J. L. (2008). *Cellular Automata: A Discrete View of the World*. Wiley-Interscience.

Schimit, P. H. T., Pattni, K., and Broom, M. (2019). Dynamics of multi-player games on complex networks using territorial interactions. *Physical Review E*, 99(3):032306.

Schmid, L., Chatterjee, K., Hilbe, C., and Nowak, M. A. (2021). A unified framework of direct and indirect reciprocity. *Nature Human Behaviour*, 5:1292–1302.

Schuster, P. and Sigmund, K. (1981). Coyness, philandering and stable strategies. *Animal Behaviour*, 29(1):186–192.

Schuster, P. and Sigmund, K. (1983). Replicator dynamics. *Journal of Theoretical Biology*, 100(3):533–538.

Schuster, S., Kreft, J. U., Schroeter, A., and Pfeiffer, T. (2008). Use of game-theoretical methods in biochemistry and biophysics. *Journal of biological physics*, 34(1):1–17.

Schwalbe, U. and Walker, P. (2001). Zermelo and the early history of game theory. *Games and Economic Behavior*, 34(1):123–137.

Seifan, M., Hoch, E.-M., Hanoteaux, S., and Tielbörger, K. (2014). The outcome of shared pollination services is affected by the density and spatial pattern of an attractive neighbour. *Journal of Ecology*, 102(4):953–962.

Selten, R. (1965). Spieltheoretische Behandlung eines Oligopolmodells mit Nachfrageträgheit: Teil I: Bestimmung des dynamischen Preisgleichgewichts. *Zeitschrift für die gesamte Staatswissenschaft/Journal of Institutional and Theoretical Economics*, 121(2):301–324.

Selten, R. (1975). A reexamination of the perfectness concept for equilibrium points in extensive games. *International Journal of Game Theory*, 4:25–55.

Selten, R. (1980). A note on evolutionarily stable strategies in asymmetric animal conflicts. *Journal of Theoretical Biology*, 84:93–101.

Seymour, R. M. and Sozou, P. D. (2009). Duration of courtship effort as a costly signal. *Journal of Theoretical Biology*, 256(1):1–13.

Shakarian, P. and Roos, P. (2011). Fast and deterministic computation of fixation probability in evolutionary graphs. In *CIB '11: The Sixth IASTED Conference on Computational Intelligence and Bioinformatics*. IASTED.

Shakarian, P., Roos, P., and Johnson, A. (2012). A review of evolutionary graph theory with applications to game theory. *Biosystems*, 107(2):66–80.

Sharkey, K. J., Kiss, I. Z., Wilkinson, R. R., and Simon, P. L. (2015). Exact equations for SIR epidemics on tree graphs. *Bulletin of Mathematical Biology*, 77(4):614–645.

Sharkey, K. J. and Wilkinson, R. R. (2015). Complete hierarchies of SIR models on arbitrary networks with exact and approximate moment closure. *Mathematical Biosciences*, 264:74–85.

Sharpe, F. R. and Lotka, A. J. (1911). A problem in age-distribution. *The London, Edinburgh, and Dublin Philosophical Magazine and Journal of Science*, 21(124):435–438.

Sherratt, T. N. and Mesterton-Gibbons, M. (2015). The evolution of respect for property. *Journal of Evolutionary Biology*, 28(6):1185–1202.

Shirokawa, Y. and Shimada, M. (2020). Synchronized emergence under diatom sperm competition. *Proceedings of the Royal Society B: Biological Sciences*, 287(1936):20201074.

Sigmund, K. (1993). *Games of Life: Explorations in Ecology, Evolution and Behaviour*. Oxford University Press.

Sigmund, K. (2007). Punish or perish? retaliation and collaboration among humans. *Trends in Ecology & Evolution*, 22(11):593–600.

Sigmund, K. (2010). *The Calculus of Selfishness*. Princeton University Press, Princeton, NJ, USA.

Sigmund, K., De Silva, H., Traulsen, A., and Hauert, C. (2010). Social learning promotes institutions for governing the commons. *Nature*, 466(7308):861–863.

Sigmund, K. and Holt, R. D. (2021). Toward ecoevolutionary dynamics. *Proceedings of the National Academy of Sciences*, 118(9):e2100200118.

Similä, T. (1997). Sonar observations of killer whales (*Orcinus orca*) feeding on herring schools. *Aquatic Mammals*, 23:119–126.

Simmons, L. W. (2001). *Sperm Competition and Its Evolutionary Consequences in the Insects*. Princeton University Press, Princeton, NJ, USA.

Simon, P. L. and Kiss, I. Z. (2005–2020). On bounding exact models of epidemic spread on networks. *Discrete and Continuous Dynamical Systems - Series B*, 23(5):74–85.

Sims, D. W., Southall, E. J., Humphries, N. E., Hays, G. C., Bradshaw, C. J. A., Pitchford, J. W., James, A., Ahmed, M. Z., Brierley, A. S., and Hindell, M. A. (2008). Scaling laws of marine predator search behaviour. *Nature*, 451(7182):1098–1102.

Sinervo, B. and Lively, C. M. (1996). The rock-paper-scissors game and the evolution of alternative male strategies. *Nature*, 380:240–243.

Sirot, E. (2000). An evolutionarily stable strategy for aggressiveness in feeding groups. *Behavioral Ecology*, 11(4):351.

Skyrms, B. (2004). *The Stag Hunt and the Evolution of Social Structure*. Cambridge University Press, Cambridge, UK.

Skyrms, B. (2008). Evolution of signalling systems with multiple senders and receivers. *Philosophical Transactions of the Royal Society B: Biological Sciences*, 364(1518):771–779.

Smallegange, I. M. (2007). *Interference Competition and Patch Choice in Foraging Shore Crabs*. Ph. D. thesis.

Smallegange, I. M. and van der Meer, J. (2009). The distribution of unequal predators across food patches is not necessarily (semi) truncated. *Behavioral Ecology*, 20(3):525–534.

Sobel, J. (2020). Signaling games. In *Complex Social and Behavioral Systems: Game Theory and Agent-Based Models*, pages 251–268. Springer.

Soler, M., Martinez, J. G., Soler, J. J., and Møller, A. P. (1995). Chick recognition and acceptance: a weakness in magpies exploited by the parasitic great spotted cuckoo. *Behavioral Ecology and Sociobiology*, 37(4):243–248.

Sood, V., Antal, T., and Redner, S. (2008). Voter models on heterogeneous networks. *Physical Review E*, 77(4):041121.

Southwell, R. (2009). *Reproducing Graphs and Best Response Games*. PhD thesis, University of Sheffield.

Southwell, R. and Cannings, C. (2009). Games on graphs that grow deterministically. In *2009 International Conference on Game Theory for Networks*, pages 347–356. IEEE.

Southwell, R. and Cannings, C. (2010). Some models of reproducing graphs: III game based reproduction. *Applied Mathematics*, 1(05):335–343.

Southwell, R. and Cannings, C. (2013). Best response games on regular graphs. *Applied Mathematics*, 4:950–962.

Southwell, R., Huang, J., and Cannings, C. (2012). Complex networks from simple rewrite systems. arXiv preprint: 1205.0596.

Souza, M. O., Pacheco, J. M., and Santos, F. C. (2009). Evolution of cooperation under n-person snowdrift games. *Journal of Theoretical Biology*, 260(4):581–588.

Sozou, P. D. and Seymour, R. M. (2005). Costly but worthless gifts facilitate courtship. *Proceedings of the Royal Society B: Biological Sciences*, 272(1575):1877–1884.

Spear, L. B., Howell, S. N. G., Oedekoven, C. S., Legay, D., and Bried, J. (1999). Kleptoparasitism by brown skuas on albatrosses and giant-petrels in the Indian Ocean. *The Auk*, 116(2):545–548.

Spencer, H. (1864). *The Principles of Biology, Vol. 1.* Williams and Norgate, London.

Spencer, R. and Broom, M. (2018). A game-theoretical model of kleptoparasitic behavior in an urban gull (*Laridae*) population. *Behavioral Ecology*, 29(1):60–78.

Stahl, S. (1998). *A Gentle Introduction to Game Theory.* The American Mathematical Society.

Staňková, K. (2019). Resistance games. *Nature Ecology & Evolution*, 3(3):336–337.

Staňková, K., Brown, J. S., Dalton, W. S., and Gatenby, R. A. (2019). Optimizing cancer treatment using game theory: A review. *JAMA Oncology*, 5(1):96–103.

Steele, W. K. and Hockey, P. A. R. (1995). Factors influencing rate and success of intraspecific kleptoparasitism among kelp gulls (*Larus dominicanus*). *The Auk*, 112(4):847–859.

Stephan, W. and Langley, C. H. (1989). Molecular genetic variation in the centromeric region of the X chromosome in three *Drosophila ananassae* populations. I. contrasts between the vermilion and forked loci. *Genetics*, 121(1):89–99.

Stephan, W., Wiehe, T. H. E., and Lenz, M. W. (1992). The effect of strongly selected substitutions on neutral polymorphism: analytical results based on diffusion theory. *Theoretical Population Biology*, 41(2):237–254.

Stephens, D. W. and Clements, K. C. (1998). Game theory and learning. In Dugatkin, L. A. and Reeve, H. K., editors, *Game Theory and Animal Behavior*, pages 239–260. Oxford University Press, New York, NY, USA.

Stephens, D. W. and Krebs, J. R. (1986). *Foraging Theory*. Princeton University Press, Princeton, NJ, USA.

Stillman, R. A., Goss-Custard, J. D., and Caldow, R. W. G. (1997). Modelling interference from basic foraging behaviour. *Journal of Animal Ecology*, 66(5):692–703.

Stockley, P., Gage, M. J. G., Parker, G. A., and Møller, A. (1996). Sperm competition in fishes: the evolution of testis size and ejaculate characteristics. *The American Naturalist*, 149(5):933–954.

Straffin, P. D. (1993). *Game Theory and Strategy*. The Mathematical Association of America.

Straffin Jr, P. D. (1980). The Prisoner's Dilemma. *UMAP Journal*, 1:101–103.

Stratton, M., Campbell, P., and Futreal, P. A. (2009). The cancer genome. *Nature*, 458(7239):719–724.

Street, G. M., Erovenko, I. V., and Rowell, J. T. (2018). Dynamical facilitation of the ideal free distribution in nonideal populations. *Ecology and Evolution*, 8(5):2471–2481.

Stroh, S., Debatin, K.-M., and Westhoff, M.-A. (2014). Darwinian principles in cancer therapy. *European Oncology and Haematology*, 10:116–120.

Sugden, R. (1986). *The Economics of Rights, Co-operation and Welfare*. B. Blackwell, Oxford; New York, NY.

Sumaila, U. R. (1999). A review of game-theoretic models of fishing. *Marine Policy*, 23(1):1–10.

Summers, K., Speed, M. P., Blount, J. D., and Stuckert, A. M. M. (2015). Are aposematic signals honest? A review. *Journal of Evolutionary Biology*, 28(9):1583–1599.

Sumpter, D. J. T. (2006). The principles of collective animal behaviour. *Philosophical Transactions of the Royal Society B: Biological Sciences*, 361(1465):5–22.

Sumpter, D. J. T. (2010). *Collective Animal Behavior*. Princeton University Press, Princeton, NJ, USA.

Sun, S., Broom, M., Johanis, M., and Rychtář, J. (2021). A mathematical model of kin selection in floral displays. *Journal of Theoretical Biology*, 509:110470.

Sun, S., Leshowitz, M., and Rychtář, J. (2018). The signalling game between plants and pollinators. *Scientific Reports*, 8(1):1–8.

Sutherland, W. J. and Parker, G. A. (1985). Distribution of unequal competitors. In Sibly, R. M. and Smith, R. H., editors, *Behavioural Ecology Ecological Consequences of Adaptive Behaviour*, pages 255–274. Blackwell Scientific, Oxford, UK.

Sykes, D. and Rychtář, J. (2015). A game-theoretic approach to valuating toxoplasmosis vaccination strategies. *Theoretical Population Biology*, 105:33–38.

Szabo, G. and Fath, G. (2007). Evolutionary games on graphs. *Physics Reports*, 446(4-6):97–216.

Szabó, G. and Szolnoki, A. (2009). Cooperation in spatial prisoner's dilemma with two types of players for increasing number of neighbors. *Physical Review E*, 79(1):016106.

Szabó, G., Vukov, J., and Szolnoki, A. (2005). Phase diagrams for an evolutionary prisoner's dilemma game on two-dimensional lattices. *Physical Review E*, 72(4):047107.

Számadó, S. (2011). The cost of honesty and the fallacy of the handicap principle. *Animal Behaviour*, 81(1):3–10.

Számadó, S. and Penn, D. (2015). Why does costly signalling evolve? Challenges with testing the handicap hypothesis. *Animal Behaviour*, 110:e9.

Szolnoki, A. and Perc, M. (2011). Group-size effects on the evolution of cooperation in the spatial public goods game. *Physical Review E*, 84(4):047102.

Szolnoki, A., Perc, M., and Danku, Z. (2008). Towards effective payoffs in the prisoner's dilemma game on scale-free networks. *Physica A: Statistical Mechanics and its Applications*, 387(8-9):2075–2082.

Szolnoki, A., Perc, M., and Szabó, G. (2009). Topology-independent impact of noise on cooperation in spatial public goods games. *Physical Review E*, 80(5):056109.

Takasu, F. (1998a). Modelling the arms race in avian brood parasitism. *Evolutionary Ecology*, 12(8):969–987.

Takasu, F. (1998b). Why do all host species not show defense against avian brood parasitism: evolutionary lag or equilibrium? *The American Naturalist*, 151(2):193–205.

Takasu, F., Kawasaki, K., Nakamura, H., Cohen, J. E., and Shigesada, N. (1993). Modeling the population dynamics of a cuckoo-host association and the evolution of host defenses. *The American Naturalist*, 142(5):819–839.

Tang, C., Li, X., Cao, L., and Zhan, J. (2012). The σ law of evolutionary dynamics in community-structured population. *Journal of Theoretical Biology*, 306:1–6.

Tanimoto, J. (2015). *Fundamentals of Evolutionary Game Theory and Its Applications*. Springer.

Tarnita, C. E. (2016). Mathematical approaches or agent-based methods?: Comment on "Evolutionary game theory using agent-based methods" by Christoph Adami et al. *Physics of Life Reviews*, 19:36–37.

Tarnita, C. E., Antal, T., Ohtsuki, H., and Nowak, M. A. (2009a). Evolutionary dynamics in set structured populations. *Proceedings of the National Academy of Sciences*, 106(21):8601–8604.

Tarnita, C. E., Ohtsuki, H., Antal, T., Fu, F., and Nowak, M. A. (2009b). Strategy selection in structured populations. *Journal of Theoretical Biology*, 259(3):570–581.

Tarnita, C. E., Wage, N., and Nowak, M. A. (2011). Multiple strategies in structured populations. *Proceedings of the National Academy of Sciences*, 108(6):2334–2337.

Tavakoli, F., Sartakhti, J. S., Manshaei, M. H., and Basanta, D. (2020). Cancer immunoediting: A game theoretical approach. *In Silico Biology*, 14(1-2):1–12.

Taylor, C., Fudenberg, D., Sasaki, A., and Nowak, M. A. (2004). Evolutionary game dynamics in finite populations. *Bulletin of Mathematical Biology*, 66(6):1621–1644.

Taylor, C. and Nowak, M. A. (2006). Evolutionary game dynamics with non-uniform interaction rates. *Theoretical Population Biology*, 69(3):243–252.

Taylor, C. and Nowak, M. A. (2009). How to evolve cooperation. In Levin, S. A., editor, *Games, Groups, and the Global Good*, pages 41–56. Springer-Verlag, London, UK.

Taylor, P. and Jonker, L. (1978). Evolutionarily stable strategies and game dynamics. *Mathematical Biosciences*, 40:145–156.

Taylor, P. D., Wild, G., and Gardner, A. (2007). Direct fitness or inclusive fitness: how shall we model kin selection? *Journal of Evolutionary Biology*, 20(1):301–309.

Templeton, A. R. and Lawlor, L. R. (1981). The fallacy of the averages in ecological optimization theory. *The American Naturalist*, 117(3):390–393.

Thapar, V. (1986). *Tiger: Portrait of a Predator*. London: Collins

Thomas, B. (1985). On evolutionarily stable sets. *Journal of Mathematical Biology*, 22(1):105–115.

Tildesley, M. J., Bessell, P. R., Keeling, M. J., and Woolhouse, M. E. J. (2009). The role of pre-emptive culling in the control of foot-and-mouth disease. *Proceedings of the Royal Society B: Biological Sciences*, 276(1671):3239–3248.

Tinbergen, N. (1963). On aims and methods of ethology. *Zeitschrift für Tierpsychologie*, 20(4):410–433.

Tkadlec, J., Pavlogiannis, A., Chatterjee, K., and Nowak, M. A. (2019). Population structure determines the tradeoff between fixation probability and fixation time. *Communications Biology*, 2(1):1–8.

Tkadlec, J., Pavlogiannis, A., Chatterjee, K., and Nowak, M. A. (2020). Limits on amplifiers of natural selection under death-Birth updating. *PLOS Computational Biology*, 16(1):e1007494.

Tkadlec, J., Pavlogiannis, A., Chatterjee, K., and Nowak, M. A. (2021). Fast and strong amplifiers of natural selection. *Nature Communications*, 12(1):1–6.

Tomlinson, I. P. (1997). Game-theory models of interactions between tumour cells. *European Journal of Cancer*, 33(9):1495–1500.

Torices, R., Gómez, J., and Pannell, J. (2018). Kin discrimination allows plants to modify investment towards pollinator attraction. *Nature communications*, 9(1):1–6.

Traulsen, A. and Hauert, C. (2009). Stochastic evolutionary game dynamics. In Schuster, H. G., editor, *Reviews of Nonlinear Dynamics and Complexity*. Volume 2, pages 25–61. Wiley Online Library.

Traulsen, A. and Nowak, M. A. (2006). Evolution of cooperation by multilevel selection. *Proceedings of the National Academy of Sciences*, 103(29):10952–10955.

Traulsen, A., Shoresh, N., and Nowak, M. A. (2008). Analytical results for individual and group selection of any intensity. *Bulletin of Mathematical Biology*, 70(5):1410.

Triplet, P., Stillman, R. A., and Goss-Custard, J. D. (1999). Prey abundance and the strength of interference in a foraging shorebird. *Journal of Animal Ecology*, 68(2):254–265.

Trivers, R. L. (1971). The evolution of reciprocal altruism. *The Quarterly Review of Biology*, 46(1):35–57.

Trivers, R. L. (1972). Parental investment and sexual selection. In Campbell, B. G., editor, *Sexual Selection and the Descent of Man*. Routledge, New York.

Tucker, A. W. and Straffin Jr, P. D. (1983). The mathematics of tucker: A sampler. *The Two-Year College Mathematics Journal*, 14(3):228–232.

Turnell, B. R., Shaw, K. L., and Reeve, H. K. (2018). Modeling strategic sperm allocation: Tailoring the predictions to the species. *Evolution*, 72(3):414–425.

Turner, P. E. and Chao, L. (1999). Prisoner's dilemma in an RNA virus. *Nature*, 398(6726):441–443.

Turner, P. E. and Chao, L. (2003). Escape from prisoner's dilemma in RNA phage $\varphi 6$. *The American Naturalist*, 161(3):497–505.

Tyler, J. A. and Gilliam, J. F. (1995). Ideal free distributions of stream fish: a model and test with minnows, *Rhinicthys atratulus*. *Ecology*, 76(2):580–592.

Ujvari, B., Gatenby, R., and Thomas, F. (2016a). Transmissible cancers, are they more common than thought? *Evolutionary Applications*, 9(5):633–634.

Ujvari, B., Gatenby, R. A., and Thomas, F. (2016b). The evolutionary ecology of transmissible cancers. *Infection, Genetics and Evolution*, 39:293–303.

Vahl, W. K. (2006). *Interference Competition among Foraging Waders*. PhD thesis, University of Groningen.

Vahl, W. K., van der Meer, J., Weissing, F. J., van Dullemen, D., and Piersma, T. (2005). The mechanisms of interference competition: two experiments on foraging waders. *Behavioral Ecology*, 16(5):845–855.

van Baalen, M. and Rand, D. A. (1998). The unit of selection in viscous populations and the evolution of altruism. *Journal of Theoretical Biology*, 193(4):631–648.

van Boven, M. and Weissing, F. J. (1998). Evolution of segregation distortion: Potential for a high degree of polymorphism. *Journal of Theoretical Biology*, 192:131–142.

van Boven, M., Weissing, F. J., Heg, D., and Huisman, J. (1996). Competition between segregation distorters: Coexistence of 'superior' and 'inferior' haplotypes at the t complex. *Evolution*, 50:2488–2498.

van Damme, E. (1984). A relation between perfect equilibria in extensive form games and proper equilibria in normal form games. *International Journal of Game Theory*, 13(1):1–13.

van Damme, E. (1991). *Stability and Perfection of Nash Equilibria* (2nd edn) Springer-Verlag, Berlin.

van der Meer, J. and Ens, B. J. (1997). Models of interference and their consequences for the spatial distribution of ideal and free predators. *Journal of Animal Ecology*, 66(6):846–858.

van der Meer, J. and Smallegange, I. M. (2009). A stochastic version of the beddington-deangelis functional response: modelling interference for a finite number of predators. *Journal of Animal Ecology*, 78(1):134–142.

Van Dooren, T., Durinx, M., and Demon, I. (2004). Sexual dimorphism or evolutionary branching? *Evolutionary Ecology Research*, 6(6):857–871.

van Doorn, G. S., Hengeveld, G. M., and Weissing, F. J. (2003a). The evolution of social dominance I: two-player models. *Behaviour*, 140(10):1305–1332.

van Doorn, G. S., Hengeveld, G. M., and Weissing, F. J. (2003b). The evolution of social dominance II: multi-player models. *Behaviour*, 140(10):1333–1358.

van Lenteren, J. C. (1981). Host discrimination by parasitoids. In Nordlund, D. A., Jones, R. L., and Lewis, W. L., editors, *Semiochemicals: Their Role in Pest Control*, pages 153–180. John Wiley & Sons, New York, NY, USA.

van Schaik, C. P. (2000). Infanticide by male primates: the sexual selection hypothesis revisited. In van Schaik, C. P. and Janson, C. H., editors, *Infanticide by Males and Its Implications*, pages 27–60. Cambridge University Press, Cambridge, UK.

van Schaik, C. P. and Janson, C. H. (2000). *Infanticide by Males and Its Implications*. Cambridge University Press, Cambridge, UK.

van Veelen, M. and Nowak, M. A. (2012). Multi-player games on the cycle. *Journal of Theoretical Biology*, 292:116–128.

Varga, T., Garay, J., Rychtář, J., and Broom, M. (2020a). A temporal model of territorial defence with antagonistic interactions. *Theoretical Population Biology*, 134:15–35.

Varga, T., Móri, T. F., and Garay, J. (2020b). The ess for evolutionary matrix games under time constraints and its relationship with the asymptotically stable rest point of the replicator dynamics. *Journal of Mathematical Biology*, 80(3):743–774.

Vehrencamp, S. L. (1983). Optimal degree of skew in cooperative societies. *American Zoologist*, 23(2):327–335.

Vickers, G. T. and Cannings, C. (1987). On the definition of an evolutionarily stable strategy. *Journal of Theoretical Biology*, 129:349–353.

Vickers, G. T. and Cannings, C. (1988a). On the number of stable equilibria in a one-locus, multi-allelic system. *Journal of Theoretical Biology*, 131(3):273–277.

Vickers, G. T. and Cannings, C. (1988b). Patterns of ESS's I. *Journal of Theoretical Biology*, 132:387–408.

Vickery, W. L., Giraldeau, L. A., Templeton, J. J., Kramer, D. L., and Chapman, C. A. (1991). Producers, scroungers and group foraging. *The American Naturalist*, 137(6):847–863.

Vincent, T. L. and Brown, J. S. (2005). *Evolutionary Game Theory, Natural Selection and Darwinian Dynamics*. Cambridge University Press, Cambridge, UK.

Vincent, T. L. and Gatenby, R. A. (2005). Modeling cancer as an evolutionary game. *International Game Theory Review*, 7(03):331–346.

Vincent, T. L. and Gatenby, R. A. (2008). An evolutionary model for initiation, promotion, and progression in carcinogenesis. *International Journal of Oncology*, 32(4):729–737.

Vincent, T. L., Van, M. V., and Goh, B. S. (1996). Ecological stability, evolutionary stability and the ESS maximum principle. *Evolutionary Ecology*, 10(6):567–591.

Viossat, Y. and Noble, R. (2021). A theoretical analysis of tumour containment. *Nature Ecology & Evolution*, 5(6):826–835.

Voelkl, B. (2010). The 'Hawk-Dove' game and the speed of the evolutionary process in small heterogeneous populations. *Games*, 1(2):103–116.

Voelkl, B. and Kasper, C. (2009). Social structure of primate interaction networks facilitates the emergence of cooperation. *Biology Letters*, 5(4):462–464.

Volterra, V. (1926). Variazioni e fluttuazioni del numero d'individui in specie animali conviventi. *Mem. AccaD. Sci. Lincei*, 2:31–113.

von Neumann, J. (1928). Zur Theorie der Gesellschaftsspiele. *Mathematische Annalen*, 100(1):295–320.

von Neumann, J. and Morgenstern, O. (1944). *Theory of Games and Economic Behavior*. Princeton University Press, Princeton, NJ, USA.

von Stackelberg, H. (1952). *The Theory of the Market Economy*. Oxford University Press, New York, NY, USA.

Voorhees, B. (2013). Birth–death fixation probabilities for structured populations. *Proceedings of the Royal Society A: Mathematical, Physical and Engineering Sciences*, 469(2153):20120248.

Voorhees, B. and Murray, A. (2013). Fixation probabilities for simple digraphs. *Proceedings of the Royal Society A: Mathematical, Physical and Engineering Sciences*, 469(2154):20120676.

Vukov, J., Szabó, G., and Szolnoki, A. (2006). Cooperation in the noisy case: Prisoner's dilemma game on two types of regular random graphs. *Physical Review E*, 73(6):067103.

Wakano, J. Y. and Yamamura, N. (2001). A simple learning strategy that realizes robust cooperation better than Pavlov in iterated prisoners' dilemma. *Journal of Ethology*, 19(1):1–8.

Walker, P. (2008). An outline of the history of game theory. Discussion paper 9504, University of Canterbury, Department of Economics.

Walsh, G. R. (1985). *An Introduction to Linear Programming*. John Wiley & Sons, New York, USA.

Wang, J., Wu, B., Ho, D. W., and Wang, L. (2011). Evolution of cooperation in multilevel public goods games with community structures. *Europhysics Letters*, 93(5):58001.

Wang, N., Zhou, W., and Wu, Z. (2019a). Equilibrium distributions of populations of biological species on networks of social sites. *Journal of Biological Dynamics*, 13(S1):74–98.

Wang, Q., Gosik, K., Xing, S., Jiang, L., Sun, L., Chinchilli, V., and Wu, R. (2017a). Epigenetic game theory: how to compute the epigenetic control of maternal-to-zygotic transition. *Physics of Life Reviews*, 20:126–137.

Wang, S.-C., Yu, J.-R., Kurokawa, S., and Tao, Y. (2017b). Imitation dynamics with time delay. *Journal of Theoretical Biology*, 420:8–11.

Wang, X.-j., Gu, C.-l., and Quan, J. (2019b). Evolutionary game dynamics of the wright-fisher process with different selection intensities. *Journal of Theoretical Biology*, 465:17–26.

Waxman, D. (2011). Comparison and content of the Wright-Fisher model of random genetic drift, the diffusion approximation and an intermediate model. *Journal of Theoretical Biology*, 269(1):79–87.

Waxman, D. and Gavrilets, S. (2005). 20 questions on adaptive dynamics. *Journal of Evolutionary Biology*, 18:1139–1154.

Weatherhead, P. J. (1979). Do savannah sparrows commit the concorde fallacy? *Behavioral Ecology and Sociobiology*, 5:373–381.

Webb, J. N., Houston, A. I., McNamara, J. M., and Szèkely, T. (1999). Multiple patterns of parental care. *Animal Behaviour*, 58(5):983–993.

Wedell, N., Gage, M. J. G., and Parker, G. A. (2002). Sperm competition, male prudence and sperm-limited females. *Trends in Ecology & Evolution*, 17(7):313–320.

Weibull, J. (1995). *Evolutionary Game Theory*. MIT Press, Cambridge, MA, USA.

Weiling, F. (1986). What about RA Fisher's statement of the "too good" data of J. G. Mendel's Pisum paper? *Journal of Heredity*, 77(4):281–283.

Weinberg, W. (1908). On the demonstration of heredity in man. In *Boyer, S. H.* , trans. (1963). *Papers on Human Genetics*, pages 4–15. Prentice Hall Englewood Cliffs (NJ).

Weissing, F. J. (1996). Genetic versus phenotypic models of selection: can genetics be neglected in a long-term perspective? *Journal of Mathematical Biology*, 34:533–555.

Werner, B., Lutz, D., Brümmendorf, T. H., Traulsen, A., and Balabanov, S. (2011). Dynamics of resistance development to imatinib under increasing selection pressure: a combination of mathematical models and in vitro data. *PLOS One*, 6(12):e28955.

Werner, B., Scott, J. G., Sottoriva, A., Anderson, A. R., Traulsen, A., and Altrock, P. M. (2016). The cancer stem cell fraction in hierarchically organized tumors can be estimated using mathematical modeling and patient-specific treatment trajectories. *Cancer Research*, 76(7):1705–1713.

West, S. A., Griffin, A. S., and Gardner, A. (2007a). Evolutionary explanations for cooperation. *Current Biology*, 17(16):R661–R672.

West, S. A., Griffin, A. S., and Gardner, A. (2007b). Social semantics: altruism, cooperation, mutualism, strong reciprocity and group selection. *Journal of Evolutionary Biology*, 20(2):415–432.

Whitmeyer, M. (2021). Strategic inattention in the Sir Philip Sidney game. *Journal of Theoretical Biology*, 509:110513.

Wiley, R. H. (1991). Lekking in birds and mammals: behavioral and evolutionary issues. *Advances in the Study of Behavior*, 20:201–291.

Wilkinson, G. S. (1984). Reciprocal food sharing in the vampire bat. *Nature*, 308(5955):181–184.

Wilson, D. S. (1975). A theory of group selection. *Proceedings of the National Academy of Sciences*, 72(1):143–146.

Wölfl, B., te Rietmole, H., Salvioli, M., Kaznatcheev, A., Thuijsman, F., Brown, J. S., Burgering, B., and Staňková, K. (in press). The contribution of evolutionary game theory to understanding and treating cancer. *Dynamic Games and Applications*, pages 1–30.

Wolfram, S. (1986). *Theory and Applications of Cellular Automata*. Advanced Series on Complex Systems, Singapore: World Scientific Publication.

Wolfram, S. (2002). *A New Kind of Science*, volume 1. Wolfram Media Champaign, IL.

Wright, S. (1930). Evolution in Mendelian Populations. *Genetics*, 16:97–159.

Wu, A. (2015). *Emergence of chemotherapy resistance in cancer: microenvironments, genomics, and game theory approaches*. PhD thesis, Princeton University.

Wu, B., Arranz, J., Du, J., Zhou, D., and Traulsen, A. (2016). Evolving synergetic interactions. *Journal of the Royal Society Interface*, 13(120):20160282.

Wu, B., Bauer, B., Galla, T., and Traulsen, A. (2015). Fitness-based models and pairwise comparison models of evolutionary games are typically different—even in unstructured populations. *New Journal of Physics*, 17(2):023043.

Wu, B., Traulsen, A., and Gokhale, C. (2013). Dynamic properties of evolutionary multi-player games in finite populations. *Games*, 4(2):182–199.

Wu, J.-J., Zhang, B.-Y., Zhou, Z.-X., He, Q.-Q., Zheng, X.-D., Cressman, R., and Tao, Y. (2009a). Costly punishment does not always increase cooperation. *Proceedings of the National Academy of Sciences*, 106(41):17448–17451.

Wu, T., Fu, F., and Wang, L. (2009b). Individual's expulsion to nasty environment promotes cooperation in public goods games. *Europhysics Letters*, 88(3):30011.

Wu, T., Fu, F., and Wang, L. (2009c). Partner selections in public goods games with constant group size. *Physical Review E*, 80(2):026121.

Wynne-Edwards, V. C. (1962). *Animal Dispersion in Relation to Social Behaviour*. Oliver and Boyd Edinburgh.

Xia, C.-Y., Meloni, S., and Moreno, Y. (2012). Effects of environment knowledge on agglomeration and cooperation in spatial public goods games. *Advances in Complex Systems*, 15(supp01):1250056.

Xu, F. and Cressman, R. (2016). Voluntary vaccination strategy and the spread of sexually transmitted diseases. *Mathematical Biosciences*, 274:94–107.

Yagoobi, S. and Traulsen, A. (2021). Fixation probabilities in network structured meta-populations. *Scientific Reports*, 11(1):1–9.

Yamauchi, A. (1995). Theory of evolution of nest parasitism in birds. *The American Naturalist*, 145(3):434–456.

Yates, G. E. and Broom, M. (2005). A stochastic model of the distribution of unequal competitors between resource patches. *Journal of Theoretical Biology*, 237(3):227–237.

Yates, G. E. and Broom, M. (2007). Stochastic models of kleptoparasitism. *Journal of Theoretical Biology*, 248(3):480–489.

Ydenberg, R. C. and Dill, L. M. (1986). The economics of fleeing from predators. *Advances in the Study of Behavior*, 16:229–249.

Yee, K. K. (2003). Ownership and trade from evolutionary games. *International Review of Law and Economics*, 23(2):183–197.

Yi, M., Huo, L., Koenig, K. B., Mittendorf, E. A., Meric-Bernstam, F., Kuerer, H. M., Bedrosian, I., Buzdar, A. U., Symmans, W. F., and Crow, J. R. (2014). Which threshold for ER positivity? A retrospective study based on 9639 patients. *Annals of Oncology*, 25(5):1004–1011.

Yi, T., Cressman, R., and Brooks, B. (1999). Nonlinear frequency-dependent selection at a single locus with two alleles and two phenotypes. *Journal of Mathematical Biology*, 39(4):283–308.

Yiu, S. W., Keith, M., Karczmarski, L., and Parrini, F. (2021). Predation risk effects on intense and routine vigilance of burchell's zebra and blue wildebeest. *Animal Behaviour*, 173:159–168.

Young, B., Conti, D. V., and Dean, M. D. (2013). Sneaker "jack" males outcompete dominant "hooknose" males under sperm competition in chinook salmon (*Oncorhynchus tshawytscha*). *Ecology and Evolution*, 3(15):4987–4997.

Zahavi, A. (1975). Mate selection–a selection for a handicap. *Journal of Theoretical Biology*, 53(1):205–214.

Zahavi, A. (1977). The cost of honesty (further remarks on the handicap principle). *Journal of Theoretical Biology*, 67(3):603–605.

Zahavi, A. (1979). Parasitism and nest predation in parasitic cuckoos. *The American Naturalist*, 113(1):157–159.

Zeeman, E. C. (1980). Population dynamics from game theory. In Nitecki, Z. and Robinson, C., editors, *Global Theory of Dynamical Systems*, volume 819 of *Lecture Notes in Mathematics*. Springer-Verlag, Berlin, Heidelberg.

Zeeman, E. C. (1981). Dynamics of the evolution of animal conflicts. *Journal of Theoretical Biology*, 89:249–270.

Zermelo, E. (1913). Über eine Anwendung der Mengenlehre auf die Theorie des Schachspiels. In *Proceedings of the Fifth International Congress of Mathematicians*, volume 2, pages 501–504.

Zhang, B., Li, C., De Silva, H., Bednarik, P., and Sigmund, K. (2014). The evolution of sanctioning institutions: an experimental approach to the social contract. *Experimental Economics*, 17(2):285–303.

Zhang, B.-Y., Fan, S.-J., Li, C., Zheng, X.-D., Bao, J.-Z., Cressman, R., and Tao, Y. (2016). Opting out against defection leads to stable coexistence with cooperation. *Scientific Reports*, 6(1):1–7.

Zhang, C. Y., Zhang, J. L., Xie, G. M., and Wang, L. (2011a). Coevolving agent strategies and network topology for the public goods games. *The European Physical Journal B*, 80(2):217–222.

Zhang, H.-F., Liu, R.-R., Wang, Z., Yang, H.-X., and Wang, B.-H. (2011b). Aspiration-induced reconnection in spatial public-goods game. *Europhysics Letters*, 94(1):18006.

Zhang, J., Cunningham, J., Brown, J. S., and Gatenby, R. A. (2017). Integrating evolutionary dynamics into treatment of metastatic castrate-resistant prostate cancer. *Nature Communications*, 8(1):1–9.

Zheng, X., Cressman, R., and Tao, Y. (2011). The diffusion approximation of stochastic evolutionary game dynamics: Mean effective fixation time and the significance of the one-third law. *Dynamic Games and Applications*, 1(3):462–477.

Zollman, K. J. S., Bergstrom, C. T., and Huttegger, S. M. (2013). Between cheap and costly signals: the evolution of partially honest communication. *Proceedings of the Royal Society B: Biological Sciences*, 280(1750):20121878.

Zukewich, J., Kurella, V., Doebeli, M., and Hauert, C. (2013). Consolidating birth-death and death-birth processes in structured populations. *PLOS One*, 8(1):e54639.

Index

Printed in the United States
by Baker & Taylor Publisher Services

Printed in the United States
by Baker & Taylor Publisher Services